SPORTS ENGINEERING AND COMPUTER SCIENCE

Advances in Sports Engineering and Technology

ISSN: 2377-2085
eISSN: 2333-8067

Volume 1

PROCEEDINGS OF THE 2014 INTERNATIONAL CONFERENCE ON SPORT SCIENCE AND COMPUTER SCIENCE (SSCS 2014), SINGAPORE, 16–17 SEPTEMBER 2014 & THE 2014 INTERNATIONAL CONFERENCE ON BIOMECHANICS AND SPORTS ENGINEERING (BSE 2014), RIGA, LATVIA, 24–25 OCTOBER, 2014

Sports Engineering and Computer Science

Editor

Qi Luo
*School of Sports Engineering and Information Technology,
Wuhan University of Sports, Wuhan, China*

CRC Press is an imprint of the
Taylor & Francis Group, an **informa** business

A BALKEMA BOOK

CRC Press/Balkema is an imprint of the Taylor & Francis Group, an informa business

© 2015 Taylor & Francis Group, London, UK

Typeset by V Publishing Solutions Pvt Ltd., Chennai, India
Printed and bound in the UK and the US

Published by: CRC Press/Balkema
            P.O. Box 11320, 2301 EH Leiden, The Netherlands
            e-mail: Pub.NL@taylorandfrancis.com
            www.crcpress.com – www.taylorandfrancis.com

ISBN: 978-1-138-02650-6 (Hbk)
ISBN: 978-1-315-75571-7 (eBook PDF)

Sports Engineering and Computer Science – Luo (Ed.)
© 2015 Taylor & Francis Group, London, ISBN 978-1-138-02650-6

Table of contents

Section 3: Computer science and applications in sport

Preface

SSCS 2014 was an integrated conference focusing on Sport Science and Computer Science. SSCS 2014 was held September 16–17, 2014 in Singapore.

The goal of the conference was to provide researchers from Sport Science and Computer Science fields with a free exchange forum to share new ideas, new innovation and solutions with each other. In addition, the conference organizer invited famous keynote speakers to deliver a speech at the conference. It was thought helpful for all participants that they would have a chance to discuss with the speakers face to face.

The main objective of the papers concerns Sport Science and Computer Science. The collection is intended to serve as resource material for scientists working on related topics in many disciplines, including Sport Science, and Computer Science. It is hoped that the contents of this volume will prove useful for researchers and practitioners in developing and applying new theories and technologies in Sport Science and Computer Science.

In the proceedings, we can acquire much more knowledge about Sport Science and Computer Science of researchers all around the world. The main role of the proceedings is to be used as an exchange pillar for researchers working in the aforementioned fields. In order to meet the high standards of CRC Press/Balkema (Taylor & Francis Group), the organization committee has made an effort to do the following. Firstly, poor quality papers have been refused after the reviewing process by anonymous referee experts. All accepted papers of SSCS 2014 have been strictly selected for quality and relevance to the conference. Secondly, about five periodical review meetings have been held between the reviewers for exchanging reviewing suggestions. Finally, the conference organization held several preliminary sessions before the conference. Through efforts of different people and departments, the conference was successful and fruitful.

During the course of organizing, we received help from different people, departments, and institutions. Here, we would like to thank the organization staff, the members of the program committees and the reviewers. They have worked very hard in reviewing papers and making valuable suggestions for the authors to improve their work. We also express our gratitude to the external reviewers, for providing extra help in the review process, and to the authors for contributing their research results to the conference.

We hope that the participants can give us good suggestions to improve our working efficiency and service in the future. And we also hope to receive your continuing support. Next year, we look forward to seeing all of you at SSCS 2015.

The 2014 International Conference on Biomechanics and Sports Engineering (BSE 2014) was held on October 24–25, 2014, in Riga, Latvia. BSE 2014 was a comprehensive conference focused on the various aspects of advances in Biomechanics and Sports Engineering. Our conference provided a chance for academic and industry professionals to discuss recent progress in the area of Biomechanics and Sports Engineering.

The goal of this conference was to bring together researchers from academia and industry as well as practitioners to share ideas, problems and solutions relating to the multifaceted aspects of Biomechanics and Sports Engineering.

Sports engineering is a field of engineering that involves the design, development and testing of sports equipment. Sports Engineers are typically involved in the following activities: Equipment design, Lab experiments and testing, Computational modeling, Field testing, Working with athletes and so on.

The call for papers attracted 50 papers, which were submitted by authors from 8 countries and 3 continents. The paper evaluation this year was both very thorough and highly selective. Each paper was peer reviewed by at least three members of the program committee. After an extensive peer review, 18 papers were selected for presentation at the conference and inclusion in this proceedings volume.

The conference featured invited as well as contributed talks, with morning plenary sessions followed by parallel sessions, covering the following topics:

Sports Science
Computer Science
Education Science
Other Related Topics.

The accepted papers relate to conference topics. The proceedings include the papers presented at the 2014 International Conference on Biomechanics and Sports Engineering (BSE 2014).

On behalf of the organizing committee, we gratefully acknowledge the technical support from CRC Press/Balkema (Taylor & Francis Group). We also thank all the members of the Program Committee and the external referees. Their work in evaluating the papers and comments during the discussions was essential to the decisions on the contributed papers.

Finally, we warmly thank all invited speakers and all the authors of the submitted papers. Their efforts were the basis of the success of the conference. We hope you enjoyed the conference and look forward to seeing you next year at BSE 2015.

Sports Engineering and Computer Science – Luo (Ed.)
© 2015 Taylor & Francis Group, London, ISBN 978-1-138-02650-6

SSCS 2014 organizing committee

KEYNOTE SPEAKER AND HONORARY CHAIR

Gerald Schaefer, *Loughborough University, UK*

GENERAL CHAIRS

Yanwen Wu, *Huazhong Normal University, China*
Ming Ma, *Singapore NUS ACM Chapter, Singapore*

PROGRAM CHAIRS

Prawal Sinha, *Department of Mathematics & Statistics, Indian Institute of Technology Kanpur, India*
Hong Tan, *Wuhan University, China*

PUBLICATION CHAIRS

Weitao Zheng, *Wuhan University of Sports, China*
Qi Luo, *Wuhan University of Sports, China*

COMMITTEE

Hongcheng Zhou, *Nanjing University of Aeronautics and Astronautics, China*
Linjie Wei, *Xian University of Architecture & Technology, China*
Dawei Cao, *Huaibei Normal University, China*
Xuemin Han, *Hainan University, China*
Jiabao Ye, *Tianjin University, China*
Bin Xu, *Northeast Dianli University, China*
Jianping Shu, *Chengdu Sports University, China*
Bin Yang, *Sichuan University, China*
Xiudi Yang, *Shandon Sport University, China*
Lingling Zhang, *Wuhan University of Science and Technology, China*

Sports Engineering and Computer Science – Luo (Ed.)
© *2015 Taylor & Francis Group, London, ISBN 978-1-138-02650-6*

BSE 2014 organizing committee

KEYNOTE SPEAKER

Gerald Schaefer, *Loughborough University, UK*

GENERAL CHAIRS

Dehuai Yang, *Huazhong University of Science and Technology, China*
Minli Dai, *Suzhou University, China*

PUBLICATION CHAIRS

Weitao Zheng, *Wuhan University of Sports, China*
Qi Luo, *Wuhan University of Sports, China*

INTERNATIONAL COMMITTEE

J. Herold, *Heidelberg University, Germany*
M. Buchner, *Heidelberg University, Germany*
K. Daunoravičienė, *Vilnius Gediminas Technical University, Lithuania*
J. Žižienė, *Vilnius Gediminas Technical University, Lithuania*
J. Griškevičius, *Vilnius Gediminas Technical University, Lithuania*
D. Kowalik, *University of Technology and Humanities of Radom, Poland*
M. Siczek, *National Research Institute, Poland*
A. Zhanxia Wu, *Shanghai University of International Business and Economics, China*
Yue Sun, *Northeast Dianli University, China*
Zhao-long Zhang, *Yunnan University, China*

Section 1: Education, physical education and training

Sports Engineering and Computer Science – Luo (Ed.)
© 2015 Taylor & Francis Group, London, ISBN 978-1-138-02650-6

Improving college students' ability of appreciating sports artistic through movies and teleplays

Xiao Gong, Yanqiu Guan, Xiaoying Zhuang & Haoyuan Xu
Beijing Film Academy Modern Creative Media College, China

ABSTRACT: College physical education should take use of "sports culture theory" instead of "physical exercise theory". This requires college physical education to not only teach students with methods and skills of physical exercise, but also to cultivate students' ability of appreciating sports artistic. Sports movies and teleplays have irreplaceable status and roles in improving college students' ability of appreciating sports artistic.

Keywords: Sports movies and teleplays; Sports artistic appreciation; College Students

1   INTRODUCTION

Students' ability of appreciating sports artistic refers to people's understanding of physical exercise, physical and mental health, sports culture, and people's sports view and value etc. It includes sports knowledge, sports consciousness, sports skills, sports personality, sports morality and physical behavior and other aspects. To enlarge students' horizon in field of sports culture, we must use a variety of ways. In these ways, teaching through movies and teleplays is obviously one of the most effective ways. During the long time of sports teaching, the author deeply appreciates the important effects of movies and teleplays visual in order to improve college students' ability of appreciating sports artistic.

China has rich resources in sports movies and teleplays. Chinese sports films appeared very early, the famous director Sun Yu directed "Queen of Sports", in1930s. After the establishment of new China, "Women's Basketball No. Five", "Spring and Autumn above Water", "An Unfinished Chess", "Seagull", "Shaolin Temple", "Huo Yuanjia" and other sports movies have caused great repercussions at that time. These movies played an important role in improving college students' awareness in sports cultural, and encouraged young athletes to train hard to win medals glory for our country. In sports TV plays, our current materials are more abundant. According to incomplete statistics, by the end of 2009, there are a total of more than 60 sports TV channels, wherein there are 17 sports Satellite TV channels with sports as the main broadcast content. It means that we can see sports TV programs in every corner of our country.

Those rich sports movie resources are extremely favorable to college physical education. In fact, in virtual college physical education, many teachers take good use of numerous movie materials, but it is rare to summarize the role of movies and teleplays in physical education teaching. This paper tries to put forward some immature opinions in this regard, for colleagues and experts reference.

2   INCREASE COLLEGE STUDENTS' SPORTS KNOWLEDGE THROUGH MOVIES AND TELEPLAYS

The inadequacy of college students' ability of appreciating sports artistic is first reflected in the limited sports knowledge. Although many college students know physical education is important, they don't think it does matter for exercise when they are young, and won't pay attention to the ways of keeping good health until they are in middle or old age. Some students think that Chinese martial art doesn't belong to the range of sports, as well as social dancing and other kinds of dancing; some students think physical education is simply exercising their body by playing balls and running, and they think it is not necessary for them to learn the basic theory and basic knowledge of sports; some students even don't know the full name and the origin of Olympic Games, it is also needless to follow the rules of basketball games and volleyball games. The lack of these basic sports knowledge reflects the poor ability of appreciating sport artistic. In fact, physical exercise theory in the advanced countries has already gone out of fashion. They pay more attention on sports culture theory nowadays.

Under the heavy pressure of study, employment, and examination, many college students in our country not only have no interest in sports culture, but also ignore body exercise. The lack of physical knowledge is a significant factor affecting college students' physical quality in our country. The author pays more attention to teach students basic knowledge in the long-term teaching practice. At the final exam in each semester, the author not only tests students' physical skills, but also tests students' sports knowledge through close-exams. A good way to teach students sports knowledge is to make them watching sports film. Students can master sports knowledge when they watch these sports films, such as basketball techniques and rules, technical rules for volleyball and table tennis and so on. By watching movies and television, students can remember very well.

3 IMPROVING COLLEGE STUDENTS' INTEREST IN SPORTS AND SPORTS SKILLS THROUGH MOVIES AND TELEPLAYS

At present, various sports competitions and live broadcast have attracted a large number of sports enthusiasts. But at university, students have little free time to watch athletic contests under the pressure of study and time. If teachers can play some sports movies in physical class, it would enhance students' interest in sports greatly, expand their horizon, improve students' sport artistic appreciation, and improve their ideological level and knowledge level. Teachers can arrange students to watch a higher level live sports games at a particular time, so that they can carefully enjoy the world level athletes' excellent skills. Generally, colleges have their own school sports teams, such as dance team, football team, basketball team, volleyball team and so on. The gym teachers can ask players to watch televised live matches during their usual training, talk about their experience after watching, and find the differences and deficiencies to improve their competitive level. Teachers can also cultivate students' patriotism through arranging students to watch sports films, such as "Wu Linzhi" and "Huo Yuanjia".

Anyway, the outstanding sports movies and the extremely rich TV sports programs in our country can be used in university sports teaching. As an advanced teaching technology, movies and teleplays play an irreplaceable role. In many cases, the effects of movies and teleplays are far better than the theoretical knowledge. Watching movies and teleplays about sports can help them understand the development and other aspects of sports and improve their sports skills at the same time. Sports movies and teleplays are the live teaching materials for training students' patriotism and collectivism. Urged by strong sense of mission and responsibility to win for school and motherland, some excellent athletes adheres the spirit to win without pride, until they win the honor for schools and countries.

The thing which should be paid attention to is that most physical education classes in China are extremely limited. This requires teachers to focus on efficiency when taking use of movies and teleplays. The teachers should try to compress the time of playing films and teleplays in classrooms. Students can use extra-curricular time to watch movies and teleplays about sports. Students will benefit from physical culture when students have rest and entertainment at the same time, increase their knowledge of sports culture. Above all, it will not occupy their valuable time. In addition, teachers should try to select the most suitable sports video.

REFERENCES

Anderson J.R. and G.J. Julien, A Ikegami, R., D.G. Wilson,. 1990. *Active Vibration Control Using NiTiNol and Piezoelectric Ceramics,* J. Intell. Matls. Sys. & Struct, 20(2):189–206.

Hoffer R. and D. Dean. 1996. Geomatics at Colorado State University, *presented at the 6th Forest Service Remote Sensing Applications Conference*, April 29–May 3, 1996.

Inman D.J. 1998. Smart Structures Solutions to Vibration Problems, *in International Conference on Noise and Vibration Engineering,* C.W. Jefford, K.L. Reinhart, and L.S. Shield, eds. Amsterdam: Elsevier, pp. 79–83.

Margarit K.L. and F.Y. Sanford. March 1993. Basic Technology of Intelligent Systems, *Fourth Progress Report*, Department of Smart Materials, Virginia Polytechnic Institute and State University, Blacksburg.

Mitsiti M. 1996. *Wavelet Toolbox,* For Use with MALAB. The Math Works, Inc., pp. 111–117.

Sports Engineering and Computer Science – Luo (Ed.)
© *2015 Taylor & Francis Group, London, ISBN 978-1-138-02650-6*

Research on current situation and countermeasures of physical training for elite male athletes of Dragon Boat Racing

Bing Li

School of Physical Education, Northeast Dianli University, Jilin City, China

ABSTRACT: After the reform and opening up, the dragon boat project has been developing rapidly, and is widely spread across the world. According to the Olympic spirit, the dragon boat race has gradually transformed from a traditional folk sport into a competitive one. The dragon boat competition requires athletes to possess good physical basis, so as to gain more competitive skills, and to achieve excellent results in competitions. After 30 years of development, thanks to the unremitting exploration of coaches, athletes and scientific research personnel, the standard of Chinese Dragon Boat Sports has been greatly improved. But considering other ability levels, there is still a large gap. [1] Through a questionnaire survey and some expert interviews, causes of the gap are as follows: 1) Influence of the traditional view about dragon boat sports; lack of knowledge of the Dragon Boat Festival; insufficient attention. 2) Lack of a unique scientific system in physical training; incomplete system of training mode. 3) Use of modern scientific methods; unbalanced understanding of the important role played by scientific equipment in the training. 4) Single composition structure of athletes; weakness of cultural knowledge; limited physical training of comprehension; inefficient training effect. 5) Unqualified coaches; imbalance between special physical training programs and the technical characteristics of dragon boat movement; special physical fitness of low level athletes.

1 THE MAIN PROBLEMS OF OUR ELITE MALE ATHLETES CONFRONTED BY PHYSICAL TRAINING

1.1 Insufficient attention to physical training

Through access to China HowNet and investigations on athletes, the current research status of Chinese athletics dragon shows that researches are more concentrated in the dragon boat rowing technique rather than in Athletes' physical condition. Content is the only index to study the characteristics of physical mechanism. Still, there is a lack of research on physical fitness training system. Obviously, now the competitive dragon boat training idea is about light weight technology physical. This reflects the widespread view: as to the dragon boat industry, technology is the only factor that determines the outcome of the game. So, every time we see special camera boat of the scientific research personnel in the dragon boat competitions, those videos only overemphasize research technical characteristics and ignore the great effect of physical play in matches. Through investigations and interviews, athletes on the importance of physical training is also far from enough, which argues that physical training is so important that accounts for 20% of the total survey. The general attitude accounts for 47.7% of the total survey. Almost

half of the players do not pay enough attention to the physical training. The survey also reflects that the training content of various sports teams is not ideal, mainly because there's no scientific training system. Training plan, content and strength are determined merely by the coach's subjective experience, which is lacking in certain scientific evidences. Creating of the training content is not reasonable. Special training is not prominent and the training effect is poor. According to the theory of energy supply, dragon boat racing is a physical fitness project. However, there's no physical fitness as the dominant action in technology and perfection is not likely to come out to play. Physical fitness and technology are the two important factors of the dragon boat racing and they are complementary. It must be corrected if we ignore one side of the training, as it breaks the laws of motion. [2]

1.2 Less time for physical training

The boat is not the national focus during the development of athletics. It has no state system of security or professional athlete. Through the investigation on the players, most of them are farmers or freelancers, as their lives are so busy that they could only participate in the training in their leisure time. They have less training time. Athletes who have trained for 2–3 times account for 22.1%

of the total survey, those who hardly train account for 42.3% of the total survey. It means there's little time for physical training, which is unable to meet the need of the increasingly fierce competition. With the application of modern training techniques, dragon boat technology is also experiencing constant innovation. Time for technical training increases prominently, which has broken the traditional training content and method. Physical training time is occupied and physical training is gradually reduced. Paying attention to technical training only and ignoring the training is a wrong idea that must be corrected in future training.

1.3 The lack of specialized training methods

Dragon boat racing is a folk activity in China with a history of more than 2000 years. As the event is to be carried out after the reform and opening up, dragon boat athletes with physical training methods of similar projects account for 41.8%, the special physical stamina training boat accounts for only 14.4%. At present, some contents and methods of training our boat inherit the traditional ones, but they cannot adapt to the need of dragon boat sports physical demands, which is increasingly high. It negatively affects the development of dragon boat sports. At the same time, due to the late start of modern dragon boat racing, it has no special strength training method of its own, instead, it mainly uses other project training and physical training for reference. This method can have a certain effect on physical training in a short time. However, due to different projects, the physical characteristics of energy supply and muscle force approaches are also different, which will deviate from the actual requirements for physical dragon boat project long time after training. Therefore, we must change the special training which does not meet the need of special sports practices. We should combine it with the characteristics of the project itself and explore suitable solution for its own special strength training method.

1.4 The lack of modern scientific and technological means of physical training

The development of modern science and technology provides scientific help for physical training. Some physiological and biochemical instruments can accurately detect the athlete's functional level. They have realistic guiding meaning to the physical training. Physical training athletes who have no effective detection equipment account for 77.2 of the total survey, those who can satisfy the physical training monitoring need only account for 7% of the total survey. In China, physical training of athletes has no effective training system. Therefore,

with the help of scientific instruments, the coaches can have control of good in the process of training. In the future physical ability training, the coaches can adjust training according to various physiological, biochemical indexes of strength and content, and avoid blind excessive training effect. [3] At the same time, according to the index of the feedback, which has been gradually worked out for the athletes' physical training and the training load intensity method, some important indexes of training help to summarize a set of specialized physical training of dragon boat system.

2 RESEARCH ON IMPROVING THE FITNESS LEVEL OF OUR ELITE MALE DRAGON BOAT ATHLETES

Improving physical fitness is the effective way to achieve excellent results. If athletes participate in the competition without good fitness, the match will be not competitive at all. While the modern dragon boat racing promotes the traditional culture, it also pays more attention to the dragon boat competition level. With the popularization of dragon boat project, there are more international competitions of dragon boat racing, the technical level of national sports also increases by years, which brings a challenge to us. Compared with other foreign powers, athletes of Chinese dragon boat racing have been the major factor that hinders the development of the project. Only by improving the physical fitness level of the athletes and the level of Chinese dragon boat sports can we make a difference. Therefore, as research indicates, we should spare no effort to ensure our traditional strengths, enhance the overall competitive level, and further strengthen the physical fitness of training.

2.1 To ensure scientific training

Science is an objective law which effectively reflects the development of things. The objective laws of scientific physical training process are followed by physical training. The high physiological needs of modern dragon boat racing are shaped by its high intensity. Therefore, the training mode of dragon boat athletes should be revolved around the standard system of scientific training. First of all, the physical training of dragon boat ought to be conducted in a scientific way. The time arranged for technical training, physical training and refreshing is far from enough, which is the direct factor that damages the health of athletes. Therefore, we must carry on the reform of training process. Meanwhile, we should strengthen technology training and increase the proportion of physical training, making sure that substantial supports are arranged for physical training. Secondly, the physical training of dragon boat should be conducted in a scientific way. Considering modern dragon boat on Athletes' physical requirements and human biological characteristics, training based on those principles and a treasonable raining content system are the embodiment of scientific training level. Therefore, we need to reform the contents of physical training and arrange some targeted and practical training content of special physical fitness. The contents of physical training are scientific, rational and efficient. Still, the dragon boat athletes training methods should be scientific as well. The development of modern science and technology, has provided a good platform for human to understand themselves. Through the physiological, biochemical instruments and the monitoring of physical training, people can better understand the change of their functions, which will provide timely and accurate feedback for physical training. Hence, we should vigorously promote the application of high-tech, applied technologies and the use of scientific instruments. Meanwhile, supervision and guidance on training are necessary to avoid the blindness and randomness in training, enhance physical training operation and improve the efficiency of physical training.[4] Finally, physical training of Dragon Boat athletes should receive scientific guide. At present, training is arranged by coaches. Coaches should be responsible for the technical training and the physical training. They are under great pressure for taking care of two items. Therefore, we should make reference to other excellent projects and hire professional and knowledgeable coaches, specifically responsible for the physical training. The physical ability level of athlete can be greatly improved if their training follows the objective law.

2.2 *To ensure the harmonious physical development*

The athletes' training includes two aspects: the general physical training and physical training. General physical training is a special foundation. Special physical training is one of it, which is inseparable in the process of physical training. We must coordinate its development. Physical training is aimed to improve the dragon boat sport. Therefore, professional training must be in accordance with the actual requirements of the dragon boat competition, pay close attention to the dragon boat sport characteristics of energy supply, technical and tactical arrangements. Without special physical training, we cannot improve the athletic level or meet the need of modern dragon boat competition. However, of organ system functions of any special physical training have their own limitations.

Contents and means of General physical training can constantly be changed. It can make up the deficiency of the special physical ability training, laying a solid foundation for the special physical training. Meanwhile, the general physical training can adjust if special physical training fails. The general physical training and physical training are two important factors in physical training, ignoring or canceling either side will lead to less efficient training, or even failure. Therefore, we must combine the two to make sure that they develop harmoniously.

2.3 *Pay attention to the scientific selection*

The scientific material is a key link in the training of high level sports talent. Practice has proved that everyone cannot become an elite athlete after years of training. Only those who exercise most can make it. At present, Chinese dragon boat athletes are composed of farmers and free occupations, some of them are not congenitally suitable for the Sports, which hinder the development of dragon boat racing. Therefore, in the future, we should strengthen the scientific material of dragon boat and choose the best athletes for the racing. We can base on the three aspects of human body (body shape, movement quality, body function) to utilize natural science and social science knowledge and methods with the help of instruments and equipment related. Through objective indicators and measurement data, we can compare the comprehensive evaluations of scientific and comprehensive analysis; select the more suitable three aspects to meet the actual needs of the athletes. The so-called "empirical method", "observation" and the single method can be more scientific and accurate for the next training selected material with the help of coaches.

2.4 *Improve the comprehensive quality of coaches*

Top athletes are cultivated by top coach. It is universal law in the sports circle. Due to the late start of Chinese dragon boat racing, most coaches in the dragon boat club are from the traditional dragon boat team. Their overall cultural quality is not as well as their professional skills. They have little knowledge about training basic theory, weak awareness of physical training, and low coaching level. It means they cannot fully understand the real meaning of the modern dragon boat racing, which, to some extent, hinders the development of dragon boat racing. Therefore, to cultivate a group of highly educated, competent and devoted couches enlarge the coach team, the science training idea, way, method should be applied to the dragon boat training and then new momentum can be injected to the development of dragon boat racing. Therefore, we can learn from other projects to implement the "absorbing and going outside" strategy. We can absorb those basic physical training theories and the scholars' work ethic and then apply them to our coaches, making them the most professional coaches. Combining their rich physical theories with the experience together to achieve the training target. Now, some couches are "going outside", taking full time or distance classes to learn training theories and methods. They attend seminar of theory study to strengthen their physical training theories and enhance their teaching level.

2.5 *Summary*

Through the research on the physical training of Chinese male dragon boat athletes, there are several factors affecting their physical development, among which the most eminent are the Modern Dragon Boat Sports Cognition attention, scientific training system, training mode, the modern scientific method, scientific research equipment, players' use structure and cultural knowledge, the teaching level of the coaches and so forth. Special physical training programs are not implemented because of the technical characteristics of dragon boat racing, which results in low level of special physical fitness of athletes. But of the male athletes' physical training in our country at the present stage, physical training gets insufficient attention and time. The lack of specialized training equals the lack of physical training, modern means of science and technology and other major issues. Besides, putting forward some countermeasures to improve the male athletes' physic level can ensure the scientific physical ability training as well as a coordinated development of physical fitness. Moreover, we need to pay attention to the scientific selection and improve the comprehensive quality of coaches.

ACKNOWLEDGEMENT

Fund Project: The doctoral scientific research foundation of Northeast Dian li University (bsjxm-201335); Academy of Social Science Fund Project in Jilin province (2013B238).

Brief introduction of the first author: Li Bing (1972-), male, professor and doctoral candidate.

REFERENCES

[1] Zheng Wenhai, Yang construction. The present situation and development strategy [J], of traditional dragon boat race and the modern competitive dragon boat sport combined with China's dragon boat racing and Sports Journal of Xi'an Physical Education University, 2007, 09, 25.

[2] Yang construction. Our traditional festival sports research status and development of [D]. doctoral dissertation of Shanghai Institute of Physical Education, 2007.04.

[3] Yang Luosheng. Dragon Boat Race research status and literature review [J]. Journal of Lingling college, 2004, 07, 15.

[4] Cao Jingwei for the 2008 Olympic Games Chinese Canoeing (hydrostatic) theory and empirical research on the scientific selection of athletes [D], Doctoral Dissertation of Beijing Sport University, 2004, 05.

Sports Engineering and Computer Science – Luo (Ed.)
© *2015 Taylor & Francis Group, London, ISBN 978-1-138-02650-6*

Study on English teaching strategies for majors of Physical Education

HaiYan Zhao
College of Foreign Languages, Northeast Dianli University, Jilin, China

ABSTRACT: In recent years, English plays as an international language. With an increasing number of international sports events and frequent exchanges between countries in the field of sports, English is becoming more and more important for majors of Physical Education (PE majors). The author is in view of the weak foundation, low interest of PE majors towards English and the monotonous teaching method. Thus, this paper puts forward that teachers should combine situational teaching strategy, task-based teaching approach, cooperative method and communicative strategy cultivation. By doing so, they can stimulate students' interest and initiative in learning English. As a result, using the method can improve English teaching effect, promote the development of China's sports undertakings and adapt to the trend of sports internationalization.

1 INTRODUCTION

With the further development of globalization, international exchanges and cooperation have become increasingly frequent in many fields, which holds true for sports. Therefore, Sports majors' universities have gained rapid development and great importance has been attached to English teaching for PE majors in recent years. However, having spent most of their time on sports training, PE majors are poor at academic studies, especially English. Due to insufficient vocabulary reserve, poor mastery of grammar, coupled with the complexity and abstract nature of English, PE majors would gradually lose their enthusiasm, which is bound to deepen the difficult with English teaching. Therefore, it is utterly necessary to explore suitable English teaching strategies for PE majors.

2 ANALYSIS OF CURRENT STATE OF ENGLISH TEACHING FOR PE MAJORS

2.1 *Analysis of PE majors' English level*

PE majors are different from students doing other majors. PE majors mainly come from two sources. Some students are admitted to college after having sat for the College Entrance Examination. Others are graduates from vocational schools or sports schools. It is not because they want to major in PE for the former group of students to choose PE major, but because they are poor in academic performance. Because of the low requirement for the scores of academic courses in entrance examination, these candidates spent a lot of time and effort

in training while neglecting academic courses, especially English. It needs to carry a lot of time for listening, speaking, reading and writing. The latter group students are even poorer in academic courses since they did not do a rigorous three-year study in senior middle school. Therefore, PE majors are generally inferior to other non-English majors in both language competence and language performance. On the other hand, many colleges and universities place a low requirement on PE majors in English course. These factors all contribute to the low interest and low pressure in learning English.

2.2 *Disadvantages of teaching methods*

English course is offered to PE majors in all colleges, but it is rarely affected by the wave of college English teaching reform. English teachers still adopt traditional teaching methods, which followed a fixed procedure. Firstly, students read new words after the teacher; secondly, the teacher explains new words by giving examples; and then the teacher himself/herself or the students will translate the text sentence by sentence. Throughout the teaching process, the teacher is in the dominant position and students are just passive recipients of knowledge. There seldom exists interaction among students. It is evident what kind of effect will come out of such a boring teaching process. In view of the specificities of PE majors, English teachers should adopt a variety of teaching methods to stimulate students' interest in learning English so as to effectively improve English teaching effect.

3 AN EXPLORATION OF ENGLISH TEACHING STRATEGIES FOR PE MAJORS

3.1 Application of situational teaching strategy teaching process aided by computer technology

Situational approach, which came into being in 1920s, refers to a teaching process during which the teacher purposefully introduced or created vivid and specific scenes to activate the context of teaching materials. So that students feel integrated with it, so as to help students understand and acquire knowledge. It is a collaborative teaching method, which depends on the cooperation between the teacher and students as well as between students and students.

The crucial function of situational approach is to stimulate students' emotions and directly improve students' motivation to learn. As a result, students will consider learning English as a fascinating experience and actively engage in learning activities. Take the passage entitled "A Busy Weekday Morning, taken from Unit 2, Book One of New Horizon College English", as an example. Before the class, the teacher may ask students to decorate the classroom to make it a living room. Three students are designated to play the roles of father, mother and daughter respectively. After the performance, the other students may be asked to analyze the root of conflicts between parents and daughter and then evaluate the performance of the 3 actors or actresses. In this way, students may feel so relaxed that they are willing to engage in learning activities and they can be impressed by knowledge more deeply.

3.2 Application of task-based approach in combination with communicative approach

Task-based teaching approach emphasizes the learning of language by implementing tasks. Communicative approach stresses the key position of students and the communicative activities in learning process. The combination of the two strategies can ensure the combination of language learning and language use. In the meantime, the state of affairs that the teacher alone is responsible for designing learning tasks can be changed. Under the new model, the teacher can make requirements and students can put forward suggestions. It is during the cooperation and interaction between teachers and students that English is acquired and the students-centered philosophy is realized. In addition, the teacher can give guidance and provide feedback.

The teacher can organize the teaching process in the following steps: (1) making clear the teaching objective of the lecture; (2) dividing the whole class into several groups and assigning tasks to each group; (3) asking students to design communicative tasks according to requirements; (4) regrouping students on the basis of each group's performance and putting well-performed and poorly-performed students together so as to mobilize all the students' initiative to engage in activities; (5) evaluating students' performance and asking students to make self-assessment and mutual assessment so as to prepare for next task.

3.3 Application of cooperative teaching approach

Cooperative teaching method relies on the division of labor and cooperation among team members, who share learning resources. Besides, groups are encouraged to compete with each other to enhance the effectiveness of learning. On the one hand to make learning opportunities more equal, on the other hand makes a strong motivation to learn more. On the one hand, learning opportunities become more equal; on the other hand, students have a much stronger motivation to learn more. For example, when teaching a lesson related to sports, the teacher may ask each group to prepare some sports tips, sports star stories, sports adage etc. before class.

One representative will stand for each group to give a presentation in class. Thus, students can learn both English and sports knowledge.

3.4 Cultivating students to apply communicative strategies

Communication strategy refers to the way the speakers turn to when they encounter difficulties in communication due to lack of the necessary language knowledge. It includes two major types: achievement strategies and reduction strategies. In order to promote students' language skills, the former is advocated to be put in use, which can be subdivided into paraphrase way (such as using synonyms, and alternative), non-verbal strategies (such as mimicry and gestures), asking for help strategies (such as questions and queries) and time-delaying strategy (such as the use of fillers and repetition of words).

The teaching of communicative strategies can be preceded in 4 steps. Firstly, students are provided with samples of communicative strategies. Take the paraphrase way as an example, whose sample can be stated as "the thing(s) that is (are) used to/ for …, It is something you do/say when …, something that you can … with, it is a kind/sort of …, they are the people who …" Secondly, students practice using communication strategies. A variety of ways can be adopted. For example, the teacher can prepare some cards with new words on them.

Two students come to the front of the classroom, one of them explains the words on the cards in English and the other tries to guess what the exact words are. Thirdly, feedback exercises are used to improve students' ability to use communication strategies. The teacher records and plays students' output from the second step, and then ask students to make self-assessment and peer assessment. This step is designed for students to find out the proper communicative strategies used by themselves and others as well as whether some better strategies can be used. Fourthly, students expand knowledge through reflection. After the successful use of relevant communication strategies, students should grasp the knowledge they did not know by consulting dictionary or related information through this knowledge.

Learning and application of communication strategies depend on communicative contexts. Students in China have fewer opportunities to communicate with native English speakers. Schools and teachers should try to create opportunities for students to speak English. On the other hand, communication strategies are taken by students to accomplish communicating process when they lack necessary language knowledge. In this way, students can accomplish infinite communicative purposes with limited knowledge.

Therefore, some students may depend so much on them that they do not want to learn new knowledge. The teacher should take note of the problem and use appropriate methods to guide students acquire new knowledge by using communication strategies.

4 CONCLUSION

In current society, interdisciplinary talents have more potential to become successful. A solid foundation in English and a good command of English will not only promote the personal development for PE majors, but also promote the international development of sports undertakings in China. Therefore, the university should pay attention to the teaching of English for PE majors. English teachers should improve their teaching methods in an effort to stimulate students' interest and initiative in learning English, thereby improving their comprehensive English competency.

REFERENCES

Chen, Qi & Zhang Jianwei. 1998. Constructivism and Teaching Reform. Educational Research and Experiment: 153–154.
Liu, Qiaoyan. 2006. The Application of Cooperative Teaching Method in Sports English Education. Jounal of Huaihua University: 165–166.
Meng, Pei. 2012. Task-based Teaching in Sport English Education. Cultural Education: 218.
Wang, Cairen. 1996. Communication Theory in English Language Education. Nanning: Guangxi Education Press.

Sports Engineering and Computer Science – Luo (Ed.)
© 2015 Taylor & Francis Group, London, ISBN 978-1-138-02650-6

Study on the correlation between personality traits and physical health of college students

JinSuo Ren

Department of Physical Education, North China Electricity Power University, Beijing, China

ABSTRACT: Interrelated analyses on personality characteristics and constitution health of freshmen have been conducted by Eysenck Personality Questionnaire [EPQ]. According to the result, the author shows that there is a difference between male and female university students on P and N dimensionality of personality characteristics. In addition, it turns out to be a distinct difference between personality characteristics and constitution health. Reality, the persons who have more extrovert trend [E] and emotion stability [N] on personality characteristics than others have better condition in constitution health of freshmen.

Keywords: Eysenck Personality Questionnaire [EPQ]; constitution health; university students

1 INTRODUCTION

University and college education stage is the period when undergraduates' character and personality characteristics gradually mature and fix. Besides, it is also a key period when laying a solid foundation for physical quality. Numerous information shows that there exists a obvious conflict in the personality development of university and college students. Furthermore, the stage is a course of dividing, integrating and unifying. Meantime, university and college education is the last school education stage before most students take up an occupation. It is also the key stage of character formation, thus during which it's malleable for students both physically and mentally. If personality characteristics and physical quality are strongly developed during this period, it will mean too much for students to adapt to the competitive society and improve future living quality.

This article proves that there is an association. The association discusses between university students' personality characteristics and physical health in university and college education period, and that introvert and unsteady characteristics adverse to good physical foundation. So, as university and college physical education workers, apart from passing on sports knowledge and skill, we should also pay attention to every student's character feature in psychological view and make pointed references in teaching method and tools. Teach as well as foster, and make students to have both healthy physique and healthy personality characteristics. Thereby, it can lay a firm foundation for entering the society, adapt to it and improve future living quality.

Correlational researches on university students' personality characteristics and constitution health tests contribute to deeply understanding of students' constitution and physical quality. The purpose is to teach students in accordance of their aptitude and characters to achieve the aim of educating students. Eventually, the students can gain health psychology and sound physique. This article attempts to provide reference basis for university physical education teaching through the discussion on relevant factors of freshmen's personality characteristics and constitution health evaluation.

2 RESEARCH OBJECT AND METHOD

2.1 *Research object*

Based on 2016 senior students and 2173 junior students of NCEPU, Eysenck Personality Questionnaire [EPQ] survey and constitution health evaluation are conducted at the beginning of their entrance to university.

Chart 1. Research object distribution.

Research object	Amount	Male	Female
Senior	2016	1256	760
Junior	2173	1212	961

2.2 Research method

2.2.1 Literature review method

According to the research content, the author refers to periodical literature and works about personality characteristics and constitution health evaluation at home and abroad.

2.2.2 Psychology measurement method

This research chooses the adult edition of EPQ redesigned mainly by Chen Zhonggeng as a tool to evaluate students' personality characteristics. EPQ is a questionnaire with recognized high reliability and validity. Totally 85 questions, and each of them is prepared with 'Yes' and 'No' two alternative answers. Testees choose the answer according to their own situation. Experimenters respectively calculate testees' raw scores of each rating scale according to the scoring question number in Appendix 1.

It's noticeable that when testees choose 'Yes' in some questions they score 1 point while some questions are contrary, that is, when testees choose 'No', they get 1 point. For example, in E scale, question 1,5 and so on, option 'Yes' scores 1 point; question 26 and 37, option 'No' scores 1 point. In other words, if they choose 'No' in question 5, or they choose 'Yes' in question 26, they score nothing. Respectively add up the obtained score of each rating scale, then comes the raw scores of testees in this rating scale. And the raw scores can be converted into standard scores according to Appendix 2. So, every testee's rating scores and standard scores in each rating scale can be worked out.

Test result is analyzed mainly in terms of standard scores. EPQ is a self-report personality questionnaire involving 3 dimensions and 4 rating scales (as in Appendix 1):

E rating scale: 21 entries, scoring interval [27, 75] for male, [28, 79] for female;

N rating scale: 24 entries, scoring interval [28, 80] for male, [26, 78] for female;

P rating scale: 20 entries, [31, 93] for male, [32, 100] for female;

L rating scale: 20 entries, [27, 62] for male, [16, 73] for female.

Based on statistics theory, standard scores between 40 and 60 cover about 68.46% norm group, between 30 and 70 cover about 95.45% norm group. It is generally acknowledged that if one testee's standard score is higher than 60 or lower than 40, he is considered having high marks or low marks features in the rating scale. In addition, if the score is higher than 70 or lower than 30, these features are more distinctive.

L rating scale: test testees' dissimulation, pretext or self-concealment, or the plain and childishness level of sociality. High-marks people imply dissimulation, or maturity and ripeness, and itself stands for a stable personality function.

2.2.3 Instrument measurement method

The paper adopts cstf-2000 type tester developed by China tongfang-height and weight testmeter, vital capacity testmeter, gripe dynamometer, long jump testmeter and step testmeter-to measure. Measurement index is as follows: height and weight, vital capacity, grip strength, step test, and long jump indexes. To minimize the random error and guarantee the testing effects, in the whole test procedure, all is conducted in terms of the following requirements:

1. Unifying test conditions (room temperature, test frequency, etc.).

Chart 2. High marks and low marks features of three dimensions in EPQ.

Dimensionality	Features	
	High marks features	Low marks features
E	Extrovert, may desire excitement and adventure; easy to express emotion, impulse; enjoy taking part in lively gatherings, social; cheerful.	Introvert, quiet, live alone, full of introspection; cold to others except for close friends; dislike excitement, adventure and impulse, like regular life-style, rarely attack, steady mood.
N	May often feel anxious, nervous, worried, blue; Emotion fluctuate greatly, easy to react strongly, even without reason when facing excitement.	Tend to slow and slight emotional reactions. Even if emotion is aroused, easy to recover calm, generally perform steadily, moderate, and good at self-control.
P	May be lonely, tend to be alone, indifferent to others, hard to adapt to external environment, lack of sympathy, blunt, hostile to people, cannot get along well with others, obstinate, stubborn, like to provoke, aggressive, and neglect dangers.	Get along well with others, adapt to environment well, mild, not rude, considerate.

2. Operate in strict accordance with test detailed rules and instrument operation specifications.
3. The whole test procedure is carried out by trained teachers.

2.2.4 *Mathematical statistics method*

Study with correlation analysis and other methods. Conduct by social sciences statistic software package SPSS10.0.

3 RESEARCH RESULT AND ANALYSIS

3.1 *Test on university students' personality characteristics otherness*

From Charts 3 and 4, we can see personality characteristics of male and female has significant difference in P&N dimensionality. But in E dimensionality, there is no significant difference level. This result basically accord with the result of study and survey on gender in PE major and non-PE major by Chen Yuxia, etc. Male and female university students have no significant difference in E dimensionality (tendentiousness). It may be related to following factors:

1. Heredity. Genetic factor is the material base of the formation and development of character, and it is the concurrent result of heredity, education and environment. In terms of genetics, male and female have the same congenital quality and thinking model.

Chart 3. Descriptive analysis on university students' personality characteristics.

| Dimensionality | Senior | | Junior | |
	Male $\overline{X} \pm S$	Female $\overline{X} \pm S$	Male $\overline{X} \pm S$	Female $\overline{X} \pm S$
P	55.15	50.12	56.47	49.16
	7.23	6.58	6.91	5.14
N	48.74	51.16	48.12	51.46
	5.66	7.28	5.49	6.46
E	56.3	55.95	55.86	54.39
	5.57	4.49	6.13	6.07

Chart 4. Test on university students' significant difference in P, N, E dimensionalities

Dimensionality	Variance	F numbers	P
P	98.15	5.69	0.021*
N	86.69	5.36	0.042*
E	91.23	3.45	0.056

*Shows there is significant difference.

2. Students have close cognition in ability, skill and other things under the same culture atmosphere, social background and study environment. Male and female university students have the same chance to take part in public activities and almost similar life-style, that's the very result. In P&N dimensionality there is also significant difference. Male is higher than female. This is because university students are in transformation period, and they undertake the responsibility and pressure from all aspects. And male is always regarded as the hard core of undertaking the responsibility. It may impel male to rise in great vigor, be absorbed in personal achievements and ignore contact with external environment and other classmates. But this may be the result of higher psychological pressure than that of female.

3.2 *Comparison of different constitution health condition of university students in P&N&E dimensionality*

Statistical result of test on remain samples after rejecting samples of high L rating scale scores and existing dissimulation, pretext or self-concealment is as Chart 5 shows.

Constitution health evaluation is the inter-comparison of excellent, good and passed three groups. We can find from Chart 5 and Chart 6, university students with highly valued constitution health is different from those who have low evaluation in N&E dimensionality.

When N shows the low marks features (<40), meanwhile, E shows high marks features (>60), it indicates the characteristics of extrovert:

Desiring excitement and adventure;
Being easy to express emotion, impulse;
Enjoying taking part in lively gatherings;
Being social and cheerful;
Emotional reactions tend to be slow and slight;
Being easy to recover calm, moderate, and good at self-control even if emotion is aroused.

Chart 5. Descriptive analysis on excellent, good and passed university students in P, N, E dimensionality.

	P $\overline{X} \pm S$	N $\overline{X} \pm S$	E $\overline{X} \pm S$
Excellent	53.15	37.73	63.3
(86~100)	7.45	5.36	6.17
Good	57.64	55.81	52.82
(76~85)	5.12	6.29	7.77
Passed	55.48	65.19	35.72
(60~75)	7.17	4.99	6.06

Chart 6. Mean value comparison of excellent, good and passed university students in P&N&E dimensionality.

	Excellent (86~100)			Good (76~85)			Passed (60~75)		
	P	N	E	P	N	E	P	N	E
Excellent	–	–	–	1.34	0.025*	0.044*	0.89	0.019*	0.031*
Good	1.34	0.025*	0.044*	–	–	–	0.97	0.039*	0.041*
Passed	0.89	0.019*	0.031*	0.97	0.039*	0.041*	–	–	–

All these are the personality characteristics of university students with good constitution. Sports can promote the development of constitution health and affect character development, which is connected with people's nervous activities, competition and overcoming of various difficulties. Practice confirms that taking exercise regularly can change introvert character and get rid of quiet. Besides, lonely activity features especially cultivate the willpower, and engage one to be grave, self-restrain, resolute, stubborn, self-disciplined and other character features.

Participant students are required to possess the rapid reaction, stable emotion, stability and moldability in movements, and strong adaption ability as well. During long-term physical exercise, some factors of his character have several positive changes, which reflects their self-improvement and helps to better adapt to and serve the society, and that is main trend. The study also found that the influence that physical exercise has on their character development exists no difference in P dimensionality. It prompts that physical exercise may have different effects on different university students, but the effects are limited in some aspects of character, which has a small effect range.

4 CONCLUSION AND SUGGESTION

1. *Compared to female, male has the personality characteristics of independence and stable emotion.*

2. *There is significant difference between character features and constitution health level. Students with extrovert trend (E) and stable emotion (N) personality characteristics have higher level of constitution health condition while students with opposite situation have the contrary health level.*

REFERENCES

[1] Chen Yuxia, Sun Jie. Personality characteristics difference of university students in PE major and non-PE majors. Journal of Physical Education. 2002(3):99–101.
[2] Study and research group of students' constitution health standard. Exercise handbook of students' constitution health standard. People's Education Press., 9, 2002.
[3] Chen Zhonggeng. Project analysis on Eysenck Personality Questionmaire [J]. Psychology Journal, 1983(2).
[4] Wang Shulan, Yang Yongming. Youth psychics introduction. Xi'an: Shanxi University Press, 1986. 120–122.
[5] General Administration of Sport of China. National Civil Constitution Monitoring Center.2000 Civil Constitution research report. People's Education Press, 2003.1:57.
[6] Qiu Yijun. Reviews of some Questions on Sports Psychological Theory [J]. Foreign Psychological Science, 1982(2).
[7] Zhou Zhihua, Huang Haiping: Personality Characteristics and Suitable Material of Wrestlers. Chengdu Physical Culture Institute Journal. 2000(4):55–57.
[8] Ma Qiwei, Zhang Liwei. Sports Psychology. Zhejiang Education Press, 1998, first edition in May.

Sports Engineering and Computer Science – Luo (Ed.)
© 2015 Taylor & Francis Group, London, ISBN 978-1-138-02650-6

On the relationship between lifelong physical education and school Physical Education

HongSheng Zhao & Hong Zhang
College of Physical Education, Northeast Dianli University, Jilin, China

ABSTRACT: School PE is not only an important part of lifelong PE, but also the foundation for lifelong PE. However, lifelong PE is the continuation and development of the school PE. Only when this interface between them is recognized, the task of school PE can be accomplished to improve the students' ability for lifelong PE. We should be fully aware of the importance of lifelong PE in school PE as well as the important role of school PE in lifelong PE. Therefore, school PE should be reformed from many aspects to meet lifelong PE.

1 THE POSITION OF LIFELONG PE IN SCHOOL PE

Lifelong PE refers to a person who performs lifetime physical activity and receives physical education, growing and developing under the influence of lifelong education ideology. PE should be accompanied by a person's life. We divide the physical exercise activities into several stages. Then school PE lies in the middle stage of connecting family PE with community PE in the whole lifetime sports. School PE is part of the school education. It is an important stage in the implementation of lifelong physical education. In the stage, the school PE will accompany everyone through their career from their entering the school. The life in school is the most guaranteed stage during which a person can receive the system of physical education for the longest time. In addition, school PE enables the students to recognize school PE. School PE is not the end of a person's physical education and sports practice process, but the foundation stage of lifelong physical activity of a person and a round of a lifelong PE. Thus, students are improved on the knowledge of the school PE eventually, and become the beneficiaries of lifelong physical exercise and the guide of self-workout.

2 THE INFLUENCE OF SCHOOL PE ON LIFELONG PE

1. *Firstly, school PE lays the foundation for lifelong PE.* Children and adolescents in school are in a critical period of physical and mental development. Hence following the law of physical and mental development, and imposing a good physical education can effectively promote the normal development of the students. The aspects include the students' body shape, function and quality, improve students' health and enhance students' physical fitness. It is conducive to the healthy growth of students during school. School PE is to make them energetically committed into learning. Besides, it is to lay a solid physical foundation for the students' learning, work, and engagement in lifelong PE after their immersion in the society.

2. *Secondly, school PE tries to train students for life awareness, habits and the ability to lifelong PE.* The school PE is a targeted and planned process of physical education. During the process, the students can improve themselves by learning and mastering the system of sports science knowledge, skills and scientific principles and methods of physical exercise. On one hand, students' physical health can be promoted and physical fitness enhanced. On the other hand, students' awareness of lifelong physical, habits and abilities can be cultivated.

3 HOW SCHOOL PE FACES LIFETIME PE

3.1 *Guiding physical education reform through the use of lifetime PE*

School PE is the entry of lifelong PE. On the contrary, lifelong PE ideas have a profound impact on the reform of school PE. Physical Education is a central part of school PE, as well as the focus of the reform. In physical education the instructors enable students to master the sport knowledge, technology and skills, and develop awareness of physical exercise, interests and habits. By this way, the students can enhance physical fitness, and strive to promote the harmonious development of body and mind as the goal. The instructors should also make full use of the ideas of lifelong physical education to guide the teaching. The purpose is to allow every educated student to hold the consciousness of maintaining and enhancing the health. As a result, Physical Education adapts more responsively to the needs of social development.

3.2 *Changing concepts, focusing on students interested in sports, training habits and ability*

The students took to the community can make the conscious physical exercise and adhere to engage in physical exercise. After that, the key is to make the students receive physical education process to develop sports interests and habits. Secondly, we should strengthen the ability of students to sports training. In the end, students in school sports master certain theoretical knowledge, certain skills, technology, and autonomous rational exercise. School sports should cross out of school. The thing could happen after the students embark on society, still adhere to consciously exercise, until the whole process throughout its life and lay a good foundation.

3.3 *Lets students choose sport*

Lifelong physical implementations rely on students' interest in sports and physical skills necessary, which is the direction of physical education reform. Improving students' interest and physical skills is the key to lifelong physical implementation. In real life, skills and personal interest in sports and good habits have a causal relationship. As long as there is this interest, students can be able to master certain skills in this area. Similarly, with the skills they will generate interest. Schools should be more difficult to set those large lifelong sports. Those sports are too big, easy to carry. Besides, they are the number of unrestricted energy for adults, the elderly. Finally, students can enhance the physical practicality of large lifelong sports such as martial arts, swimming, power lifting running, fitness and bodybuilding, dance, table tennis, tennis and so on.

3.4 *Reform of teaching methods, students play the main role*

At present, China's middle school physical education is in terms of teaching ideas, forms of organization or teaching methods. The school physical education has yet to break the shackles of traditional. It is a closed model of physical education. Besides, this teaching is teacher-oriented, based on the student's physical condition. In addition, other factors are the semester, seasonal arrangements learning content, student activities in accordance with the provisions of the model. The simple pursuit of exercise order, exercise mode is also included. Therefore, the way in PE teaching reform of PE teaching should be to the student. Students become sports center in the teaching process. Teachers adopt a "heuristic" teaching, "small groups" teaching. The method can increase classroom fun, lively, and allow students to feel warm, happy and satisfied. As a result, the active input, active exercise, learning will become "passive" to "active", play the initiative of students. The purpose is to encourage students to actively participate in physical exercise.

3.5 *Improve teachers' professional quality, and strengthen the construction of sports teams of teachers*

PE teacher is to teach students the knowledge and skills of sports direct mentors. Their ideas, knowledge levels will have a tremendous impact on students. The factor is directly related to consistently achieve lifelong physical ideas, and even the "National Fitness Program" promotion. So, as a sports teacher cannot simply think: "I just have a lesson, and give good demonstration to complete the schedule of lesson." PE teachers should do many things: a) adopt a positive attitude; b) accelerate the transformation of ideas; c) strive to update their knowledge; d) enrich themselves; e) keep the "charge", lapping materials; f) and choose suitable teaching content combinations. As a consequence, the teaching process can interest students. They will be willing to engage in physical activity consciously. [2].

3.6 *Emphasis on sports theory of education, highlighting the sport to teach the students for life science knowledge*

When we choose materials, teaching materials should reflect on science and seek to propose effective, targeted guidance and times and other features. We should consider the reality. Moreover, we should focus on the future, the main content of the foreseeable lifetime protection for student health and exercise science. Furthermore, we also consider the need to work in the future for life knowledge. The content includes the basic law of sports, value and function of sport, physical development quality theory, simple common sports injury prevention and treatment formed. Finally, as the main content of the theory of knowledgeable

University, the main content should make students meet the future needs of the students.

3.7 *Grasp extracurricular sports activities, the effect of the consolidation exercise*

Extracurricular physical exercise is a continuation and supplement classroom teaching sports. It can improve and consolidate what they have learned in PE teaching contents. Besides, it can also cultivate students' interest, hobby. The important measure is to improve students' ability to exercise and exercise habits. Physical exercise is the ability to participate in sports activities in the form of physical exercise in extracurricular activities, organizational forms of competition events. In the end, physical exercise is to increase students 'opportunities to practice. In addition, it helps to improve students' sports awareness and the ability to consolidate the effect of exercise.

REFERENCES

[1] Liu Wei, Li Chuanzhu. The lifelong sports on. Journal of Qiingdao University Teachers College, 2000, 10.
[2] Zhao Zijian. A modern concept of survival—lifelong physical education. Cambridge University, 2000,09.
[3] Gao Yanhua, lifelong physical education and sports teaching reform in Colleges and universities (J). China School Physical Education Journal, 2002, 2.

Sports Engineering and Computer Science – Luo (Ed.)
© *2015 Taylor & Francis Group, London, ISBN 978-1-138-02650-6*

Research on the evaluation index system of university students' physical education self-regulated learning

Long Zhang
Department of Physical Education, Liupanshui Normal College, Liupanshui, P.R. China

ABSTRACT: In order to improve the self-regulated learning ability of university students, this paper is intended to establish a set of scientific, comprehensive and systematic self-regulated learning evaluation index system for university students. Besides, the method is unitizing approaches as theoretical analysis, experts' consultations and Analytic Hierarchy Process (AHP). There is the application of self-design, self-monitoring, self-assessment together with the combined utilization of the 4 first-class indexes, 12 second-class indexes and 30 third-class indexes. In addition, it is providing positive practical value in conducting tracking and evaluation on the current status and longitudinal changes of the self-regulated learning for university students.

Keywords: university students; physical education self-regulated learning; index system

1 INTRODUCTION

Self-regulated learning (autonomy) is originated from the Greek word "autonomia", which means self-government, self-government rights, and freedom and so on. As a brand new learning concept, it emphasizes the subject consciousness and creative spirit during the learning course of physical education of students[1].

Physical education self-regulated learning refers the following things. In the learning course of physical education, students determine the learning target and select the learning contents, and monitor the learning course. Besides, they evaluate the learning results by themselves under the necessary directions, guidance and assistance from teachers[2]. Faced with the society where the knowledge is under the constant renewal, we need to obtain the lifelong learning concept and learning awareness. Hence self-regulated learning is acting as the pre-condition. Moreover, it's clearly pointed out from "*National Medium- and Long-Term Plan for Education Reform and Development (2010–2020)*" that, by 2020, our country will basically form into the learning society. In addition, "*The Teaching Program for National College Physical Education Course*" released by the Ministry of Education also indicates that the role of learning subject should be fully exerted. Under the directions of teachers, the self-learning ability and self-practice ability should be enhanced.

At present, some domestic scholars are taking the perspectives of the conations, influences, impacting factors and investigations of physical education self-regulated and so on. For example, Deng Guoliang[3] and Pang Weihua[4] have performed theoretical analysis and intervention study,

joined with other scholars adopted the perspective of the physical education self-regulated scale. However, they haven't scoped physical education self-regulated learning into the target of education evaluation. As a result, there is a lack of respective method of evaluation index system and practical directions for physical education.

Therefore, the establishment of the evaluation system for physical education self-regulated learning will improve students' learning methods on physical education and teachers' teaching quality. In addition, it will provide further references to the reform on physical education classes and management works for normal universities.

2 THEORETICAL ANALYSIS OF THE INDEX SYSTEM

The author collected research materials about physical education self-regulated learning both at home and aboard. Finally, this paper conducts classification and comprehensive analysis on these materials in terms of definition, properties, features, influences of self-regulated learning and other aspects. The purpose is to be aware and grasp the current status and latest research achievements of this area. Based on the above, by unitizing the 4 first-class indexes, this paper initially defines the evaluation index system of students' physical education self-regulated learning in normal universities. The way includes self-design, self-monitoring, self-assessment and innovation.

2.1 *Self-design*

As indicated from the research of Dickinson in 1992, self-regulated learning refers that learners are able to identify the teaching objective of the teachers. They know what to learn, how to set up the study target for themselves and re-establish the teaching target. As a result, the students can satisfy the constant changes by themselves[5]. The famous western scholar on self-regulated learning, Zimmerman also pointed out that efficient management and arrangement plan on study schedule is an extremely crucial feature for self-regulated learners. That is to say, before starting self-regulated learning, students should clearly set up the learning objective, work out the learning schedule. Furthermore, they should also inspire the learning motivations and interests.

2.2 *Self-monitoring*

Through the study in 1985, Holec found out that self-regulated learning is the ability for learners to manage learning. It includes the setup of learning objective, self-monitoring and self-assessment and so on[6]. Self-monitoring refers to the adjustment and control performed by students on the external environmental factors. The factors are learning places, learning materials and learning time as well as internal factors such as learning habits and learning emotions and so on. It is done when the students overcome the physical and psychological barriers and monitor themselves in the course of learning. The main purpose is to make efforts towards the pre-defined learning objective.

2.3 *Self-assessment*

In the study of 1990, Chamot & O'Malley found out that after a task is completed, it's better to leave certain time to learners. Actually, only by this, can the learners reflect their learning activities and evaluate the utilization conditions of the learning strategies applied in completing the learning tasks. Besides, it's required for enough time to evaluate if the pre-defined learning objectives are reached or not[7]. Self-assessment refers that students perform assessment on the matching degree between learning objective, learning process, learning approaches and learning effects. This will assist students to think about their expectations from the study. Meanwhile, this will also improve their learning motivations and allow them to make objective rewarding and punishing assessments for practices or contests. As a result, it consequently will better make them understand themselves, grasp the learning rules and lay a solid foundation for future self-regulated learning.

Through study (2012), Wu Benlian discovered that physical education self-regulated learning is of unique value in cultivating the innovation ability for university students on difficult and beautiful sports items[8]. While the students' physical education innovation ability is not only confined to the discovery and invention, but also refers to a comprehensive

quality and learning ability. It includes the spirits and quality for innovative learning, the courage of independent thinking, innovation in learning ability, learning approaches and learning methods, and the ability. Finally, it can scientifically set up fitness plan (exercise prescription) and entertainment plan. Besides, it can handle sports injury correctly as well as the ability to apply the sports knowledge and skills learned to future career development and so on. Such ability is not only the objective of physical education self-regulated learning, but also is the detailed manifestation of the students' physical education self-regulated learning ability when it develops into higher stage.

Therefore, in the course of physical education self-regulated learning, it's required to always take the active and comprehensive development of students. His first priority is to pay attention to the subject role of the students and cultivate their self-innovation awareness. Moreover, he also made creative thinking ability and to inspire their learning motivations and interests. Furthermore, he allowed them to learn how to study through the course of physical education regulated-learning. Only through this way, will the university students develop the habit of insisting lifelong physical exercises.

3 PHYSICAL EDUCATION SELF-REGULATED LEARNING EVALUATION INDEX SYSTEM

3.1 *Principles of establishing the evaluation index system*

3.1.1 *Comprehensiveness of the evaluation index*
Principle of comprehensiveness requires the index system to fully reflect the integrity of the learning contents and structure of physical education self-regulated learning. At the same time, it is with emphasis and highlight on essential indexes. Meanwhile, this principle does not exclude to remove those indexes that are reflecting the minor issues. As a result, it can ensure the briefness of the system and achieve the effect of non-repetition and non-omission.

3.1.2 *Objectivity of the evaluation system*
The establishment of the physical education self-regulated learning evaluation index system is intended to perform comprehensive and objective analysis and evaluation. The establishment is based on taking relative evaluation theories and the actual conditions of normal universities students' physical education self-regulated learning into consideration. By doing so, it can ensure the objectivity, authenticity and rationality of the evaluation system.

3.1.3 *Operability of the evaluation process*
The various indexes in the self-regulated learning evaluation system should fully consider the

specialty of self-regulated learning evaluation. The system is different from the evaluations for learning types. Finally, it should allow students to make choices and controls on their own in various learning aspects. The main purpose is to design the evaluation process that is in accordance with the real circumstances.

3.1.4 *Principle of consistency between the evaluation index and the objective*

Evaluation indexes set should be consistent with the evaluation objective, namely to reach the expected target.

3.1.5 *Principle of development*

Evaluation result should not be based on one-time only evaluation. But it is the result generated based on multiple rounds of evaluations on the progress obtained, in order to assess whether students are developing towards the expected direction.

3.2 *Filtering process and analysis for evaluation indexes*

The filtering of university students' physical education self-regulated learning evaluation index is the crucial link to establish the physical education self-regulated learning evaluation index system. The research consults more than 40 first-class physical educational teachers. These teachers have been working in physical education teaching, sports and exercises, learning evaluations and teaching management from Guizhou University, Guizhou Nationalities University, Kaili University and Liupanshui Normal University together with 10 experts and scholars. As per research requirements, this research has designed expert consultation table and conducted two rounds of consultations on the index contents. Besides, by utilizing expert consultation approach and AHP approach, this research has preliminarily selected the evaluation indexes for qualitative and quantitative analysis.

3.3 *Reliability analysis on the experts' consultation results*

3.3.1 *Consistency degree of the experts' advice*

Coordinate coefficient is intended to reflect if there are giant divergences among the evaluation advices for each index. The index is given by the experts within the experts' panel. Through two round of consultation, the first, second and third-class experts' consultation coordinate coefficients are computed as 0.917, 0.901 and 0.891, respectively. All these are all next to 1, meaning the experts' advices are relatively consistent. In addition, through the significance test on all the coordinate coefficients of the indexes (P < 0.05), we can

observe the evaluations or predictions of experts' consults are of relatively high reliability.

3.3.2 *Experts' authority weight q*

The authority of the experts is exerting a tremendous impact on the reliability on the research results:

a. Authority degree q equals to the average value of the weight value of the experts' academic level ($q1$);
b. Weight value of the index judgment ($q2$);
c. Weight value of the familiarity degree of the investigation questions ($q3$), namely $q = (q1 + q2 + q3)/3$.

The average coefficient for the experts' authority weight of this research is >0.75. Generally, coefficient for the experts' authority with weight $q > 0.70$ is considered as acceptable, namely the evaluation authority is relatively high and reliable.

3.4 *Confirm the contents of the various indexes of the evaluation system*

On basis of the initial set up of the evaluation index, two rounds of consultations are performed by handing out consultation sheets to the experts. This paper adopts the advices and comments from the experts as well as the contents modification and number adjustments on the index system. Thus, this paper eventually sets up the physical education self-regulated learning evaluation index system of normal university as 4 first-class indexes, 12 second-class indexes and 30 third-class observation points.

4 ANALYTIC HIERARCHY PROCESS (AHP) AND THE ESTABLISHMENT OF THE EVALUATION MODEL

AHP is used for carrying out the decision analysis combining with qualitative analysis and quantitative analysis. It is based on the orderly hierarchies (goals, criteria, index and program levels), which are resolved from the complicated issues. Through comparing the factors pairwise and determining the superior degree index, it realizes the total order of the relative importance of the decision scheme on the goals. In solving the multi-goal decision scheme, compared with other methods, it is obviously simpler and more practical.

4.1 *Establishment of the hierarchical structure model*

According to the understanding of the ordinary university students' physical education self-regulated learning ability, the evaluation hierarchical structure system of the physical education self-regulated learning ability in the universities is determined (Fig. 1).

The evaluation index system of the university students' physical education self-regulated learning			
	Self-design A1	Goal setting B1	To understand the physical education goals of the university（C1）
			To understand the classroom learning goals and requirements proposed by the physical education teachers（C2）
			To formulate reasonable learning goals according to your own needs（C3）
		Reasonable planning B2	To flexibly arrange activity time according to the physical education learning requirements（C4）
		Motivational beliefs B3	To be confident about your own physical education learning（ C5）
			To be interested in physical education learning activities（C6）
			To actively participate in the physical activities organized by the university（C7）
	Self-monitoring A2	Strategy selection B4	To have your own physical education learning method which is commonly used and suitable for yourself（C8）
			To be good at making full use of various resources for physical education learning（C9）
			To consult others when facing difficulties in the physical education learning（C10）
		Content selection B5	To reasonably arrange the content of the physical education self-regulated learning（C11）
			To select the appropriate sports for exercise（C12）
		Emotion regulation B6	To take the initiative to　overcome the interference from the outside world（C13）
			To objectively treat criticism and praise（C14）
			To keep a good mood in the physical education learning（C15）
		Creating environment B7	To reasonably arrange and adjust the time of the physical education learning（C16）
			To select learning partners for cooperative learning（C17）
			To reasonably arrange comfortable places for the physical education learning（C18）

Figure 1.　(*Continued*)

		To establish the role model, and to actively seek help from teachers and classmates (C19)
	Self-reflection B8	To timely adjust the physical education learning plan according to the actual situation (C20)
		To observe and understand your own behavior and process of the physical education learning (C21)
		To correctly estimate your own level of the learning goals you have achieved (C22)
Self-assessment A3	Self-estimation B9	To summarize and evaluate the physical education learning at any time (C23)
		To conduct the attribution analysis of the sports technology, and then to find out the problems (C24)
	Self-reinforcement B10	To be self-satisfaction when obtaining good results of the physical education (C25)
		To positively adjust the strategy instead of negatively reacting to it when your physical education study has dragged behind that of the other students (C26)
	Innovation ability B11	To have the innovative learning spirit, and to have the courage of the independent thinking (C27)
		The innovation of the skills, use methods and means of the learning (C28)
Innovation and application A4	Application ability B12	To apply the learned sports knowledge and skills in the future career development (C29)
		To scientifically develop the fitness plan, and to have the ability of correctly dealing with the sports injury (C30)

Figure 1. The evaluation index hierarchical model of the university students' physical education self-regulated learning.

4.2 *Constructing the pair-wise comparative judgment matrix*

Table 1. The weight values partition table.

The relative importance	Equally important	Adjacent	Slightly important	Adjacent	Fundamentally important	Adjacent	Really important	Adjacent	Absolutely important
Saaty scaling	1	2	3	4	5	6	7	8	9

4.3 Calculating weight and the consistency test

With the judgment matrix, as for calculating the corresponding eigenvector of the largest eigenvalue of the judgment matrix, the power method or the approximate power method could be used. Besides, for computing the index weight, the AHP could be used. The paper uses the square root method to calculate the largest eigenvalue of the matrix. If a_{mn} is defined as the element of the m line and the n column in the k order of the judgment matrix, the specific operation is shown as follows.

4.3.1 Calculating the product of all the elements of each line in the judgment matrix

$M_m = \prod_{n=1}^{k} a_{mn}$, $m = 1, 2, \ldots k$; (k is the order of the judgment matrix).

4.3.2 Calculating W_m, the kth square root of M_m

$W_m = \sqrt[k]{M_m}$, $m = 1, 2, \ldots k$; (k is the order of the judgment matrix).

4.3.3 Standardizing $W = (W_1, W_2, \ldots W_k)^T$

$W_m = W_m / \sum_{m=1}^{k} W_m$, $m = 1, 2, \ldots k$; (k is the order of the judgment matrix).

The result of $W = (W_1, W_2, \ldots W_k)^T$ is the approximate value of the required eigenvector, namely, the weight of the element.

4.3.4 Calculating the largest eigenvalue of the judgment matrix

$\lambda_{\max} \approx \sum_{m=1}^{k} (AW)_m / K W_m$, $m = 1, 2, \ldots k$; (k is the order of the judgment matrix).

Wherein, (AW) is the mth element of vector (AW).

4.3.5 The consistency test

Due to the complexity of the objective reality or the one-sided understanding of the material objects, the consistency test and the random test are needed. The purpose is to verify whether the eigenvector (the weight) obtained through the constructed judgment matrix is reasonable: a) CR value is calculated as follows: $CR = CI / RI$, $CI = [1/(k-1)](\lambda_{\max} - k)$; b) wherein, CI is the consistency index of the judgment matrix;

c) $\lambda_{\max}$ is the largest eigenvalue; d) k is the order of the judgment matrix.

RI is the mean random consistency index of the corresponding judgment matrix (Table 2).

4.3.6 Calculating the index weight of each group

According to Table 1, the primary index weight, the largest eigenvalue and the random consistency ratio (Table 3) can be obtained from the above calculation process (3.3.1–3.3.5). Similarly, based on the judgment matrix of four secondary indexes and twelve tertiary indexes in the layer, the weight and the consistency test could be calculated respectively. Due to space limitations, the paper only constructs the judgment matrix of $A1$, $A2$, $A3$, $A4$ on A. The results of the weight are obtained as follows:

$$\lambda_{\max} = \sum_{i=1}^{n} \frac{(AW_i)}{nW_i}, CI = \frac{\lambda_{\max} - n}{n-1}, CR = \frac{CI}{RI}$$

The value is substituted into the formula, and the results are shown as follows:

$\lambda_{\max} = 4.1342$; $CI = 0.0502$; $RI = 0.90$; $CR = 0.0558 < 0.10$

Because of $CR < 0.10$, it is indicated that the judgment matrix of the primary index of the evaluation index of the university students' physical education self-regulated learning possesses is higher consistency. That's to say, the index weight setting is reasonable, and the hierarchical ranking is effective. The weight order for the above evaluation of the physical education self-regulated learning is: $W1 > W2 > W3 > W4$, which shows that, with the contribution rate of 78.13%, self-design and self-monitoring are the main factors of the evaluation of the physical education self-regulated learning. With the contribution rate of 12.57%, self-assessment is the important factor of the evaluation of the physical education self-regulated learning. On the basis of the above formula, it can also be concluded that all the results of the following judgment matrix ($A1 \sim B_i$, $A2 \sim B_i$, $A3 \sim B_i$, $A4 \sim B_i$, $B1 \sim C_i$, $B2 \sim C_i$, $B3 \sim C_i$, $B4 \sim C_i$, $B5 \sim C_i$, $B6 \sim C_i$, $B7 \sim C_i$, $B8 \sim C_i$, $B9 \sim C_i$, $B10 \sim C_i$, $B11 \sim C_i$, $B12 \sim C_i$) pass the consistency test.

Table 2. The mean random consistency index (M is the order).

M	1	2	3	4	5	6	7	8	9
RI	0.00	0.00	0.58	0.90	1.12	1.24	1.32	1.41	1.45

Note: When $CR < 0.10$, the consistency of the judgment matrix is satisfied, otherwise, the judgment matrix should be adjusted to make the consistency satisfied.

Table 3. A: A_i compared with the overall goal, the comparison of the importance between various primary indexes.

A	$A1$	$A2$	$A3$	$A4$	Weight coefficient W
$A1$	1	3	4	5	0.5474
$A2$	1/3	1	3	2	0.2339
$A3$	1/4	1/3	1	2	0.1257
$A4$	1/5	1/2	1/2	1	0.0930

4.3.7 The evaluation index weight system of the university students' physical education self-regulated learning

Based on the weight calculation from the previous section, the complete evaluation index weight system of the university students' physical education self-regulated learning is made below (Fig. 2).

4.3.8 The process of four-in-one evaluation system

The evaluation of the students' physical education self-regulated learning is completed by teachers, peers, students themselves and teaching management personnel. It can not only urge the students to change the routine, scheduled and mechanical learning method, but also no longer make exams the only evaluation of the students' study results in the physical education. And in the meanwhile, the students do not regard the teachers as the only source of knowledge any more. This evaluation focuses more on the procedural evaluation, including attendance, homework, class performance, learning attitude, emotion, progress margin and so on.

In the process of evaluation, the teacher's role has gradually shifted from the "soloist" in traditional teaching to the "accompanist". The teachers are also the beneficiaries in the evaluation process, because a large number of evaluation information can provide the specific and accurate feedback information for PE teachers. Finally, it is to promote teachers to perfect their teaching, and constantly improve their business ability and teaching level.

On the basis of four aspects in the "Tyler Rationale", the teachers' evaluation process is: a) determining the learning goals; b) quantifying the evaluation indexes and weights; c) organizing the implementation of the teaching plans; d) observing the students' learning state; e) evaluating the students' self-regulated learning state; f) integrating the information, summarizing the feedback; g) improving the teaching strategies.

Students' self-assessment of their own learning behavior enables students to understand the degree of their own knowledge and skills. The knowledge and skills are that they have mastered and discover problems in their learning. At the same time, it can also urge the students on to make constant progress and study positively, mobilize students' independence and autonomy of the learning, as well as improve the self-competence. All these help to improve the students' self-reflection and promote the development of physical education teaching activities.

The students' evaluation process is: a) confirming the goals and the evaluation indexes of the physical education self-regulated learning; b) using the evaluation criteria to judge the learning activities at any time during the study, students' self-assessment, peer-assessment, teachers' evaluative feedback, reflection; c) adjusting the learning strategies.

In the peer-assessment process, the students can learn from each other, deepen the self-understanding, and conduct the self-regulation. Peer-assessment can make up for the one-sidedness of PE teachers' unilateral evaluation and establish the integrity awareness among students. Finally, they can fully ensure the fairness of the evaluation results of the physical education self-regulated learning.

The peer-assessment process is: a) confirming the individual responsibilities and goals; b) making the practical appraisal according to the classmates' learning performance.

Teaching management personnel's evaluation can provide teachers and students with the information about their learning from another perspective. As a result, it is helpful to arouse the enthusiasm of teachers and students. Besides, it also contributes to developing the students' ability to fully understand and evaluate themselves and others as well. Through the evaluation, the good cooperative relationship between teachers and students is established, which can promote the further development of the students.

The teaching management personnel's evaluation process is: a) understanding the teaching goals, mastering the evaluation indexes; b) observing the teaching progress; c) observing the students' self-regulated learning state; d) examining evaluation and summary, feedback, reflection on the teaching strategies; e) adjusting the teaching management strategies.

5 CONCLUSION

1. *This research takes the concept, property and features of the university students' physical education self-regulated learning as the entry*

The evaluation index system of the university students' physical education self-regulated learning			
	Self-design 0.5474	Goal setting 0.3740	To understand the physical education goals of the university（0.0294） To understand the classroom learning goals and requirements proposed by the physical education teachers（0.0983） To formulate reasonable learning goals according to your own needs（0.2464）
		Reasonable planning 0.1088	To flexibly arrange activity time according to the physical education learning requirements（0.1094）
		Motivational beliefs 0.0639	To be confident about your own physical education learning（0.0400） To be interested in physical education learning activities（0.0087） To actively participate in the physical activities organized by the university（0.01573）
	Self-monitoring 0.2239	Strategy selection 0.1381	To have your own physical education learning method which is commonly used and suitable for yourself（0.0408） To be good at making full use of various resources for physical education learning（0.0821） To consult others when facing difficulties in the physical education learning（0.0140）
		Content selection 0.0441	To reasonably arrange the content of the physical education self-regulated learning（0.0154） To select the appropriate sports for exercise（0.0307）
		Emotion regulation 0.0255	To take the initiative to overcome the interference from the outside world（0.0166） To objectively treat criticism and praise（0.0063） To keep a good mood in the physical education learning（0.0036）
		Creating environment 0.0163	To reasonably arrange and adjust the time of the physical education learning（0.0050） To select learning partners for cooperative learning（0.0080） To reasonably arrange comfortable places for the physical education learning（0.0013） To establish the role model, and to actively seek help from teachers and classmates（0.0026）
	Self-assessment 0.1257	Self-reflection 0.0800	To timely adjust the physical education learning plan according to the actual situation（0.0539）

Figure 2. (*Continued*)

		To observe and understand your own behavior and process of the physical education learning（0.0181） To correctly estimate your own level of the learning goals you have achieved（0.0033）
	Self-estimation 0.0132	To summarize and evaluate the physical education learning at any time（0.0033） To conduct the attribution analysis of the sports technology, and then to find out the problems（0.0099）
	Self-reinforcement 0.0325	To be self-satisfaction when obtaining good results of the physical education（0.0065） To positively adjust the strategy instead of negatively reacting to it when your physical education study has dragged behind that of the other students（0.0260）
Innovation and Application 0.0930	Innovation ability 0.0155	To have the innovative learning spirit, and to have the courage of the independent thinking（0.0039） The innovation of the skills, use, methods and means of the learning（0.0116）
	Application ability 0.0775	To apply the learned sports knowledge and skills in the future career development（0.0258） To scientifically develop the fitness plan, and to have the ability of correctly dealing with the sports injury（0.0517）

Figure 2. The index weight figure of the evaluation system of the university students' physical education self-regulated learning.

point. And then based on the above, through literature review and experts' consultations, this research invites experts to perform filtering and modifications on the indexes. Meanwhile, with AHP method, this paper establishes the physical education self-regulated learning evaluation index system. Moreover, it also discusses about the implementation procedure of the four-in-one physical education self-regulated learning evaluation index system for the university students.

2. The physical education self-regulated learning evaluation index system is composed by 4 first-class indexes. The indexes include self-design, self-monitoring, self-assessment and innovation and application together with 12 second-class indexes and 30 third-class indexes. According to their weight from large to small, the first-class indexes can be arranged as: self-design, self-monitoring, self-assessment and innovation and application. Among these, self-design and self-monitoring are acting as the major factors

in the physical education self-regulated learning evaluation with a contribution rate as 78.13%. Self-assessment is the crucial factor in the physical education self-regulated learning evaluation with the contribution rate as 12.57%. This system has provided referential basis for the scientific evaluation on the physical education self-regulated learning of university students. Moreover, it has offered framework for the future development direction of the physical education self-regulated learning of university students, which is both highly scientific and reliable.

3. Due to the wide scope of the physical education self-regulated learning, educators are adopting different perspectives and emphasis in the research. As a result, it's of great difficulty and complexity to establish a relatively thorough physical education self-regulated learning evaluation index system. Hence, this research is just a preliminary exploration, which requires for further study.

FOUNDATION PROGRAM

Guizhou University of Outstanding Technological and Innovative Talents Support Program
(Scientific Research Program of Guizhou Education Department [2012] number 100.

REFERENCES

[1] Xin Li. Based on the network of middle school
    students' autonomous learning ability of the survey
    and research [D]. Harbin Normal University.
[2] W.U. Ben-lian, J.I. Ji. Discussion on the Self-regulated
    Learning in New Physical Education Curriculum [J].
    China School Report (6): 56–58.
[3] Guo-liang Deng. Investigation and Research on
    the Status Quo of College Students. Ability of
    Autonomous Learning in Physical Teaching [J].
    Journal of Beijing Sport University. 2007, (02):
    245–246.
[4] Pang Weihua, Zhou, Xuerong. Research on the
    Several Issues on Physical Education Self-regulated
    Learning [J]. Shandong Sports Technology. 2007,
    (01): 50–52.
[5] Dickson L. Learning Autonomy 2: Learner Train-
    ing for Language Learning [Z]. Dublin, Ireland:
    Authentic Language Learning Resources Ltd, 1992.
[6] Holec H. On autonomy: some elementary concepts
    [J]. Discourse and learning. 1985, 985: 173–190.
[7] O'Malley J.M., Chamot A.U. Learning strategies in
    second language acquisition [M]. CUP, 1990.
[8] W.U. Ben-lian. Experimental Study on the Influ-
    ences of Self-regulated Learning of Physical Educa-
    tion on the Innovation Capacity of College Students
    [J]. Journal of Beijing Sport University. 2012, (04):
    99–104.
[9] Saaty T.L., Alexander J.M. Thinking with models:
    mathematical models in the physical, biological, and
    social sciences [M]. Pergamon Press, 1981.

Sports Engineering and Computer Science – Luo (Ed.)
© 2015 Taylor & Francis Group, London, ISBN 978-1-138-02650-6

On the theoretical basis of visual symbols used in physical education and training

HongGang Qu
School of Physical Education, Northeast Dianli University, Jilin City, Jilin Province, China

ABSTRACT: By means of relevant literature, expert interviews, logical deductions and detailed comparisons, we elaborate some theories of the sense of vision sign and use in the sports teaching and the movement training. The study shows that the theoretical basis is plentiful in the sense of vision sign, which can be applied to sports and movement training and this way is scientific and can become teaching method.

Keywords: sports teaching and training; teaching method; sense of vision sign

1 INTRODUCTION

In the process of recognizing the objective world to human, there are two ways: one is sensing it directly; the other is knowing it by receiving education. In real life, human's cognition of the objective world is mainly from education, so when taught, what students in the objective world faced with is the information carrier of the objective world rather than texts, graphics, and symbols. The purpose of education is to enable students to understand the objective world and improve practical operation abilities from observation, information extraction, analysis and deduction. From semiotics perspective, the knowledge system is a series of symbols that are arranged by specific rules, the representation of human's cognition about objective world and the meaningful expression about objective world. And human's cognition of the objective world is the only medium [1]. In the education process, students are faced with a variety of symbols and in most subjects, what the students face is the abstract objections so it requires them to have a rational understanding of symbols. When teaching, teachers use rational ways to explain the cognitive object. However, sports teaching has its own particularity, that is, teachers use visual symbols because of the perceptual cognition of objects, namely perceptual symbols.

2 THEORY OF SEMIOTICS

Since ancient times, to express human's understanding of nature and their emotions, they have created graphics, languages, behaviors and a series of communication tools. Besides, we have accumulated from the work experience and our emotions and gradually formed a symbol. The history of mankind's development is the history of creation and the use of symbols; the development of people's consciousness is a symbol of the world's progress. When human could walk upright, they learned to use a variety of symbols to record and express ideas such as cave paintings and keeping records by tying knots. Before the use of body language, the most important way is the use of symbols, which become a strong evidence of civilization transmission. The French philosopher Maritain Thomas sent a letter in 1957 and he said: "No elephants and marks have the questions as complex and basic as the relationship between human and civilization. In the whole field of marks and human knowledge and life, it is a common tool, just as the physical world is always in motion." (quoted in Dille, (ed.), 1986, 51). Symbols as expressions of human civilization stand for the human spirit. The Study of mark problems is called semiotics, which is a popular and contemporary research in western social science. When some western researchers analyze the influence of philosophy, linguistics and modern natural science, they pay more attention to symbols. The British philosopher and scientist C. Beards once said: "In a broad sense, semiotics is undoubtedly the contemporary philosophy and the core of many other fields' thoughts." Semiotics is formed in the early twentieth century and has now become one of the most important fields of science over the last century. The essence of symbolic activity is a kind of thinking activities of human internal conversion. First one should understand the appearances of things, then the image will store in the brains recording to the formation of concepts, and then what relates these concepts with the outside world is symbol. Thinking can be regarded as the symbol of the selection process-composition, transformation, regeneration and thinking. The ultimate goal is to practice better. From the form point of view, human's knowledge system is composed of a series of symbols with a specific rule. In contrast, from the content point of view, the knowledge system is a process of the reproduction of human's cognition of the objective world, that is to say, the symbol is the carrier of objective world of human meaning. This determines the nature of symbols: on the one hand, symbols are the external form of the meaning of the world of objects; on the other hand, to the objective world carrier, symbols and human significance are the content of the human mind manifestation. Therefore, what people's cognition faces with is two worlds: the objective reality of the world and the shows of the objective symbol world [2]. Learning process is cognizing process. Those who face is not the objective world, but on behalf of his symbol system and people use symbols to cognize the objective world, that is to say, the objective world can be recognized with symbols. The symbol system is a medium to cognize the objective world. From the perceptual cognition to rational cognition, it can be completed by coding activities for symbolic cognition, which endows the meaning of symbols and codes corresponding to perceptual materials and the perceptual information into symbol information. In rational thinking, first the carried information is obtained by symbol, and then a kind of symbol manipulates and changes in activities, namely symbolic thinking. In the expression of cognitive results, cognitive achievements have been obtained. Express something bravely so that it will help people understand and have a social value system of knowledge. Therefore, the symbol is no longer just a manifestation of the meaning of the objective world. And what more important is that the symbol is becoming the medium of human's cognition of to the objective world. And it's even the only medium, which creates the use of symbols. The symbol world is not only with lively reactions but also embodied as a social subject and the nature of human existence.

In the teaching process, each links symbols with the material carrier of information throughout the whole teaching. In the view of modern semiotic, each symbol itself is just a self existence and is to become a symbol, which must be in a certain process, that is, some relationship to work. Therefore, to achieve its function, the teaching process as symbols may be provided. At the same time, the symbols in the teaching of this dynamic process actively promote the development of students' thinking. Semiotic theory thinks, the main memories are to provide many possibilities for thinking, because most people use symbols to help them think. The symbols used gradually deepens on the further development of the symbol and gives it a certain creativities. The establishment of a fixed link will sign with certain content and form the symbol signifier and signified. This symbol reflects economic liberation of people's thinking and in fact also forms the external form of thinking, storing with a large number of potential of thinking so as to think further [3]. One aspect of this is virtually the process of education and in line with the education, teaching, at the same time imparting knowledge and the education goal of all-round development.

3 THE BASIC IDEA OF MODERN TEACHING THEORY

With the rapid development of science and technology, all kinds of advanced scientific theories have been cited in education: the educational theory and educational practice, the continuous development and improvement. Theories that are cited mostly are recognized the most theory. These three theories are in fact complementary. Only when the system exists, can it transfers information, then realizes the controlling process and then improves the system. The main form of education is the transmission of human civilization and the whole process is to form a larger system. Slag is B in "science and education system": education is a diverse, open, comprehensive and big system and science education should be a large complex system. Baban J Ki in one book *the teaching process optimization, general teaching theory* completes exposition of the basic ideas about how to use the system to study and structure its optimization problems of teaching. Sun Shaorong in the "educational information theory" pointed out: "The essence of education is the transmission of a message". Modern teaching theory is that students "understanding of non mature" because they are engaged in learning of the leading role of the teacher's, which determines that students must be in the explicit form and make feedback on their acceptance of information. This in theory for the controlling of the education process provides a foothold. The modern teaching about "three" based on the education is regarded as a control system, according to certain teaching objectives and certain elements, which constitutes a form of organization. To achieve the implementation of certain education functions they think that the teaching process is a complex information exchange system []. In this process, information and education are the most basic elements. If the implementation of the function of the educational system is to reach the aim of education and educators must use the information to control the educated (see Fig. 2). Li Chengzhong and Wang Xusun, in the "education" in control theory tutor that controlling on psychological and social system is achieved by control of education. As the carrier of information, symbol has played the role of the medium in the whole control process and is the indispensable ingredient of education controlling. According to the information receiver, the information is divided into visual, auditory, tactile, olfactory and gustatory information. Large numbers of experiments confirm that people are realized mainly through vision and auditory. Therefore, in the process of teaching, teachers and students are major recipients of the visual information and auditory information and transmission is the visual symbol and auditory symbol. The students in the process of cognizing, the visual and auditory perception through thinking symbol, finishing and deductive reasoning finally form the cognitive object of cognition. Because of the characteristics of physical education, visual symbols show cognitive object through the human bodies' movements and therefore, through visual symbols strengthening students' cognitive process is the main features of sports teaching.

4 BASIS OF PSYCHOLOGICAL, PHYSIOLOGICAL VISUAL SYMBOLS OF PE TEACHING IN COLLEGES

4.1 The psychological basis of visual symbols used in physical education teaching in Colleges and Universities

4.1.1 Cognitive

Cognitive psychology in a narrow sense is the so-called information processing psychology. Thinking that the cognitive process is information receiving, encoding, storage, exchange, operation, retrieval, extraction and use process and this process is divided into four system models: sensory system, memory system, the control system and the response system, which emphasizes the decisive role of knowledge and knowledge structure takes on his behaviors and current cognitive activities and people in various shapes of things will cognize objects stored in three different locations and then can use the sound, shape and righteousness as three different ways to retrieve the memory.

Visual symbol is with visual stimuli and cognitive of visual symbols. The form code is stored in memory and shown at the right time. The cognitive process is the visual symbol of extraction, observation definition, inductive, deductive reasoning, a series of cognitive activities of continuous process expression and practice.

4.1.2 Delayed learning

Studying delay is a concept of psychology. It refers to people by the formation of learning form new patterns of behaviors and not immediately reveal, but it needs to be delayed until the proper time. Spirit of the environment and atmosphere of public opinion may inspire the normal accumulation of memory. Visual symbols retained in the students' memory by cognitive, using a longer time to feel, digest and understand memory symbolic form and content. Studying delay helps students to cognize the behaviors in willing basis and the behaviors of internalized significant efforts are great.

4.1.3 Consciousness

The consciousness is the sum of a personal psychological experience and a whole system to psychological process based. Among them, emotion and ration are the dominant aspects of consciousness, especially that language and thinking are the core consciousness. Language is a symbol of all the members of the society of common understanding. In people's cognition, in addition to the cognitive process of the little outside, sign language is the most important means for human cognitive activities. In sports teaching, teachers with action, gesture, the visual symbol matching language so that students obtain visual perceptual knowledge through a large number of perceptual knowledge, which gradually rises to rational knowledge, namely the formation of thinking. Thinking is a cognitive process and the key to forming the consciousness. In psychology, thinking the indirect and general is an important characteristic in human thinking process. The so-called indirect is to know the objective things through other media, namely the knowledge and experience to understand or grasp those have no direct perception or impossible to perceive and predict and infer the process of development of things. The so-called general is the same kind of things abstracts the common features and the essential characteristics of the memory summary. Characteristics of substantial thinking are two ways of cognition: from the features of thinking, you can clearly see the sign of the shadow. According to the theory of semiotics, mankind cannot think and communicate without symbols and the consciousness process. There is a symbol of the process, so thinking basically is the operation for a process of combination, transformation and regeneration of symbols. The visual symbol's application in the teaching of physical education and training is mainly to strengthen the transformation process of visual perception and thinking and promote the rapid formation of the tactic consciousness.

From the cognitive interpretation and evaluation of science point of view, many cognitive scientists think that symbol system has strong explanatory power. It can explain the higher cognitive processes, such as thinking, problem solving, speech understanding and understand the cognitive process of the lower, such as perception, pattern recognition and the body movement. I think this is another kind of sports teaching process and teaching using semiotics theory. For example, in the basketball teaching, the training process is usually continuously through the perception of motion itself, which is to enable students to form a rational thinking and obtaining good basketball skills.

4.2 Physiological basis of visual symbols used in teaching of basic basketball tactics

The receiver receives the information from the outside of the main body is the visual, auditory, tactile and other three systems, but more visual and auditory utilization. According to the study, visual people receive the information volume of 80% above, therefore, in the process of human's cognition, vision is one of the main ways that people communicate with the outside world. Physiologist Ba P Love found that the human higher nervous activity has two signal system theory. He said: the most basic activities of the cerebral cortex are the signal activity and in essence, the signals can be divided into two categories: One is the reality and the attributes of the specific signal are the first signal; the other is that the reality of abstract words is signal (second signal) and cortex system can react to the first signal and second signal known as the first signal and the second signal system. In order to promote the formation and development of basketball skills, the visual, auditory and tactile input information such as receptor channels are necessary. Foci formed foci and proprioceptive that formed in the cortex of a temporary connection, strengthening repeatedly by neural activity and produced trace effect which will store information, formed a strong memory and established dynamic exercise to consolidate the setting. Therefore, according to the receiving way of information, the information carrier is divided into visual, auditory and tactile symbols. The study shows that, in people's cognition, cognitive activities through vision accounted for 25% of all people's cognitive activities and in teaching, body movement is the main object of teaching and cognitive activities through the vision are 65% above [4]. Symbologists Cahill said that sign function was giving the object form and concept, and made the objects more easily recognized than in

the pure nature and symbolic importance, not only because it is a part of the material world, but he can enter the world's meaning. Presentational symbol includes visual symbols and auditory symbol, but the visual symbol is far superior to the hearing in the information transmission. Therefore, the visual symbol is the main information communication mode besides language symbol. A kind of symbolic form of visual symbols with the language can not achieve in the expression of human emotional field.

5 SUMMARY

A lot of visual symbols used in sports teaching and training in the form. Scope is very broad, but has a different name in the using process. The specific methods used are not the same. In this paper, from the perspective of semiotics, it introduces the concept of visual symbols in physical education and training and hopes that the theory in the application of visual symbols has a unified understanding. Based on elaboration, analysis and using the related theory of visual symbols involved in physical education and training, we can see very clear that visual symbols in sport are the support of their solid theories. Symbol theory is consistent with the law of education and teaching, which can reflect the mental body exercise and the law of physical objective. Visual symbols for various forms of physical education teaching and training in the use of a unified concept are supplements and development of the methods. Training of sports science means system is in line with the objective law. In the future, sports training will gradually increase the visual symbol's application for promotion efforts and make it become the new teaching methods.

REFERENCES

[1] Yuan Zhenguo. New ideas of education [M]. Beijing: Educational Science Publishing House, 2002.
[2] Sun Yiliang [J] Journal of Chengdu Sport University 2003 to explore the visual ability and its training of basketball players, 29 (2).
[3] Wang Zhe, Xiao Jin Shan. The vision and the application of [J] nonverbal signals in the teaching of physical education in Liaoning Teachers College Journal 2003, 5 (2).
[4] Wang Xiaodong. A preliminary study on nonverbal symbols in the process of transmitting sports information [J]. China sports science and technology, 2003, (9).

Sports Engineering and Computer Science – Luo (Ed.)
© 2015 Taylor & Francis Group, London, ISBN 978-1-138-02650-6

Research on physical fitness training methods of football athletes

Bin Xu
College of Foreign Languages, Northeast Dianli University, Jilin, China

ABSTRACT: Physical fitness, an important part of the overall structure of competitive ability of athletes, is the sum of all kinds of necessary physical ability of athletes in order to enhance the movement technique and tactic level and create excellent results. Reviewing some literature material and previous studies, this article briefly analyzes other methods of soccer movement characteristics of energy supply, and finally sorts out the factors of physical ability level of football players and explores a kind of physical training method for them.

1 INTRODUCTION

Physical fitness, an important part of the overall structure of competitive ability of athletes, is the sum of all kinds of necessary physical ability of athletes in order to enhance the movement technique and tactic level and create excellent results. The main purpose of the physical training is to achieve the athlete's body potential, make the sports ability of the organism reach the highest level and lay a good foundation for the successful completion of the training process. The level of training is not only the primary material basis for athletes to master complex technique and tactic and the important assurance for modern athletes adapting to the training and competition requirements, but also a kind of cultivation of the athlete's tenacious will which can increase the sports life of athletes.

2 FOOTBALL MOVEMENT CHARACTERISTICS OF ENERGY SUPPLY

The energy in which all kinds of daily life, the work, the life and muscle activities was required from fat, carbohydrate oxidation and protein decomposition. When the body is in the state of motion, energy consumption increased. The increase in the intensity and motion to the motion depends on the length of accounting for 18.9% of energy fast sprint "direct source is Adenosine Triphosphate (ATP), the ultimate source of muscle activity energy is matter of aerobic oxidation (sugar, fat)". ATP has three different energy supplies. The first starts from the phosphate energy supply system (atp—cp), the next is lactic acid energy supply system and the last is aerobic oxidation function system. So there are different requirements about football on three energy supply systems, which is in order to finish anaerobic metabolism and aerobic mixed energy movement. The most of the energy in playing the football is from ATP, provided by cp system; but the anaerobic energy production is also very important. It is the requirement of high strength movement because there are about 200 times 3s sprint to top players in the race, which is provided by the anaerobic system energy. "The concentration of lactic acid in the football and part of the track and field project are different; glycolysis intensity is low, more focus on aerobic energy and non lactic anaerobic energy supply". Therefore, the football match energy supply is concentrated in the aerobic and anaerobic energy supply, and non lactic acid and anaerobic glycolysis ability are relatively low. The athletes have not strong acid resistant ability.

3 PHYSICAL TRAINING METHODS CURRENTLY 2 FOOTBALL PLAYERS

In order to meet the requirements for the quality of the game and ensure that the technical and tactical athletes can fully play the game, players need a high level of physical fitness training. So the physical training of the training plan is an important part. According to the different requirements of the body function based on matches, physical training can be divided into three parts, namely the aerobic training, aerobic training and specialized muscles training.

3.1 *Aerobic training*

Football players' aerobic training is to improve the quality of game action, prevent the drop and reduction induced by fatigue in the match. Aerobic training enhanced oxygen transport capacity of the cardiovascular system, the ability to recover after intensive exercise, the ability of muscle using oxygen and oxidation of fat in the long time exercise. Aerobic training includes the recovery training, aerobic small strength training and aerobic training. For example, jogging, playing the role of resumed training in the game and after the exercise can help athletes recover to normal body level. Aerobic endurance training can be completed through continuous passing, shooting or equal number of combat exercises.

3.2 *Anaerobic training*

In the game, players need to practice frequently, such as running at full speed, quickly changing direction and fast motion. Elevated blood lactate in motion also suggests the existence of a large number of glycolysis. Anaerobic training can improve the activity of the creatine kinase and glycolytic enzyme. Anaerobic training aims to enhance anaerobic potential athletes who can adapt to high intensity training and competition. It can be divided into training of speed and training of speed endurance. The speed training can be divided into: the reaction speed, movement speed and displacement velocity. Improving the speed of reaction is mainly to improve the ability of response to selection of moving objects. The main methods to improve movement speed and displacement velocity are repeating and transforming strength. Speed endurance training can be obtained by high intensity training race and ball training.

3.3 *Muscle training*

Muscle specific training is of great significance to improve performance. There are many action in the football such as body confrontation, jumping, speed, turning, the strength and power combination; and football players have to train high level muscle strength required by the development of specialized muscle strength training to get the game. Selection method of strength training should be considered when coaches and athletes need a proper one among different methods to improve the maximum strength, explosive force, muscular endurance and other different types of muscle strength.

4 CONCLUSION

What coaches need to pay attention to is the physical fitness training and special training of athletes in training together. Do not blindly to emphasize physical fitness test requirements, and arrange the training effectively, solve the energy relationship in quality training, physical training, physical training, competitions, good physical training and technology. Tactical training, generally, requests the athlete to have conscious, long and uninterrupted physical quality training. Coaches will not view general physical training as the main content of courses for training; but in the technical and tactical training, the training requirements reflect throughout the training process, so that players will not feel bored and dull, and coaches can achieve the purpose of physical training.

REFERENCES

[1] Qin Zhihui: "to explore the" problems of endurance training in football players, "Journal of Beijing Sport University" 2002.25 (1):136–138.
[2] Geng Jianhua: "analysis and training countermeasures" structure of physical training at the present stage of football players in China, "Journal of Wuhan Sports Institute", 2004.38 (5): 111–113.
[3] Chen Ming: "physical training and eliminate fatigue" football "Journal of Beijing Sport University", 2006.29 (2): 206–208.
[4] Liu Fuli, Zhou Yi: "analysis" means occupation football team trainingsystem and its components, "Journal of Guangzhou Sports University", 2002.22 (1): 68–71.

Sports Engineering and Computer Science – Luo (Ed.)
© 2015 Taylor & Francis Group, London, ISBN 978-1-138-02650-6

Comparative study of Beijing elite female softball players batting training

HuiZhi Zhang

Department of Physical Education, North China Electric Power University, Beijing, China

ABSTRACT: The paper analyzes the specific hitting training of female softball in Beijing and discusses the rationality and effects of new training method. On purpose, it aims to provide a reference for the future training and match of Chinese Softball Team. The new specific hitting training method is used in the daily training of female softball team. Then the method is analyzed in detail and studied comparatively by methods of documentation, observation and record, data statistics and tracing comparison, and so on. The batting average of Beijing Team is 0.23 ± 0.03 in the league matches in 2010 and is 0.37 ± 0.02 in 2011, increasing 73.4%. In a word, the specific hitting training is the effective way to improve the hitting speed and batting average.

Keywords: softball; hitting; bat swing; batting

1 INTRODUCTION

In softball game, offensive starts when a batsman hits the ball in the shots box. Only when the batsman hits the ball successfully, he can create opportunities on his base and then create further scoring opportunities for himself or teammate. One of the key factors that determine the offensive ability is the hitting ability, which is the ability of hitting the ball accurately to combat. Offensive is the key to victory, without the offense to score, it is impossible to win only by perfect defense. Only when offensive was played in a high level, the team can win the final victory. By continuous threat of offensive, not only opponents can we depress, but also inspire our teammates. "Offense is the best defense" also proved the principle. To win a game, superior offensive capability is an indispensable condition. However, a strong offense is based on the tough and effective training. Scientific training is the essential element. Correct training methods are more effective approach. We will make analysis and evaluation on the new swing speed training of the individual training.

1.1 *Traditional methods of combat training*

In the process of hitting, swing "quickly, accurately, and heavily" is the three most important factors: the speed and orbit of swing and the hitting position of the ball, real-time speed of swing. The factors are the determinants that result in hitting effect. Excellent physical, reasonable swing action, hitting timing and accurate hitting position

are the fundamental conditions to complete the process of hitting. In order to achieve good results of hitting, the above requirements for athletes are indispensable.

Main points of hitting: swing as fast as possible, motor coordination to achieve reasonable swing orbits, bat on the ball with the power as large as possible. The points mentioned above require excellent physical fitness. The hitting timing was not only related to the casting lines by the pitcher, but also related to individual swing skills (speed, trajectory and hitting position). These are not just physical issues, and more importantly, skills issues, which need multiple attempts to accumulate their experience gradually. Hence, it requires a large amount of exercise as a guarantee.

Traditional training methods have been mainly to strengthen an individual, such as continuous airswing for 500 times. But there is no reasonable collocation and schedule for individual training as well as strengthening the links between the various individual training. On the basis of the original training content, Beijing team carried out the reforms by a large number of experiments. Aimed at the characteristics of each training, Beijing team carried out scientific analysis and considerations and developed a new training method, which increased emphasis on the relevance of individual practice and the link between the various individual training.

1.2 *Study objectives*

Alan M. Nathen, a physicist at the University of Illinois pointed out that: to make a good hit, the speed when the ball leaves the bat is the key factor. The factors determine the initial velocity of the ball to leave the bat, in addition to the casting speed. Within them, the factor that can be controlled by the batsman is nothing more than swing speed, as well as the touch point of the softball and the bat. According to physical principles, in the case that the quality of ball and bat was fixed, the faster the bat was swing, the faster the initial velocity will be (in the case of the same point of contact). How to give full play to the body's physical strength and skills to maximize the swing speed is the most important factor of improving the initial velocity at the premise of insuring the accuracy and timeliness. It is for this reason that all the softball teams have attached great importance to increase swing speed training in the combat training. How to improve the swing speed and increase the ball's flight distance involves many factors, such as force mode, swing time, hit points, the highest swing speed and swing speed before hitting the ball and so on. Due to the limitations of experimental conditions and other factors, it is quite difficult to make quantitative analysis and subtle description for the status of swing speed, here in the studies of swing techniques. This paper only take Beijing team as a sample to analyze the effectiveness of the training programs that developed and implemented to improve the swing speed and increase the ball's flight distance.

2 RESEARCH OBJECT AND RESEARCH METHODS

2.1 Research object

Nine key players of Beijing softball team, training age range from 5 to 10 years, exercise levels are excellent.

2.2 Research methods

2.2.1 Documentation method
Statistical investigations on the swing speed and technique of the 9 key players.

2.2.2 Observation
Take site records of Beijing team. Observation time: 2009–2011 season.

2.2.3 Comparative method
Track the swing speed and batting average before and after this program of Beijing Team athletes, compare the data before and after to define the existence of significant differences.

3 ANALYSIS AND DISCUSSIONS

Take statistical analysis on the indicators of the 9 key players' swing and number of battings. Tables 1–4 are the statistical data of each indicator before and after the improved training program of Beijing team.

3.1 Analysis on swing technical of part players of Beijing softball team

The tests of the swing speed in 2010–2011 season show that the new training methods applied by Beijing softball team has a significant effect for improving swing speed, among which, 2 players increased 10 km/h.

3.2 Statistical analysis on techniques of Beijing team in 2010–2011 season

Beijing team won the Championship in the 2010 season. Although the batting average is not very high, but there were many key scores. In the 9 games in statistics, there are 9 players batting in each round. The total batting number of all team is 275, the actual number is 254 and the total number of bingles is 55. The player with the highest and lowest number of bingles is 13 and 3. With the batting average is 0.448 and 0.125, while the team batting average is 0.217.

Zhang Yin was awarded the Home run award and Batting award. In the 9 games in statistics, there are 9 persons batting in each round. The total batting number of all team is 308, the actual number

Table 1. Average swing speed, swing speed increase and batting direction of five players of Beijing softball team in the season of 2010 and 2011.

Name	Year of birth	Swing count	2010 (km/h)	2011 (km/h)	Increase	Batting direction
Xue Shen	1984	10	105	115	9.5%	Right
Zhang Yin	1973	10	105	113	7.6%	Left
Lei Donghui	1984	10	103	113	9.7%	Left
Chen Jiawen	1983	10	102	110	7.8%	Right
Jiang Jing	1985	10	100	107	7.0%	Left

Table 2. Statistics of batting number of Beijing team in National Softball Championship in 2010.

Name	Position	Number of hitting	Number of batting	Number of bingles	Number of bagger	Batting average
Wang MiaoMiao	8	36	30	7	10	0.223
Zhang Yin	DH	36	29	13	17	0.448
Xue Shen	2	38	31	8	10	0.258
Chen Jiawen	6	37	30	6	8	0.200
Wang Xiaoqing	3	31	24	3	3	0.125
Lei Donghui	9	39	35	9	10	0.257
Yin Jie	7	34	29	5	5	0.172
Zhang Jing	4	25	17	3	3	0.176
Jiang Jing	5	33	29	7	9	0.241

Table 3. Statistics of batting number of Beijing team in National Softball Championship in 2011.

Name	Position	Number of hitting	Number of batting	Number of bingles	Number of bagger	Batting average
Wang MiaoMiao	8	28	21	9	15	0.428
Zhang Yin	DH	37	35	16	22	0.457
Xue Shen	2	38	32	11	16	0.343
Chen Jiawen	6	36	34	13	20	0.382
Wang Xiaoqing	3	32	27	8	11	0.296
Lei Donghui	9	35	31	12	17	0.387
Yin Jie	7	37	32	10	14	0.312
Zhang Jing	4	31	26	9	12	0.346
Jiang Jing	5	34	28	12	13	0.428

Table 4. Statistics of batting number of Beijing Team in National Softball Championship in the season of 2010 and 2011.

Athletes	2010	2011	Increase
Wang MiaoMiao	0.223	0.428	91.9%
Zhang Yin	0.448	0.457	2.0%
Xue Shen	0.258	0.343	32.9%
Chen Jiawen	0.200	0.382	91.0%
Wang Xiaoqing	0.125	0.296	136.8%
Lei Donghui	0.257	0.387	50.6%
Yin Jie	0.172	0.312	81.4%
Zhang Jing	0.176	0.364	96.6%
Wang MiaoMiao	0.241	0.428	77.6%
Average	0.233	0.375	73.4%

is 266 and the total number of bingles is 100. The player with the highest and lowest number of bingles is 26 and 8. With the batting average is 0.457 and 0.296, while the team batting average is 0.376.

The average batting rate of Beijing softball team in 2010 is 0.233. After the applying of new training method in 2011, the result has been greatly improved. The batting average in 2011 had increased by 73.4% compared with 2010, which

resulted in significant differences in statistics. In other words, the swing speed of Beijing team members is significantly higher than that in 2010, which shows that the new training method is effective. Thus the special combat training method of Beijing team is worth learning and promotion.

3.3 Analysis on the training of batting technique of Beijing Team athletes

3.3.1 Various kinds of special combat training

1. Throw hit: i vertical throw 50×6 group; ii oblique throw 50×5 group; iii fast throw hit 30×5 group;
2. Free hit: i inside the cage (≥ 200); ii free hit (middle hit) in court (≥ 100);
3. Air swing: i 30×5 group; ii 50×10 group; iii 20×10 group;
4. Air swing and thong swing alternately with the proportion of 7:5;
5. Oblique throw: inside the cage 50×10 group;
6. Swing middle hit: i in court 30×60; ii in the cage 50×10 group;
7. Hit machine ball: inside the cage, linear matches with changes, coaches can decide what kind of ball to be used according to the specific circumstances of each athlete;

8. Hit handball: complete hitting practice in a game mode, the hitting number ranges from 40 to 60 each day, the whole number reaches about 183 one year.

Among the above, coaches can build up different training method at different phases and cycle with concerning the specific circumstances of each athlete, and execute alternately. The purpose is to enable that the training content can help improving the game effect in greatest extent.

3.3.2 Various speed quality training

Speed quality includes reaction speed, action speed and movement speed of the shift in space.

1. A variety of base running practice;
2. Running hitting practice of interior angle, exterior angle, curve, rise, fall, speed ball;
3. Doing action speed training with visual and auditory as signals;
4. Short-distance running (20 m–50 m);
5. Middle-distance running (150 m–300 m);
6. Jumping exercises (such as back step run, hop, frogs jump, squat jump, bench jump, and barrier jump).

Through the analysis of the relationship between hitting speed and batting average of the two years, Figure 1 and Figure 2 show that the batting average was positively correlated with the hitting speed. Actually it means that the higher the hitting speed is, the higher rate of forming bingle will be. Reversely, if the hitting speed decreases, the rate of forming bingle will decrease. However, the attention must be paid that batting is not only related to hitting speed, but also related to direction of flight after being hit and flight path. Therefore, hitting speed and batting average are just a relationship of trend. However, it will undoubtedly increase the time of judgment of the flight path through increasing the swing speed. If the hitting point is the same, the speed of the ball

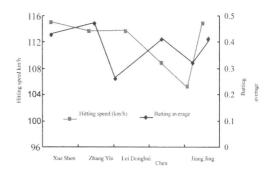

Figure 2. Relationship between hitting speed and batting average of Beijing team in 2011.

will also increase, which will undoubtedly improve the hitting effectiveness, thereby improving batting average.

4 CONCLUSIONS

1. *By the new method of offensive training, the swing speed of Beijing softball team in 2011 has significantly increased 8.33±0.54% compared with the traditional training method in 2010. And the new method has made the stick speed increased 8.6±0.6 km/h, which has a significant difference compared with the traditional training method in 2010. It shows that the new training method does improve the swing speed of Beijing softball team players.*
2. *By analyzing the data of National Softball League in 2010 and 2011, after the implementation of the new training method for a year, the batting average of Beijing softball team in 2011 has significantly increased 73.4% compared with 2010. The new combat training was developed and implemented to increase the swing speed. In return, increasing the swing speed can improve batting average in games so as to guarantee the victory from the offensive side in a competitive race.*
3. *In the relevant relationship between stick speed and batting average, batting average and increasing of swing speed shows a positive relationship. It shows that increasing of swing speed does improve the batting average. Compared with the traditional method, we find that the new training method not only improved the individual capabilities, like swing speed, but also thereby improved the results of the game. Surely, the new special offensive training helps to improve the results of the game, so the new training method is proved to be scientific and effective.*

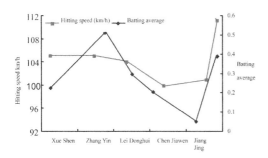

Figure 1. Relationship between hitting speed and batting average of Beijing team in 2010.

REFERENCES

[1] Jiang Minghong compiled. Baseball learning skills. [M] Big Kun bookstore distributor, 2001.

[2] Ishii Fujiyoshi Lang, Chiharu Sato, western tatsume forever. Real baseball. [M] compiled United Advertising Book Company, 1999.

[3] Ishii Fujiyoshi Lang, Chiharu Sato, western tatsume forever. Baseball classroom. [M] compiled United Advertising Book Company, 2002.

[4] Wang Hui, Zhou Jiaying, Sun Boqing, Zhong Caoyue, Zhu Xiyou. Analysis of motion on swing technology of partial outstanding baseball players in Chinese. [M]. Xi'an Physical Education University, 1995, 12 (2), 1–4.

[5] Sun Boqing, Zhou Jiaying, Wang Hui, Yao Kangle, Chen Fang, Zhong Caoyue. Analysis on swing speed. [M]. Xi'an Physical Education University, 1995, 12 (1), 30–34.

[6] Zhou Jiaying, Wang Hui, Zhao Guofeng. A study and analysis of countermeasures on the swing technique of current Chinese baseball players. [M]. Xi'an Physical Education University, 1996, 13 (1), 35–38.

[7] Wang Pai. Swing-(b) combat exercise: Published by the department of physical of Chinese Culture University for the fourth, 1970, p. 8.

Sports Engineering and Computer Science – Luo (Ed.)
© 2015 Taylor & Francis Group, London, ISBN 978-1-138-02650-6

Technical analysis of Beijing's elite first baseman

HuiZhi Zhang

Department of Physical Education, North China Electric Power University, Beijing, China

ABSTRACT: Data analysis, references and observation were used to study first baseman defensive position in this paper. The data of catching the ball and positioning was employed as the main content analysis. Main conclusion is that first baseman's defensive positions are flexible according to different situation in defensive stance. He should start fast in all directions. His movements contain starting around accurately, reacting quickly, turning back to home. He is quite familiar with each place and he can go to the base and touch it accurately. He catches up the balls, especially difficult ones. He can play in different conditions with defensive consciousness. The paper emphatically analyzes the first baseman's defensive position with technical details for Beijing baseball team as reference of the first baseman defensive position skill.

Keywords: Beijing first baseman softball team; defense position; technical analysis

1 INTRODUCTION

Ever since softball has been carried out in China, we learn a lot from playing games in other countries and have made considerable progress after long-term research studies. After long time study, China improves a lot of skill. However, comparing with the world-class to team U.S.A., China still has some needs to improve to reach the top team. Take the 2008 Beijing Olympic Games for example, the Chinese women's softball team had worked hard, but the end result was still not ideal, which exposed the gap between teams. Outstandingly, offensive is important, also a solid tight defense is the basic foundation. The team, which not only have offensive skill, but also has defensive skill, could be the best one.

From practical experiences, we realized the importance of position technology. Also, people have to be good in mastering the correct basic technology firstly in order to play a skilled technical position. The basic technology is a shared technology, and location technology is both common and personalized. Determining a team defensive system and strengthening contingency is based on the defense team's technical style and the talent that meet with a variety of opponents to play the game needs. Position to an athlete, it is important that the position should be met with the athlete, that is, considering the technical and psychological characteristics of players (including the pace of movement, arm strength, fitness, flexibility, ability of judge, courage, etc.) and also consider the wishes and interests of player, on this basis to finalize the team's defensive position.

Position technology follows the change of defensive technology; position of the first baseman is one of many defensive positions. In order to enhance technology for the location, the need for good basic skills, but training is both effective ways to turn. Basic skills of softball defense are the foundation for the project, not here as a major research topic. Position technology is the most important factor in the decision of defensive ability; the softball's position technology is varying in different location. For example, a first baseman requires the most comprehensive location technologies. A good baseman also plays an important role in the game. A good first baseman not only good at her own position's skill, but also could work will with different tasks so that the team could reduce their mistakes. She can link infield and outfield, and establish the overall defense system.

After years of practice, we find that the ability to improve the defense technology needs to start from positional skill, pick up the skills fully and correctly. Only each position skill of a first baseman can be used skillfully is the high performance, which shows the importance of position defensive skill in practice.

2 THE OBJECT AND METHODS OF STUDY

2.1 *The object of study*

The first baseman defensive position technology skill of Beijing Women's softball team.

2.2 The methods of study

2.2.1 The reference analysis

Read the relevant literature of softball, comprehensive analysis and collection the basic theory of the position technology, as the basis of thesis study, provide the basis for writing paper.

2.2.2 The expert interviews

Interviews of many national softball experts discuss the relevant technical issues. For example, national team coach Mr. Huang Weigang, former national team first baseman Ms. Wang Ying who were named the world's star players.

2.2.3 Observation

Watching the technology videos of Beijing team national tournament and championship, which are a total of six games, focusing on the movement techniques of both first baseman observation stations, back to base and catch the ball. And watch the technology videos about the group within the field of national team to observe the position technical details of the first baseman.

2.2.4 Mathematical statistics

By analyzing the defensive data from 2008–2011 of Beijing softball tournament games, we find that the first baseman is not only the position that uses more back to base to catch the ball, but also the position of a high success rate defending.

3 ANALYSIS AND DISCUSSION

3.1 Analysis of first baseman's catching technique

According to Table 1, comparing first baseman put-out number and entire unit put-out number of Beijing team from 2008 to 2011, first base average put-out number occupies one third of entire unit average put-out number in these four years, first base use more back base catching technique. In view of the contrast of the first baseman's fielding average and entire unit fielding average, the first baseman has higher fielding average and less

muff, manifests the first baseman has stable catching defensive technology and solid basic skills, Suitable to be the first baseman.

Table 2 manifest first baseman has the most put-out number in entire unit of each team (List Beijing, Shanghai, and Sichuan these three teams). Emphatically analyzes the first baseman's catching data of Beijing team, according to the different styles of pitcher, the effects of balls will be different too. The put-out number of first baseman of Beijing team is 56, which means that there were many grounder in infield of Beijing team. The first baseman can get more point when the game has more grounder. According the date, there were 56 put-outs, 50 of them are put-outs after back base, coming balls are all in normal catching scope.

Two of them are double play back to first base put-out. Four of them are one-jump-ball, two of these four are teammate's passing problem, and the other two are the position's problem. Standing a little bit further make the back base distance too long to arrive ahead of time. So get back to base quickly is a manifestation of first baseman's catching skill. The more quickly to get back to base, the more time people could prepare for catching, and the more obviously the object will be, then the higher teammates' passing precision will be.

3.2 First baseman position of Beijing softball team

First baseman is required grasp of the most comprehensive technology throughout the game, this location has divided into left and right. In terms of position needs, left-handed players have an advantage in defensive because the first baseman's defensive position is in right field. Here's take left-handed players for example. The type of technology is as follows: back to base ball technology, back to the base tag technology, ground ball techniques, batting touch technology, the ball pass the base technology, playing pull the ball cross-sectional technique, complement-bit number.

Table 1. Statistical table of first baseman defensive skills of Beijing softball team.

Year	First baseman put-out number	Entire unit put-out number	First baseman fielding average	Entire unit fielding average
2008	53	166	97%	93.5%
2009	44	170	98%	92.3%
2010	65	173	100%	96.1%
2011	56	168	100%	94.6%

Picks from 2008–2011 year national technology statistics.

Table 2. Statistical table of defensive put-out technology in every position of part teams of National female softball championship tournament in 2011.

Team	First baseman's put-out number	Second baseman's put-out number	Third baseman's put-out number	Shorts top's put-out number
Beijing team	56	9	12	14
Shanghai team	39	11	11	12
Sichuan team	48	6	8	16

Picks from 2011 national technology statistics.

The normal position of first baseman is 5–8 meters before the pad, but first baseman will have different stations in different situation. There were many types of situations as follows: according to the hitter, if no one on the base, left hitter hit the ball in the case of precordium site position to move forward 1–2 meters in order to trigger a faster defense before the district to hit the ball. Right hitter could not move forward, but the person should have anti-sense to hit the ball. When someone in first base, the first baseman's defensive position should forward relative to some other hitter in the exposure tactics, obviously fight bunt, they can move forward to 8 meters away from the home and make anti-bunt preparation. This is a special station that the player can choose which is only using in this particular case. The station is normal when someone is in second base. Specific location is changing with different hitter and situation just pay little attention to the direction. When someone in third base, the player should make preparation for all kinds of balls according to defensive location. The reason is that the other offensive players have many ways to launch attack. It is the most passive defensive situation when some people in first or second base or more, at this time other hitter has many options. Therefore, the normal defensive location for first baseman can defense variety of coming ball. According to the batter to make the same movements to predict in advance and focus on the defense direction, the base defensive position is generally with closer distance, while increasing the difficulty of assaulting the grounder quickly.

3.3 Analysis of first baseman defensive

3.3.1 Back to base and catch the ball

Back to base and catch the ball is a basic technology of the first baseman. After the hitter hit the ball, the player runs to the first base as fast as she could.

The infielder pass the ball to the first base for kill the hitter, at the same time, the first baseman turns back to the base quickly and faces where the ball comes generally 5–8 meters before the first base, and shouts loudly to prompt his teammates. When the player turns back to the base, she suppose to make counterclockwise rotation, and the left foot (the left hand passing player) teach the inner side of the pad to ensure the foot will not leaving the base pad when her leg stretch, while the right foot as far as possible has always been direction towards the ball with the body to build the most long-distance, which could shorter the distance to the ball's flight, also faster seal kill hitter. The player should get back to the bit after the passing of the ball according to different circumstances.

The ball within the normal defense is the best. However, if ball go to side, the first baseman has to give up base pad to catch the ball firstly. So, we can see that controlling the game is skill of a good first baseman. Back to the base of the steps required that turning fast, running quickly, and arriving position correctly. A good first baseman also requires getting base bag position without looking down.

3.3.2 Diversion tactics and tag out

When there is someone at first base, receiving signal sent diversionary tactic, first baseman needs to get back to the base quickly, after catching the ball, she should "killing" the first base runner. The whole process shows that the important technology depends on the first baseman receiving the signal. After the player throws the ball, she needs to go back as quickly as possible, get the station, and turn back to catch the ball so that the team could get this point. Touch first base back to base to "kill" the runner. Requirements turned back to the base speed, the card position is correct, consistent tag moves fast. However, the basic diversionary tactic is the second baseman into the bit, and took over the completion of tactical coordination, not to first baseman back to the position as the main tool, only as one of the technologies, without specific analysis.

3.3.3 Bunting defense

Bunting is a mean for hitter gets on base, which is an important defense mission for the first baseman besides back to base and catch the ball. The main task of the first baseman is to react according to hitter's behavior, towards ball, catching ball, passing ball to first base, and "kill" the hitter.

The feature of bunting defense mainly action of player that after the pitcher pitch the ball and then hit it, she will make some move. With careful watching, after the bunting action is made, the first baseman runs fast to the ball. Place soon after the emergency stop, the card bit the ball, pass the ball to first base, blocked hitter, action sooner the better. In the course of the defense in advance of the pre-sentence, starting the most important. Because it is and take over, pitcher, third baseman at the same time bunt defense, so the ball in play at the same time, while starting forward, side to cry, according to the direction of the ball before the meet and determine the speed of the ball defender. Ball players wear to kill to complete the task, did not catch the rest of the task team needs to fill the seats of their completion, command, and other various tasks. Although the defense is a personal technology, is the only skilled with the orderly defense. Is a reasonable allocation of time for the hitter to block key.

3.3.4 *Ground ball defense*
Ground ball is the defensive basis of all players and it need the player defensive front, a fast ground ball between the first base edge second base, flat circles inside and outside the ball and fly ball, player need about fast bowling when the ball flutter then the peace. There is between the right and ground ball pitcher. This skill shows the first baseman's personal ability, the greater the range of the ball relative to the stronger defense. Conversely the greater the scope of each person's defense, the team's overall defensive ability is stronger.

3.3.5 *Baseman passes*
Passing the ball depends on the base runner, when no one on base that the hitter need to block first base. After the first baseman catch the ball, the ball does not need to be passed. When many runners get different stations, the first baseman has to make decision according to the situation. The player needs proficient skill of passing the ball to second base, third base, home plate, and throwing the pace of home plate.

3.4 *Requirements of first baseman defensive position*

The ability of sensitive decides the anticipation of the game. First baseman defensive awareness in the following two aspects, especially in playing pull cut.

3.4.1 *Pull fight cut*
Outfielder passes the ball to home base, first baseman needs to enter the relay in position, so that the fielder can pass to the first baseman's right glove

as target, and according to the receiver's command to intercept the ball, after get the ball, passing it to ready station and then intercept the ball since the situation of the field. Generally, when second base runner is not on base, the fielder shout out "cut!", and the first baseman tag her out according to the position of second base runner running. The following situations are most common:

1. The reason of the first baseman's cutting is because the ball passed to home plate, which is distorted. The first baseman caught the ball early so that she could adjust the ball, and then pass the ball to home plate. Therefore, they can "kill" the runner who ran from the second base.
2. At the receiving of the command to intercept the ball, the second base runner ran pass the third base, stopped between the 3rd plate and home plate. The first baseman changes direction quickly when catch the ball. According to the runner's purpose, the first baseman passes the ball to third baseman or home plate. When the runner is standing between two fixed base, first baseman needs to hold the ball after receiving it, and runs from home plate to third base in a clockwise direction driving other runners, also force on the runner run in direction of the 3rd plate, and then tag out the runner with her teammates.
3. Make determination according to the time, when the runner running from second base to home plate is safe, the first baseman should cut the ball, give up catch off the runner, change the pace, and catch off other runner. As the case may be, player can directly pass the ball to second base and return the ball to first base. Resulting in putout between first and second base, tag the runner out, and control the situation, stop the situation continues to expand.

From relay and a range of technologies follow-up, we can see a first baseman's level and experience.

Then, let us use specific case to analysis. Example: 2009 National Championships Beijing, Jiangsu, Jiangsu team had runners in second base, hit to right field, first baseman play came out to home plate and pulled cut. Second base runner ran to home plate, when the home plate was obviously safe, the first baseman cut the ball and return home plate, then passed the ball to the second base. To prevent the runner went to second base hit back.

4 CONCLUSION

The first baseman is not only skilled in position skill, but also good at basic catching technique. In the other words, the player should turn rapidly, go back position quickly; catch ball ready, and also has the outstanding ability to catching difficulty ball.

The first basemen have different choice in defensive stances and she can catch the ball correctly according the different situations. The player also need have her own unique defensive habits and tactical coordination so that she can play out all kinds of technical skills flexibly in cooperating with other team members.

The first baseman can make a correct decision in cable cut-off and control the situation. A quick response to make a treatment way is formed in repeated cooperation.

REFERENCES

[1] Li Bin. Baseball [M]. Beijing. Higher Education Press. 1996.

[2] Chen Wenyuan. Baseball Technology Magazine Weekly. 1993(2).

[3] Li Minkuan. Professional Tactical in Baseball. Baseball Technology Information. 1985.3.

[4] China Baseball Association. Baseball Album. 1980. 10(18).

[5] National Women's Baseball championship. Defense Technology Statistics. Organizing Committee Records. 2006–1009.

[6] National Women's Baseball championship, Championship Video. Recorded by National Team. 2009.

[7] National Women's Baseball Infield Group Defense Technology Video. Recorded by National Team. 2007.

Sports Engineering and Computer Science – Luo (Ed.)
© *2015 Taylor & Francis Group, London, ISBN 978-1-138-02650-6*

Research on the problems in college aerobics teaching

WenJie Zhu
Suzhou Art and Design Technology Institute, Jiangsu, Suzhou, China

ABSTRACT: Based on the educational function and value of aerobics course, this article analyzes the current problems in college aerobics teaching. In addition, it also conducts to promote the establishment and development of aerobics in colleges and universities.

Keywords: college; aerobics; teaching problem

1 INTRODUCTION

As a cardio program, aerobics takes physical exercise as basic means and develops health, strength and beauty as features. Surely, it makes shaping, improves health and entertainment as main targets. With its strong sense of the times, rhythm and rhyme, it has gradually become one of the most important parts of the university sports curriculum and very popular with students especially girls. Eventually, it turns into a hot item in college students extracurricular sports activities. How to face up the main problems has become an important research subject in College PE teaching.

2 THE EDUCATIONAL VALUES OF AEROBICS IN COLLEGE PHYSICAL TEACHING

2.1 *Fitness value*

Since the sports aerobics teaching at colleges and universities is very popular, its fitness value has been well reflected in teaching practice. Students through long-term aerobics exercises can have a positive influence on the human movement system, cardiovascular system, respiratory system, urinary system, nervous system and endocrine system. The purpose appears to improve joint function, avoid the abnormal development of bone and enhance the ability of capacity and self-motion.

At the same time, when doing aerobics exercise, it does good to increase the myocardial contractile protein and myoglobin quantity. Besides it can also accelerate the blood circulation and reduce the likelihood of various cardiovascular diseases. In addition, long-term aerobics exercise can also increase the trainer's respiratory muscles and enhance the exchange. At last, the exercise

increases storage capacity of oxygen respiratory gases and promotes the improvement of human glandular structure and functions and maintains normal glucose homeostasis.

2.2 *Entertainment value*

First of all, aerobics has the sense of beauty formally. This is mainly showed in two aspects of dance and music. In virtue of the figure and sound beauty from dance and music, it can help students get more enjoyment in the aerobics class, make emotional and mental state harmonious and obtain double music and movement pleasantness. Then reach the purpose of entertaining students physically and mentally. Also, aerobics has autonomy, openness and compensatory in relationships to let the students fully demonstrate the beauty of their own in class. Finally, they can get psychological satisfaction and share the collective motion of happiness.

2.3 *Educational value*

Aerobics, as a means of college physical education, is a collection of music, gymnastics and dancing. It has a positive effect on students' aesthetic ability, morality, intelligence and emotion attitude. For example, aerobic exercise can let students express their bad emotions in motion and establish friendship with the others. Then they realize the coordinated development of students' physical and psychological qualities.

3 THE MAIN PROBLEMS EXISTING IN COLLEGE AEROBICS TEACHING

As a whole, the College Aerobics Teaching model still follows the traditional classroom teaching model at present. Normally, the teacher gives students one-way standardized explanation and demonstration. At the same time, students are passive recipients of knowledge and keep practicing mechanically. No extracurricular coaching makes the aerobics course interest greatly decreased. Thus, students' participation enthusiasm along with deep teaching is difficult to maintain. The improvement of aerobics course teaching quality is effected. Specifically, the college aerobics teaching problems mainly lie in the following points:

3.1 Theoretical aerobics teaching curriculum is relatively less

Through investigation, it is not difficult to find that the college aerobics teaching mainly take the forms of practical teaching. However, theory teaching content is relatively small, basically covering 2–4 hours. This is mainly because many teachers and students think that non-sports professional students do not need too much to develop the theory of aerobics course. Obviously, it is unreasonable. On the one hand, due to the lack of theoretical courses, many students are unable to fully understand the aerobics movement project. For the cognizance of its exercise value and educational function is not so clear, the students are unable to establish correct aerobics learning goals. At last, it affects the future result of practice teaching. On the other hand, the class hour is less theory class. Therefore, teachers are hard to combine the college physical education with health education. As a result, it also hard to cultivate students' consciousness of lifelong sports for effective integration, referring to the characteristics of aerobics sports. So that many students simply take PE class in order to learn aerobics, which is not conducive to develop students' comprehensive physical quality. In addition, many students do not master enough theory knowledge and the practice in the class lacks theoretical guidance. As a consequence, it not only greatly influences the aerobics training result, also even causes damage to the health of students. What's more, it is clearly not consistent with the sports teaching idea of "Health first".

3.2 The whole class time is not enough and little effective guidance after class

On the whole, the number of teaching aerobics curriculum in colleges is relatively low. Basic arrangements are at the 30–36 lessons. These lessons also include students' theory course, physical fitness test and final examination time. Eventually, the real condition makes the real implementation of the aerobics teaching time limited to physical education teachers. In order to deliver on time

or ahead of the completion of the teaching task, some teachers have to blindly drive teaching schedule and ignore the ability of students to accept and training effect. Even some students reflect that although the choice of aerobics class is very exciting, they find the task heavy, the training difficult and aerobics class atmosphere not active. Thus, the situation affects the aerobics confidence and initiative of learning to a great extent. In reality, some students want to consolidate the knowledge in class or develop self-practice. But due to the lack of effective guidance, the students can only give up outside self-practice and promotion opportunities. The restriction on the class hours is the main character of the courses. Due to it, a lot of students, both in the aerobics class and the aerobics extracurricular class, find it difficult to fully meet their aerobics learning needs.

3.3 The aerobics teaching mode and method are single

At present, the college aerobics teaching mainly takes the class demonstration method. Usually, teachers do the action demonstration in class and students do mechanical imitation and training, not requiring student aerobics thinking basically. Therefore, this way of teaching is teacher centered, a one-way indoctrination, oppression and cramming teaching methods. It is difficult to guarantee the main body of students in sports class. So it is not good to stimulate the students' learning enthusiasm and interest. From the perspective of the characteristics and essence of aerobics, aerobics has plenty of exercise value and educational function. Simple imitation and mechanical practice are obviously difficult to reflect the educational value of aerobics. Whereas, the method adds some teachers' relatively backward aerobics teaching philosophy, the lack of aerobics teaching professional skills and little knowledge in the modern scientific teaching methods. This will inevitably affect the college sports aerobics curriculum teaching result.

3.4 The teaching evaluation is not so scientific

First of all, because some aerobics teachers make the teaching evaluation system more relaxed, the assessment of student learning outcomes often has very strong subjectivity. The real condition makes the assessment methods prone to error and omissions so that many students doubt on the existing teaching evaluation system. Finally, they think it restrains the full play to students' subjective role.

Secondly, the existing aerobics teaching evaluation mode only treats the students simply as evaluation objects. Students are always in the assessment, management and evaluation status. The standard, method, content and principles of teaching evaluation are formulated by the teacher unilaterally. It is difficult to mobilize the students' participation enthusiasm in the teaching evaluation and affects the students to play subjective initiative in aerobics class.

Thirdly, a lot of PE teachers mainly make disposable summative assessment to students when doing aerobics teaching evaluation work. They only use the result of final one-time examination as a basis evaluation of Aerobics learning outcomes for students. But in teaching practice, we cannot rule out some students who study hard and trained with regularity. Probably because of examination stress or physical discomfort caused by disorders play, their aerobics final examination scores are not ideal and even affect the final total score and individual scholarship recipients. So, the way of disposable summative assessment may ultimately affect the students for the aerobics course of long-lasting study interest and the learning enthusiasm. Besides, it is not conducive to form the students' consciousness of lifelong sports. In addition, some aerobics teachers make the evaluation too much over the standard. Most of them take aerobics sports skills as the only reference. Meanwhile, they are lack of evaluation for the learning process, emotional attitude and other aspects. Eventually, they lead some students who have poor coordination movement quality to feel hard to achieve the teacher requirement in limited time. Even when it get more serious that students produce certain psychological barriers of aerobics, causing students' lack of motivation.

3.5 Aerobics teachers team construction leaves behind the demands of students' development

First of all, the number of college aerobics teachers cannot meet the learning and development needs of students. In recent years, due to the impact of high enrollment policy, the number of college students appeared to have a multiple type of growth. But the number of Aerobics teachers did not achieve the corresponding growth. Especially in the big background of college sports curriculum reform, the aerobics course is the basic course to students. Actually, girls become aerobics main classroom personnel structure. And the team of college sports male teachers who understand aerobics teaching is very lack. Female physical education teachers become difficult to adapt to the strong sense of rhythm and large amount of exercise aerobics teaching because of the growth

of the age. Then it forms a great contrast that students' learning demand increases and the teaching task becomes heavier while the number of teacher is insufficient.

Secondly, when aerobics teachers are in the process of construction, some schools overemphasize the high degree. As a result, it leads a lot of new teachers to be lack of teaching experience. There are many deficiencies of aerobics professional teaching ability. For example, a lot of Aerobics teachers' information quality is low. And they have not carried out the consciousness and ability of aerobics education network basically. At last, it results that the class is very difficult to carry out effective learning guidance to students. Thus restricting the aerobics course teaching effect.

4 CONCLUSION

Including all the aerobic fitness function, such as improving the body quality, aerobics improve cardiopulmonary function, muscle endurance, and promote the coordination of the body's tissues and organs. Can make human body to play the best skills, have good fitness results. With the school sports education supplement, after-school aerobics exercises comprehensively enhance the students' physique. And it also guides students to the pursuit of health Sig, positive, energetic, resist the decadent, the erosion of ideological trend of campus culture.

ACKNOWLEDGMENTS

2013 years of Jiangsu Province Education Science "The 12th Five Year Plan" planning issues (Grant NO. T-C/2013/050).

REFERENCES

[1] ZhaoHui Hu, Research of Universities and Colleges in Jilin province Aerobics Curriculum Reform [J]. Heilongjiang Science and Technology Information, 2010, (19):11–13.
[2] Lei Kang, Tao Yan, Na Zhao, LiJie Zhang, Explore the College Aerobics Teaching Problems [J]. Exam Week, 2009, (43):23–25.
[3] FanHua Meng, Research on the Value, Development and Utilization of Fitness Aerobics in National Fitness [J]. Journal of Nanjing Sport Institute (Natural Science Edition), 2013, (50):22–24.

Sports Engineering and Computer Science – Luo (Ed.)
© 2015 Taylor & Francis Group, London, ISBN 978-1-138-02650-6

Research on the innovative strategies in college aerobics teaching

WenJie Zhu
Suzhou Art and Design Technology Institute, Jiangsu, Suzhou, China

ABSTRACT: Aerobics, as a combination of music, gymnastics and dancing, has a strong educational function and exercise value. In recent years, aerobics courses have been widely opened in universities. The main reason is that the students can achieve the goal of fitness shaping and attitude adjustment simply by themselves or using body-building equipment. Therefore, this article finally comes up with the effective innovative strategies to improve aerobics teaching quality.

Keywords: college; aerobics; teaching; innovation

1 INTRODUCTION

With its strong sense of the times, rhythm and rhyme, it has gradually become one of the most important parts of the university sports curriculum and very popular among students, especially among girls. As a result, it turns into a hot item in college students' extracurricular sports activities. How to innovate the traditional aerobics teaching mode and promote the benign development of aerobics in PE teaching is worth studying.

2 EFFECTIVE INNOVATIVE STRATEGIES TO IMPROVE THE QUALITY OF COLLEGE AEROBICS TEACHING

2.1 *Adhere to the learner-centered and develop the consciousness of main body character*

The main objective of college aerobics curriculum is to promote the coordinated development of students' physical and mental quality, make them realize the importance of physical exercise and develop lifelong sports consciousness. Therefore, there is no doubt that students dominate in the aerobics class. So it requires physical education teachers in Colleges and universities to change the traditional ideas of teaching aerobics and stick to the student-oriented. By playing the correct guidance function its own, they will lead students to participate in aerobics teaching activities actively and enthusiastically. Besides, in order to fully stimulate students' main body consciousness in aerobics class, aerobics teachers are supposed to establish harmonious relationship between students and create a relaxed, pleasant atmosphere of classroom teaching. So that students can be kept for a long time attention and emotion in the classroom. More communication and exchange between teachers and students should be strengthened in teaching and training process. Teachers should make the students know how to dialogue and cooperation with other players in a team environment. Meanwhile they also help them realize the coordinated development in moral, intelligence, body, beauty and other comprehensive quality.

2.2 *Set clear teaching objectives*

The teaching objective refers that the students achieve the expected learning results and standard after teaching activities. An explicit college aerobics teaching objective is beneficial to carry out the system of teaching management activities and make aerobics teaching more systemic, targeted and scientific. Under the background of quality education in college aerobics teaching, targeted adjustments in the teaching objective should be made. Apparently, it means to weaken competitive aerobics teaching and focus on the development of college students' comprehensive quality. Besides, it also needs to increase health education teaching and pay attention to develop students' interest in sports aerobics, emotion and personal qualities. In this case, students can establish the correct concept of sports and health and form sound knowledge structure and sports thinking ability. For example, referring to domestic and foreign development of physical education theory in recent years, College Aerobics teaching goal should at least cover sports skill, movement participation goal, physical health goal, mental health goal and social adaptation goal. Surely, the 5 goals must be put into the aerobics teaching practice deeply so that the continuous improvement in college aerobics teaching quality will be seen.

2.3 Innovate the traditional model of aerobics teaching

In the traditional College Aerobics Teaching mode, the teaching model is too uniform and backward. The teaching content mostly selects the basic type. Besides, the teaching type is one-way, ignoring the full combination of aerobics teaching school specific conditions and regional characteristics and lacking new ideas of teaching aerobics. It requires that college aerobics teaching select various and targeted teaching methods. Furthermore it also demands to adapt to the teaching reality and the needs of students when strengthening the construction of aerobics teaching model. At the same time, teachers should attach importance to layered teaching strategy and happy teaching strategy in aerobics teaching. Then fully adapt to and meet the aerobics students' knowledge and learning needs. According to the actual situation of aerobics class in college students, they ought to take fitness teaching as the main line. However, treat the development of students' physical quality as the fundamental basis. Certainly, mainly center on fitness knowledge, fitness concept and fitness learning and develop students' awareness of lifelong sports.

2.4 Make optimized settings of the aerobics teaching content

According to the aerobics teaching experiences of many years in summary, the aerobics content system for student has been explored basically. First of all, in order to help the students start from the most basic steps and gradually form a complete set of movements, some basic aerobics teaching content is supposed to set up. Through opening theoretical courses and explanation for students, their basic knowledge of the aerobics sports system can get learned and strengthened, which lays a good theoretical foundation. Secondly, combine with the students' interests and learning needs, choosing some teaching content with strong sense of times, such as hip hop aerobics, Latin Aerobics, jazz aerobics and other distinctive features of aerobic exercise. So that the existing university aerobics curriculum content system will be more perfect, practical and scientific. This will further stimulate students' interest in aerobics and the aerobics fitness quality will be improved gradually.

2.5 Establishing diversified teaching evaluation system

The new era college aerobics teaching needs to weaken the competitive sport examination under the guidance of scientific development view. Because college students are not aerobics professional athletes and their main learning goal is to entertain and keep fitness. So the assessment proportion of health education, learning process and sports emotion should be increased in the teaching evaluation system. In addition, it is necessary to build a diversified aerobics teaching evaluation system, so as to give full play to the functions and role of teaching evaluation and promote college students' learning methods and teaching methods. In the practical and diversified construction on teaching evaluation system, teachers should mainly persist in formative assessment and process assessment with summative assessment supplement. Then the teachers also should guide the students to do self-evaluation and mutual evaluation in aerobics class. Absolutely, the purpose is to make multi-dimensional and diversified aerobics teaching evaluation system come true and further improve the science and effectiveness of the aerobics teaching evaluation.

2.6 Increase the aerobics teaching input and improve the comprehensive quality of aerobics teachers

Faced with the various problems in current college aerobics teaching, colleges need to expand the enrollment scale, at the same time, they should notice to carry on the conformity to the aerobics teaching resources. Combined with the actual needs of teaching aerobics, they ought to increase aerobics teaching input constantly and create good aerobics teaching environment. Since time and space have a relatively small influence on aerobics, increasing the input in teaching should focus on the introduction and training of high-quality aerobics teachers. The teacher is the organizer, guidance and implementation of aerobics teaching Moreover, the teacher is responsible for the teaching objective system, implementation of teaching objectives and the management of teaching objectives. So each college must firstly ensure the number of aerobics teachers is sufficient and adjusted the aerobics teacher ratio to a reasonable state. Besides, the university should pay attention to the aerobics teachers' comprehensive quality promotion and constantly improve themselves in the aerobics teaching skills and coaching skills. To update existing knowledge structure, teaching ideas and teaching skills positively is also one of the important measures.

3 CONCLUSION

To sum up, at this stage of the aerobics teaching in Colleges, still there are great problems and disadvantages. On the basis of strengthening aerobics teachers' team construction, colleges must actively change the aerobics teaching and the traditional concept of education. Besides, it is also a great way to adhere to the students as the main body, take health education theory as guidance and further enhance the interesting, openness and efficiency aerobics teaching in colleges. In addition, colleges should give full play to the value and functions of aerobics curriculum and realize the coordination and unification of the aim of college physical education. Improving the comprehensive quality of contemporary college students can still play an important part in aerobics curriculum.

ACKNOWLEDGMENTS

2013 years of Jiangsu Province Education Science "The 12th Five Year Plan" planning issues (Grant NO. T-C/2013/050).

REFERENCES

[1] QiuMei Zheng, Study on the Problems of Psychological Training in Aerobics Sports—Taking Northwest Normal University Aerobics Team and Professional Group as an Example [J]. Northwest Adult Education Journal, 2013, (03):15–17.

[2] Yan Li, Present Learning Evaluative Situation and Development Strategy of Aerobics Curriculum in General Undergraduate colleges in Jiangsu Province [J]. Fight (Sports Forum), 2013, (05):23–26.

[3] Feng Du, Qian Li, HongBing Zhou, Experimental Study of Aerobics Athletes Rhythm [J]. Journal of Jiangsu Teachers University of Technology, 2013, (02):19–21.

Sports Engineering and Computer Science – Luo (Ed.)
© 2015 Taylor & Francis Group, London, ISBN 978-1-138-02650-6

Training and innovation research of Chinese youth football talents

Bin Cao

Shandong Women's College, Shandong, Jinan, China

ABSTRACT: Our youth football is the future development of Chinese football. Therefore, only more scientific, systematic and innovative football system, and selection and training of youth players can make young players grow better.

Keywords: youth; football; talent training; innovation

1 INTRODUCTION

Football is the NO.1 sports in the world. The football match with big venue, long time, fierce competition and strong antagonism requires players to have good and more comprehensive physical fitness. Especially for youth players, without very good physical fitness, they cannot play technical and tactical level. But they do have to win the game and guarantee the successful completion of the training content. Thus the rapid improvement of technical and tactical level is affected.

National team and coaches are trained with the same methods and policies to achieve common development goal. Implementation of the policy "trinity" can not only maintain the development of each echelon of youth football players, but also guarantee a substantial improvement in the level of the national team players. When guaranteeing the improvement of coaches at all levels, all those three are complementary. Comparing with Japanese strengthening system policy "Trinity", the system and policies of Chinese youth football are still not perfect. It is important to coordinate development of youth players and coaches and mutual assistance degree between organizations and units. However, it still needs a very great improvement.

2 COMPARISONS OF CHINESE-JAPANESE YOUTH FOOTBALL SYSTEM

Through comparisons of Chinese-Japanese youth football policies, league system and records in recent years, a lot of suggestions for Chinese youth football are obtained. Japan Football Association has made their football development goal as "the world's football powers" in 1980s. To achieve this goal, Japan Football Association has formulated a series of corresponding policies and measures. The most influential one is the famous strengthening policy "Trinity".

3 THE TRAINING OF YOUTH FOOTBALL TALENTS MEANS A LOT TO THE DEVELOPMENT OF CHINESE FOOTBALL

1. The training of our youth football talents is currently in a restructuring phase, and gradually develops from the original oneness to now diversity.
2. In our youth football competitions, issues of age falsification in athletes are very outstanding. Charging standards of the training of our youth talents are high. Subordinations of our youth football talents are very complex. So there is a great restrictive function if talented youth football talents want to change the training unit for better training and development.
3. At present, football fields of sport talent training for our youth football talents are in short supply. Now the number of youth football players showed a gradual decline in recent years. Although the number of coaches in football is steadily increasing, the level of football coaches still needs a great improvement.

4 HOW TO STUDY YOUTH CHARACTERISTICS ON THE BASIS OF THE COMBINATION OF CAMPUS FOOTBALL AND SEMI-PROFESSIONAL, PROFESSIONAL YOUTH FOOTBALL

In accordance with youth training and competition tasks, and physical development characteristics in different ages, daily training and means needs to be arranged reasonably. Surely, that will play a very positive role in the improvement of youth fitness.

4.1 *Development of speed quality*

Development period of speed quality of Chinese male youth is fastest between ages 9–15, top at age 19 and basically stable after age 20. Therefore, youth period is the best period for the improvement of speed quality. We should take full use of this opportunity to obtain the training of speed quality.

4.2 *Development of power quality*

The youth in different ages have sensitive periods of growth and development of muscle groups in different parts. Therefore, we shall use sensitive periods of development of power quality and make proper and reasonable training content, then better training effect will respond.

4.3 *Development of endurance quality*

From the physical fitness test results of our young boys, between ages 10–14, can do endurance training with high intensity. Meanwhile boys, between ages 19–23, have reached the maximum work capacity of endurance. We use heavy exercise and period training of super-compensation to arrange endurance training of youth players reasonably. In accordance with endurance increase periods of youth players and the principle of progressive training, the desired training effect will be achieved.

5 THE ROLE OF PSYCHOLOGICAL AND EMOTIONAL STATE OF THE YOUTH IN GAMES

There are big differences between adult and youth football players in aspects of physical fitness, psychological quality and understandings of technique and tactics. So we need to start from the most basic of all aspects of the youth.

5.1 *Psychological training in youth football*

5.1.1 *Independent emotional control*
Psychological training of the youth shows immaturity because of their age limit, entirely from inner process of self mobilization and encourage. So in order to effectively achieve the best results, our coaches should seize the youth to improve their ability of emotional stability.

5.1.2 *Motivation training for independent training*
The importance of motivation selection for football training is obvious. Good or bad training motivation of the youth has very direct relationship with their sport performance. During training, training and inspiring of motivation is the important basis to promote the youth to train actively and hardly. Therefore, they can win excellent results and glory for the collective.

5.2 *Thinking of technical and tactical training of youth football*

With the professionalism of Chinese football, competition of football games is fiercer and the technical difficulty and skills are also higher and higher. Thus in order to fit the needs of football game, it requires the participating players to constantly improve their own quality and technical level and reasonably use technique.

6 MAIN IDEAS AND INNOVATIONS OF OUR YOUTH FOOTBALL TALENT TRAINING

The selection and training of youth football players shall not only follow the formation rules of football skills, but also take targeted training as youth characteristics of physical and psychological development. Youth players shall master basic football technique and tactics. Besides, for youth players with different physical characteristics, different training content, methods, tools and strength shall be taken by following growth and development characteristics of the youth.

1. Nowadays, many youth football coaches do not have high cultural quality and do not follow processes of physical and psychological maturation of the youth. They only pursue their current achievements to power the youth blindly, which makes their physical development deformed and affects the late development. How can we design a suitable training and competition program for our youth players? At the same time, we should maximize current capabilities of the youth and do not affect future adult athletic performance?
2. How can we select the most outstanding football seedlings in current youth? There are few minority athletes in our Super League and that cannot be explained as insufficient level of our minority players but insufficient opportunities given to them.

   How can the selection ratio between Han and minority be balanced to give more performance opportunities to players of minority? Now the number of minority youth studying in the Mainland is growing and many of them are talented football players. Due to national culture, many minorities have a blood advantage, and diet structure and the tradition of loving sports builds their strong physique. From the view of national character and cultural traditions, many minorities have outgoing personality, overflowing enthusiasm and worship of heroes. And all those are important qualities to play good football.

3. The current football system in china is bad for our players, especially for youth players. The youth players often start formal training around the age of 10, plus the too high expectations of coaches and parents. They only care about their results but frequently ignore the importance of physical development and cultural qualities. Many of them from our province served as top players in national junior team or national youth team. But they got quick elimination in participation of adult competitions for reasons of insufficient physical development and cultural qualities.

4. Our football system should have bold reforms, just like the American NBA basketball. We can develop the campus football. On one hand the body and volitional quality of the youth can be exercised. And on the other hand, college students are mature in both mind and body. Then students with outstanding achievements in football can be selected to continue football career in professional teams. However, students with insufficient level can get the degree to choose other careers. In that way, not only players will have high culture but also the eliminated youth can also have their degree and majors to choose other careers.

ABOUT THE AUTHOR

Cao Bin (1981-) male, master, national level football player, lecturer with Shandong women's college, research direction: Physical Education.

REFERENCES

[1] Zhu Ning. Youth training mode in JFA strength system "Trinity" [J]. Sports and Science, 2002, 23.

[2] Jiang Yong. Professional influence on the development of Chinese youth football [J]. Sports and Science, 2003 (4).

[3] Yao Zhenguang, Feng Wenchang. Status and countermeasures of youth coaches in prefectures of Henan Province [J]. Journal of Jiaozuo Teachers College, 2006.

[4] Wu Chenghui. Research of training systems of Chinese football reserve talents, The Wealth of Networks 2009 (24).

[5] Sun Qinghai, Zhan Guoxiang. Thinking of overage phenomenon in Chinese youth football players at present [J]. Liaoning Sport Science and Technology, 2004 (1).

Sports Engineering and Computer Science – Luo (Ed.)
© *2015 Taylor & Francis Group, London, ISBN 978-1-138-02650-6*

Research on strength training of the elderly under the vision of "nerve accommodation"

Hong-Chun Pu
Sports Department, Chengdu Sport University, Sichuan, China

ABSTRACT: By referring to a great deal of literature and data, this paper analyzes the current situation of the fitness of the elderly in China and clarifies the characteristics of the neuromuscular system and the influence of strength training on the elderly. According to the existing researches and reports at home and abroad, it discusses the importance of the muscular strength training of the elderly in terms of nerve accommodation, draws the conclusion that the elderly need appropriate strength stimulation to spend the remaining years in comfort and brings forward the suggestion that the elder should strengthen the training of fast muscle fiber, easily weakening muscle group and flexibility during the training of muscular strength, aiming to stimulate the public fitness to attach importance to the training of the strength of the elderly.

1 WHAT IS NERVE ACCOMMODATION

The change of the accommodation of nervous system caused by a kind of training called nerve accommodation[1]. Just like the adaptive hypertrophy and hyperplasia of muscle during the training of strength, the nervous system also has such feature. As early as in 1988, Eno[2] argued that during the training of strength, especially during the first several weeks after the training starts, the growth of strength is caused by the change of the accommodation of nervous system to a great degree.

2 CHARACTERISTICS OF THE NEUROMUSCULAR SYSTEM OF THE ELDERLY

As people growing elder, the brain will be gradually degenerated and atrophied, while both the surface area of the cerebral cortex and the cerebral blood flow will be correspondingly decreased. The degeneration of the dendrite (process of nerve cell) in the nervous tissue of the elderly will impact the connection between nerve cells. Furthermore, the stability decline of nervous system appears, the conversion between excitation and inhibition slows down, the formation of new conditioned reflex connection becomes more difficult, the hearing and memory become weaker, and the response to stimulus becomes longer. Since the age of 60, the quantity of the peripheral nerve motor units of the elderly will decrease, and the transduction between distal motion and sensation of axon will become slower. The nerve cells shall have weak endurance.

People tend to feel tired and it will take more time to allay tiredness or get recovered, and the main symptoms includes amyotrophia, decrease of muscular strength, slowed speed of action, weakened elasticity of ligament, osteanabrosis and osteoporosis[3].

It is reported[4] that the muscular mass of the elderly aged 60 to 70 will decrease by 15% and the muscular strength will decrease by 30% in every 10 years, which is dominated by the atrophy of fast muscle fiber, decreasing from 60% at the age of 60 to below 30% at the age of 80. In foreign countries, a series of symptoms caused by the growth of age, including the constant decrease of muscle, gradual decrease of the muscular strength and the degeneration of human body structure and functions, are collectively called as "sacropenia"[5]. The denervation of fast muscle fiber and conversion to slow muscle fiber is an important factor, which causes the decrease of the maximum strength and explosive power level of the elderly[6]. The aging-related function degradation and death of nerve controlling muscle are the important reason of the weakening of skeletal muscle. From 25 to 70, the skeletal muscle decreases by 25% to 50%. Especially after the age of 60, the decrease of nerve will be accelerated[7].

3 INFLUENCE OF THE TRAINING OF STRENGTH ON THE PHYSICAL FITNESS OF THE ELDERLY

According to a number of researches, the scientific training of strength of the elderly can prevent the outbreak of osteoporosis, improve the balanced capacity, postpone the degenerative changes of muscular mass and strength, relieve symptoms of many chronic diseases like chronic heart disease, arthritis and type II diabetes mellitus, improve the quality of sleep, lessen depressed mood[8] and even function as an important measure to treat hypertension and heart disease[9].

For example: the research on postmenopausal women conducted by Pruitt[7] et al. have showed that the lumbar BMD (Bone Mineral Density) of the group that has undergone 9 months of weight lifting training and sports increase, while that of the control group decreases. Narici et al.[10], through the strength training of the elderly aged from 60 to 97 and the young, have discovered that both the muscular CSA (Cross Section Area) and muscular strength of the elderly group evidently increase, while the daily increasing proportion of muscular CSA is even higher than that of the young group. Japanese scholars have carried out the aerobic training and strength training of women aged from 65 to 75 for as long as 3 years (140 weeks), and found out that the physical fitness of the elderly is evidently improved, and the chronic diseases[9] like hypertension are effectively controlled. Besides, the elderly, especially the retired ones, often suffer the psychological disorders like void and loneliness. Frequent participation in some training of strength can not only maintain the physical fitness, but also eliminate the negative psychological impacts and improve the quality of life.

4 ANALYSIS OF THE IMPORTANCE OF NERVE ACCOMMODATION AND TRAINING OF STRENGTH OF THE ELDERLY

4.1 Influence of the training of strength on the balancing function of the elderly

The high frequency of falling occurred on elderly has already become an important social problem, which is mainly caused by the degeneration of the balancing function. To maintain the balance of human body mainly depends on the information input of vestibule, sight and proprioceptive sensation systems and the central nervous system's integration of such information and control of action effectors. The training of strength is an important method to improve the balancing function and prevent the falling of the elderly.

Keijio Hakkinen et al[11] have evaluated the functions of neuro-skeletal muscle of the elderly after the training of strength. The elderly under test was made to sit on a 40 cm tall chair. And then, the minimum time they spent in finishing 3 times of standing-up action in succession was measured.

The result shows that after the training of strength, the time required to finish standing-up action is decreased by 24.8% (P<0.001) for women and 25.4% (P<0.001) for men, while those of the control group does not change. It means that the evident improvement of the function of neuromuscular system for both men and women under test, after the training of strength may result from the shortening of response time and increasing of muscular strength the elderly under test. It further explains that the training of strength may improve the response capacity of the elderly. Through researches, Lord has discovered that the strength and response time of quadriceps femoris and surae muscle group are significantly correlated to the five gait parameters that reflect the balancing capacity, which further illustrates that the training of strength of the elderly can improve the balanced capacity. During a 12-week training of strength, in form of load knew extension, of the community residents aged 56 to 72, Zhu Lingqing et al.[12] have discovered the significant improvement of the muscular strength and balanced capacity of the lower limbs, which proves this viewpoint as it turns out.

Of course, the improvement of balanced capacity may also result from the improvement of the functions of peripheral proprioceptors, and the consequent strengthened control of the tensing and relaxing of agonistic muscle and antagonistic muscle due to the training of strength. It can be well proved by PNF (proprioceptive neuromuscular facilitation) training, which can improve the balancing capacity[13] of human body.

To sum up, appropriate training of strength can improve the balanced capacity of the elderly by enhancing the functions of nervous system.

4.2 Influence of the training of strength on the intelligence level of the elderly

Through the comparative study on the response time, memory span and muscular strength of 187 aged persons, who are classified into physical training group and control group, Williams et al.[14] have discovered that the memory span of physical training group improves as the muscular strength and response time improve. Tsutsumi et al.[15] have randomly classified 42 healthy aged persons who are sedentary without exercise into three groups, namely group of small quantity of medium—and high-impact exercises, group of large quantity of low-impact exercise and group of no exercise, and carried out the training of strength for 12 weeks. The result shows no significant change of cognitive function. It means that the training of strength will help to maintain the cognitive function and to postpone the degeneration of cognitive function. It suggests that the training of strength can not only improve the function of sports system, but also enable the nerve center that controls sports to be more strongly stimulated and exercised, stimulate the blood circulation of brain and postpone the degeneration of the functions of motor center in cerebral cortex.

4.3 Influence of nerve accommodation on the muscular strength of the elderly during the training of strength

4.3.1 Nerve accommodation stimulates the gaining of strength

The muscular strength does not only rely on the muscle mass. Otherwise, when the muscle (group) of the same mass is activated to the greatest extent at any time, the peak value of strength should be roughly the same, while the study on the cross section area of muscle discovers that the muscular strength is only about 50% correlated to the mass. Obviously, the muscular strength also depends on other factors. As a matter of fact, the improvement of strength exercises and coordination exercises is not correlated to the growth of muscle mass. Such inconsistence largely results from the accommodation of nervous system[16]. For example, with the control of weight, the lifting athletes can still gain the growth of strength year on year. Therefore, the muscular strength does not only depend on the mass of correlated muscle, but may also depend on the capacity of nervous system to reasonably activate muscle.

Bilateral (cross) transfer refers to the change of strength or sport skill of the contralateral non-exercising limb caused by the exercise of unilateral limb. It has been proved that such transfer of strength does exist in various trainings and studies. Based on the data sorting and analysis of 29 research reports on bilateral transfer[17], scholars have discovered that the muscular strength of both hands and/or both arms will change after the strength training of one hand and/or one arm. In the all the researches, the strength of the exercised limbs have undergone growth of different degrees. Except for the slight negative growth of the contralateral limbs occurred in 3 groups of experimental data, the rest researches all show the significant strength growth, lower than that of the exercised limbs, and the approximated positive correlation of strength growth between the bilateral limbs.

The above-mentioned researches have sufficiently proved gaining of nerve accommodation stimulates strength during the training of strength.

4.3.2 Nerve accommodation of the gaining of strength of the elderly

Presently, there are few literatures about the direct researches on the relationship between the training of strength and strength gaining of never accommodation of the elderly. However, Aniansson and Custafson (1981) have discovered that after prescribed training of strength, the elderly men aged 69 to 74 have almost gained the same strength growth as the young men. Besides, a lot of experiments[18] show that the muscular strength of the elderly could be significantly improved (higher than 100% for many cases) after systematic training of strength, and the growth of the cross section area of muscle (about 10%) is highly inconsistent with the growth of muscular strength.

Therefore, from the perspective of nerve accommodation, the strength training of the elderly can stimulate the growth of muscular strength, compensate for the decrease of muscular strength of the elderly caused by the decrease of exercises, and improve the quality of life.

5 CONCLUSION AND SUGGESTION

5.1 Conclusion

The degenerative change of neuromuscular system is a kind of manifestation of the aging process. In terms of the system theory, for human body, which is an inseparable organic whole, change of any part is correlated with the whole body. Besides the aging of the muscle itself, all the degenerative changes of neuromuscular system, including the action features of the nervous system, features of the endocrine system, psychological factors and decrease of physical exercises, may impact the degeneration of muscle. Furthermore, each factor is correlated with each other. The physical exercises of the elderly, especially the scientific and appropriate training of strength may activate all the systems of the human body and produce good nerve accommodation, thus postponing the aging-related degeneration of the functions of the main center, improving the balancing capacity, enhancing the muscular strength and maintaining the high working efficiency of the body. Therefore, the elderly have more needs of appropriate training of strength.

5.2 Suggestion

Besides following the general fitness rules, the elderly should well completed the following tasks according to the features of the neuromuscular system:

1. The design of the exercise prescription of the elderly should take the combination of aerobic exercise and strength training into consideration, and appropriately increase the proportion of strength training to respond to the decrease of fast muscle fiber. Special attention should

be paid to the exercises of the easily weakening muscles, such as longus colli, scalene, trapezius, oblique abdominis, rectus abdominis, gluteus and so on to avoid the occurrence of humpback, hernia and so on.

2. To prevent the elderly from falling, emphasis should be laid to the strength training of lower limbs. At the same time, it is advised to combine the recently popular core strength training method to strengthen the core stability.

3. The neuro-muscle of the elderly is fatigable. After strong-impact exercises, the muscle needs more time to be recovered, so special attention should be paid to the relaxing after exercises. The adoption of PNF retraction method can not only sufficiently relax the muscle, but also improve the flexibility, joint motion range and elasticity of the ligament of the elderly.

4. Taking advantage of the cross transfer of strength, carry out the strength recovering exercises of the elderly.

Besides, the sports science scholars should strengthen the research on the physical fitness of the elderly and establish safe and effective fitness guidelines for the elderly in accordance with the current situation of China.

REFERENCES

[1] Enoka, R.M. Muscle strength and its development: new perspectives. *Sports Medicine.* 1988, 6(3):146–168.

[2] Morley J.E, Baumgartner R.N, Roubenoff R. et al. Sarcopenia. *J Lab Clin Med*, 2001, 137(4): 231–243.

[3] Hurley B.F. Age, gender, and muscular strength. *J Gerontol,* 1995, 50: 41–44.

[4] Pruitt L.A., et al. Weight-training effects on bone mineral density in early postmenopausal women. *J Bone Miner Res,* 1992(7):179–185.

[5] Narici, M.V., Reeves, N.D., Morse, C.I. et al. Muscular adaptations to resistance exercise in the elderly. *J Musculoskel Neuron Interact,* 2004, 4(2):161–164.

[6] Toshiaki Kato, Youichi Kurosawa. Effect of long-term exercise on physical function and medical examination in elderly women. *Yonago Acta Med*, 2002, 45:75–88.

[7] Jensen J., Lundin-Olsson L., Nyberg, L., et al. Fall and injury prevention in older people living in residential care facilities. Acluster Randomized Trial. *Issue of Annals of Internal Medicine,* 2003, 136:733–741.

[8] Hakkinen, K., and Pakarinen, A. Muscle strength and serum hormones in middle-aged and elderly men and women. *Acta. Physiol. Scand,* 1993, 148:199–207.

[9] Lord S.R, Lloyd D.G, Li S.K. Sensori, motor function, gait patterns and falls in community-dwelling women. *Age Ageing,* 1996, 25(4):292–299.

[10] Williams, P., Lord, S.R. Effects of group exercise on cognitive functioning and mood in older women. *Aug N Z J PublicHealth,* 1997, 21(1): 45–52.

[11] Tsutsumi, T., Don, B.M., Zaichkowsky, LD. Physical fitness and psychological benefits of strength training in community dwelling older adults. *Appl Human Sc,* 1997, 16(6): 257–266.

[12] Narici, M.V. Hoppeler, H. et al. Human quadriceps cross sectional area, torque and neural activation during 6 months of strength training. *Acta Physiologica Scandinavica.* 1996, 157:175–18.

[13] Laidlaw, D.H., Kornatz, K.W., Keen, D.A., et al. Strength training improves the steadiness of slow lengthening contractions performed by old adults. *Journal of Applied Physiology.* 1999, 87:1786–1795.

[14] Yue, G., Cole, K.J. Strength increase from the motor program: comparison of training with maximal voluntary and imagined muscle contractions. *Journal of Neurophysiology,* 1992, 67:1114–1123.

[15] Higbie, E.J., Cureton, K.J., Warren, G.L., et al. Effects of concentric and eccentric training on muscle strength, cross-sectional area, and neural activation. *Journal of Applied Physiology.* 1996, 81:2173–2181.

[16] Hakkinen, K., Katlinen, M., Izquierdo, M. et al. Changes in agonistantagonist EMG, muscle CSA, and force during strength training in middle-aged and older people. *Journal of Applied Physiology.* 1998, 84:1341–1349.

[17] Admiridis, I.G., Martin, A., Morion, B. et al. Co-activation and tension regulating phenomena during isokinetic knee extension in sedentary and highly skilled humans. *European Journal of Applied Physiology.* 1996, 73:149–156.

[18] Charette S.L. et al. Muscle hypertrophy response to resistance training in older women. *J Applied Physiology,* 1991, 70:1912~1916.

Sports Engineering and Computer Science – Luo (Ed.)
© 2015 Taylor & Francis Group, London, ISBN 978-1-138-02650-6

Study on the reform of teaching modes of physical education curriculum in colleges and universities

Feng Wang
Department of Physical Education, China University of Petroleum, Huadong, P.R. China

ABSTRACT: Through investigation, the conventional teaching mode is still mainly used in physical education curriculum in most colleges and universities, which no longer meets the requirement of "Health education". The "Three-self" classroom teaching form proposed in the current guiding outline of public physical education curriculum in colleges and universities issued by the Ministry of Education, which can bring students' principal role into full play, expand time and space of physical education curriculum, and benefit the implementation of "Health education". In this article, the survey method, the mathematical statistics method, the experimental research method and so on are taken to carry out empirical research of the "Three-self" teaching form, demonstrate the necessity and feasibility to implement the "Three-self" teaching form in colleges and universities.

1 INTRODUCTION

At present, it is desired that the ideas about physical education of the school can be updated. Physical education has been developed in the direction of "Health education". Throughout the modes of physical education curriculum in colleges and universities today, in spite of some improvements and changes, there is still certain restriction, and it still can not meet the demand of increasingly developed physical education in colleges and universities. The development and diversification of the modes of physical education curriculum in colleges and universities not only represent the development level of the ideas about physical education in colleges and universities, but also symbolize the integration of modernized educational thoughts into the ideas about physical education. It is the necessary trend for the development of modern educational thoughts.

Through making survey and referring to a large number of references, it is indicated that there are just the following several kinds of modes of physical education curriculum existing in colleges and universities: compulsory courses of physical education; optional courses of physical education; elective courses of physical education and so on. Although these courses have made the reform of physical education teaching step forward to a great extent, there is a common feature among them, that is, with physical education teachers as the leading role, it is still the teaching mode in which the teaching of teachers, the learning and practice of students are taken as the major form. Importance is attached to teaching sports technique and improving special sports ability of students, which does not benefit the cultivation

of students' innovative ability and the stimulation of students' interest. It is still conventional physical education teaching mode. In addition to this, is it possible to establish new physical education classroom modes which more comply with the idea of "Health education", benefit the cultivation of students' consciousness of "Lifelong physical education", and take students as the leading role? In current guiding outline of public physical education curriculum in colleges and universities issued by the Ministry of Education, the classroom teaching form of "Students should have the freedom to select courses, teachers and class hours by themselves" is proposed for the first time. The "Three-self" teaching system can bring students' principal role into full play, expand time and space of physical education curriculum, and lay the foundation for the implementation of "Health education". In this article, the questionnaire survey method, the mathematical statistics method, the experimental research method and so on are adopted to carry out empirical research of the conditions to implement the "Three-self" teaching and of the method to determine the "Three-self" teaching contents and make conclusions. The research will play a driving role in the implementation of "Three-self" teaching in physical education curriculum of colleges and universities in a better way.

2 METHODOLOGIES

2.1 Research object

Undergraduates of Year 2008 and Year 2009 in 16 colleges and universities in Beijing and Shandong are taken as the research object.

2.2 Research method

2.2.1 Questionnaire survey method

The questionnaire survey method is used to make survey and research of students' interest, hobbies and feeling about the "Three-self" teaching. In 2008, the first survey was made to freshmen. 1250 questionnaires were distributed, and 1099 valid questionnaires were taken back; in 2009, 600 questionnaires were distributed for the second time to make survey about reliability. 448 valid questionnaires were taken back. Among the survey objects, there are 220 boys and 228 girls.

2.2.2 Expert interview method

The expert interview method is used to ask for experts' opinions.

2.2.3 Mathematical statistics method

SPSS mathematical statistics method is used to makes statistical analysis of the obtained data.

2.2.4 Experimental research method

Follow-up survey is made in the process of public physical education curriculum, study and improvement is continuously made according to students' feedback opinion.

3 RESULTS AND ANALYSIS

3.1 To adhere to the educational idea of "human-centered" is the ideological basis to establish the "Three-self" teaching system

The basic function of education is to improve the overall development of people, while "human-centered" is the core of modern educational values. "Human-centered" puts emphasis on bringing students' potential into full play and improving the healthy development of students mentally and physically. The value orientation of education in "human-centered" is to impel everyone at his best as possible as he can be in the way of respecting students' personality, feelings and so on. The formation of knowledge-based economy in today's world makes knowledge the important factor of production, while the main carrier, disseminator and creator of knowledge are human. The essence of college education is the activity to make individual socialized through transmission, internalization, selection and innovation of profound culture as well as the cultivation of senior specialized talents with subjective sprit required for social development. The stage of higher education is the important stage for talent cultivation. At this stage, it is required not only to grasp profound specialized knowledge and skills, but also to set up health consciousness and grasp knowledge and skills to improve and maintain health, and has a healthy constitution accordingly. Physical education curriculum just takes the improvement of overall healthy development of body, mentality and social adaptive ability of students as the major objective. Through learning sports knowledge and practicing sports events, students are made to grasp effective methods for bodybuilding, form the habit of taking exercise on their own initiative, and promote the sustainable healthy development mentally and physically.

However, the tendency to put too much emphasis on sports skills and ignore students' interest and specialties. It is ever existed in previous sports teaching. Since such physical exercise is not out of students' own interest, it is very hard to stimulate students' internal motive to take physical exercise actively and form the good habit to take physical exercise. There are just a small number of students who will insist on taking physical exercise after graduating from colleges and universities. In order to change the situation, it is required to carry out the idea of "human-centered", respect students' will and individual difference, and stimulate students' passion to participate in physical exercise on their own initiative.

The "Three-self" teaching system is just established based on the idea of "human-centered". The so-called "Three-self" teaching refers to the classroom teaching form to make students "have the freedom to select courses, teachers and class hours by themselves". In such teaching form, students can select sports teaching contents suitable for themselves according to their own interest, hobbies and demand. The implementation of "Three-self" teaching will bring students' principle role into full play, expand time and space of physical education curriculum, and plays the important role in helping students form the healthy life style and form the habit of taking physical exercise lifelong. Such teaching form fully respects students' will and individual difference. It is a complete reflection of the educational idea of "human-centered" in physical education teaching.

3.2 To know students' interest and hobbies is the prerequisite to determine "Three-self" teaching

In order to make the contents of "Three-self" teaching closer to the actual demands of students, the survey about students' interest and hobbies should be made first before teaching contents are designed. The teaching contents of physical education are classified, and students make choice in various courses according to their respective hobbies, so that good preparations are made for the arrangement of teaching contents.

For example, it is known in the survey that among big-ball sports, what makes boys interested most is basketball and football, accounting for 82.9%. When the teaching class is arranged, the number of teaching classes for basketball and football will be increased properly, so that students' demand can be met to the fullest extent. Take another example, it is known in the survey that the events in which girl are interested among dancing events include fitness dance (31.3%), bodybuilding (28.7%), ballroom dance (21.5%), sports dance (18.5%). In a word, it is very important to make survey about students' interest and hobbies in every term in order to meet students' demand to the fullest extent.

3.3 The condition of instruments at the field is the material guarantee to carry out "Three-self" teaching

The condition of instruments at the field is the important material guarantee to implement "Three-self" teaching. In order to meet students' diversified demands in real sense, it is necessary to have sufficient field facilities as guarantee.

For example, we have learnt from the survey that of entertainment and bodybuilding, which boys are interested among events include swimming (33.4%) and roller skating (24.3%). The case is the same with girls: swimming (36.6%) and roller skating (34.5%).

It can be seen that both boys and girls like swimming. However, the event is not available in our school due to the class hour arrangement and the restriction of field at present, which obviously affects the effect of "Three-self" teaching.

In the survey, we also know that among small-ball events, the first three events that boys like most are table tennis (43.3%), badminton (36.8%) and tennis (15.6%). For girls, they are badminton (51.2%), table tennis (34.7%) and tennis (10.6%).

It can be seen that badminton is an event that both boys and girls like. However, due to the restriction of field, the demand can not be completely met when teaching contents are arranged in most schools. Obviously, the condition of field facilities is the important guarantee to carry out "Three-self" teaching in the implementation of teaching.

3.4 To set up teaching contents which can meet students' diversified demands and have a teaching staff that can undertake the teaching task is an important guarantee to carry out "Three-self" teaching

With social progress, people attach much more importance to health than ever. Various bodybuilding methods generated at the right moment are also continuously updated and developed. Newborn bodybuilding methods are widely promoted via news media. Therefore, in the design of teaching contents, if it is only restricted to conventional events in physical education curriculum, and no fashionable fitness methods are introduced, it will not meet students' diversified demands.

In the implementation of "Three-self" teaching, the condition of teaching staff is a basic guarantee to carry out "Three-self" teaching smoothly. In the execution of "Three-self" teaching, we should not be restricted to conventional events in previous physical education curriculum. Instead, fashionable bodybuilding methods that have newly emerged at home and abroad and are available for the field conditions in our school should be introduced.

Nevertheless, in order to set up new events, there must be teachers who grasp new events for teaching, or these emerging events can not be implemented. In order that the teaching staff can meet the requirement of the "Three-self" teaching contents, it is required to tap teachers' potential. Teachers need to continuously learn and make progress, so that they can not only undertake their own specialty events, but also undertake the teaching work of other emerging events. Besides, the way in which teachers go out for learning and those who grasp new events should be engaged should be taken. By doing so, the smooth implementation of "Three-self" teaching can be ensured. It can be seen that the establishment of the teaching staff to undertake teaching task in "Three-self" teaching is an important guarantee to carry out "Three-self" teaching.

3.5 To carry out matching teaching management system is an important condition to carry out "Three-self" teaching

Physical education curriculum is the compulsory course in colleges and universities. In compulsory courses, if the fixed class hours arranged by the school previously are changed to that selected by students themselves, it is required to carry out the matching teaching management system, or there is possibility to have no way of implementing the "Three-self" teaching form. The conditions may happen that students can not participate in the physical education courses in the period when they are arranged since students have other courses, or no physical education courses are arranged when students are free. It can be seen that in order to carry out "Three-self" teaching in real sense, there must be the teaching management system matching to it in teaching management. For the purpose of establishing the system, the teaching competent authorities of the school should give sufficient importance and support first and create favorable conditions for that; second, the uniform network management system of the school should be set up, so that there is also equal possibility for physical education curriculum as other courses for time selection; third, when teaching is arranged, the time for classes should be prolonged as possible as it can be when teaching is arranged so that students are open to more alternatives.

In order to know the trend that students select time for physical education curriculum, so that the teaching time can be arranged properly, we have made survey about students' selection of class hours (Table 1). It is known from survey that there is the greatest number of people who select to have physical education course at section 7 and section 8 of class hours. Therefore, on the condition that field instruments and teachers are available, physical education courses should be arranged at section 7 and section 8 of class hours, which can meet students' demand in a better way.

In addition, it is also available to prolong the term for selection of physical education curriculum to facilitate students' choice. In some terms, students fail to select courses that they like due to other reasons, and they can select them in another term. Altogether, in order that students can select class hours on their own, the matching teaching management system must be carried out in teaching management. It is an important condition to carry out "Three-self" teaching.

3.6 Analysis of students' approval of the "Three-self" teaching form

In order to know students' opinion about the "Three-self" teaching form, we made survey among students: in the questionnaire about whether you like such teaching form to select teacher, time and event on your own, those who like it very much account for 26.4%, who like it account for 57.2%, who think it passable account for 15.5%, who don't like it account for 0.7%, and who refuse it account for 0.2%. On the whole, boys and girls who like the teaching form of "Three-self" teaching account

for 83.6% in total, those who think it passable or don't like it only account for 16.4%. Therefore, it turns out that "Three-self" teaching is very popular among students.

In the questionnaire about what kind of teaching form you are satisfied in the two modes of learning via self-choice and learning according to regulations of the school, 83.2% select the former, 14.6% select the latter, and other accounts for 2.2%. From the two questions, it can be seen that students still like the way to have the decision-making power to select event, teacher and time they prefer.

4 CONCLUSIONS

1. To adhere to the idea of "human-centered" is the theoretical basis for the implementation of "Three-self" teaching, to know students' demand is the basic prerequisite for the implementation of "Three-self" teaching, to have necessary field facilities is the material guarantee for the implementation of "Three-self" teaching, to have a teaching staff who can undertake various tasks in physical education teaching for the implementation of "Three-self" teaching, to provide the matching teaching management system is an important condition for the implementation of "Three-self" teaching.

2. In the content system of "Three-self" teaching, it is required to carry on the conventional events in the physical education curriculum regulated in the school that students like very much, and introduce bodybuilding events fashionable at home and abroad and available in colleges and universities, so that students' diversified demands can be met.

3. It can be known from students' feedback result that 83.6% of the students think it necessary to carry out "Three-self" teaching. Obviously, "Three-self" teaching is a teaching form very popular among students.

4. Only 26.1% of the students are very satisfied or relatively satisfied with the field facilities required for the implementation of "Three-self" teaching, and a large proportion of students just think it passable or feel unsatisfied. It can be seen that the improvement of the facility conditions at the sports field is a problem to be solved urgently in the process of the implementation of "Three-self" teaching.

Table 1. Survey of students' intention about time selection of physical education curriculum.

Class	Section	Time	Percentage
01–02	07:30–09:20	7.2%	5
03–04	09:40–11:30	20.4%	2
05–06	14:00–15:50	17.6%	3
07–08	15:10–17:00	37.4%	1
09–10	18:30–20:20	14.5%	4
11–12	20:40–22:30	2.9%	6

REFERENCES

Bao Zhen W. 2010. The Shandong public sport resource present situation research in national constitution. *Journal of Shandong Sport Institute*, (7): 17–21.

Cheng Liu & Guo Gui Wu. 2009. The study of using colleges and universities sports resources to develop community sports activities. *Journal of Harbin Institute of Physical Education*, 27(6): 35–37.

Feng Wang. 2010. Notification of the Ministry of Education on Printing and Distributing the Guiding Outline of Physical Education Curriculum. *Teaching in Colleges and University*, 48(10): 789–791.

Feng Yan L. 2006. Study of the "Eleventh Five-Year Plan" of sports business in Shandong province. *Master Thesis, Beijing Normal University.*

Guang Feng Yuan. 2003. The study of the status and development and the configuration countermeasures of Shandong urban area sport resource. *Capital Institute of Physical Education,* 15(4): 25–27.

Jian Zhang. 2011. Investigation and research on college sports resources opening to the surrounding communities. *Journal of Harbin Institute of Physical Education*, 29(3): 30–33.

Jian Qing Zhu & Hu Ping C. 2006. The structure of sport resource and the choice of optimal collocation manner of common colleges in east area. *China Sport Science*, 42(6): 104–110.

Jie Yuan. 2011. College sports resources service city community sports development countermeasures. *Journal of Nanjing Institute of Physical Education (Natural Science)*, 10(3): 100–111.

Jie Dong & Yong-Jun. Zhang. 2006. The public investigation and analysis of sport resource in Shandong province. *Journal of Shanghai Institute of Physical Education*, (5): 20–23.

Jin Ping Xiong & Shun Yi Li & Dan Chen. 2011. A study on the sports resources' allocation status and operation countermeasures in Jiangxi colleges and universities. *Liaoning Sport Science and Technology*, 33(1): 23–26.

Jing G. 2009. The university as the center for regional sport development research. *Master thesis, Southwest Jiao Tong University.*

Kai Gao. 2007. Interactive development research of Shandong province community sport and college sport. Master thesis, Nanjing Normal University.

Li Hong Yu. 2011. The strategic choice of the status quo and optimization of allocation of sports resources in higher education institutes—A case study of higher education park in Hangzhou, Zhejiang Province. *Journal of Beijing Sport University*, 32(2): 92–96.

Li Cheng Tang & Li Hui Tang. 2010. On the status of university sports resources service the fitness for all. *Sichuan Sport Science*, (2): 82–85.

Lian Ming Zang. 2001. Apraisal about the utilization efficiency of sports resource in colleges and universities. *Journal of Physical Education*, 8(4): 76–78.

Ling Hua R & Yu Shan T. 2007. Synergetic outlook of social sports resources under pan-resource background. *Journal of Shanghai University of Sports*, (2): 1–5.

Ming Li. 2006. Status study development and utilization of resource in Hubei province. *China Outstanding Doctor/Master Dissertations Full-text Database.*

Ping Zhou & Shao Bo X & Zheng-Qian Z. 2011. Research on the current situation and countermeasures of college sports resources serves of rural community sports. *Journal of Guangzhou Sport University*, 31(1): 34–37.

Qiao Ling Li & Hong Ling Hu. 2008. Simply analysis the concept and definition of university sports resources. *Journal of Nanjing Institute of Physical Education*, 22(1): 84–86.

Rong Li & Zi Xing Hang. 2001. Discussion on development and disposition of resource of university P.E. education[J]. *Journal of Nanjing Institute of Physical Education*, 15(6): 29–30.

Rui Xia Huang. 2004. Discussion on construction of the common service system of P.E. resource in colleges and universities. *Fujian Sports Science and Technology February*, 23(1): 57–58.

Shui Sheng P et al. 2005. Research on the status quo of sports industry in Shandong province and its strategic objective for development and its countermeasures in 2010. *Shandong Institute of Physical Education and Sports*, 21(3);19–21.

Tie Min S & Chang Zheng Z. 2005. Paid service present situation and feasibility study of college sport stadiums in Shandong province. *Journal of Shandong Sport Institute*, (6): 51–52.

Tong Yan Li & Yun Hong Z. 2009. Tentative research on the development of competitive sports resource. *Journal of Chengdu Sport University*, 35(6): 14–15.

Xiao Ping Liu & Yu Liu Tao. 2007. Research on the universities sports resources and their social sharing in China on the basis of resource deployment theory. *Journal of Sports and Science*, 28(3): 94–96.

Xin Hong Wang. 2009. Integration implementation mechanism research of Shandong province sport human resource. *Master thesis, Shandong Normal University.*

Yan Fang Geng. 2009. The study of the impact on the sustainable development of competitive sports in Shandong province laid by the eleventh national games. *Master Thesis, Qufu Normal University.*

Yang Li & Yong Gen J & Yu Ying S. 2000. Investigation on rational distribution of P.E. resources in Shanxi colleges and universities. *Journal of Physical Education*, (2): 76–79.

Yi Lao. 2005. Cultivation of the primary social sport instructors with P.E. resources of colleges and universities. *Journal of Shenyang Physical Education Institute*, 24(2): 40–41.

Yi Xiang Liu & Xin Feng. Social allocation of college sports resources in China[J]. *Journal of Wuhan Institute of Physical Education*, 2009, 43(1): 70–72.

Zhen Li. 2009. Sport stadium's present situation investigation and the idea of the use of integrated reformation in Shandong province. *Master thesis, Shandong Normal University.*

Zhen Li. 2009. Investigation on status quo of Shandong sports facility resources and assumption on the integrated reform of the operations. *Master Thesis, Beijing Normal University.*

Sports Engineering and Computer Science – Luo (Ed.)
© 2015 Taylor & Francis Group, London, ISBN 978-1-138-02650-6

Research on the construction evaluation index system of China university extracurricular sports activities—Beijing ordinary university as an example

WenZhong Li, YunHua Cao & HuiZhi Zhang
Department of Physical Education, North China Electric Power University, Beijing, China

ABSTRACT: In the "on strengthening juvenile sports and improve the health of young people's opinion" and "National Ordinary College Sports Curriculum Teaching Instruction Summary" and other documents, point out clearly the importance of students' extracurricular sports activities. In this article, Beijing college students' extracurricular sports present situation investigation found, students in extracurricular sports activities the overall situation is good, can reach the national standard of college students. However, the college students' extracurricular activities vary greatly, affect the overall effect of colleges and universities in Beijing. In order to solve the problems, the establishment of Beijing college students' extracurricular physical activity evaluation index system is very necessary. This article uses the literature material law, expert interview law, according to the analytic hierarchy qualitative evaluation and quantitative evaluation principles, developed in Beijing college students' extracurricular sports evaluation indexes, including: 3 first level indexes, 10 level two indexes and 24 level three indexes, and on the part of college students extracurricular sports activities were pre evaluation, evaluation of effect is good by the majority of teachers and students recognition.

Keywords: school physical education; college students extracurricular sports; evaluation index

1 INTRODUCTION

The Ministry of education, the State Sports General Administration, the Communist Youth League Central Committee in order to further improve the physique and health level of the students, at the end of 2006, jointly launched the "nationwide Sunshine Sports" hundreds of millions of students, schools at all levels in the positive response, extensive and in-depth development of sunlight sports. So that all sectors of society have a more profound understanding to improve the physical health of students, with more attention to the school sports work, to mobilize the broad masses of young students to participate in physical exercise passion. In this great in strength and impetus of the sports activities of the students in extracurricular sports activities, the major colleges and universities students obtained the rapid development, the college extracurricular sports activities organizing content rich and colorful, diverse forms.

Beijing university school sports work has excellent traditions, decades of mass sports activities to flourish, become a common practice, has formed a fine tradition and culture, it has become a kind of campus culture, is an organic part of school, have been awarded the advanced unit of the national sports work. In order to better implement the party's education policy, adhere to the "health first", the educational idea of "people-oriented", and strive to build the school sports in colleges of Beijing city "boutique", at the same time the fine tradition in order to better keep the school mass sports activities, I develop this study.

2 RESEARCH OBJECT AND METHOD

2.1 Research object

In this study, 8 schools in Beijing city as the object of study; according to the research needs, investigated a total of 1200 students, each school were randomly selected 150 test, the competent students extracurricular activities of 16 people, each school 2 people, a total of 1216 people were in the test personnel.

2.2 Research methods

Literature method: through access to the Internet and many core periodical literature, data.

Method of investigation: the extraction of all kinds of education experts and a total of 23 people were interviewed, the famous 2 professors, 5 professors, 13 associate professors, 2 lecturers, and 1 assistant; these experts are all teachers. The results of the interviews are interviews and finished.

The complex system construction method: establishment of extracurricular sports activities in Beijing colleges and universities evaluation indexes at all levels, and to filter the index and standard.

Using the analytic hierarchy process; determining the evaluation index of extracurricular sports activities in ordinary colleges and universities in Beijing city specific weight.

3 THE RESULTS AND ANALYSIS

3.1 *Investigation of the present situation of extracurricular sports activities of college students in Beijing city*

Before I building the index system, the first of Beijing college students sports activities after the investigation, by using the method of questionnaire, issued a total of 1200 student questionnaires, 1131 were recovered, the recovery rate is 99.02%, of which 727 boys and 404 girls. After a week in the students in the selected 80 people, the retest reliability coefficients, obtained from the two income data, R = 0.935, P < 0.01 level, is of high reliability. The questionnaires were distributed to 16 copies, 16 were recovered, the recovery rate is 100%.

3.1.1 *Student participate in the extracurricular sports attitude*
Attitude is the psychological process of a comprehensive, is the formation of comprehensive based on cognition, emotion, motivation and other psychological process. Attitude value tendency to cognition and emotion as the main content, in order to have the motivation and emotion as the main content of the motivation. So, the student to participate in the extracurricular sports attitude is mainly manifested in three aspects: whether there is interest in extracurricular activities? What is participating in the sports activities motivation? How to understand the extracurricular sports activities?

From Figure 1, you can see students' extracurricular sports activities like, 73.11% of the total number, there were 17.81% students' extracurricular physical activities like. This school organization every semester extracurricular activities such as competitions, games, there is a direct relationship. School extracurricular sports activities in the campus culture construction, make students aware of the benefits of participating in physical activity,

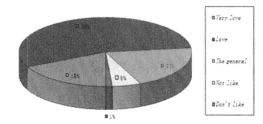

Figure 1. The interest of students in extracurricular sports activities distribution.

and create a good environment for students to participate in extracurricular sports. Students participating in extracurricular sports activities motivated college students to participate in physical exercise motivation, motivation oriented active students participating in extracurricular sports activities in order to stimulate students interest in participation, can induce students to participate in the initiative.

It is seen from Table 1, boys and girls are to enhance the health in the first place, respectively is 88.69% and 67.73%, indicating that the boys and girls to recognize the importance of strengthening the personal physique, this is one of the main motivations to participate in extracurricular sports activities. Second, in order to enrich school life, male, female students are ranked second, 63.33% and 57.27% respectively; in addition to the psychological aspects of girls is ranked third in the 42.01%, the boys came in fourth to 40.39%; however, in the "exam" this option in the proportion of boys was only 8.64%, ranked ninth the girls, and the ratio of 38.08%, ranked fourth. From the analysis of the option, students participate in extracurricular sports activities in order to meet girls physical examination, this phenomenon should arouse enough attention.

3.1.2 *Students participating in extracurricular activities*
Students participate in extracurricular sports activities is one of important index of frequency response of extracurricular sports. According to the survey: in 1–2 sports college students accounted for 53.8%, to participate in the 3–4 sports college students accounted for 38.4%, participate in more than 5 sports college students accounted for 3.1%, not to participate in the physical training college students accounted for 4.7%, often take part in physical exercise of students accounted for 95.3%, said the extracurricular activities of colleges and universities students, the development of Beijing city the overall situation is good, the students to

Table 1. Student participating in extracurricular sports activities motivation survey.

Motivation	Boys (727)			Girls		
	People	Ratio/(%)	Qua	People	Ratio/(%)	Qua
Enhance physical	645	88.69	1	274	67.73	1
The technical level	254	34.89	5	57	14.10	7
Enrich your life	460	63.33	2	231	57.27	2
Social networking platform	15	2.02	11	21	5.09	9
Psychological adjustment	294	40.39	4	170	42.01	3
The exercise will	133	18.26	7	27	6.69	8
Hobbies	424	58.32	3	122	30.23	5
The exam	63	8.64	9	154	38.08	4
Shape	51	7.03	10	84	20.78	6
For honor	200	27.46	6	18	4.36	10
Habit	77	10.66	8	12	3.05	11
The other	25	3.47		16	4.07	

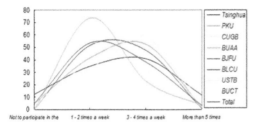

Figure 2. Sports activities frequency distribution.

participate in extracurricular sports activities. So how do the students of the school extracurricular sports activities? (For Fig. 2).

See from Figure 2 simple can be, the school students to participate in extracurricular sports activities frequency distribution is not uniform, the cumulative frequency method of testing, the frequency of students extracurricular sports activities in $P < 0.05$ level, as shown in Table 2:

From Table 2 results we also clearly to understand, do not participate in any activities and a number of students participating distribution of more than 5 times the very uneven. Although the extracurricular physical activity of college students in Beijing city to carry out in good condition, but the difference of each school is bound to affect the whole of Beijing college students extracurricular sports activities, this phenomenon should pay great attention to.

The student extracurricular activity time and strength are also one of the important indicators of quality reflects the extracurricular sports. Through the statistics of Beijing college students extracurricular sports activities for more than 2 hours accounted for 17.8%, 1 hours to 2 hours accounted

for 27.3%, 30 minutes to 1 hours 30 minutes following accounted for 46.1%, accounted for 8.8%. Take part in extracurricular sports activities often sweating accounted for 57.5%, sometimes sweating accounted for 34.4%, don't sweat accounted for 8.1%. From this we can see that the extracurricular physical activity time and intensity of exercise are ideal, which help to improve the physique and health level of the students, so the school should pay more attention to the students in extracurricular sports activities.

3.1.3 Factors affecting the students to participate in extracurricular sports activities

It can be seen from the above survey, extracurricular physical activity of college students in Beijing city to carry out the overall situation is good, the internal problems, mainly the college student participation in extra-curricular sports activities differences. Specific performance in the following aspects: (1) the school management of students' understanding of extracurricular sports activities. (2) the school extracurricular sports activities of students lack of support. (3) the number of faculty of extracurricular sports activities in college teacher shortage. (4) college students in extracurricular sports activities can not guarantee the time. (5) lack of facilities, seriously affecting the extracurricular sports school. So, from the school of management, to establish a correct concept of sports, through the physical education and extra-curricular physical training, trains the student to sports and health knowledge and attitude correct; suit one's measures to local conditions to carry out extra-curricular sports activities; to improve school facilities, create a good campus sports culture atmosphere, and establish a scientific student extracurricular physical activity evaluation index system.

Table 2. Students extracurricular sports activities in different test frequency.

School referred	$D = cf_1 - cf_2$	x^2	$x^2_{0.05}$ $P < 0.05$	Comparison of results
Tsinghua	0.113	6.982		Sig. difference
PKU	0.123	8.276		Sig. difference
CUGB	0.076	3.162		Not sig.
BUAA	0.047	1.227		Not sig.
BJFU	0.069	2.626	$x^2_{0.05 = 5.991}$	Not sig.
BLCU	0.125	8.559		Sig. difference
USTB	0.170	15.91		Sig. difference
BUCT	0.188	19.27		Sig. difference

3.2 Construction of the evaluation index system of extracurricular sports activities in ordinary colleges and universities in Beijing city

3.2.1 Evaluation of extracurricular sports activities in ordinary colleges and universities in Beijing city

Through the establishment of the evaluation index system, in order to further implement the "national construction plan", "the school sports work regulations", "national student physical health standard", "National Ordinary College Sports Curriculum Teaching Instruction Summary", the spirit of the document. The promulgation and implementation of national education policy and implementing the policies to basic education department. But also for the development of condition monitoring school sports work, check the school sports work level. In order to improve the physical health of students, inhibition of students' physical health decline, cultivating good sports consciousness, further strengthen the school sports work. At the same time, but also to promote the construction of campus sports culture, to create a good learning and communication platform between the students, between teachers and students.

Principles of establishing evaluation index system. The selection of evaluation indexes should not only take into account the basic situation of the school development, can reflect the state of development of the students of the school extracurricular sports activities, should also take into account the actual situation of each school. Fully implement the national education policy. At the same time should follow the scientific, representative principle, feasibility principle, qualitative and quantitative indicators, the principle of consistency, and comparability of the 6 basic principles. Only by adhering to the above principles, can be more effective in establishing a suitable evaluation index system, the evaluation result can reflect the original appearance of things.

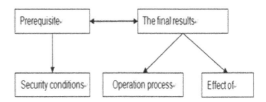

Figure 3. Extracurricular sports activities evaluation model.

3.2.2 The theoretical model constructed evaluation index system

To establish a scientific evaluation index system is the first to have a theoretical model, in this model framework, choice, judgment, series, set up the index, can be in a model to determine the direction of. At present, the most widely used evaluation index system of school sports in China is "preconditions—the development of the model" theory model, the paper from two aspects: first, evaluation of school physical education is the full extent of society and schools for school sports development provided conditions; second is to get students sport interest and equity of abundance.

According to this model, constructs the framework of evaluation index system of extracurricular sports activities in ordinary colleges and universities in Beijing city, including the departments of extracurricular physical security conditions, management, college extracurricular sports activities of extracurricular sports activities of three parts (Fig. 3).

3.3 Determine the evaluation index system of extracurricular sports activities in ordinary colleges and universities in Beijing city

Because the factors affecting students' extracurricular sports activities is more complex, so are determined by using multi-index comprehensive evaluation of complex. According to the students in ordinary colleges and universities in Beijing city of extracurricular sports activities of the survey results, and the results of a large number of literature and expert interview, determined by the study's sports college extracurricular activity evaluation index system for evaluation and theoretical model, the evaluation index system of indicators for the pre-selection. Qualifying a 3 level indicators, 10 level two indexes, 35 level three indexes, the formation of the evaluation index system of experience.

3.3.1 *The selecting evaluation index*

The establishment of the panel of experts, the expert members of a total of 23 people, at the first expert will experience index system made questionnaire, the index validity evaluation. After the first round of the meeting of experts and experts give opinions and suggestions, the extracurricular sports activities of students in Beijing colleges and universities evaluation index system, identified as

the 3 level indicators, 10 level two indexes, 24 level three indexes and 1 additional indicators (Table 3).

3.3.2 *To determine the weight of evaluation index*

Using hierarchical analysis software Yaahp V0.5.3, analyze and judge the hierarchical relationship between the index and calculates the index of 3 one class, weight 10 two level indexes and 24 three level index, complete the extracurricular sports activities of students in Beijing colleges and universities evaluation index system of evaluation form. The analytic hierarchy model of software Yaahp based V0.5.3.

3.3.3 *To complete the construction of evaluation index system of extracurricular sports activities in ordinary colleges and universities in Beijing city*

Through the above result, the assembly of experts' expert ranking vector data using the weighted geometric mean method, complete the evaluation index system of extracurricular sports activities in Beijing colleges and universities summary.

Table 3. Extracurricular sports activities in Beijing colleges and universities evaluation indicator screening results.

One class index	Two level index	Three level index
I1. Conditions of extracurricular physical security	II 1 Policy formulation	III 1 The measures for the implementation of the extracurricular sports activities
		III 2 Plans to carry out extracurricular sports activities
		III 3 Management of extracurricular sports activities
	II 2 Active site	III 4 Site type
		III 5 Site area per capita
	II 3 Funds	III 6 The proportion of per capita sports funds
		III 7 Budget
	II 4 Management	III 8 Business training and assessment
		III 9 Work attitude
		III 10 Student evaluation
I2. Management of extracurricular sports activities	II 5 System management	III 11 The implementation of file of extracurricular sports activities of proportion
		III 12 Complete the project of extracurricular sports activities
		III 13 The organization and management of Sports Association
	II 6 Organization and management	III 14 The number of weekly extracurricular sports activities
		III 15 Incentive and reward measures
	II 7 Counseling course	III 16 Counseling course
		III 17 The course of extracurricular sports Coaching
I3. Effect of extracurricular sports activities	II 8 Students' physical fitness	III 18 Student physical health standard rate
		III 19 The qualified rate of sunlight long-distance race
	II 9 The effect of exercise	III 20 Intramural sports competition performance
		III 21 Reap the rewards
		III 22 Participation in group activities in Beijing City
	II 10 The influence of	III 23 The big game organization
		III 24 The propaganda effect
Additional score (not to exceed the maximum 10 points)	International sports tournament of 3 plus 6 points; 4 to 8 plus 4 points; 9 to 16 plus 2 points; 17 to 32 plus 1. The national sports games. 3 plus 4, 4 to 8 plus points, 9 to 16 plus 1 Beijing City sports tournament of 3 plus 2 points, 4 to 8 plus 1. Athlete grade level 3/two; 2/three 1/person. The referee Level 3/two; 2/three 1/person.	

Table 4. Evaluation index system of college students' extracurricular sports activities in Beijing summary table.

One class index	Wt	Two class index	Wt	Three class index	Wt
I1. Conditions of extracurricular physical security	0.2254	II 1 Policy formulation	0.0502	III 1 Implementation measures for the extracurricular sports activities	0.0179
				III 2 Extracurricular sports activity programs implemented	0.0155
				III 3 Extracurricular activities Manager	0.0168
		II 2 Active site	0.0593	III 4 Site type	0.0261
				III 5 Site area per capita	0.0332
		II 3 Funds	0.0631	III 6 Proportion of per capita sports funds	0.0408
				III 7 Budget	0.0223
		II 4 Management	0.0528	III 8 Business training and assessment	0.0207
				III 9 Work attitude	0.0155
				III 10 Student evaluation	0.0166
I2. Management of extracurricular sports activities	0.2871	II 5 System management	0.1002	III 11 Implementation of extracurricular sports activity percentage of files	0.0378
				III 12 Completion of extracurricular sports activity programs	0.0309
				III 13 Management of sports associations	0.0315
		II 6 Organization and management	0.1012	III 14 Weekly hours of extracurricular sports activities	0.0718
				III 15 Incentive and reward measures	0.0294
		II 7 Counseling course	0.0857	III 16 Counseling course	0.0530
				III 17 Sports Coaching courses	0.0327
I3. Effect of extracurricular sports activities	0.4872	II 8 Students' physical fitness	0.2099	III 18 Rates of physical health standard for students	0.1108
				III 19 Rates of long-distance race	0.0991
		II 9 Effect of exercise	0.1560	III 20 Intramural competition results	0.0664
				III 21 Rewards	0.0460
				III 22 Beijing community activity participation	0.0436
		II 10 Campus impact	0.1213	III 23 The big game organization	0.0653
				III 24 The propaganda effect	0.0560
Additional grade		International sports tournament of 3 plus 6 points; 4 to 8 plus 4 points; 9 to 16 plus 2 points; 17 to 32 plus 1. National sports games top 3 and 4 Add 2 points; 4 to 8,9 to 16 plus 1 point Beijing sports before the match 3 and 2, 4 to 8 and 1. Athlete grade level 3/person secondary 2/person; three 1 minute/person. Referee Level 3points/person; II 2/person; three 1 minute/person.			Maximum 10 points

4 CONCLUSION

4.1 The results of the survey extracurricular sports and students' physical health of students

The overall situation of Beijing city college students extracurricular sports activities and physical health level of students is good, can reach the national standard of College students. But the college students to participate in the extracurricular activity differences. In order to solve the problem, we should start from the school's management personnel to establish the correct concept of sports, physical education and college through extracurricular sports training, trains the student to sports and health knowledge and attitude correct; suit one's measures to local conditions to carry out sports activities to enrich the content of extracurricular sports activities; to improve school facilities, the construction of campus sports culture good atmosphere, students build a scientific evaluation index system of extracurricular sports activities is necessary.

4.2 Quantitative and qualitative evaluation

Students in extracurricular sports activities is a complex process consisting of a multiple factors, there are many cases can only be evaluated by subjective judgment, therefore only qualitative and quantitative analysis, if only to emphasize the quantization, will make the evaluation result is unilateral, does not have a comprehensive and scientific. Therefore, to evaluate the organization of extracurricular sports activities and the management, to follow the principle of combining quantitative and qualitative, the only way to fully grasp the true situation in.

4.3 To promote the development of students' extra-curricular activities in ordinary colleges and universities in Beijing city

The students of Beijing colleges and universities extracurricular sports activities of the investigation result and 23 experts inside and outside the school, after the three meeting of experts to discuss college extracurricular sports activities, evaluation index system established, with guide and pertinence, index system by "extracurricular physical security conditions", "extracurricular sports management", "extracurricular sports activities effect", 3 aspects constitute a level indicators, two indicators to 10 indicators, three indicators of 24 indexes. Using AHP software for statistical analysis of questionnaire, the weight of each index was calculated, and the consistency check

of a questionnaire, finally completes the evaluation index system. The evaluation index system of appraisal of the college extracurricular sports activities. The evaluation results of the department are satisfied, can reflect the development situation of college students extracurricular activities.

4.4 Set up an appraisal system has a positive role to promote the school sports work

Extracurricular sports activities in Beijing ordinary colleges and universities better overall situation, student activities frequency, time, physical health condition of students are with the national level, but between the various departments within the school of relatively large differences. In order to further improve the students in extracurricular sports activities, should start from each department, improve the grass-roots level of the extracurricular activities. A measure of the quality of college extracurricular activities evaluation, so as to improve the overall level of the school extracurricular sports activities.

REFERENCES

[1] Education department.[J] outline of the National College of physical education teaching of school physical education in China, 2002, (6).
[2] CPC Central Committee and the State Council. On the strengthening of the youth sports and enhance their physical views in [S]. [2007] 7.
[3] Week economy. To strengthen the college physical education, Chinese college students fitness craze [R]. Set off in the National University Sports Forum on the speech, 2004.8.
[4] Chen Zhili. Strengthening school sports work, improving the health of adolescents with [R]. Speech at the National Conference on school sports, 2006.12.
[5] Ren Hai. Sports [M]. popular abroad. Beijing: Beijing Sport University press, 2003.8.

Sports Engineering and Computer Science – Luo (Ed.)
© 2015 Taylor & Francis Group, London, ISBN 978-1-138-02650-6

Theoretical study on core strength training and traditional training

XiuDi Yang
Shandon Sport University, China

ABSTRACT: In this paper, by using the method of literature and logical analysis, based on the synthetical analysis of previous research results, the author analyzes the characteristics and the existing problems of traditional strength training. Besides, he also analyzes the core strength training from the perspective of sports training, sports physiology, biology and other aspects. Core strength training has the characteristic of "instability". In the view of the author, this instability is the essential difference between core strength training and traditional strength training. However, the added unstable factor not only increases the difficulty of strength training, but also adds a fresh factor to the traditional strength training. Therefore, the introduction of the concept of core strength training updates the traditional idea of strength training. This new concept has the following effects. First, it innovates the training methods. Second, it makes up for the deficiency of the traditional strength training in the improvement of players' coordination, sensitive, and balance ability. Third, it can produce effective prevention of sports injury in training practice.

Keywords: traditional strength training; core strength training; theoretical study

1 INTRODUCTION

In recent years, core strength training in sports training has aroused the interest of many people. But the study on core strength training for athletes is a new research topic in the field of sport training. In the recent years of the development of athlete strength training, the development of athlete strength training is centered on the muscle strength of the core strength training. This has become a research hotspot of many foreign experts and coached in physical training. Since 2003, there have been many articles on the research of the method of core strength training published on NSCA, the magazine of physical training.

Chinese sports coaches, swimming and other magazines have also published articles on the use of fitness ball equipment to develop the core strength. Core strength training has become a hot spot of physical training. It is very effective for the overall and balanced development of core muscle strength and body central equilibrium and stability. And it is also an effective way to improve exercise capacity and injury prevention. At present, there is a lot of research on the core strength training methods. But theoretical research is relatively rare. Foreign sports workers have put the "core power" into the track and field, ball games, swimming, gymnastics, weightlifting and other sports training. And the training has been regarded as parts of the training program me. A year after

Nadle SF et al. (2006) took on the core strength training of the American NCAA university sports team waist, abdominal muscles and hip extensor muscle, it is found that the athletes' hip extensors strength balance is improved. And the lower back pain associated with this (LBP) incidence rate also decreases. A stable support is very important for players. Studies have confirmed the following effects of the core strength training. First, it will enable the body to respond quickly to changing needs in the process of movement. Second, it can help the athletes to control the body acceleration, deceleration and stability in the competition. Third, it can improve the body balance and muscle sensory perception. Fourth, it can save energy consumption in the movement process, relieve fatigue and reduce sports injury.

2 THE CHARACTERISTICS OF TRADITIONAL STRENGTH TRAINING IN CHINA AND ITS PROBLEMS

From the biological point of view, we know that the core stability is improved through strengthening core muscles. At present, domestic strength training mainly focuses on resistance training, the overcoming of the elastic object and external environment resistance and other 7 main means of training practice. These exercising form and means have common characteristics, that is, in the process of strength training, body centre of gravity is in relative balance state. This balance is realized through the stable support, which is provided by the instrument or the ground reaction force. And we call this kind of training as the strength training under the condition of the steady state. Strength training in steady state has a certain effect in improving the strength of motor muscle in core part. But the training has two disadvantages. First, during exercise, athletes' body posture will be in a state of constantly changing and even in the state of loss of balance. This unstable state destroys the forces that we have cultured under steady state conditions. Therefore, this strength can hardly play the best. Second, it is difficult to realize the strength training methods to improve the small muscles in the core parts. So we must make a breakthrough and innovation in the traditional training methods and means.

3 STUDY ON THE CORE STABILITY TRAINING

The USA, Germany and Norway and other countries have walked in front of the world in terms of study on physical training. As early as 1990s, scholars in these countries began to expand the training method for fitness and rehabilitation to the field of sports training. And the training of core stability is one of the results. Core strength exists in all kinds of sports. All sports action is a sport chain centered with core muscle group. A strong core muscle of the body plays a stable and supportive role in motion posture, movement skill and special skill. Any technical movements of athletic project are not completed by a single muscle. Instead it must mobilize much muscular coordination. The core muscles bear the responsibility of stabilizing centre of gravity, and are also a main part of the overall force. It plays a pivotal role on the upper and lower limb coordination and integration of force. In the core stability training, people innovate many training methods and means, for example, suspension training is one of the main methods. In suspension training, the sling hangs up part or all of the body to force the body to constantly adjust the unstable state in order to improve proprioceptive neuromuscular function. Research shows that the core strength training helps to improve athletic performance. For example, 12 Norwegian footballers were trained with the stability-oriented suspension for eight weeks. After the training, it has been found that sleeping while standing on one leg athletes' body weight transient shaking speed decreases. And the average difference of the instantaneous legs shaking lowers from 51% to 3%.

The kicker maximum speed of players increases significantly, and the pelvic rotational stability has also improved significantly. Therefore, the conclusion can be drawn that stability training in special unstable suspension ropes can obviously improve the static balance ability.

Stabilizing the spine and pelvis on core areas is like the convergence of the lower and upper body of the bridge. Its importance is like the foundation of the house, which not only affects limb movements, but also responsible for the control of the body posture. The coordination activity between stabilizing muscle and movement muscle of the spine realizes its function of stability and movement. On the contrary, our traditional strength training exercises more on the surface of movement muscle. They ignore the training of deep stabilizing muscles. According to this concept, we believe that the first core stability training should be the core stability muscles under the dynamic proprioceptive training. The force point is based on an unstable support surface when exercising. And the body is completed under the unstable support surface. Second, core stability training is the strength training of proprioceptive. When facial muscle group of proprioceptive training is carried out, weight training is also carried out at the same time. It improves the large muscle groups' core strength while improving the stability of the deep spinal muscular strength.

4 THE COMPARISON BETWEEN CORE STRENGTH TRAINING AND TRADITIONAL STRENGTH TRAINING

Core strength training has the characteristic of "instability". This instability is the essential difference between core strength training and traditional strength training. However, the added unstable factor not only increases the difficulty of strength training, but also adds fresh factor into the traditional strength training.

Updating the idea of strength training, based on kinematic chain theory, core strength training participates the limb together into a "chain" in the completion of movements. And each part of the body is a link in the chain. The completion of the skilled movement is achieved through momentum transfer in various links. The core strength plays a vital part in the process of momentum transferring in the power chain. Core stability can effectively transfer forces from the ground to the upper limbs to achieve the maximum acceleration or deceleration towards the upper limb or the instrument. It can also transfer the momentum from the upper limb to the lower limb, adjusting the effect of the lower extremity muscle on the ground. By doing so, it can improve the efficiency of the coordination of actions between the lower and upper limbs or technical movement. Therefore, core strength training highlights the transfer of power, the improvement of coordination and the capability of muscle control. It reflects the new concept of the systemic holistic, multi-muscle group simultaneously taking part in multiple dimensions.

Core strength training makes up for the deficiency of the traditional strength training in improving coordination, agility, and balance ability. The physiological mechanism to maintain the stability of human body is a process. In this process nervous system continually receives signals coming from the vestibular, visual center and focuses on the muscles, tendons, ligaments, joint capsule, and skin proprioception. Body balance is adjusted through activation and the controlling of limb muscles maintaining body stability. This movement mechanism strengthens the recruitment and exciting ability of deep muscles. And it is advantageous to improve muscle coordination, agility and balance ability. This makes the supplement for the traditional strength training in improving the explosive strength, speed strength, strength endurance and so on.

Core strength training means the innovation of the methods and means of strength training. At present, domestic strength training mainly focuses on resistance training, the overcoming of the elastic object and external environment resistance and other 7 main means of training practice. These exercising form and means have common characteristics, that is, in the process of strength training, body centre of gravity is in relative balance state. This balance is realized through the stable support, which is provided by the instrument or the ground reaction force. And we call this kind of training as the strength training under the condition of the steady state. However, in the actual movement, the instability of body destroys the condition in which the power is cultured under the steady state. So the power hardly plays the role in human movement state. This can probably explain why some athletes can show great strength in equipment strength training, but they fail to perform it in the field. Because of the unstable factor added into the strength training, there appear two main core strength training methods: suspension training and vibration training.

Core stability training not only innovates the training idea, but also differs from the traditional strength training in the training apparatus. Most of the instruments used abroad are exercise balls, balance boards, medicine balls, elastic bands, mini-trampolines, stability balls, half foam roller, slide board, CorDisc and so on. These practitioners are mostly standing weight-bearing or non-weight-bearing equipment for strength training. This update of the training equipments makes training from standing on "Earth" to stand on "balloon". It can be seen that core stability training makes a breakthrough of the country's existing training concepts and methods. Therefore it is a supplement and development to China's physical training concepts and methods.

Core strength training also means sports injury prevention. At present, the study on the relationship between core strength and damage are common. For example, Devlin's literature review of rugby League players' injuries points out that muscle fatigue is a factor of hamstring injuries. Summer's report also reveals that in the case of fatigue, athletes will appear femur adduction and internal rotation movements when jumping. These position changes are related to injuries. He maintains that it is because athletes' glottal, hamstring and abdominal muscles cannot produce enough torque to contend with external forces that the hip and knee bear. Knee value is due to the declined body's position control ability caused by the weakness of waist and hip muscles. Zeller and some other people completed the studies of kinematic and EMG activity. They have found kinematic differences caused by gender tend to be more related to hip muscles, rather than the different quadriceps muscle activation levels as is previously thought.

5 CONCLUSION AND SUGGESTIONS

To sum up, core stability training is the dynamic proprioceptive training of stable muscle, and the dynamic physical training of core exercise muscle. This train gets rid of the disadvantages of getting support for the body from external force in traditional strength training. Neuromuscular function continuously adjusts the unstable state, which in return realizes the improvement of stable muscle function and the strengthening of the movement muscles' strength. Therefore, it has the incomparable superiority than traditional physical training. But core stability training also follows the principle of combining with special training. So athletes are supposed to combine the special characteristics with the core stability training. They should not follow the others blindly. It can be seen that the combination of core stability training and specific training needs further investigation in the future.

REFERENCES

Anderson, J.R. and G.J. Julien, A Ikegami, R., D.G. Wilson. 1990. *Active Vibration Control Using NiTiNOL and Piezoelectric Ceramics,* J. Intell. Matls. Sys. & Struct, 20(2):189–206.

Hoffer, R. and D. Dean. 1996. Geomatics at Colorado State University, *presented at the 6th Forest Service Remote Sensing Applications Conference*, April 29–May 3, 1996.

Inman, D.J. 1998. Smart Structures Solutions to Vibration Problems, *in International Conference on Noise and Vibration Engineering,* C.W. Jefford, K.L. Reinhart, and L.S. Shield, eds. Amsterdam: Elsevier, pp. 79–83.

Margarit, K.L. and F.Y. Sanford. March 1993. Basic Technology of Intelligent Systems, *Fourth Progress Report,* Department of Smart Materials, Virginia Polytechnic Institute and State University, Blacksburg.

Mitsiti, M. 1996. *Wavelet Toolbox,* For Use with MALAB. The Math Works, Inc., pp. 111–117.

Sports Engineering and Computer Science – Luo (Ed.)
© 2015 Taylor & Francis Group, London, ISBN 978-1-138-02650-6

Study on the construction of China's college cheerleading development situation and training content system

LingLing Zhang
Department of Physical Education, Wuhan University of Science and Technology, Wuhan, Hubei, China

ABSTRACT: Cheerleading is a subject that has attracted large group of young people for its strong appeal and fascinating glamour. As for the colleges and universities, it has a great effect on its cultural construction. As for the students, it contributes to physical enhancement, development of independent characters, implantation of strong fortitude and the sense of team. Accordingly, cheerleading has played a significant part in spreading the quality-oriented education program of colleges and universities in China. This paper adopts the methods of the literature data, questionnaire survey, and mathematical statistics as the prerequisite. By inheriting, improving the current foreign cheerleading training contents, we can establish a feasible cheerleading training framework system with diverse functions. This will be of great significance in promoting the extension and development of cheerleading in colleges and diversifying the school cultural life.

Keywords: cheerleading; current development situation; training; system

1 RESEARCH TARGETS AND METHODS

This paper has referred to relevant sports documents, data, core academic journals of our country and collected data about cheerleading. To clarify its development and current situation in colleges and universities and to seek the thinking and means of its project study; 30 questionnaires have been sent to partial team coaches and referees who have participated in The 2013 National Cheerleading Championships and The 2014 World Cheerleading Championships. Consequently, the effective questionnaires are 26, with its ratio 93.3%.

2 RESULTS AND ANALYSIS

2.1 *Analysis of the current development of cheerleading in China's colleges and universities*

Cheerleading is a new sport in our country. Since the lecture given by the President of the American Cheerleading Association in 2000 in Guangzhou, China students Aerobics Gymnastics Association has vigorously promoted and facilitated this sport during the short 14 years. It has scored visible achievements and been funded by a great number of youngsters and college students. At present, cheerleading teams are set up in numerous universities in China, which regard this sport subject as competitions. Many sports colleges and universities have also opened cheerleading courses.

2.1.1 *Analysis of participating people, group settings and regional distribution*

There are 36 teams participating in the 2011 National Cheerleading Championships, with 582 players in total. In 2012, the number of the teams has increased into 49, and the number of the players is 766. In 2013, these figures have been added to 58 teams and 1051 players. Group settings include primary school, middle and high school, colleges and universities, which contained the routine and optional routine at the same time according to every grades. Moreover, contents of cheerleading have been boomed, with different dance styles such as jazz, hip-hop and folk dances.

Regions of participating units are mostly from Southeast coast areas, Northeast developed places. Although there are also participating teams in Henan, Shaanxi, Heilongjiang, Guizhou provinces, marking a new regional distribution of participators, the participating scale of undeveloped regions should still be further improved and enlarged.

2.1.2 *Analysis of the selection and sources of athletes*

Competitive sports in our country, especially the difficult type, group type sports, are generally choosing the athletes from children. Due to the short development history in this sport, there is a certain degree of difficulty in selecting the players professionally. However, according to the acceptable principle of selecting the people through item groups, new players can be chosen from the big

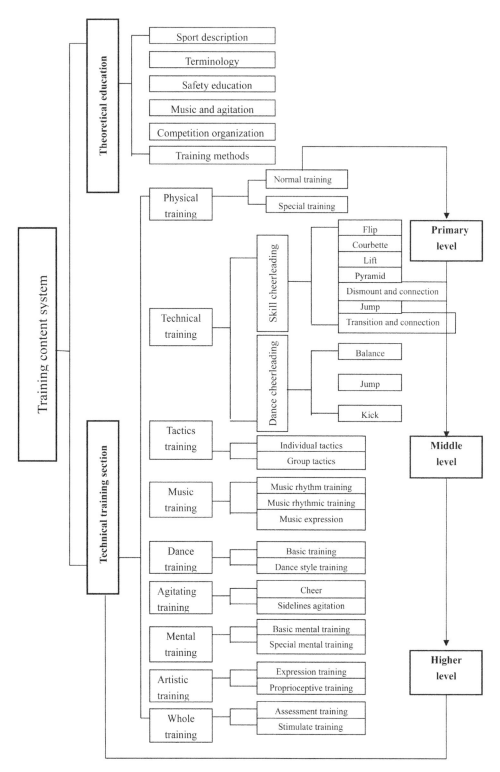

Figure 1.　Training content system of cheerleading.

item of the same subject group. Statistics show that the events, such as aerobics, gymnastics, acrobatics, martial arts, eurhythmics, and dances, have close relation with the cheerleading, thus a large number of our cheerleaders have come from these items.

2.1.3 *Analysis of the instructing level of coaches*

The understanding of cheerleading has direct effects on its development in China. Our cheerleading coaches have a relevant higher academic level. According to the survey, most of them have more than 5 years coaching experience and the minority has 1 or 2 years. Among the surveyed 26 coaches, more than 60% take the aerobics, other 20% taking gymnastics and skills, which indicate the coaches become professional. Among these coaches, more than 60% of them have been trained more than 3 times, acquiring the higher level degree, with less 30% being the middle level.

2.1.4 *Analysis of referees*

In 2013 National Cheerleading Championships, 11 of all 20 referees are senior professional referees, and the rest is intermediate professional referees. This figure generally reflects the referee level of this sport.

2.2 *The construction of cheerleading training system of China's colleges and universities*

2.2.1 *Construction principles*

According to the questionnaires from specialists, the main consensus of conducting cheerleading training must comply with four principles: principles of safety, scientific, measurable and pertinence. Safety principle is of the utmost important among them all. Because of the hazardous characteristic of cheerleading, endurance of the personal safety is the prerequisite of exercising training smoothly. As a result, the construction of training system must be based on the principle of safety, avoiding injury during the exercises.

Scientific principle requires that every steps of overall system should meet the scientific demands. Measurable principle allows the index data to be demonstrated with accurate figures. All the index system can be illustrated in the trial run, testing the accuracy of index figures. Pertinence principle requires that each index in system should contain multiple training levels and stages of colleges and universities. This helps the coach to choose the targeted index and conduct the instruction.

2.2.2 *Training goal and mission positioning*

The training of cheerleading is to win the games, and promotes its popularization and expansion. Also, it aims to strengthen the physical quality of young teenagers, build up their independent character, diversify the quality of the oriented education, and promotes the construction of school culture.

2.2.3 *The basic framework of China's college cheerleading team training content system*

This paper has referred to plenty of writings about sports talent science, structure learning, and sociology. It discusses about the future development of China's cheerleading training, the quality and talents through collecting generous data about training systems. And the basic framework of China's college cheerleading team training has been primarily pictured and readjusted by relevant professors all over our country towards the framework design. After being appraised twice, the content of cheerleading training system index has come into being, consisting of two main sections: the theoretical education section and technical section. (Fig. 1)

3 CONCLUSIONS AND SUGGESTION

3.1 *Conclusion*

1. *As cheerleading expands around China, colleges and universities have established their own cheerleading team one after another. There are more cheerleading events in China, with larger scale, wider effect and annually booming number of their participators.*
2. *Group settings become more appropriate, mainly grouped by schools. Regions of participating units are mostly from Southeast coast and Northeast developed places. The athletics level of Guangdong, Guangxi areas is higher than that of the central and western regions. The following are the reasons for this situation—the restricted economic condition, their teachers' being devoid of accepting new things, school's unwillingly paying the high competition spending and lack of professional cheerleading coaches. All these factors result in the regional disparity among different areas.*
3. *Cheerleaders are mostly from the events, such as aerobics, gymnastics, acrobatics, martial arts, eurhythmics and dances, who have some of the physical quality. Aerobics, gymnastics and other subjects also bear lots of traits similar with cheerleading. So it is an important requirement for coaches to select athletes.*
4. *Coaches, conducting the cheerleading training, have a relevantly higher academic level; most of them have Master or Bachelor degrees and age 25–35. They are generally senior level and major in cheerleading and gymnastics.*
5. *The majority of referees are levels of senior or middle, basically severed by the physical education*

instructors from colleges and universities. The referee group still need a constantly improvement due to its small scale, low academic level and instability of a professional team.

6. *Principles of safety, scientific, measurable and pertinence are the basic principles of the construction of cheerleading training system of China's colleges and universities.*

7. *The content of cheerleading training system of China's colleges and universities are divided into theoretical education section and technical section.*

The former part includes training methods, security education, competition organization and judgment, music and agitation, terminology and sport description. The latter one contains nine training parts such as physical, technical, artistic expression, team, dance, music, psychology, agitating and tactics.

8. *The three levels of training system for colleges and universities constructed by this research have provided a reference for current cheerleading training of our country. The study is still not so perfect due to the limitation both in time and condition. Its training system should be further improved and optimized.*

3.2 *Suggestions*

1. *Though cheerleading has developed rapidly in our country, with its range spreading and effect expanding, the history of this sport is still short and inexperienced coaches are still here. Therefore the suggestion must be launched here that physical departments of universities should gather relevant experts to map a suitable cheerleading training system for our higher leaning schools. They should lead a theoretical and exercising guidance to cheerleading trainings in our campuses.*

2. *Though promoting cheerleading functions in building a diverse school cultural life, this subject will become an attracting point in school culture and physical activities. School should adopt the reward and punishment system and combine them with cheerleading activity, which can stimulates enthusiasm of students. Cheerleading courses should also be set up. Besides, a variety of cheerleading activities should be progressed in campus.*

3. *Our national cheerleading championships will be further developed through integrating with TV and printed media. Communication and promotion in our northwest regions should be strengthened and the academic level of both coaches and referees should be improved. Meanwhile, we should enhance the cooperation and exchange with foreign cheerleading organizations, vigorously driving the internationalization of cheerleading sport movement.*

REFERENCES

[1] Du Liping Research Status and Development Strategies of College Athletes [J] Chengdu Institute of Physical Education, 2007 (3): 120–123.

[2] College sports teams' aerobics Status Quo and Development Strategies compliance Li Hua. Fujian Province [J]. Jilin Institute of Physical Education, 2006 (2): 117–123.

[3] On behalf of Yongsheng. Hubei Province, a high level of education and training college sports teams Situation and Countermeasures [J]. Sports Adult Education, 2007 (5): 47–48.

[4] Status Survey and Coaches of College Competitive Aerobics team [J] Shandong Institute of Physical Education, 2006 (6): 117–119.

[5] Section long wave research organizations to manage and control the college level sports teams training [J] Wuhan Institute of Physical Education, 2003 (7): 158–159.

[6] Yang Jian, Sun Rende constraints college building high performance sports teams and developing countermeasures [J] Hebei Institute of Physical Education, 2007 (2): 62–65.

[7] Thinking Tang Zhiming. High level sports teams in the Running System and the Reform and Development of Our College [J]. Beijing Sports University, 2007 (1): 87–89.

Sports Engineering and Computer Science – Luo (Ed.)
© *2015 Taylor & Francis Group, London, ISBN 978-1-138-02650-6*

The special characteristics and the prospect forecast of cheerleading

LingLing Zhang

Department of Physical Education, Wuhan University of Science and Technology, Wuhan, Hubei, China

ABSTRACT: This paper discusses the special characteristics and prospects of cheerleading with the help of literature material and expert interview. It analyzes the skills, tactics, physical ability, psychology and environment to provide a theoretical basis for the further development of the cheerleading in China.

Keywords: cheerleading; the special characteristics; the prospect forecast

1   INTRODUCTION

Cheerleading, a newly burgeoning sport with its original name cheerleading in english represents a modern sport event emerging from the US. It refers to the completion of superb cheerleading skills and the combination of a variety of dance movements conducted by the athletes. This sport event embodies the youthful, healthy team spirit and pursues the honor of team. After being introduced from abroad, cheerleading has promoted leapfrog advancement in China within a few years, with all kinds of games correspondingly carried out. As for the training of the competitive sports, however, first, the accurate management of the sport characteristics and training rules is the prerequisite. Then, the cognition of the event plays a key role in the improvement of the training level. Then, the understanding and grasp of special training is the fundamental principle of improving training efficiently and maintaining the quality-oriented training for long time. This paper discusses the special characteristics and prospects of cheerleading with the help of literature material and expert interview. It analyzed the skills, tactics, physical ability, psychology and environment to provide a theoretical basis for the further development of the cheerleading in China.

2   SPECIAL CHARACTERISTICS OF CHEERLEADING

2.1   *The description of special characteristics*

Special characteristics, under the allowed game rules with the goal of maximum movement efficiency, refer to the main sport characteristics featured by Physics, Biology, and so on. Usually the special characteristics can be divided into a variety of aspects, such as tactics, physics, mental and environment, and so on. Each of them has also been composed of some heterogeneous elements (Fig. 1).

2.2   *Technical characteristic*

2.2.1   *Competitive and skill*
Cheerleading is a strong penetrability movement, which combines many tricks from gymnastics, trampoline, aerobics with many folk dance elements from Latin, jazz, HIP-HOP and ballet. This combination gives cheerleading the different styles and themes. Cheerleading has currently been divided into technical cheerleading and dance cheerleading. The former has its main content of somersault, juggling, lifts, Pyramid, dance movements, transitional or connected slogans. These actions are

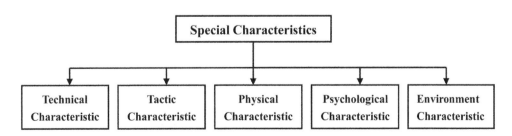

Figure 1.   Special characteristics of cheerleading.

difficult and dangerous, requiring higher competitive and technical skills. Cheerleading athletes must have supreme physical and mental quality as the basis to complete them. However, dance cheerleading is mainly composed of dance movements, including balanced swivel, jumping and kicking. It's relevantly less difficulty than technical one, ensuring that it can be promoted in regular colleges and universities.

As cheerleading sport develops in China, a variety of cheerleading games have been launched. In the annually National Cheerleading Competition organized by China students Aerobics Gymnastics Association, as the selection game of the cheerleaders for the 2012 Olympic Games, this sport has been heatedly popularized across the state with its own styles. In addition, places in anywhere and from all levels are carrying out the competitive games. With innovative, spectacular ideas and compositions, the competitive and skillful characteristics of cheerleading have been further displayed.

2.2.2 *Team coordination and artistic expression*

In addition of completing the difficult movements, team coordination and fantastic artistic expression are also the important technical characteristics. The team of the group events is formed by certain principles and organizations. Compared with the individual sports, the sole characteristic of team is to rely on the collective strength and explore the intelligence in training and contest. Therefore, in group events, the group overall strength is the main aspect of estimating their competitive ability. And the whole strength enhances the possibility that they bear strong cohesive power and cooperating spirit. Whether cheerleading can have powerful and stunning visual effects lies in its cooperation. A plenty of difficult movements and poses can not be shaped regardless of the coordination from every group member. In light of this, cheerleading game is also a competition of team cohesion and cooperating spirits. In every event of 2012 Olympics Games, cheerleaders shouted loudly their encouraging slogans though utilizing props and performed enthusiastically their limb activities accompanied by the bright and happy music. Their powerful appeal touched every nerve of the spectators, pushing the environment to the highest point. They released their youthful energy, passion and sweat, getting the viewers involved into the sports atmosphere to enjoy the happiness brought by the games.

Hence, cheerleading is quite similar to other difficult movement projects. It stresses the fitness of competition sports, uses the means of flexion, extension, jumping and body straightening and combines elements such as poses, gestures with

expressive force harmoniously. It takes the characteristics such as fitness, contest, team cooperation, entertainment and creation. Therefore, it contributes to building a healthy body and readjusting the mental work.

2.3 *Tactic characteristics*

Just as other difficult movement projects, cheerleading requires no more tactic characteristics than the combat sports. Being sports of pursuing the highest team honor, the group tactics shares the main parts in cheerleading contest. Its tactics mostly lies in organization and innovation of movements, choosing music and edition, designing slogans, costume styles and novel color, athlete spirits and great team coordination.

2.4 *Physical characteristics*

In macro respect, physical characteristics include two parts: physical fitness and physical condition. Physical level of one athlete relies on the development of her body shape, physiological function and sports quality. Specifications of cheerleaders mainly rest in the height, weight, body shape, and physiological function. Cheerleading requires not strictness in physical fitness of their players. Only with their body symmetrical, healthy, and their attitude positive, energetic they can participate in this event. For this, cheerleading also has the merits of universality and mass publicity.

Physical condition depends on the level and potentiality of cheerleaders, mainly including flexibility, strength, speed, endurance and coordination. Flexibility and coordination play the key parts in the physical condition. With high action range and difficulty, those who have supreme flexible body and coordinating ability can finish the movements perfectly. At the same time, speed, strength and endurance can not be underestimated because cheerleading stresses the powerful force, high speed and precise location. It's difficult to finish the big range movement and highest difficult actions forcefully and correctly under the quick, changeable rhythm. If the cheerleaders want to achieve this goal, the quality of great speed, strength and endurances is the prerequisite for the cheerleaders.

2.5 *Psychological characteristics*

As has been proved, whether physical training and tactical level can be enhanced and displayed or not, it has great connections with the psychological training. In light of this, master of the psychological traits of cheerleading has the significant meaning of progressing our national cheerleading event level as a whole. The sports traits of cheerleading are mainly shown by the confidence and strain force.

In this performing project, cheerleaders must have the passion for performing, which makes confidence important. Such kind of confidence is emerged from a confident, fearless mental state of a specific sport player. This psychological condition can highlight the passionate mood and fantastic performance. And the inner enthusiasm can make setting movements full of attraction. The spirit springs from the self confidence. There are also various urgent situations confronted by the player in the game court, which can be tackled easily once the cheerleaders respond positively. If the player reacts nervously or negatively, performance in the game would be seriously affected. The improvement of confidence and strain force need the players to gain their courage and accumulate experience. In addition, encouragement and support from coach are also essential.

2.6 Environment characteristics

Environment characteristics of cheerleading include costumes and props, rules and field, coach and spectator, and so on. Cheerleading game costumes are mostly made from tight and elastic fabrics. Female players can dress conjoined outfit skirt, split vest and short skirt; male players can dress split short (long sleeves) and pants. Props include sigh board, microphone, flag, banner, bouquet and inflatable stick, and so on. *The international all star cheerleading Competition scoring rules (version 2006–2009)* published in 2006 stipulates that: height of game field is 80–100 cm, with background culling behind; area of game stage must be more than 13×13 m^2. Gymnastic board or carpet can be used in the game field, clearly marking the area of 12×12 m^2.

The advancement and development of a sport have numerous connections with the relevant coaches. Being the main organizer and instructor, a coach is responsible for selecting the materials, planning the training program, organizing the implementation and guiding the game. Then, the players are likely to hold high grade. The understanding of cheerleading sport, the theoretical ability, the master of cheerleading skills along with the qualification is directly related to the cheerleading development within our country. Since the first cheerleading coach training class was founded by CSARA in November 2006, our national cheerleading coach group has constantly expanded. For being this case, coaches of our country are mostly transferred from other events. It indicates that the cheerleading training is still infant, and the overall strength and popularization of our cheerleading sport is not as mature as the internationals. One of the different

traits of cheerleading is that the spectators play a great part. Field passionate atmosphere touches the viewers to shout the slogans together with the players, enjoying the game thoroughly. Their enthusiasm has also in turn influenced the players, encouraging the players to be preoccupied and fully release their youth and passion.

3 THE PROSPECT FORECAST OF CHEERLEADING

3.1 The popularity of perfection and development

The special characteristic of cheerleading sport has rapidly promoted its own popularity and extension around the world. International Federation of cheerleading, IASF, USASF and other associations are forcing the further extension of cheerleading, making this sport more and more a eye-catching subject in worldwide. However, as an emerging sport, the history of cheerleading is not long, and there are still some ways to go in its rules, contests, and training. An overall theoretical system should be built, just like other sports. Since cheerleading was imported from abroad, our country has invited experts from the US, Japan to organize the training classes, educating the coaches for cheerleading sports. At present, cheerleading teams in primary, middle school, colleges and universities have been launched. Some other institutions of high learning have set up the cheerleading courses, making this during the games. The cheerleading team is a charming view whether in the campus or during the games.

3.2 Competitive development

Nowadays, as international cheerleading game increases and competition scale enlarges, the competitive level has grouped explicitly. With the age of the participators extends, the number of them has also grown and so has their skill level and the athletic level. China now has striven to establish the cheerleading games, and actively participated in international contest to keep up our ability with the world level. In recent years, there has emerged successively a variety of cheerleading team to represent China to go to America for attending the world cheerleading competition and score excellent achievements, such as teams in Wuhan Institute of Physical Education, Guangzhou Institute of Physical Education and Nanning NO.26 Middle School. Therefore, cheerleading has a great potentiality for its competition development in China. Coaches of our country have vigorously researched numerous training means, promoting the movements' difficulty and innovating our styles, so as to move cheerleading forwardly in China.

3.3 *Commercialization development*

Commercialization movement of sports is nothing new. In the UK and the US, cheerleading has already come into the sport market operation, getting the paid performances and titling sponsorship for entrepreneurs during the games. It has brought visible commercial benefits of its own. China steps into the market economy. And professional sports teams transforms into club system alone with the professionalization of athletes. All have spurred more and more national players to attend foreign marches. Especially the establishment of the 2012 Olympic Games made China sports, as well as its momentum gradually in line with the rest of the world. The competitive sports in China, from contest institutions and mechanisms to systems and methods, have implemented the overall reform systematically and comprehensively.

With sports competition market soars, functions of professional cheerleading team have been highlighted. Cheerleading performances are generally injected into the middle break of the games. Its main purposes are to create a heated game atmosphere, mobilize audience enthusiasm, and add the appreciation and entertainment of the sports programs. As early as 1999, the first professional cheerleading team came into being in China and performed in the break of the CBA game. Recently, however, professional cheerleading teams in China are mostly based on schools, aiming at attending the contests, not operating as club system into market. More and more international events are hosted in China, and games mode becomes closer to the foreign countries. Therefore, market requirement of cheerleading teams will be stimulated and its commercialization tendency will be boomed, picturing a bright and promising prospect of cheerleading development.

REFERENCES

[1] Jiang Yuhua Research Campus Sports Culture [J]. Southwest University for Nationalities, 2004 (1): 443–445.

[2] Chang paint column. Status of group cohesion and direction of movement of the study [J]. Wuhan Institute of Physical Education, 1994 (3): 64–66.

[3] Wu Rongrong. Cheerleading sports groups cohesion factors [J]. Quanzhou Normal College, 2006 (6): 126–128.

[4] Sun Tie-min, Li Hui-Juan. Investigation of the Current Situation of Cheerleading [J]. Xi'an Physical Education College, 2005 (4): 85–87.

[5] Wang Lijuan. Theory of the necessity and feasibility of Cheerleading Sports [J]. Shenyang Institute of Physical Education, 2005 (6): 117–118.

[6] Classification and technical characteristics Xuzhong Qiu. Cheerleaders analysis [J]. Sport Science & Technology, 2006 (10): 5–6.

[7] Sun Tie-min. Investigation of the Current Situation of Cheerleading [J]. Xi'an Physical Education College, 2005 (4): 85–87.

[8] Wang Hao. Cheerleading dance class athletes should have the basic qualities and training methods [J]. Hubei Radio and TV University, 2007 (2): 159–160.

[9] International Association all-star cheerleading validation international all-star cheerleading competitions scoring rule (2006–2009 edition) [S].

Sports Engineering and Computer Science – Luo (Ed.)
© 2015 Taylor & Francis Group, London, ISBN 978-1-138-02650-6

The influence of physical exercise on the 3–4 grade students' Self-Control Ability

Yan Rao
Wuhan Sports University, Wuhan, Hubei, China

ABSTRACT: The primary school stage is an important phase in the development of the individual. Self-control ability is not only taking an important position in the physical and mental development of primary school students, but also exerting a profound influence on their future development. We conduct a research on the self-control ability of pupils in the first primary school of Shuiguohu, Wuchang District of Wuhan by literatures, questionnaire, experiment and statistics. We found that 3–4 grade in a primary school is a main phase in the development of Self-Control Ability (SCA). In this period, the development of SCA declines a bit, not that remarkable. And girl's SCA is a little better than boy's. The 4 grade student's SCA and all-dimensional development are a little worse than the 3 grade, whereas the grade difference is not noticeable. Pupil's physical exercise has a positive correlation with SCA. In short, those who are doing the physical exercise have a higher SCA than those who don't. Besides, continuant physical exercise for some time can enhance pupil's SCA.

1 INTRODUCTION

According to Xinhua News website, reporting in January 27th, 2011, an article named *Self-Control Ability in Childhood Can "Predict" One's Road to Success* wrote about the experiments of some researchers' from Duke University. They made a random selection of 1000 newborn babies who would consistently bear a test of their self-control ability. When the subjects were at the age of 32, the researchers made an analysis of their health, wealth and family, trying to find out their relationship with their IQ, self-control ability as well as their social economic status. And they have found that children with low control ability might be more likely to have some health problems, be incapable of financial management or even have a criminal record in their adulthood. According to the experimental results, if children's self-control ability get improved with time passing, they will live a better life. In other words, one's future can be altered or changed by improving his self-control ability. However, children' parents often neglect improving their babies' self-control ability, which results in the problems in their children's later life. Therefore, it is necessary for everyone to improve his self-control ability, which will play a vital role in one's life.

The past studies pointed out that one's self-control develops relatively slow when they are pupils. Some studies have said that doing physical exercise has an active effect on lifting one's self-control ability. This paper is trying to analyze the relationship between physical exercise and pupil's self-control ability through experiments and researches in order to provide a new thinking and way for the pupils to have a sensible physical exercise and promote their self-control.

2 SUBJECTS AND METHODOLOGY

2.1 *Subjects*

3–4 grade pupils in the first primary school of Shuiguohu, Wuchang District of Wuhan are selected at random. Two classes of each grade are selected and in total, 4 classes in sum, 124 students as subjects. 60 are for boys and 64 for girls, of whom 30 boys and 30 girls are set as experimental group and another 30 boys and 34 girls as control group.

2.2 *Methodology*

2.2.1 *Literature*

We collected and sorted out a large number of papers, books and relevant documents on pupil's self-control and physical exercise's effects on it to get a comprehensive understanding of current situation and cutting-edge studies in this field. So we can classify and analyze them to obtain some theories and findings related to this paper.

2.2.2 *Questionnaires*

A total of 40 days, from October 18th, 2013 to November 26th, 2013, are spent in conducting a questionnaire research for the selected students and for the selected classes in a collective anonymous way 5 minutes early before the class begins, and then the questionnaires will be timely collected. Also, we have contacted leaders and head teachers or other teachers.

Based on the real situation of the school and our research, we looked for relevant authoritative questionnaires on self-control ability. Here we choose *self-control ability questionnaires* edited by Wang Hongjiao and Lu Jiamei in 2004 to test the students' self-control. The questionnaire has totally 36 questions, of which 10 are for positive questions and 26 for negative question. We use the five points scale that 1–5 stands for totally disagree to totally agree. We adjust the questionnaire to our test for the pupil's characteristics and delete some questions unrelated to the pupil's real situation. Eventually we keep 23 questions including three-dimensioned emotional self-control, behavior self-control and thinking self-control. Among them, the percentage of behavior self-control multiplies 2 and the negative questions should be transformed to positive questions to be calculated. The Cronbacha alpha coefficient in all dimension stands between 0.49 and 0.78 and it is 0.81 in the total scale, which indicates the questionnaire is reliable and has a fairly good construct validity.

2.2.3 *Experiments*

A total of 124 questionnaires of *self-control ability questionnaires* of 2004 edition are given out and 118 of collected questionnaires are valid with a validity percentage of 95.16%.

1. Group of experiments

   All the subjects should be divided into experimental group, students with intervention in their physical exercise, and control group, students usually without intervention in their physical exercise. Sports planned for the experimental group mainly focuses on the track and field with support of other sports. Sports duration lasts from 45 to 60 min, in which warm-up takes up 10 min and after-sport relaxing accounts for 10 min. These subjects will be tested of their heart rate during their exercise so as to monitor their cardiac load. The load intensity will be kept within 130–150 beat/min for heart rate. Sports time is in the afternoon and 3–4 times a week will last for 2 months. After that, the independent sample of basic situation of the two groups will go on a T examination.
2. Arrangement for experimental phases

   In the first phase, the pre-experiment will use *self-control ability questionnaires* edited by Wang Hongjiao and Lu Jiamei to test all students of two groups. In the second phase, the control group will undergo the arranged physical exercise for 2 months. And they will do it 3–4 times a week. On the contrary, the control group will not carry out any physical exercise and attend their classes as usual. In the third phase, the data of the two groups then will be compared at the same time.

2.2.4 *Statistics*

First and foremost, we are going to add up all the valid questionnaires and count the number of them, and then make a regular statistics with the SPSS.

3 RESULTS AND ANALYSIS

3.1 *Current situation of self-control ability of the students in the first primary school of Shuiguohu, Wuchang District*

From Chart 1, we can observe that the subjects in this research are generally getting relatively low Self-Control Ability (SCA). And high-level SCA takes up only 26.92% and low-level SCA is 14.58%. According to a report from Chinese Association for Mental Hygiene, in recent years, most of the primary and secondary school students have been self-centered and lack solitary and cooperative spirit. There is an urgent need for an improvement of pupil's SCA. In this research, primary school girls' SCA is obviously higher than the boys'. Although all the subjects in the test do not own a high SCA, middle-level SCA is higher than the other two SCA. Therefore, we think that the subjects' SCA is relatively higher.

3.1.1 *Current situation of gender difference in the 3–4 grade pupils' SCA of Shuiguohu school*

There is a common idea that girls are mature earlier than boys in primary school. So people tend to hold the view that the development of girls' SCA is higher than boys'. It is true that with the development of pupil's self-consciousness, social cognition and gender awareness, the girls show gender behavior compatible with social and cultural requirement in their daily life. In 3–4 grade, despite a minor difference, the girls' emotional SCA and thinking

Chart 1. Analysis table for self-control ability.

	High-level SCA (%)	Middle-level SCA (%)	Low-level SCA (%)
M (n = 60)	26.25	58.13	15.62
F (n = 64)	27.55	58.84	13.61
S (n = 124)	26.92	58.50	14.58

Chart 2. Comparison tables of pupils with different genders.

Factors	Gender	M	SD	T	P
Emotional SCA	M	31.92	5.184	−1.705	0.060
	F	33.75	4.278		
Behavioral SCA	M	49.79	8.924	−2.046	0.042*
	F	53.65	8.062		
Thinking SCA	M	31.48	5.062	−2.719	0.051
	F	33.85	4.448		
SCA in sum	M	113.19	19.17	−2.871	0.007**
	F	121.25	16.788		

(Note: *stands for $P \leq 0.05$, **stands for $P < 0.01$).

Chart 3. Comparison tables of pupils with different grades.

Factors	Grade	M	SD	T	P
Emotional SCA	3 grade	33.39	4.849	1.610	0.055
	4 grade	32.02	4.624		
Behavioral SCA	3 grade	47.88	8.953	0.836	0.051
	4 grade	47.74	8.341		
Thinking SCA	3 grade	32.45	4.979	0.213	0.050*
	4 grade	31.83	4.947		
SCA in sum	3 grade	113.72	18.781	0.0612	0.005**
	4 grade	112.15	17.912		

(Note: *stands for $P \leq 0.05$, **stands for $P < 0.01$).

SCA are a bit higher than boys'; for the P-value of behavior SCA factor is lower than 0.05. As a result, we can observe the girls' behavior factor is higher than the opposite sex. Therefore, we can observe that the 3–4 grade children' SCA development has an obvious sexual difference that results from individual socialization.

3.1.2 Current situation of grade difference in the 3–4 grade pupils' SCA of Shuiguohu school

From Chart 3, we have found that 4 grade students' SCA and all-dimensioned development are only little lower than the 3 grade students and no striking difference exists. The P-value of thinking SCA factor is lower than 0.05, which means those 3–4 grade students have remarkable differences in thinking SCA. However, the P-value of emotional and behavior SCA factor is higher than 0.05, which indicates a reverse result, compared with the former. The lower one's self-control level gets, the harder the students can control their emotion and the more misbehavior will turn up. So it is worth noticing that there is a trend that the 3–4 grade students' behavior SCA is in a transitional period and is going through an unstable development. What's more, behavioral control is declining. However, for

the 3–4 grade pupils, due to a small age span, a relatively slower growing-up period and a limit on their knowledge and ability, their SCA level does not stay high.

3.2 Research on physical exercise's influence on the 3–4 grade students

3.2.1 Pre-comparison study on experimental group and control group

From Chart 4, it can be told that the P-value is higher than 0.05, so there is no noticeable difference. It means that before doing physical exercise, data of emotional SCA in both groups does not have a big fluctuation. Remarkable differences are hardly observed in behavior and thinking SCA. So far, there is no obvious difference in SCA of the two groups.

3.2.2 Comparison study on the pre and after-data of the experimental group

From Chart 5, it can be observed that after doing exercise the experimental group has a remarkable influence on the emotional, behavioral and thinking SCA ($P \leq 0.05$). The date shows that a great change takes place in the subjects' pre and after experimental SCA. The emotional self-control

Chart 4. Data pre-comparison table of experimental group (Exc. G) and control group (Con. G).

Factors	Group	M	SD	T	P
Emotional SCA	Exc. G	36.78	5.138	1.468	0.143
	Con. G	36.31	5.203		
Behavioral SCA	Exc. G	51.90	8.623	1.539	0.122
	Con. G	51.63	8.934		
Thinking SCA	Exc. G	33.67	5.159	1.425	0.164
	Con. G	33.48	5.479		
SCA in sum	Exc. G	122.35	18.92	1.731	0.071
	Con. G	121.42	19.616		

Chart 5. Comparison tables of pre and after-data of experimental group.

Factors	Data	M	SD	T	P
Emotional SCA	Pre	35.29	6.339	−2.934	0.007*
	After	37.47	5.424		
Behavioral SCA	Pre	50.73	12.839	−2.542	0.015*
	After	52.18	10.352		
Thinking SCA	Pre	31.95	6.547	−2.055	0.041*
	After	33.41	6.058		
SCA in sum	Pre	117.97	25.725	−2.849	0.006*
	After	123.06	21.834		

(Note: *stands for $P \leq 0.05$).

factor takes on a remarkable changing average value both in the pre and after experiment. Before conducting this experiment, pupils are limited to assign their attention, weak in willpower, and usually self-centered. So the average value is commonly low. However, after a moderate intervention, the pupils learn how to do their practice when they can consciously regulate and try to control their own time and activities. That is the reason why the average value comes to rise. After some proper exercise, the fall of variance demonstrates an obvious difference. The behavioral self-control factor often takes on a great casualty before an intervention. Since one-child takes up a large share of children in modern society and gets indulged by their parents, the pupils have a weak self constrain and feel not easy to distinguish the good from the bad. The behavioral self-control gains a huge improvement after it is intervened. Before doing sports, the thinking self-control factor shows itself as a dependent thinking, following suit and great dependence, which is the result from pupil's immature mental development. As a result, the average value looks low and then goes higher after the experiment. And in the experiment, proper physical exercise is good to promote pupil's SCA. SCA consists of emotional SCA, behavioral SCA and thinking SCA. Different changes of each SCA after a physical exercise also exert an effect on the change of SCA in sum. That manifests that the experiment is valid and efficient. That is to say, pupils' all round ability get improved after doing physical sports.

3.2.3 *Comparison study on the pre and after-data of the control group*

From Chart 6, in addition to normal activity time, the control group only goes through the normal teaching activities before doing sports. Pupil's SCA in sum and average value does not change dramatically, which means the normal teaching activities and activity time do not have big effects on the SCA.

3.2.4 *Comparison study on after experiment between experimental group and control group*

Some studies have proved that all kinds of sports activities are taken on the basis of mental qualities such as SCA, strong perseverance, bravery and consistent will. Therefore, some sport activities for those qualities should be adopted so as to foster a healthy and all-round character.

The mentioned contrast testifies that physical exercise can lift pupil's SCA. To further illustrate that it can enhance 3–4 grade students' SCA, we make a statistical analysis on the after experiment data of the experimental group and control group. And we try to prove the availability of experimental hypothesis.

From Chart 7, it can be told that after two weeks' strict exercise, whether from emotional SCA, behavioral SCA or thinking SCA, compared with the data of the control group, the experimental group's has a great change. Only from the average value, we can observe the value goes higher. To further prove a remarkable change, we can tell the changing trend from the variance that three self-control factors' P-values are lower than 0.05. So we draw a conclusion here that students in the control group have built their three-dimensional thinking model, leaned how to keep their own emotion and behavior under control, got rid of their image of being self-centered and join other students as a unity.

3.3 *Physical exercise's influence on 3–4 grade pupils' SCA*

Low SCA of pupils is an important internal factor for the appearance of behavioral problems. In recent years, there have been many researches on the nurturing and exercise of pupils' SCA, which all demonstrate the importance of improving their SCA. Some studies have said that physical

Chart 6. Comparison tables of pre and after-experiment data of the control group.

Factors	Group	M	SD	T	P
Emotional SCA	Pre	38.38	6.913	−1.956	0.451
	After	39.64	5.601		
Behavioral SCA	Pre	55.68	10.334	−1.138	0.258
	After	55.94	8.618		
Thinking SCA	Pre	34.73	6.328	−1.214	0.31
	After	35.85	5.934		
SCA in sum	Pre	128.79	23.575	−1.233	0.251
	After	131.43	20.153		

Chart 7. Comparison tables of after-experiment data of experimental group (Exp. G) and control group (Con. G).

Factors	Group	M	SD	T	P
Emotional SCA	Con. G	36.15	5.810	2.089	0.044*
	Exp. G	37.91	5.570		
Behavioral SCA	Con. G	50.61	8.394	2.054	0.047*
	Exp. G	54.58	10.251		
Thinking SCA	Con. G	33.29	5.334	2.083	0.045*
	Exp. G	34.96	5.965		
SCA in sum	Con. G	121.72	19.538	2.486	0.018*
	Exp. G	127.45	21.786		

(Note: *stand for $P \leq 0.05$).

exercise has a positive influence on pupil's SCA, (1) the longer they are doing physical exercise, the stronger they can self-control; (2) they need to do the exercise for some years so that it can further exert a positive influence on their SCA. We are mainly concentrating on the experiment on pupil's physical exercise, and the findings show that it is fairly conducive to their SCA.

3.3.1 *Physical exercise is good for pupil's emotional control*

Some studies said that doing sports played a role in mitigating one's stress reaction so as to decrease nervousness. Also, it can strengthen one's will and build one's physical body and mental mind. In spite of unstable emotion and ruthless impulse, pupils can be transformed to be more stable and profound in their feeling with the age getting old. Though low grade pupils come to get their emotion under control, it often appears to be unstable. Their emotion and feeling will become more stable and restrain them in their heart rather than be ruthless the moment they become high grade pupils. A variety of sport games can not only develop the student's thinking ability, but also cultivate them of a good morality and will. It promotes one's development of mental health and stir up many kinds of potentials of pupils. The students learn to develop their happy and optimistic emotion and make an all-round development of their personalities. A lot of sport games are collective. They can cultivate the student's character, put their irrational behave under control; strengthen their awareness of collectivism and spirit of team work. And then they will be likely to meet different people and be a part of the team. Sport games are a kind of competitive activity they are helpful to nurture pupils to be brave and tough when they face difficulties so that they can lift their SCA.

3.3.2 *Physical exercise is good for shaping pupil's healthy personality*

The pupil's self-centered awareness is sprouting in the bud, and his self-consciousness in the progress of development. His personality is taking shape and he tends to imitate others and full of plasticity. Doing sports makes a student's body and heart in a dynamic status and his brain is always working actively. This enhances the growth of his brain and connect his mind closely related with his body. And his personality is fully demonstrated and developed. Besides, pupils can gain happiness from success and mental satisfaction from being respected in the sports. To testify his own capability in front of his classmates, he will get over any difficulties and accomplish his goal. As a consequence, his confidence and ego gets fulfilled and then his healthy personality comes into being and get fully adjusted and developed.

3.3.3 *Physical exercise is good for enhancing pupil's capability of social adaption*

Pupils are having a weak will, poor continuity and SCA, but all of these are being developed. Through the changing environment in the sport games, pupils play varied roles and abide different rules of games, their decisiveness gets fostered. Also, they improve their social cognition by exchanging ideas among classmates, which increase chances for communications between them and other students. It is conducive to the coordination of their social intercourses. By the competitive games, they overcome their consciousness of fear and become bolder. And they raise their competitive as well as fair competition awareness and learn to go well with others and to forgive. Also, they acquire a stronger ability of social adjustment and SCA.

4 CONCLUSION AND ADVICE

4.1 *Conclusion*

1. 3–4 grade in a primary school is a main phase to the development of SCA. In this period, the development of SCA declines a bit, not that remarkable. Girl's SCA is a little better than boy's.
2. The 4 grade student's SCA and all-dimensional development are little worse than the 3 grade whereas the grade difference is not noticeable.
3. Pupil's physical exercise has a positive correlation with SCA, and those who are doing the physical exercise have a higher SCA than those who don't. Besides, continuant physical exercise for some time can enhance pupil's SCA.

4.2 *Advice*

For the pupils, 3–4 grade students are supposed to be middle grade students in a primary school. However, their SCA's development cannot be ignored, because it has a direct influence on the study achievement and social intercourse. This enlightens us that to change the current situation of pupil's SCA, we can do it in the following way:

Physical exercise contains games that are the leading activity to pupils. Games have rules and the pupils need to play different roles in the games to learn to control themselves. We promote their SCA by setting varied game modes, regulating different rules and playing distinct roles to cultivate them of a hard-working and stable learning mood.

In the daily sports games, speak highly of them rather than speak ill of them, boost their confidence, draw their attention and make them taste fun in the sports and give them more chances to show. All these are helpful to turn "force me to do sports" into "I need sports".

Different sports have different functions on cultivating one's will. Therefore, we may be able to try a variety of sports to enhance the needed ability for different types of students.

ABOUT THE AUTHOR

Yan Rao, F, born in 1974 from Wuhan Hubei, Assistant Professor, Study Exercise Training.
Tel: 13387561382.

REFERENCES

Gao Jun, Sun Jianhua & Xiao Kunpeng. Empirical Study on Sprots'Intervention in the Influence of University Students'Internet Addiction [J]. *Shenyang Sports College Journal*. 2012, 13(31): 77–78.

Jin Junqing. Brief Introduction to Self-Control's Resource Mode [J]. *Social Psychological Science*, 2007, 2(5): 45–46.

Liao Tingting & Lin Chuan. Relationship Between University Students' Time Management and Trend to Internet Addict [J]. Modern Prophylactic Medicine, 2011, 21: 4419–4425.

Liu Shujuan & Zhang Zhijun. Internet Addiction's Social—Mental—Physiological Mode and Study Prospect [J]. *Applied Psychology*, 2004, 13(6): 45–46.

Luo Ling. Self-Control Resources Mode Theory and Its Application in Its Bahavioural Control [J], *Suzhou University*, 2011, 3(2): 27–28.

Robert W. Mulder. Role of Sport's Pleasure in the relationship of physical exercise and mood [J]. Sports Scientific and Technological Information, 2002, 14(7): 35–36.

Wang Hongjiao. Relevant Study on Secondary School Students' SAC and Its Academic Achievement [D], *Shanghai Normal University Master Degree Academic Dissertation*, 2003, 5(1): 56–57.

Wang Yuxiu. Study on Influence of Emotional Adjustment of PE Teaching Mode on Mental Health of Secondary School Students [D], Academic Dissertation of Yangzhou University 2004, 5(2): 29–30.

Sports Engineering and Computer Science – Luo (Ed.)
© 2015 Taylor & Francis Group, London, ISBN 978-1-138-02650-6

Analysis of the feasibility of physical education intervention in youth network addiction

Lei Wang & Bing Li
College of Foreign Languages, Northeast Dianli University, Jilin, China

ABSTRACT: According to the reasons and the performance of Internet addiction, research in psychology, medicine and the social workers are trying to solve the problem of Internet addiction at the same time from their respective fields, which strengthens the physical education, intervention using the methods and means of physical education, in/play 0 meanings, with positive ideas and forms to influence youth, will be helpful for solving the problem of Internet addiction.

1 INTRODUCTION

In recent years, the problem of Internet addiction has become a social problem from all walks of life, especially the hot issues of common concern of parents of teenagers. The China youth addiction data report issued by China Youth Network Association in 2005 shows that at present Chinese teenagers network addiction proportion has reached as high as 13.2%, and 13% of adolescents are in Internet addiction tendency. [1] This shows that in recent years, the domestic experts and scholars pay close attention to this aroused anxiety/social disease in 0 conditions; at the same time, the means and methods of the past, the medicine, psychology, sociology and education in general, although played a certain role, but failed to stop the addiction the spread tendency. The main reason is the lack of lateral communication/between disciplines, which cannot have the effect of integration; the family, school, society, young people themselves do not form a cohesive force, led to repeated episodes of youth Internet addiction; one-sided to foreign research, ignoring the national conditions, the loss of validity 0. [2] The following topics (to take what method to regulate behavior habits, make teenagers get rid of Internet addiction) become the social issues of common concern.

2 STUDY ON DOMESTIC AND FOREIGN INTERNET ADDICTION INTERVENTION METHODS

2.1 Foreign quit addiction research

2.1.1 Cognitive behavior therapy
Cognitive behavioral therapy is a common method of psychotherapy, the representative in this aspect is

the Kimberley # yang. Yang considers social function of the network, which is very difficult to work with the traditional mode of control intervention on Internet addiction. By referring to the research on other addictions and others on the Internet addiction treatment results, Yang proposed their own treatment: A. reverse practice (Practice the Opposite), B. (External Stoppers) external check, C. goal setting (Setting Goals), D. control (Abstinence), E. (Reminder Cards), F. card personal directory (Personal Inventory), G. support group (Support Groups), H. family therapy (Family Therapy) eight steps. Yang Cong time control, cognitive restructuring and collective help point of view, the emphasis.

Therapy should be to help patients build coping strategies and effective, through the system to help the appropriate changes in patients with Internet addiction behavior.

2.1.2 Drug treatment
Drugs that are currently used to treat Internet addiction include antidepressants (antidepressants) and mood stabilizing drugs (mood stabilizers). Research results show that, although the drugs have certain impact on the treatment of Internet addiction plays, the process is still in the trial stage, the general and psychological treatment combined with the implementation.

2.2 Domestic research and treatment of Internet addiction

2.2.1 Psychology and clinical medicine
At present, in the psychology and clinical medicine, experts and scholars have put forward the cognitive therapy, behavior therapy, aversion therapy, intensive intervention method, distraction, social support, group intervention method, and physiological

feedback method; [4], [5], but most of these methods are only theoretical discussion, empirical test. Although other methods, such as the group intervention method, has been proved to have a good effect on the behavior of teenagers "network addiction"; [6] however, this method requires high psychological doctor. The lasting time is relatively concentrated, vulnerable to interference, environment variables, so, in the face of network addiction group huge would be of no avail.

2.2.2 *Aspects of education*

Experts and scholars have put forward that the students should be informed of the harm caused by Internet addiction behaviors, and the teachers ought to strengthen the education of information literacy, so that young people can consciously regulate their behavior; family and schools should establish effective monitoring system which can help the students use the Internet as the means to data access, and acquiring useful information; the Internet cafe operators should create a good network space to students. These methods have some effects, but the effect is not significant.

Professor Tao Hongkai is the first one who offers hopes to this issue. He used the talk method from the beginning of 2004 May and saved a large number of young people to make their learning and life in order. In 2005, he published the monographs 5 *children have reasons for Internet addiction* in January, and *Tao Hongkai rescued the children addicted to the Internet* in which he talked about the causes of the addiction and the measures to solve this problem. He believed that in order to solve the Internet addiction, quality education is a ready-made panacea and the social education is the only viable path. [7]

2.2.3 *Synthesis*

After Professor Tao Hongkai, Tao Ran and Li Banghe put forward that addiction treatment unit 0, [8] is the psychological treatment, drug therapy, health education, physical therapy combined therapy mode. Is a set of psychology, medicine, education means comprehensive treatment method in one. The effect of this method and whether it can be popularized in large area are still to be discussed.

2.2.4 *Research on sports intervention of Internet addiction*

Although many experts said the lack of cultural and sports activities are the main causes of network addiction, there is little systematic research on sports intervention of Internet addiction. Comparison of 6 [9] Deng Wencai's 5 Internet addiction behavior and physical addictive behavior shows that if the rich and colorful sports activities and competitions are employed, they can enrich the students

extra-curricular life, cultivate interest in sports, also reduced form of Internet time and network addiction behavior of students. [9] But the article did not deal with the effective way to interpret physical intervention of Internet addiction deeper considering the particular situations in China.

3 PHYSICAL EDUCATION INTERVENTION

Professor Tao Hongkai believes that the children have the positive minds, and in order to get rid of Internet addiction, quality education is a ready-made panacea, social education is the only feasible way. Internet addiction and drug addiction eliminating lift are different. We can ask the teenagers to be away from drug use but we can hardly ask the youngsters to be away from the Internet. We need to guide the young people to use the Internet as a tool for learning in a reasonable way. Through various forms of physical education to improve parent-child relationship, develop sports exercise interest and habit, to get rid of Internet addiction, and promote the healthy development of young people, it is a good method of persuasion. Especially the sports intervention can also be useful to large groups people suffered by the Internet addiction and the basic function and essential characteristic of sports determine the intervention effect.

3.1 *With the intervention of Internet addiction basic functions of sports*

Youth network addiction group on the Internet tends to play online games, which belongs to entertainment. Sports have the functions of body-building, entertainment, education, morality education, and society. Compared with the network, both have the functions of entertainment and virtual communication. However, sports can provide a visible world to the participants, which can make them able to enjoy the fun of the sports and be happy in this process. It can also motivate the social relationship and facilitate the sincere inter-personal relationship. Sports have a lot of functions which the Internet does not take the place. Doing sports can be helpful to the youngsters' growing up; mediate their psychology and strengthen their willpower, morality and collectivist minds. Sports, contrasted with other therapies, can be a very effective way of auxiliary effect to the Internet addicts.

3.2 *Sports' positive impacts on youngsters' personal traits*

Given the negative performances of the Internet young addicts, we can choose some group sports activities for them to participate in, such as ball games, group competitions, and field group competitions. These activities can help them improve the abilities of social communication and abiding by the social norms. The young people should choose the contents of those sports activities out of interests and physical conditions and in this way they can build their self-confidence and overcome self-contempt. The middle-distance race, the long distance walking race and swimming can help them form their willpower and overcome the lack of self-control. All of these activities can help the youngsters form good personal traits and be away from the Internet addiction.

3.3 Three forms applied in the sports and physical education

3.3.1 Physical education at home

Physical education at home can improve the body constitution of the family members and be good for their mental and physical health as well. This can help the young people form the good habits of doing sports at the very young age and it is also a good start for lifelong sports, which can facilitate the healthy personalities of the young people. Physical education at home is also the lubricant of harmonious families that can be helpful to emotional communication of the family members and the interpersonal relationship and the elimination of the negative emotions.

Recently there exist many problems in the family physical education in China. The research shows that the tense parent-children relationship is also a cause for the Internet addiction. [10] Many parents, affected by the economic conditions and the traditional culture, have little sense of doing sports. Besides, there are no enough facilities for the sports games. Therefore, the family physical education is not developing well in China. We can try to broaden the family physical education by employing many forms of activities and help the children to form good living habits and choose the appropriate forms of entertainment, which can be good for the youngsters forming interests and the stability of the families, improving the parent-children relationship, and later reaching the aim of preventing and curing the Internet addiction.

3.3.2 Physical education at school

Physical education at school can facilitate the development of intelligence, forming good moralities, improving the aesthetic qualities, forming the correct postures and improve body skills, qualities, basic sports abilities and so on. The extracurricular sports exercises and competitions are the main content. On the one hand, the students can mediate their emotions, enrich the life and lighten the

learning pressures. On the other hand, the students can be psychologically fulfilled by watching the sports matches and performances. Physical education at school is also a nice means of leisure as well as interpersonal communication for students and they can show their talents in the sports.

Professor Tao Hongkai pointed out that the examination-oriented education system is a cause of the Internet addiction and the situation is worse and worse. [11] The students are so stressed and they can easily lose the desire of acquiring the knowledge. The rote learning can make them boring and they will go the surf the Internet to find the fun.

Due to the disadvantages of school physical education, we need to improve the current situation that there are no enough gyms and other sports facilities in all levels of schools in order to meet the need of the students. In the mean time, we need to decrease the burden of other disciplines, organize more competitions and improve the professional abilities of the teachers. In this way, more students will be attracted to the sports and the Internet addiction will have less seduction.

3.3.3 Physical education in the communities

The physical education in the living communities can help the residents build a good body and satisfy their needs of leisure and entertainment, have a better communication with each other and it can also be good to the stability of the whole society.

The current situation is that the young people can hardly find the proper place to do sports for there are so few facilities. Professor Tao Hongkai observed: many domestic city lacks the children play place, the children's palace to do the training class, sports facilities fees are too high, and the children have no place to play, also did not have time to play, with Internet addiction is reasonable. Investigation of [11] in Beijing city showed, 63.4% of the street community does not have the sports venues, the remaining 36.5% of the street community have only table tennis room, chess and card room small indoor places a class; Hyundai Residence also in lack of standard of sports facilities, sports facilities to the public rarely, even a small amount of facilities, mostly business, utilization rate is low, the benefit of small. [12] Guangdong general residential district sports facilities of city residents ratio reached the standard of limit state regulations (200 per 1000 square meters) accounted for only 21136%. [13] The only sports facilities are not young people, mostly in the elderly activity room. Economically developed areas like this, one can imagine the situation in other areas.

If you have enough sports venues and facilities in the community, there are high quality sports instructors, the organization of various projects of training and competition, satisfy the need of extracurricular activities of young people, so the network game will no longer be teenagers love.

Of course, we can not arbitrarily say that physical education is the best method to solve the network addiction. However, in China under the current situation, in the face of the huge addiction group, only by using the methods such as psychological counseling, group psychological intervention, expert talk for the prevention of Internet addiction disorder of individual and small group is an utterly inadequate measure; physical education compared with other forms of education, in cultivating adolescent self-esteem, self-confidence, responsibility, self-motivated and the good aspects of behavior, at least as treatment formula of medicine, has the special function. Therefore, strengthening the physical education to many teenagers stop in the addiction, but it can reduce the degree of addiction may even released them from addiction as well.

4 CONCLUSIONS AND SUGGESTIONS

We should actively advocate family sports, perfect the physical education at school and in the communities. In the mean time, we need strengthen the construction of stadiums and sports facilities, schools and communities, and encourage the establishment of various sport clubs, increase low-cost or free sports facilities; organize various sports activities and competitions, meet the needs of the community, family and youth sports activities. The integral model of the physical education in school, at home and in the communities should be constructed and implemented. [14] We need to make each part show their merit and build a multi-source, multi-level and all-rounded network of physical education. In this way, the youngsters can be devoted to the enjoyment of doing sports and grow happily. Physical education can offer a relaxed living environment to the children and takes the place of the boring talks of the teachers and the parents. This can show the spirit of humanities and be good for the growth of the young people.

4.1 *The sports activities to exert positive influence on adolescent personality traits*

Sports psychology research proof, various sports activities need to have strong self-control. The firm faith, brave and resolute will and other psychological quality is the basis of physical education. Therefore, only taking the physical exercises as the target has serious defects, the effective measures of training perfect personality should be carried out.

In view of some negative personality traits of adolescents with Internet addiction, can choose some sports collective, such as ball games, group competitions, the field team competitions, training social exchanges and their ability to keep social norm; according to the content of adolescent self interest, ability to choose their favorite and competent, to cultivating their self-confidence, overcome inferiority complex; long distance, long distance walking, swimming endurance project selection of training their perseverance, to overcome the problem of low self-control, make them form sound personality, and the prevention and treatment of Internet addiction.

4.2 *Physical education operational intervention on adolescent network addiction*

4.2.1 *Family sports*
The family sports can enhance the family member of the constitution, the promotion of family members of mental and physical health, make teenagers form the sports exercise interest and habit since childhood, is the starting point and end result of lifelong sports, can promote the formation and the development of adolescent health personality. Family sports is family friendly lubricant, can promote the exchange of feelings between family members, form a harmonious interpersonal relationships, and enrich the modern family life, eliminate the bad mood.

At present, the family education in our country there are many family problems, research proof, the tension in parent-child relationship is one of the causes of teenagers' network addiction. [10] Is influenced by economic status and traditional culture, many parents exercise consciousness, lack of physical activity, coupled with the lack of sports venues and facilities, and thus the development of family sports is not optimistic about the situation. We can through the family sports, cultivate good habits, choice of entertainment content and form a healthy, make teenagers form physical exercise habit and interest, from promoting family stability, harmony, improve parent-child relationship, to achieve the prevention and treatment of addiction to network.

4.2.2 *School sports*
School physical education can promote intellectual development, the formation of good moral character, culture aesthetic taste; to develop the correct posture, promote growth, improve the skill level, the development of physical activity and basic ability, enhance the ability to adapt to the external environment; after-school training and competition, is an important part of students' extracurricular activities. On the one hand, students through

participation in sports activities can regulate emotion, rich life, and relieve the study because of the nervous tension and fatigue; on the other hand, the students through the watch sports competitions and performances can be obtained to meet the psychological and spiritual enjoyment. School physical education is an important means of student leisure, is to expand the important media for their social and self performance, shows an important stage of self.

Professor Tao Hongkai pointed out, / is the exam oriented education let Chinese children with Internet addiction, and more! 0 [11] in the exam oriented education under the baton of, the children are under enormous pressure, not ease, which lost the desire for knowledge, mechanical dead endorsement, is easy to produce weariness emotion, then skip class, on the Internet, indulge in the network.

The effect of school physical education environment not play situation, we should improve from primary school to university sports venues and facilities exist serious shortage situation, meet the needs of young people during the period of school classroom and extracurricular sports activities; reduce the other department to work burden, the teenage physical activity time guarantee organization; all kinds of training and competition; improve the PE Teachers' professional level and the humanities accomplishment, so that everyone can have their own love of sport, more young people are attracted to body movement, the temptation of the network power will be greatly reduced.

4.2.3 *Community sports*

Community sports can promote the health of the residents, the residents of the entertainment and leisure needs, rich cultural life of the community, to communication and the exchange of community residents, to help establish a good community relations, maintaining social stability and unity.

The current situation is, the community of the severe shortage of sports facilities and project a single, young people to find their favorite sports and sports venues. Professor Tao Hongkai observed: many domestic city lacks to the children play place, the children's palace to do the training class, sports facilities fees are too high, the children have no place to play, also did not have time to play, with Internet addiction is reasonable. Investigation of [11] in Beijing city showed, 63.4% of the street community does not have the sports venues, the remaining 36.5% of the street community have only table tennis room, chess and card room small indoor places a class; Hyundai Residence also general lack of standard of sports facilities, sports facilities to the public rarely, even a small amount of facilities, mostly business, utilization rate is low, the benefit of small. [12] Guangdong general

residential district sports facilities of city residents ratio reached the standard of limit state regulations (200 per 1000 square meters) accounted for only 21136%. [13] The only sports facilities are not young people, mostly in the elderly activity room. Economically developed areas like this, one can imagine the situation in other areas.

If you have enough sports venues and facilities in the community, there are high quality sports instructors, the organization of various projects of training and competition, satisfy the need of extracurricular activities of young people, so the network game will no longer be teenagers love.

Of course, we can not arbitrarily that physical education is the best method to solve the network addiction. However, in China under the current situation, in the face of the huge addiction group, only method psychological counseling, group psychological intervention, expert talk for the prevention of Internet addiction disorder of individual and small group is an utterly inadequate measure; physical education compared with other forms of education, in cultivating adolescent self-esteem, self-confidence, responsibility, self-motivated and the good aspects of behavior, at least as treatment formula of medicine, has the special function. Therefore, strengthening the physical education to many teenagers stop in the addiction, but also can reduce the degree of addiction may even released them from addiction.

5 CONCLUSIONS AND RECOMMENDATIONS

1. Actively advocating family sports, school sports really perfect, to the development of community sports. Strengthening the construction of stadiums and sports facilities, schools and communities, and encourage the establishment of various sport clubs, to increase low-cost or free sports facilities; to organize various sports activities and competitions, meet the needs of the community, family and youth sports activities.
2. Construction and implementation of school, family, community, sports education mode. [14] Play school, family, community sports advantages and functions, promote the participation and interaction between the three forms of physical education, social network, a multi-channel, multi-level, full range of. That young people bathed in the sports of the sea, enjoy sports happy, healthy growth.

The founder of the modern Olympic Games in 5 sports, 6 Coubertin wrote: "ah! Sport, you are beauty! You through the most direct way, enhance national physique, physical deformities corrected; Weiran in-patients. Ah! The sports, you are the progress! For the human change rapidly, physical and mental improvement at the same time to start. You provided a good living habits, asking people to guard against excessive behaviour." The face of the network, physical education can make the parents, teachers to change the endless preaching, presents to children is loose, a good living environment, which will better reflect the humanistic spirit, will be more conducive to the healthy growth of children.

REFERENCES

[1] Chinese Youth Association for network. Chinese youth addiction data report (2005) [N]. Chinese Youth Daily, 2005-11-23.

[2] Zan Yulin. Teenagers "network addiction" research [J]. Chinese youth research, 2005, (7):23–25.

[3] Gu Haigen. A study on the Internet Addiction of foreign [J]. Education in foreign countries, 2005, (9):36–39.

[4] Works, Liu Min. New social disease [J]. Internet addiction research [J]. Gansu social science, 2005, (4):42–45.

[5] Gong Yinqing, Yang Rong, Zhang Bin. And the technology methods of psychological treatment of Internet addiction. [J]. Chinese school health, 2005, (5):50–53.

[6] In 衍治. Group psychological intervention to improve the network addiction of young feasibility of [J]. Chinese clinical rehabilitation, 2005, (20):24–27.

[7] Tao Hongkai. Children have to [M]. The heart of the Changsha: Hunan people's publishing house, 2005:116–119.

[8] Tao, Li Banghe. Overview of [J]. Network addiction treatment unit China Journal of epidemiology, 2005, (8):41–44.

[9] Deng Wencai. College students' Internet addiction and physical addiction to compare [J]. Journal of physical education, 2003, (6):38–42.

[10] Li Tao. College Students Internet addiction and parental rearing pattern of [J]. In psychological science, 2004, (3):44–49.

[11] Tao Hongkai. The examination oriented education into the net addiction is the source of [N]. Xinmin Evening News, 2005-02-25:8–10.

[12] Wang Qiaojun. Sports facilities planning in city residential area of [J]. SportsScience, 2004, (2):5–8.

[13] Shao Yiqiang, Song Yunqing. Investigation of Guangdong Province, community sports facilities construction condition of mass sports activity and [J]. China sports science and technology, 2005, (1):20–24.

[14] Lin Shaona, Chen Shaoyan, et al. School, family, community sports development mode of education integration 0 [J]. Journal of Wuhan Sports Institute, 2004, (6):22–26.

Sports Engineering and Computer Science – Luo (Ed.)
© *2015 Taylor & Francis Group, London, ISBN 978-1-138-02650-6*

Research on the value of human movement science in sports practice in university

Lei Wang & Bing Li
College of Foreign Languages, Northeast Dianli University, Jilin, China

ABSTRACT: Countries in the implementation of strategic measures of rejuvenating the country through science and education gradually, comprehensive promotion of higher quality education of the students, cultivate new talents for the development of our country in the new era, to enhance the overall quality of the broad masses of the people in China. Human movement science occupies the position and plays a decisive role in the college physical education in China, higher education and training institutions in China will also develop the science of moving human body as the main task of education, it has a very important role in the development of physical education in colleges and universities of the stand, strengthen the training of human movement science knowledge, not only to provide adequate security for the sports training, colleges and universities at the same time, the direction of development of higher school education gradually toward the broad masses of the direction of development, but also provides the possibility for the healthy development of the masses.

1 INTRODUCTION

Because of the reform and opening up policy, the society has developed rapidly, people now have more advanced technology, and the living standard continues to improve, so the needs of people in higher education in every respects are gradually deepened, thus human movement science gradually enters into our daily life, which is also absorbed gradually by the college physical education. By the definition of sports science, we can infer that it mainly focus on some profession talents who can make the scientific guidance in sports education and the cultivation of these talents. The purpose of human movement science is to perform physical practice education in universities. Nowadays, human movement science becomes more important in college physical education, which targets on performing education for all-around develop and cultivating new talents. Sports science of human body mainly focuses on the researches made by physical education majors in the laboratory and improving their qualities. In the meanwhile, it intends to discover the defects of sports science of human body in college physical education and come up with the solutions to these defects. Then it can provide references for the development of college sports education practice.

2 THE OBJECTIVES OF HUMAN MOVEMENT SCIENCE IN SPORTS PRACTICE IN UNIVERSITY

The purpose of physical education in higher education is to cultivate students' physical quality, and gradually promote the comprehensive development of students, it enables students to better learning, and becomes new talents who are useful to school, society, the motherland and make contributions to the better development of the society. The objectives of college sports education are: first, to exercise the physical quality of students, and promote all-round development of students' morality, intelligence, body, and beauty. Second, to perform basic physical education, so that students can master the knowledge of sports, sports skills, basic technology and enable the student to have the interest and enthusiasm of sports education, thus students will have a better body and gradually develop the passion and habits for the sports. Third, sports education can not only promote the development of students' physical health, but also cultivate students' sentiment, and furthermore exercise students' temperament, improve the psychological quality of students. Fourth, teaching spirit can only be realized gradually in the long time study. The connotation of sports teaching, can not only teach students to fight the forces of evil, but also improve the loyalty of the students to the communism and the students' moral character.

3 THE NECESSITY OF SPORTS HUMAN SCIENCE EDUCATION OF THE COLLEGE SPORTS PRACTICE

Human movement science is a subject that focuses on the research of the human body. Students exercises most in physical classes. Therefore, the use of human movement science knowledge in the practice of sports education will not only strengthen the students' enthusiasm, but also teach the students how to protect themselves better and take care of themselves when they are injured.

Sports science of human body is not well known as a discipline by the students, people usually believe that human movement science is physical education time of it, but the main or lectures and laboratory of human anatomy. Physical education in the laboratory, is a bold attempt, often people will be considered only in the outdoor sports teaching, practice teaching in sports can also be carried out in the interior, the main research is on the human body, make human movement science curriculum in the lab, can let the students a better understanding of the human body, and a better understanding of sports. The human body in the process of movement, muscle movement, shrinkage, how to correctly exercise does not get hurt, the lecturer can in the laboratory on the human body to explain the time out, more students understanding and absorption. Sports science of human body is the human body and exercise related disciplines. Only thorough research is able to let the new process supersede the old one in sports practice. The body produces myriads of changes in the athletic process, and only by constantly new changes can better understand the movement of chemical, biological and other sports theory knowledge. The motion human body teaching in the laboratory, but also to enable students to better accept the sports teaching practice, if in the study of human muscle movement, let the students themselves boring associative learning, is more vivid understanding of the human body in the laboratory as students, so better to deepen their impression, to better grasp the theory knowledge.

3.1 Education reform

With the development of the times, the rapid economic rise, social competition is becoming increasingly fierce, the enterprises for talents of the increasingly stringent requirements, not only requires people with professional knowledge, also asked the people of ability and innovation ability can meet the needs of enterprises. Demand for high-quality talent competition in the fierce constantly, so the college students' requirements are strict constantly, physical education in colleges and universities, for the training and human knowledge of students, also in the unceasing reform and perfect. For college physical education, research and in the continuous improvement of human movement science, compensate for lack of comprehension, in order to better cultivate students' morality, intelligence, body, beauty, human knowledge is the teacher teach better, the harmonious development

of students' college life, development of sports human science education is vital the. Based on the laboratory of human movement science in universities, the introduction of advanced teaching equipment, teaching and practice in all aspects, not only can enhance student learning enthusiasm, but also can enhance students' ability to understand and. In the laboratory, college sports teaching of human body, enhance the students ability to exercise, to facilitate the development of all aspects of personnel, so as to cultivate talents to meet the demand of social development in all directions.

4 COMPONENT OF COLLEGE PHYSICAL EDUCATION

The present world situation is changing gradually, the university also in the gradual reform, colleges and universities sports also gradually go out of the school to enter the world. Now all the college sports in the constantly changing, sports teams, each is also gradually become more powerful, the ranks of advanced, gradually into the world for example: Tsinghua University, University of Cambridge rowing, diving is the count as one of the very best sports team. With the development of our society, our teaching of college sports are becoming more and more seriously, the universities of our country sports also gradually make major changes, the students "combined" teaching strategies to teach, not only to promote the development of students' quality education, but also make the students sense of competition gradually the well development. Now the world has to colleges and universities as the best training base, reserve force for the sport cause, can

5 EFFECTS OF SPORTS SCIENCE OF HUMAN BODY IN SPORTS PRACTICE IN UNIVERSITY

Concept is now teaching in colleges and universities, is to cultivate the students', to promote the healthy development of student's direction, the school will cultivate students' physical health as the first priority, especially when doing physical examination for students, will involve many fields, for example: life science, biology, medicine and nutrition, all areas for the development of students' sports make a strong backing, so that the development of college sports, forging ahead toward a more scientific, standardized direction. To make the greatest contribution to the science of human exercise also for the development of the sports practice of colleges and universities, the development of college sports needs of human movement science knowledge, study of human movement science major is to enable students to better understand the human body, understanding human bones and muscles. The new supersedes the old and so on, to enable students to better understand the limitations and explosive force own, let the students better and avoid their shortcomings in physical education teaching, to make up for their deficiencies. The college sports teaching not only teach knowledge, more important is to have a strong physique, people often say: "the body is the capital of revolution", without a strong body, how to exercise to accept physical knowledge? In college physical education, physical education teacher is not only to impart knowledge, or lead the students to carry out physical exercise instructor, the primary task for the teacher is to cultivate students' body, but not in the physical education with entertainment learning instead of physical learning. For the sports learning, not only for learning basic knowledge, more important is the practice, if not in practice and how to better learn the principles and skills taught by the teacher? The cultivation of students' sports concept on human movement science knowledge to support.

6 APPLICATION OF SPORTS SCIENCE OF HUMAN BODY IN SPORTS PRACTICE IN UNIVERSITY

6.1 Application of exercise prescription

College physical education major is to improve the physical quality of students, but for a long time, people usually have a misunderstanding, some people think sports is fitness, and fitness did not enhance the wrong idea of physical. Therefore, now the university for physical education mainly is the motor skills, sports skills to use in physical exercise to strengthen students' physique, physical exercise is different for different students' physical condition, in order to improve the physical quality of students. In fact, fitness or enhance the physical in nature and exercise or have certain difference, sports just for exercise or enhance a constitution provides material content is certain, because of improper movement and injuries. It is often seen, so, how to regular exercise? How to solve the motion after the injury problems become the main topic to research, and exercise prescription for the problem solving exercise made, pointed out the direction of science.

6.2 Application of exercise physiology

Exercise physiology is still belongs to the subject category, but only break empty talk to better applied to the real life, the research to be more practical value. Our country in the research of exercise physiology, aiming at how to improve athletic performance, how to improve the people's health has made a

lot of investigations, exercise physiology is mainly reflected in the movement in the process of sports fatigue, recovery and the athlete's body function, sport practice teaching movement, immune and other scientific research applied to in college, for students to better exercise made a strong support.

6.3 Application of sports biomechanics

In recent years, as China's reform and opening up the pace gradually thorough, the economic strength of our country is gradually rising, development of science and technology also gradually progress, science and technology content of sports have been added, study of sports biology has also made attract worldwide attention, various colleges of physical education teachers also gradually the discovery of sports biology, occupies a very important position in the sports teaching, practice in sports teaching in colleges and universities, the movement biology into knowledge, not only can improve students' ability to understand, to exercise at the same time, also can be more creative students completed the exercise needs.

7 CONCLUSION

In new period, along with the development of China's reform and opening up, China's higher education is gradually of reform, the reform in college physical education, in-depth study of the state is gradually. If the human movement science knowledge is put into practice in PE teaching in colleges and universities, the students' learning enthusiasm will be greatly improved. Professor of sports science of human body of knowledge at the same time, also can let the students lack of a better understanding of their sports, more targeted for exercise, the teaching practice of colleges and universities of our country a thorough, cultivate new talents with all-round development of morality, intelligence, a number of beauty, for the development of our future.

REFERENCES

[1] Cao Lanju. Cultivation of sports industrialization and College of human movement science talent. Journal of Henan Normal University (Philosophy and Social Sciences Edition). 2011 (6):25–28.
[2] Chen Xiaoan Chen Jing. The sharing of resources, optimize configuration, reasonable use, function development and Human Sciences Laboratory—ordinary university sports on the construction of sports space. 2013 (13):34–37.
[3] Qiao Decai Kang Daofeng Liu Xiaoli Hou Lijuan Deng Shuxun. The status quo of our college of human movement science disciplines. Journal of Capital Institute of Physical Education. 2011 (3):42–45.

Sports Engineering and Computer Science – Luo (Ed.)
© 2015 Taylor & Francis Group, London, ISBN 978-1-138-02650-6

Study of social sports specialty entrepreneurial talents cultivation

Jian Liu & LiJun Xu

School of Physical Education, Northeast Dianli University, Jilin City, Jilin Province, China

ABSTRACT: To broaden the train of thought, seek development in the process of reform, train applied talents of social sports field with the spirit of innovation, entrepreneurship ability as the goal, to "professional skills training and entrepreneurship education" of thinking, teaching, practice, education three teaching programs, tamping the specialty base, improve their professional skills, spread entrepreneurship seed, students not only can become professional to the social sports guidance and management, but also have the confidence and ability to start an undertaking.

Keywords: social sports specialty; entrepreneurship education; professional skills; applied talents

1 RESEARCH BACKGROUND

1. *In the year 1998, when the Ministry of Education revised regular college course catalog for the third time, they make the social sports specialty a formal professional.* This is good for the development of our country's economic and the improvement of the country's living standards. With the continuous increase of our sports population, the demand of talented social sportsmen is in great need. Sports instructors, as an example, as the two phase of the project outline of the nationwide body-building plan issued by the State General Administration of sports "(2001–2010) plan" pointed out that, after 10 years of efforts, our country city community, rural universal establishment of social sports guidance station (center), the number of social sports instructors will reach more than 650,000; social sports instructor has specialist degree or above accounted for 34.90%. As it can be seen, the social sports professional talents are in need.

2. *At present, in the fitness field social sports instructor posts are rarely a full-time job.* The entry to community social guidance work need to be a civil servant, and is restricted in several individual cities. The jobs for professional social sportsmen are limited. "In the Department of the Ministry of Education's Education Department of education and employment" series of 2005 college entrance examinations guide, "social sports specialty is listed as a professional of lower employment rate" [1]. Low employment rate restricted the healthy development of social sports specialty, and then lead to an atrophy trend. The data show that the number of higher colleges who has social sports specialty is 173 in the year 2007, 144 in the year 2009, and 92 in the year 2010.

On the one hand, social sports are built on a new economic platform. With China's rapid economic development and the process of modernization, demands of social sports professionals continues to grow; on the other hand, the current supply to the full-time social sports professional job is extremely limited. This poses a great challenge to our education.

2 THE TEACHING REFORM AND PRACTICE

2.1 *To strengthen the professional skills training, construction of training mode of "1-3-3-3":*

1- One goal: to cultivate outstanding social sports talents;

3- Consists of three parts: Germany, technology, can;

3- Three stages of teaching process: classroom teaching, social practice, and the new classroom teaching;

3- Three combination: class "teaching, learning, practicing the" combined, "the combination of class", and the combination of school and society.

2.1.1 *Take the social demand as the guidance, training objectives*

Taking the social demands as guidance, we can train students as a professional into social sports. They can serve the development of mass sports and serve the society.

2.1.2 *To training objectives as the basis, determine the training*

1. Morality—character, occupation accomplishment.
   Basic composition:
   ① Love. Love nation, social, dedication, love each and every one of our guidance.
   ② The sense of social responsibility. Social sports professional students must have a high sense of social responsibility, in order to achieve its culture value. Take participate in social welfare activities such as guidance, students shall put their heart and soul into service to society and make contribution to people's health and social development.
   ③ Beauty: moral beauty, image beauty. Good moral character, healthy body and mind, healthy body, forming a model effect.
   ④ The etiquette: etiquette, courtesy, master the basic etiquette knowledge, based on respect, harmonious communication.
   ⑤ Brave, tenacious, unity, passion Implementations: Classroom teaching; lectures; social activities; special training.
2. The professional and technical training, technical—
   Basic composition:
   Professional theory: sports anatomy, exercise physiology, sports psychology, sports sociology, sports market and marketing, sports management, sports nutrition, sports fitness principle and method.
   Sports: ball, winter sports, track and field, aerobics, aerobics, roller skating movement.
   Implementations: Learn professional knowledge according to the social sports specialty curriculum, learn all kinds of sports technology at the same time, select 1–2 sports technology as the major technology and expertise.
3. Can do—the cultivation of comprehensive ability.
   Basic composition:
   Problem analysis and problem solving; innovation, entrepreneurial ability, communication ability; communication; cooperation, management ability; "guidance, teaching, lead" and other professional ability.
   Implementations: Employ teaching, learning and exercise as a "dual" system. Study both learning and training skills, through the "role conversion, guidance, organization, evaluation, innovation, practice, practice" and other series of form. Then exercise in learning, learning, improvement in exercise.

2.1.3. *The implementation of three stages of teaching process: classroom teaching, social practice, and the new classroom teaching*

The first stage: lay a foundation. In the first or second year, spotlight the classroom teaching. Learn public and professional knowledge and the basic professional technology and training.

The second stage: social practice. After the third year, combine the classroom and social practice.

The third stage: targeted learning. The new classroom teaching and basic classroom teaching are different. In conclusion, it's "summarizing—studying—learning—enhancing".

2.1.4 *Three combination: the combination of "teaching, learning, practicing" in class; the combination of "in and after class"; the combination of school and society*

1. The combination of "teaching, learning, practicing" in class
   Formation: ① "Teaching and learning"; ② "Learning and practicing"; ③ Simulating practice in teaching. Objective: to master the skills, improve the ability to work, promote.
2. The combination of "in and after class"
   Formation: ① Add the club and professional associations into teaching; ② Create special association; ③ Take part in administration. Objective: To study, train and practice the ability to work.
3. The combination of school and society
   Formation: ① Research; ② Invite entrepreneurs, managers, practitioners and other in social sports field to teach and train students; ③ Performance; ④ Match referee; ⑤ Set up practice base, combine school with enterprise cooperation; ⑥ Work study.
   Objective: practice and education.

2.2 *Strengthening of entrepreneurship education*

2.2.1 *Change idea renewing the concept*

The Ministry of Education on promoting the innovation and entrepreneurship education in higher schools and students independent venture work opinions reads: innovation and entrepreneurship education are intended for all students—integrated into the training process; Strengthening the construction of curriculum system of innovation and entrepreneurship education, innovation and entrepreneurship education effectively into the cultural quality education and vocational education teaching plan and credit system; promoting innovation and entrepreneurship education work in colleges and universities. Therefore, our social sports specialty training should change the education idea, update the training concept, and reform the single professional skills model. On the basis of our professional education, improving the training objectives, Talent Training Mode and course system reformation, strengthen enterprise education and train with the high level of professional skills at the same time, also with a sense of entrepreneurship and entrepreneurial talents of entrepreneurship.

2.2.2 *Students hope to obtain the knowledge, skills and entrepreneurship education*

The results of a questionnaire survey of social sports specialty aiming at college students show that 97% of the students want to obtain entrepreneurship education, 98% of them hope to learn business knowledge in the school, 94% of the students have entrepreneurial aspirations, 99% of them don't know the policy on national support. Students have entrepreneurial aspirations, but cannot start. They don't look up the policy consciously, and don't understand the relevant information and policy. Nowadays, with the vigorously promote innovation and entrepreneurship education in higher school, our students have to come to the front of our education, education is not fully realize its function of education. We should keep pace with the times. Combining sports and students, strengthen the establishment of entrepreneurship education curriculum system. Set up the training mode of professional skills and entrepreneurship education, open up new ways of cultivating talents.

Implementations: Set up entrepreneurship education system, introduce entrepreneurship education into the school teaching. "Entrepreneurship education system should include: classroom teaching, case analysis, training simulation and field research. Central University of Finance and Economics will take the lead into entrepreneurship education and teaching. They have set up an "entrepreneurship" curriculum, established business vanguard class. These practice compose the entrepreneurship education system". [2] Entrepreneurship education is a complex and difficult process of education. Not only should one learn the theoretical knowledge, but also focus on the combination of theory and practice. These theory and practice include social entrepreneurship base, school enterprise cooperation education content, enhancement of training, strengthening practice. With only school education, it will be the examination oriented education mode of business education.

There are obstacles lie ahead, but the opportunities and challenges are coexistence. Strengthen the entrepreneurship education while cultivate talented person, build up a cultivating system of "special skills and entrepreneurship education", cultivate application-oriented talents are the way to promote social sports specialty.

REFERENCES

[1] Jia Yan specialty of social sports to the social demand condition analysis and countermeasure study [J] "Journal of Northeast Dian Li University" 2008. 05.

[2] Qi Linquan college students' entrepreneurship education is more attention to shout less good work [N] "Chinese Education Newspaper" 2006. 10. 17.

Sports Engineering and Computer Science – Luo (Ed.)
© 2015 Taylor & Francis Group, London, ISBN 978-1-138-02650-6

The training of the university students' lifelong sports consciousness, hobbies and abilities

Jian Liu & LiJun Xu
School of Physical Education, Northeast Dianli University, Jilin City, Jilin Province, China

ABSTRACT: College physical education is an important link of school sports and community sports. According to the significance of students' lifelong sports training, college physical education should aim at training of the university students' lifelong sports consciousness, hobbies and abilities, in order to make college students establish a good consciousness of lifelong physical exercise, master the more scientific method and build the concept that health is a lifelong wealth.

Keywords: sports and health; education; lifelong sports

1 INTRODUCTION

The improvement of national physique is the concern issues for whole society. It is very important for college students whether they can have a healthy body and develop a lifelong habit and skills of physical exercise. Therefore, how to cultivate college students' consciousness, interest and the ability has become an important task of College Physical and Health Education.

2 AWARENESS OF LIFELONG PHYSICAL TRAINING

From a psychological point of view, consciousness is refers to that the people reflect the objective reality of consciousness, is the subjective reflection of the objective world. And lifelong sports consciousness, refers to that person has the consciousness of physical exercise whole life long. From the teaching process of sports, sports is the unity of human and mental activities, it contains human needs, objective, motivation, knowledge, skills, ideas and ways of thinking and so on. In the university stage, the student is in an important growth and development period to build good physical quality. More importantly, it is a training time for students to develop lifelong sports consciousness. As a consequence, the sports and health education ought to include social sports and sports life, such as the principle of physical fitness, sports health carefulness and physical health theory. Let the students learn the basic knowledge of fitness, master the scientific method of exercise science, and recognize the importance and role of lifelong physical exercise, which then makes students aware that not having a strong physique and self exercise will be eliminated by the rapid development of society now and future. [1]

3 LIFELONG INTEREST IN SPORTS TRAINING

The interest is a tend for people to seek some kind of thing or to engage in an activity, besides the generation of interest is closely related to the acquired education and social practice.

If students are forced to rather than have a interest to learn and to be trained, students are not motivated and effect is not good. At the same time, interest and ability are closely linked. When students have great interest in some sports activities, they will show a positive learning attitude and determination to overcome difficulties, to master sports skills in the shortest time, and continue to practice to improve themselves on the based consolidate. This is a good foundation for the lifelong sport. Strong interest and thirst for knowledge is the power for ability formation. Therefore cultivating sports interest of students will play an important role in the lifelong physical exercise.

In the sports teaching process, the generation and strength of students' learning interest is closely linked with live and enlightening teaching, but also the students engaged in sports activities in the pleasant experience. Teacher's vivid and humorous language, graceful and accurate demonstration of movement, not only can active the class atmosphere, but also can promote the students emotion sublimation, which then can create a good learning and training environment. At different stages of the teaching, according to different teaching content, teachers' using various teaching methods is conducive to enhance the students' interest in PE, focusing the students on the teaching process. All that can make students to gain spiritual satisfaction and improvement of physical function.

4 CULTIVATING THE ABILITY OF LIFELONG PHYSICAL EDUCATION

The ability is psychological characteristics of personality which can make one a successful completion of certain activities, it refers to those abilities that conform to the requirements of the activity and the effect of activities. Ability is one of the fundamental physiological qualities, which is through education and training and then developed in practice. Improvement and enhancement of physical ability plays a very important role in the implementation of lifelong sports. Therefore the cultivation of students' sports activities in universities should use sports knowledge, fun, entertainment, fitness, grasp good sports and other features to lay a good foundation of physical education.

On physical education and health education in colleges and universities, the teacher's main task is not only to teach knowledge, skill in itself, but as a physical training instructor and guide, to help and promote students' learning. [2] With the development of society, sports and health education must be updated, the teacher must enlighteningly guide students based on the scientific knowledge education, teach students with certain aims, abandon the "examination oriented education", and cultivate ability of exercise on their own. While in the teaching of sports technology and skill, teachers also should use various forms of sports theory knowledge teaching to enable students to have the ability to use scientific knowledge to guide the physical exercise.

Including the students' ability of lifelong physical training: know exercise science principle, principle and reasonable fitness means and methods, to develop training plans; to make self supervision, self supervision and self evaluation, self rehabilitation of mastering the knowledge and skills. Scientific, diversification of sports and health teaching in enlightening guidance and teaching method is the important means and methods of training students' ability of physical education. Strengthening the students' extra-curricular guidance and help, the students' self-learning and self training, self plays an important role in improving. Cultivating the students' lifelong physical ability should have long-term plans, through the process of higher education of sports and health education, in cultivating students' general ability at the same time, pay attention to teach students in accordance with their aptitude, mining sports potential students, so that students ability to play, encourage, encourage students to master and improve the ability of lifelong physical exercise.

In the setting of college physical education and health education curriculum that offers specialized elective courses, students can chose special sports from elective course and master one to two movement technique to develop sports expertise according to their own interests, hobbies, and ability. It will play a positive role in student's lifelong physical ability. Sport is various, so the students can't learn many kinds of sports. However, students can select one or two projects which are suitable for their own conditions, interesting them a lot and worth exercising for a long time. Through the special elective courses, students constantly improve motor skills, master the scientific method of exercise special, and then can take the project as a lifelong physical exercise content, achieve physical fitness and gain benefit in the long run.

Colleges and universities are to cultivate talents for the country. Talents cultivation should be based on the health, cultivating the lifelong health consciousness, imparting lifelong health knowledge, and mastering the lifelong health ability. To accomplish the tasks, we should attach importance to the lifelong physical education, and cultivate talents with the aim that "work 50 years healthily, living a happy life".

REFERENCES

[1] Cheng Jie, Quanzheng, Our athletes training mode comparative analysis. Journal of Shenyang Sport University, 5, 2007.
[2] Liu Ming, China's elite athletes training way present situation and prospect of. Shanxi Sports Science and Technology, 2, 2007.

Sports Engineering and Computer Science – Luo (Ed.)
© 2015 Taylor & Francis Group, London, ISBN 978-1-138-02650-6

Social sports specialty talents training ski instruction of school enterprise cooperation

Jian Liu & LiJun Xu
School of Physical Education, Northeast Dianli University, Jilin City, Jilin Province, China

ABSTRACT: Ski market needs ski guide personnel because of the rapid development of skiing. Feasibility analysis of school and enterprise cooperation through cultivating talents proposed ski instruction, guidance and management of social sports specialty only combined with ski, ski equipment enterprise, integration of advantageous resources, cooperation, build a new model of the ski instruction training, in order to meet the social and market demand, broaden the field of employment, promote the benign development of professional.

Keywords: school enterprise cooperation; mass skiing ski instruction

Mass skiing is a part of social sports. Training professional guidance for the mass sports, guiding participants scientifically and promoting the scientific development of mass sports talent are some of the goals of guidance and management professional society culture.

1 FEASIBILITY ANALYSIS

1.1 *The status of mass skiing*

1.1.1 *Ski*
The ski field in China has been built, operated nearly 300, throughout the country 27 provinces and municipalities directly under the central government. South to Guangzhou, Shenzhen, Fujian, Shanghai, Jade Dragon Snow Mountain ski resort in Yunnan is China's ten largest ski in ski, only with a certain scale in Beijing and Hebei Province registered operation is a 57 home, real estate tycoon such as Vanke, Wan Da also put huge investment into "the development of ski resort style" project.

1.1.2 *Ski population*
In 1996 the popular skiing group, as a starting point Chinese skiing, skiing in the crowd began to rapid growth, an increase of 1996–2000 ski trips for 109%, an increase of 2001–2005 ski trips for 53%, 2006–2010 years ski trips a year will increase to 20%. 2010–2011 snow season, China ski trips have reached 10000000. Is expected during the 2010–2015 years, China skiing population and ski trips will continue to maintain an average annual growth rate of more than 10%.[1]

1.2 *Ski instructors*

The rapid development of mass skiing is not coordinated with the skiing real situation with the growth of the skiing population and the lack of professional guidance to China's ski. The world has one hundred years history of skiing. French, Swiss, Japan and other countries have relatively complete instructor reserves. Such as France, there are about 1000000 people participated in the skiing instructor training each year, and about 150000 people are full-time instructors in the snow. Compared with these countries, ski instructor reserves in our country is far less than 1/10000 of theirs. According to this and our country's large ski market proportion, it far cannot meet the development needs of mass skiing. "The instructor is the key to drive the skiing sports development, and is the foundation to expands the market".

1.3 *Social sports specialty*

1.3.1 *Low employment rate*
The full name is guidance and management of social sports specialty. In the current situation of our country, in the fitness field also rarely set up full-time social sports instructor posts. Community social guidance work to be implemented through the civil service exam, is only for individual city economic development. The main services of social sports specialty in sports entertainment market business and related services, provide for social sports professional jobs are limited. "In the Department of the Ministry of education's Education Department of education and employment" series of 2005 college entrance examinations

guide, "social sports specialty is the low rate of professional employment," low employment rate restricted the healthy development of social sports specialty of social sports specialty, the progressive atrophy trend. "In the Department of the Ministry of education's Education Department of education and employment" series of 2005 college entrance examinations guide, "social sports specialty is the low rate of professional employment," low employment rate restricted the healthy development of social sports specialty of social sports specialty, the progressive atrophy trend. The data show higher colleges of Social Sports Specialty: 2007 173; 2009 144; 2010 92.[2]

1.3.2 *Current situation of the development of ski guide personnel*

At present, most of the guidance and management of social sports professional training direction is still very limited, as an example, it is main for health guidance and training in my school such as aerobics (aerobics group), sports and fitness guidance (instrument for muscle training) are the required courses, students need to study all five semesters, there are 72 hours in each semester. The ice and snow sports have 36 hours as a course, the ice skating class accounted for nearly 30 hours, ski teaching time is little, the reason is that the school does not have the skiing teaching venues, equipment condition.

1.4 *Ski resorts have site, equipment condition*

In our case, our Jilin city has the national top ski Beidahu Ski Resort—Jilin City, away from the urban district only 53 kilometers, covers an area of 30 hectares, there are alpine skiing, cross-country skiing, ski jumping, freestyle skiing, modern and two sled, sledge, snow on the project site, ski shoes, plate, poles and other appliances equipment. Goods are available in all varieties.

1.5 *Analysis of adverse factors*

1. *The curriculum and time allocation is not appropriate. The students must stay at a ski resort in whole winter, which bring unnecessary trouble to the curriculum and school curriculum arrangement; make other courses to be completed with difficultly.*
2. *Allocation of the economic interests of school, enterprise and the student is the main contradiction.*

1.6 *Conclusion*

To sum up, the rapid development of mass skiing is not coordinated with the serious shortage of professional ski guide personnel, market demand for professional talents ski instruction. As the ski resorts of enterprise, there are the talent training ski equipment, appliances and other hardware; there are school guidance and management of social sports professional students which is the basic talent group, the school only joint ski enterprise can they integrate the advantage resources, overcome the disadvantage factors, collaboration, community sports training professional talents training mode of ski instruction, it is the effective talent suitable. The market not only demands as the guidance and explores new ways to cultivate talents, but also broadens the professional field of employment, improves the employment situation and promotes the sustainable development of social sports specialty.

2 THE IMPLEMENTATION OF TRAINING

2.1 *Take the social demand as the guidance, training objectives*

Taking the social demand as the guidance, training with the ski professional knowledge and skills of application of Mass Skiing outstanding leadership,

Service on the continuous development of social masses sports.

2.2 *Take the training goal as the basis, determine the training*

2.2.1 *Character, occupation accomplishment cultivation*

Basic composition:

First love: love, love of national, social, dedication, love each and every one of our guidance.

Second, the sense of social responsibility: social sports professional students must have a high sense of social responsibility, in order to achieve its culture value.

Second, the sense of social responsibility: social sports professional students must have a high sense of social responsibility, in order to achieve its culture value.

Beauty: good moral character, healthy body and mind, healthy body, forming a model effect.

The etiquette: etiquette, courtesy, master the basic etiquette knowledge, based on respect, harmonious communication.

Brave, tenacious, unity, passion.

Implementation way.

Schools: teaching seminars.

Enterprise: training.

2.2.2 *Professional and technical training*

Basic composition:

Professional theory: Sports Anatomy, exercise physiology, sports psychology, sports nutrition, sports fitness principle and method, skiing theory.

Exercise: skiing.

Implementation way:

School: Based on the social sports specialized curriculum, "teaching, learning, practicing" the combination of professional theory knowledge, learning to ski; learning skiing technique.

Enterprise: skiing, skiing technology learning practice, improve the technical level.

2.2.3 Cultivation of ability to guide

Basic composition:

The ability of analyzing and solving problems, communication ability; communication; cooperation, management ability; "guidance, teaching, band, collar" and other professional ability.

Implementation way:

Schools: teaching, learning, exercise as a "dual" system, both learning and training skills, exercising in learning and learning in the exercise through the "role conversion, guidance, organization, evaluation, innovation, practice, practice" and other series of form, Enterprise: managers, practitioners, ski training teaching; practical skiing, improve the scientific guidance ability.

REFERENCES

[1] Wang Baoheng. Present situation and development of China's popular ski instructors. Journal of Shenyang Sport University, [J]. 2005-12-30.

[2] Jia Yan. Analysis and Countermeasures of social sports specialty to the social demand. Journal of Northeast Dianli University, [J]. 2008-05.

Sports Engineering and Computer Science – Luo (Ed.)
© *2015 Taylor & Francis Group, London, ISBN 978-1-138-02650-6*

Empirical analysis of quality management assessment system of physical education in college and university

Bei Ren

Chongqing Technology and Business University, Chongqing, China

ABSTRACT: From the five factors of the teaching attitude, teaching contents, teaching methods, teaching ability and teaching effect, the paper shows to do as follows: a) set up sports teaching management evaluation system; b) design the corresponding questionnaire; c) collect and organize the documents needed for the empirical analysis through the questionnaire survey. Then, with quantitative empirical analysis using Principal Component Analysis (PCA) establishes a comprehensive evaluation model of physical education teaching quality management. Finally, it is combined with the comprehensive evaluation model policy suggestions of optimization of teaching quality management are put forward.

1 INTRODUCTION

Entering the 21st century, with the accelerating of the globalization process and the college enrollment expansion, there are more and more problems about the public physical education curriculum in colleges and universities. It is the current urgent problem to solve to improve the teaching quality of institutions of higher learning. Public physical education curriculum is an important part of the higher school undergraduate teaching. It is the inevitable requirement of higher school teaching reform and quality engineering to establish and perfect the public physical education curriculum teaching quality evaluation index system. For this reason, many academics begin to study the ordinary university public physical education curriculum teaching quality evaluation from different angles; Different index system is put forward in order to provide objective evaluation standard and effectively improve the quality of physical education teaching.

The traditional sports teaching quality in China is usually the main inspection and regular inspection on evaluation method. This evaluation method obviously is a single, cannot be the reasonable evaluation of undergraduate teaching quality. Now there have been many scholars to study the ordinary university public physical education curriculum teaching quality. For example, Hongmei Xu[1] thought that the current college physical education teaching mode has not adapted to the needs of the development of society changes. Besides, it is imminent to find out the ways to improve the quality of physical education teaching. Zijian Zhao[2] introduced the necessity and

feasibility of college physical education teaching quality management system from IS0900, and put forward to the steps and methods of the teaching quality management. Liming Xie[3] put forward the main body of teaching quality monitoring system framework and main operation way. It comes from the teaching quality monitoring mechanism of the connotation, function, theoretical basis and guarantee for colleges and universities sports teaching quality management.

However, most scholars' studies of colleges and universities sports teaching quality comprehensive evaluation are in qualitative analysis, quantitative empirical research is relatively less. In this paper, it takes the school of public physical education as research object. Besides, it collects data from the questionnaire survey and studies the main factors affecting colleges and universities sports teaching quality management with principal component analysis. The principal component factor is obtained from the empirical analysis, as it establishes the teaching evaluation model of quality management. With the factor, it puts forward policy Suggestions to improve physical education teaching quality management according to the model.

2 BUILDING OF QUALITY MANAGEMENT ASSESSMENT SYSTEM

During the physical education teaching process in colleges and universities, students are the main body. Teachers play a leading role. The students' learning method and effects are influenced by the teacher, and teaching attitude, teaching contents, teaching methods, teaching ability and teaching effect. These factors will directly affect the quality of teaching. So this paper mainly designs questionnaire from five factors: a) teaching attitude; b) teaching contents; c) teaching methods; d) teaching ability; e) and teaching effect. Meanwhile, it collects the evaluation data and establishes the teaching quality management assessment model with principal component analysis. Finally, combined with the comprehensive evaluation model, the policy Suggestions of optimizing teaching quality management is put forward.

2.1 *Build quality management assessment system*

During the comprehensive evaluation of public physical education in colleges and universities teaching quality management, first of all, the elements of evaluation must be made sure. The elements are many factors affecting the quality of physical education teaching. Then some corresponding evaluation content should be scientifically determined. In this paper, it puts forward some factors that affect physical education teaching quality management. The paper is based on the current situation of physical education teachers teaching quality and the related literature. With these factors, there builds a physical education class teaching quality evaluation index system of management, as shown in Table 1[1,3,4,5].

2.2 *Collection and processing of data*

The author designed the questionnaire for the above index system and selected the sophomore

Table 1. Quality management assessment system.

Teaching attitude	X1: Well-prepared, not to be late and not absent
	X2: Attention to extracurricular sports training
	X3: Communication between teachers and students
Teaching contents	X4: The science of sports training
	X5: The practicability of sports training
	X6: Attention to the cultivation of students' ability
Teaching methods	X7: For advice and improve the teaching
	X8: Instructional teaching
	X9: Heuristic teaching
Teaching ability	X10: Ability to teach
	X11: Health education ability
	X12: Choreographing capability
Teaching effect	X13: Attendance, discipline
	X14: Skills to master
	X15: Student feedback

of the XX University in China as the research subject. As a result, it can ensure the reliability of the questionnaire survey result. The questionnaire issue should be randomness, and questionnaires are anonymous. In the questionnaire survey there are a total of 500 questionnaires, and the recovery rate is 100%. According to the questionnaire rated conditions, there is an obvious question out of 20 questionnaires. So 480 questionnaires are valid, the efficient rate is 96%. Apparently, the recovery rate and efficient rate as you can see. The questionnaire survey from the original data is reliable, and it can carry out an empirical analysis of the physical education teaching quality management.

3 EMPIRICAL ANALYSIS OF TEACHING QUALITY MANAGEMENT

There is a principal component analysis with the above indices 15 factors of teaching quality management. The steps can do as follows: a) find out the most important factors influencing the physical education class teaching quality management (main components); b) then reorganize the data and sets up the data files; c) finally, establish the appraisal model of physical education teaching quality management with the principal component score data.

There are KMO and Bartlett's test results in Table 2. As well known KMO, the more close to 1 is more suitable for factor analysis. From the table, the KMO value is 0.809, which shows that it is more suitable for factor analysis. Bartlett test results show that the Sig value is of 0.001, less than 0.05 significant level. The results also show that there is relationship between variables. It verifies again that it is suitable for factor analysis.

The contribution rate of factor is presented in Table 3. Characteristic values of the table on the left side of the part is initial intermediate to extract the main factors as a result, on the right side of the rotation is the main factor of the result. Results show that only the first three factors of eigen values greater than 1. The sum of the eigen values of the first three factors accounts

Table 2. KMO and Bartlett's test.

Kaiser-Meyer-Olkin measure of sampling adequacy.	0.809
Bartlett's test of sphericity	
Approx. Chi-Square	198.407
df	105
Sig.	0.001

Table 3. Total variance explained.

	Initial eigen values			Extraction sums of squared loadings			Rotation sums of squared loadings		
	Total	% of variance	Cumulative %	Total	% of variance	Cumulative %	Total	% of variance	Cumulative %
1	5.531	60.107	60.107	5.531	60.107	60.107	4.391	49.374	49.374
2	2.017	18.453	78.560	2.017	18.453	78.560	3.207	26.570	75.944
3	1.270	9.341	87.901	1.270	9.341	87.901	1.975	11.957	87.901
4	0.901	3.644	91.545						
5	0.713	2.743	94.288						
6	0.574	1.410	94.698						
7	0.402	1.283	95.981						
8	0.297	1.039	97.020						
9	0.256	0.872	97.892						
10	0.193	0.611	98.503						
11	0.167	0.412	98.915						
12	0.137	0.353	99.268						
13	0.089	0.304	99.572						
14	0.067	0.257	99.829						
15	0.051	0.171	100.000						

Extraction method: Principal Component Analysis.

Table 4. Rotated component matrix[a].

	Component			
	1	2	3	4
X1	0.327	0.865	0.227	0.327
X2	−0.315	−0.797	0.261	−0.315
X3	0.208	−0.251	−0.165	0.208
X4	0.857	0.335	0.166	0.857
X5	−0.795	0.104	−0.230	−0.795
X6	0.303	0.038	0.114	0.303
X7	−0.818	0.122	0.154	−0.818
X8	0.296	−0.260	−0.283	0.296
X9	0.886	−0.097	0.226	0.886
X10	−0.337	0.217	0.873	−0.337
X11	0.229	0.238	0.097	0.229
X12	0.265	0.155	0.799	0.265
X13	−0.167	−0.076	−0.068	−0.167
X14	−0.115	0.109	0.106	−0.115
X15	0.163	−0.107	0.099	0.163

[a]Rotation converged in 6 iterations.

Table 5. Component score coefficient matrix.

	Component			
	1	2	3	4
X1	0.144	0.297	0.099	0.144
X2	−0.137	−0.266	0.112	−0.137
X3	0.098	−0.108	−0.083	0.098
X4	0.331	0.242	0.083	0.331
X5	−0.298	0.139	−0.101	−0.298
X6	0.127	0.002	0.063	0.127
X7	−0.315	0.067	0.071	−0.315
X8	0.133	−0.112	−0.121	0.133
X9	0.367	−0.038	0.096	0.367
X10	−0.150	0.094	0.332	−0.150
X11	0.100	0.102	0.007	0.100
X12	0.113	0.071	0.294	0.113
X13	−0.085	−0.005	−0.003	−0.085
X14	−0.063	0.058	0.055	−0.063
X15	0.079	−0.057	0.002	0.079

[a]Rotation converged in 6 iterations.

for 87.901% of the total characteristic value. Therefore, the three factors can be extracted as the main factor.

From Table 4, you can see that the principal component 1 is the most relevant X4, X5, X7, X9, such as the principal component 1 for "teaching content and method". In the mean time, weight coefficients are shown in Table 4. Principal component 2 is the most relevant X1, X2, such as

principal component 2 for "teaching attitude". Meanwhile weight coefficients are shown in Table 4. Principal component 3 is most relevant X10, X12, such as the main component 3 for "teaching ability". As we can see, weight coefficient is shown in Table 4.

The component score coefficient matrix calculated is given in Table 5 with the principal component factor score of data. As a result, it reflects the various survey data on the different factor score. Thus available factor score function is:

$$\begin{cases} FAC1 = 0.144X1 - 0.137X2 + \cdots + 0.079X15 \\ FAC2 = 0.297X1 - 0.266X2 - \cdots - 0.057X15 \\ FAC3 = 0.099X1 + 0.112X2 - \cdots + 0.002X15 \end{cases}$$

By the variance contribution rates of the principal component for the weight, the colleges and universities sports teaching quality comprehensive evaluation model is as shown below:

$$F = 0.49374FAC1 + 0.26570FAC2 + 011957FAC3$$

With the comprehensive evaluation model and the questionnaire survey data the sports teaching quality management of some specific university colleges and universities can be in the comprehensive evaluation.

4 POLICY SUGGESTIONS

Through the above evaluation result, it shows that physical education teaching contents and methods occupy a dominant factor in improving the quality management model (49.4%). The physical education teaching ability takes a second place (26.6%). In view of this, in order to better optimize the colleges and universities sports teaching quality management, this paper puts forward the policy Suggestions:

Firstly, optimize the sport teaching content. Teaching content is the carrier of curriculum, through the sports teaching to realize sports participation, sports skills, physical health and mental health, and so on. Therefore, when determining the physical education curriculum content, it must keep fit combined with culture and keep health first as the basic starting point. Besides, it also attaches importance to sports culture content of the course content. Combined with effectiveness, curriculum content must be rich and colorful. The content can promote the comprehensive development of students' physical and mental health and provide larger choice space for students. Combined with scientific and acceptability, it must follow the laws of the physical and mental development of college students and interests. Besides, it should also reflect the new progress and new results of the sports discipline. Combined with nationality and cosmopolitan, it must carry forward the traditional Chinese national sport and absorb excellent sports culture in the world[6,7].

Secondly, optimize physical education teaching methods. Sports teaching method is that in the process of sports teaching, teachers guide students to learn sports teaching contents in order to achieve the teaching purpose. It consists of a set of physical education teaching mode of operation strategy.

The optimization and the combination of the physical education teaching sports skill must help students to learn and promote the development of their personality, ability and creativity. In the cognitive stages of motor skill, it must select the teaching methods. In addition, the methods are given priority to with direct perception of physical education, such as the demonstration method, demonstration method and protection. Not only it can efficiently improve the quality of physical education teaching and the learning efficiency of most of the students, but also it can cultivate the students' sports ability and self-confidence, arouse their interest in learning and cultivate the students care about each other.

In the automation stages of motor skill, it must choose mainly explore sexual activity of sports teaching methods, such as discovery method and explore problem method. The activity is more advantageous to the student ability raise. With this method, it can make students to train the ability of independent analysis and to solve problem. At the same time, it can also make students to stimulate students' intrinsic motivation[6,8,9].

Finally, optimize teaching ability. Physical education teaching ability refers to the physical education teachers in teaching sports knowledge, technology, and ability of skills. The skills are action demonstration ability, password, preventive and corrective action, sports training and competition ability of organization and so on. This is the ability as a sports teacher which should be the basic skills. The sports teaching is the operation type technology teaching. The students practice body by imitating the teacher's action, which is an important medium of teachers' teaching and students learning. Thus the teachers' action demonstration must be correct, standard and skilled. Password is the medium of teaching in physical education teaching. It is the basic skill in physical education teachers, which requires: a) long and short sound must be difference; b) light is important; c) and the sound must be strong and strong rhythm. On the other hand, teachers should update their knowledge and improve teaching methods. Knowledge of teachers and students should be the relationship for "a pail of water" and "a glass of water". The pail of water should not be "dead water", and should be a long stream of "new water". Thus the teachers must pay attention to updating the knowledge, improving teaching methods and improving the ability of the business[7,10].

REFERENCES

[1] Hongmei Xu. Quality factor analysis and counter-measure research of physical education in the ordinary university, Sports World, 2011, Vol 6, P81–82.

[2] Zijian Zhao, Xiuli Zhang. Research of quality assessment system of physical education with ISO900, 2004, Vol 9, P1238–1240.

[3] Liming Xie, Analysis on Management of Teaching of Physical Education in General Higher Institution, Journal of Capital College of Physical Education, 2003, Vol 15, P64–66.

[4] Huaping Yao, Changsong Han. On College Physical Education Teaching Management. 2013 3rd International Conference on Education and Education Management, 2013, P551–556.

[5] Jianbo Dai, A study of the characteristics and methods of the management system of college P.E. teaching, Journal of Wuhan Institute of Physical Education, 2004, Vol 38, P163–164.

[6] Liu Changjiang, Liu Zhiyuan. Study on Teaching Methods of Physical Education Practice Course. Proceedings of 2012 International Conference on Education Reform and Management Innovation, 2012, P423–428.

[7] Lei Wang. Analysis of the Educational Practice of Sports Students in Henan Univer- sities. Proceedings of 2012 International Conference on Education Reform and Management Innovation, 2012, P124–130.

[8] Liu Hewang, Zhu Liming, Liu Mingchan—g. Physical Education Supervision's Influence Factors' Causes and Responses in Universities—A Case in Wuhan. Proceedings of 2012 2nd International Conference on Applied Social Science, 2012, P234–243.

[9] Runping Wang, Qiuping Jiang. Selection research of optimization method of sports teaching, Health Vocational Education, 2009, Vol 27, P31–33.

[10] Huifang HU. Research and Countermeasures on P.E. Students' Teaching Competence in Sports Universities, 2011 International Conference on Education Science and Management Engineering, 2011, P1665–1669.

Sports Engineering and Computer Science – Luo (Ed.)
© 2015 Taylor & Francis Group, London, ISBN 978-1-138-02650-6

Study on the pressure and influence factors of P.E. teachers in Jiangxi Province under the background of curriculum reform

WenTao Zhang
P.E. College, Jiangxi Normal University, Nanchang, China

Xin Lu
Mathematics and Computer Science Department, Nanchang Normal University, Nanchang, China

ABSTRACT: Using the method of literature, questionnaire survey, and mathematical statistics, the survey of 182 middle school teachers who participate in the 2011–2013 "national training plan" discuss the stressful condition of P.E. teachers, which are under the curriculum reform background. The results show that the P.E. teachers in middle school suffer a great pressure in Jiangxi province. Besides, genders, grades, educational backgrounds, the time to implement reform and other reasons had no significant effect. However, ages, years of teaching, academic title of a technical posts and other factors had a significant effect in different degree and level of P.E. teachers in the curriculum reform of pressure.

1 INTRODUCTION

The pressure of teachers has been a focus of many researchers, especially the development of global education idea renewal in recent years, the rise of education reform, and research on teacher stress are more important. However, the P.E. teachers as a special group of teachers suffer a large amount of pressure, which has not received enough attention during the process of curriculum reform. In order to promote the implementation of the new physical education curriculum reform smoothly and to promote the healthy development of physical education teachers and teaching quality, this paper conducts the research to the factors of pressure strength and influence of P.E. teachers in Jiangxi Province, so that the pressure of middle school physical education teachers in the process of curriculum reform can be discussed.

2 OBJECT AND METHOD OF STUDY

2.1 *The research object*

182 middle school P.E. teachers in Jiangxi Province are chosen as the investigation object, who participate in the "National Training Plan" in Jiangxi Province.

2.2 *Research methods*

2.2.1 *The method of literature*
By consulting literature, understanding and grasping the sports teachers' stress of influencing factors

of domestic international trends and the research situation, it provides a theoretical reference for the research of this paper.

2.2.2 *Questionnaire survey*
According to the orientation of this paper, the questionnaire is designed about the curriculum reform in the middle school's P.E. teachers who are under pressure in Jiangxi province. The questionnaire has high reliability and validity after testing, and it can detect the pressure of P.E. teachers in Jiangxi province effectively. A total of 190 questionnaires were issued, 186 were recovered, 182 were effective questionnaire, and the efficiency was 98%.

2.2.3 *Mathematical statistics*
SPSS16.0 statistical software was used for the collection, collation of data for statistical processing and analysis, and the influence factors of intensity of the pressure with the middle school physical education teachers in curriculum reform background, which provides data support for the study.

3 THE ANALYSIS AND RESULTS

3.1 *The analysis of middle school P.E. teachers' pressure strength in Jiangxi province*

From Table 1, we can deduce that 83.6% middle school P.E. teachers feel a little pressure and a lot of pressure in the process of curriculum reform, 8% middle school P.E. teachers felt great pressure in curriculum reform. In addition, the total pressure is 3.4, it is between a little pressure and a lot of pressure.

Generally speaking, this shows that P.E. teachers suffer a lot of pressure under the curriculum reform.

3.2 The factor of influencing the middle school P.E. teachers' pressure who are under the background of curriculum reform

3.2.1 The table of gender impacting factor for P.E. teachers in the curriculum reform

From Table 2, we can deduce that the independent sample T test, there was no significant difference of scores between male and female. Therefore, the pressure of curriculum reform of middle school physical education teachers' gender factors has no significant effect.

3.2.2 The influence of grade for P.E. teachers who are under the pressure of curriculum reform

From Table 3 we can deduce that with the analysis of single factor, the score between different grades

Table 1. Middle school P.E. teachers' pressure strength in Jiangxi province.

	Number of people	The proportion (%)	Average
No pressure	2	1	
Smaller pressure	13	7	
A little pressure	70	38.5	
A lot of pressure	82	45.1	
Great pressure	15	8	
A total of	182	100	3.4 ± 0.671

Table 2. The test of T in the gender factor independent sample.

Gender	Number of people	Average	Standard deviation	Values of t	Values of P
Male	136	3.32	0.812	0.887	0.236
Female	46	3.28	0.8423		

has no significant differences. Therefore, the pressures of P.E. teachers in middle school which are in different grades have no significant effect during the curriculum reform.

3.2.3 The influence of age factors for P.E. teachers who are under the pressure during the curriculum reform

From Table 4 we can deduce that the analysis of multiple factors in different age, the middle school P.E. teachers' age between ≤30 years old (M = 3.71) and 31–40 years (M = 3.31) has significant difference (P < 0.01); age between ≤30 years old (M = 3.71) and 41 to 50 years (M = 3.17) has significant difference (P < 0.01); there is no significant difference in others age.

The middle school P.E. teachers aged 30 or less feel more stressful due to the time they have the job is too short, they may lack experiences in their work. However, the idea, the way and the method of teaching have a significant difference after the curriculum reform having been carried out, and puts forward some requirements in high level. So, the P.E. teachers in these ages are hard to adjust and enhance the professional attainment of itself, and they feel stressful.

3.2.4 The teaching ages influence P.E. teachers' pressure after curriculum reforming

From Table 5 we can deduce that we have reported the multiple factors in difference teaching ages. There is a significant difference (P < 0.01) between the middle school teachers' teaching ages about 4–6 years (M = 3.87) and 7–25 years (M = 3.39). There is a significant difference (P < 0.01) between the ages about 4–6 years (M = 3.87) and 26–33 years, and there is no significant differences in others age.

Before the curriculum reform, the middle school teachers might suffer a high stress in the teaching ages of 4–5 years, because they had formed their own style of teaching gradually. However, the new ideas, ways and methods of teaching appear after curriculum reforming. So, they have to focus their attention and spend lots of time to study and practice the way of new curriculum. And their stressful are outstanding. In the teaching age of 7–25

Table 3. The analysis of variance of different grade.

The grade	Number of people	Average	Standard deviation	Values of F	Values of P
The junior school	71	3.52	0.789	1.854	0.197
The high school	64	3.27	0.836		
The part-time job in junior and high school	27	3.23	0.950		

years, the way of teachers' develop will become different: there are many teachers who have strong willpower to reform, they are brave to find the new challenges. However, another part of teachers start to doubt themselves and the continual reform, and they think it is vapidity to have the teaching life year after year. So the risk will appear in their work. The pressure in teaching ages of 26–33 years is low, and these two ages of teaching support that the new curriculum reform mean a lot to P.E. teachers, it is not only a kind of challenge, but also a value of one's work, and it can be a rare chance to wide oneself.

Table 4. The analysis of multiple factors in age.

	≤30 years old	31–40 years old	41–50 years old
31–40 years old	0.42**		
41–45 years old	0.53**	0.08	
>50 years old	0.39	−0.09	−0.15

Note: **P < 0.01, similarly hereinafter.

Table 5. The analysis of multiple factors in teaching ages.

	≤3 years	4–6 years	7–25 years	26–30 years
4–5 years	−0.19			
7–25 years	0.39	0.53**		
26–30 years	0.43	0.67**	0.12	
>30 years	0.22	0.44	−0.15	−0.22

Note: *P < 0.05, similarly hereinafter.

Table 6. The analysis of multiple comparison in different title of a technical post.

	Trainee teachers	Level two	Level one
Level two	−0.27		
Level one	0.63**	0.91**	
Senior	0.42	0.67**	−0.24**

3.2.5 *The title of a technical post that make P.E. teachers feel pressure in the curriculum reform*

From Table 6, we can deduce that the analysis includes the multiple comparisons in different title of a technical post. In the area of pressure, the trainee teachers and level one teachers have significant difference in middle teachers, and trainee teachers advanced than level one teachers, the level one teachers, level two teachers and senior teachers have significant difference, and level two teachers advanced than others; level one teachers and senior teachers have significant difference, and level one teachers are advanced than senior teachers.

There are two aspects that can contribute to the disparity of the trainee teachers in the reform of curriculum: First, for those trainee teachers, most of the faculties and schools of the P.E. physical education pay enough attention to the current reform of the physical curriculum, so that those P.E. teachers who have just graduated don't really understand the current situation of the P.E. curriculum reform. Second, for those assistant P.E. teachers who have just entered the job position, because they are lack of teaching experience, they just have so many pressures in the stage of the trying process of the new curriculum. With the rise of the middle school teacher's title of a technical of pressure, I think it's because a higher title of a technical post can represent a better P.E. teaching abilities and a more practical teaching experience.

3.2.6 *The influence of P.E. teachers' pressure in education background during the curriculum reform*

From Table 7, we can deduce that there is no significant difference in different education background after the single factor analysis of variance. Therefore, there is no significant influence in P.E. teachers who have different education background during the curriculum reform.

3.2.7 *The influence of reforming time in school in P.E. teachers during the curriculum reform*

From Table 8, we can deduce that there is no significant difference in different time of curriculum reform scores. Therefore, there is no significant difference in P.E. teachers in different time of curriculum reform.

Table 7. Variance analysis of different education levels.

Education background	Number of people	Average	Standard deviation	Values of F	Values of P
Polytechnic school	6	3.18	0.875	2.081	0.102
Junior college	52	3.23	0.838		
Undergraduate course	120	3.51	0.862		
Postgraduate student	4	2.72	0.952		

Table 8. The table of curriculum reform factor's independent sample T test.

Time	Number of people	Average	Standard deviation	Values of t	Values of P
1–2 years	64	3.29	0.871	−0.37	0.687
>2 years	118	3.33	0.859		

4 CONCLUSIONS

1. *The middle school P.E. teachers in Jiangxi Province feel stressful during the background of curriculum reform.*
2. *In the factors causing P.E. teachers' stress, there are some factors of no significant influence during the curriculum reform, such as, the different age, the different grade they are teaching, the different education background they have.*
3. *There is a significant difference between the age ≤30 years old and 31–40 years old, 41–50 years old, and they feel stressful now. There is a significant difference between the ages ≤30 years old and >50 years old, and the ages who are ≤ 30 years old feel more stressful, besides, there is no significant difference in others age.*
4. *In the different teaching ages, we have done the comparison; there is a significant difference between teaching ages 4–6 years old and 7–25 years old. There is no significant difference in others age.*
5. *In the aspect of the title of a technical post, there is a significant difference between trainee teachers and level one teacher, and there is also a significant difference between level two teachers, level one teachers and senior teachers. There is a significant difference between level on teachers and senior teachers, and the pressure will be reduced with the upgrade of a technical post.*

REFERENCES

Ji Liuwang, Xiao Zan. The teaching method of high school new curriculum of physical education and health [M]. Beijing: Higher Education Press, 2005.

Tian Qiang occupation plight of physical education teachers in middle schools and self development [J]. Journal of physical education, 2006 (6): 136–138.

Yao Lixin. Teachers' pressure management [M]. Hangzhou: Zhejiang University press, 2005.

Sports Engineering and Computer Science – Luo (Ed.)
© *2015 Taylor & Francis Group, London, ISBN 978-1-138-02650-6*

Study on Chinese health preservation and university students' self-management of health—from the global perspective of health

Yan Zhang & ZhiHua Lin
Department of Physical Education, Wuhan University, Wuhan, Hubei, China

ABSTRACT: As issues concerning health have already transcended the traditional concept of health, studies on Chinese health preservation and university students' self-management of health have taken an important position from the global perspective of health. By means of consulting relative references, logical analysis and interviewing experts, it is found that the conception of physical fitness is in line with the traditional health preservation, in which the fusion of Confucianism, Buddhism, Taoism and Medicine provide a theoretical and technical basis for self-management system of health. Physical factors, psychological factors and social factors are perfectly integrated. Integrating traditional health preservation from the perspective of health can help university students to formulate their self-management system of health.

1 INTRODUCTION

Health plays an important role in the development of humanity. The development of China is best marked by the improvement of the physical quality of all Chinese people. Hu Jintao, President of P.R. China, once said: "health is the foundation for achieving comprehensive development, concerning the happiness of thousands of families." The improvement of people's awareness of health is the basis for improving people's physical quality. Only with healthy body and mindset can they build the harmonious society and enjoy happy life. Global health features great inclusiveness, and integration of different disciplines, covering social security, laws, economy and remote sensing. Human's physical activity, mental activity, personal networking and even the whole process of living are closely related with health. The traditional Chinese culture contains abundant and profound health preservation philosophy, with generated scientific theories and methods for health preservation in such underdeveloped conditions. These theories and methods have withstood the test of time. It is of great significance to understand the traditional wisdom of health preservation for people in modern times to manage their health. As a special group at the transmission from campus to society and with vibrant ideas, university students are able to understand and accept the traditional health preservation. It will have important and positive influence for their college life, life-long health and for society if given the traditional health preservation education, and establishing a self-management system of health.

2 "ONE HEALTH"

This concept focuses on global issues such as zoonotic diseases, healthy breeding and environment change, etc. It is devoted to advocating the harmonious co-existence of human beings, animals and environment. It was first proposed in 400 B.C by Hippocrates, the father of medicine, in his work *On Airs, Waters, Places*. It was believed that the public health largely depends on clean environment. One Health refers to a development strategy for cross-discipline collaboration and communication aiming at the health of human beings, animals and environment. It calls for bringing together human health care practitioners, veterinarians, and public and environmental health professionals to address the important problems such as the newly discovered epidemics and environment change. Up to now, many countries in the world have carried out this concept in different ways. After 2000, World Health Organization has been very proactive with topics of health developing beyond the traditional sphere of health, and into the agendas of the leaderships of many countries. The problem of global health has become a "non-traditional security threat", and is turning into more and more a connection between different countries and regions. Strengthening our cultivation of talents in global health will not only create a sound public opinion environment for China's peaceful rise, but also provide China's national security with a strategic protection. This subject is with strong communicating ability. It will make full use of Wuhan University's full range of disciplines, and its advantage of early-development in internationalization to cultivate versatile and international talents capable of theoretical research and policy assessment of global health, integration of international health resources, and global disease control.

3 RESEARCH METHOD

This research is conducted from the perspective of global health, involving hygienics, Traditional Chinese Medicine (TCM), philosophy, sociology, management and sports science. Comprehensive methods, including generalization and comparison of references and interviewing experts, are used to verify.

4 RESEARCH RESULTS

4.1 Cultural background and theoretical source of traditional health preservation

Traditional Chinese health preservation contains an extensive and profound ideological system. With the development of history, different ideas and cultures have been absorbed by or transformed into health preservation theories or methods, which, at the same time, evolved into a kind of ideology and culture. Every dynasty has seen many practitioners in this field and many thinkers have contributed to its development. Tao Te Ching, Zhuangzi, Liezi and Wenzi of Taosim, I Ching, Analects and Mencius, and even the scholar official class in the Song dynasty all established their theory on the basis of human body and mind. From Huang Di's Classic of Internal Medicine, the classic of TCM, to medicines in Qing dynasties, few medical books or doctors have treated disease without considering health preservation. Either Dao Zang, a classic of Taoism, or thoughts of famous Buddhist monks talked a lot about respiratory and physical exercises and mind training. The fusion of Confucianism, Taoism, Buddhism and medicine is the main content of Chinese health preservation culture. People's pursuit for and thinking of happy life gave birth to the traditional health preservation system: Taoism advocates to follow the rules of nature and to extend life with the harmony between human and nature; Confucianism focuses on society, upholding the belief of realizing values of life in society; Buddhism concerns itself with the nature of human beings, namely to get wisdom with one's harmonious mental state; medicine is based on organs and *jingluo* of human body, striving for the balance of yin and yang of

the body by adjustment. It is in this comprehensive culture that China's health preservation has evolved into a complete system with philosophical and scientific factors.

Health can never be preserved without giving consideration to the environment, to the society, and to human body and mind. First of all, the relation between human and nature is the most important aspect to maintain human life. Without nature, nothing can survive. This is how the survival of human beings is related to nature, as well as an issue concerning human beings' sustainable development and the precondition of personal health. Secondly, people who cannot get along well with the society in which they live are prone to suffer physical and mental diseases. Both the relation between human beings and nature, and people's social relationship have much to do with the physical harmony, mental harmony and the harmony between body and mind.

Harmony between human beings and nature comes to the first. Human should not give all their attention to their own interests or plunder the nature regardless of the rules of nature. Human can only achieve long-term and sustainable development with harmonious co-existence with nature and other species. Otherwise, insurmountable disasters like El Nino would fall upon human beings. In the nature, human beings and other species form a system of circulation, in which each depends on and constrains another. The extinction or mass reproduction of one species may disturb the system and lead to the doom of human beings. From the perspective of natural humanity, Taoism calls for regressing to nature so as to bring human life back to naturalness and achieve the harmony of life.

The second is the harmony within human society, namely, harmony among human beings. From the angle of a social man, Confucianism hopes to build a harmonious world by means of *Ren*—humaneness. With one common rite, men can live harmoniously together to bring the best benefit to the society. Seemingly a constraint, rite is actually good for the maximum mutual freedom in co-existence. Without such constraint, individual's freedom is a real myth, for that the maximum freedom of one may turn out to be a great barrier for others'. For a community, only with common rules of proper behaviors can everyone enjoy the best freedom. Similarly, only with a harmonious social relationship can each individual develop healthily. Confucianism addresses the problem of harmonious personal relationship in the society.

Then is the internal harmony of a man. Any conflicts with the outside world would affect the internal harmony of a man and the quality of his life. Working on human nature, Buddhism starts with transforming men's cognition, the basis of health system, is to observe one's mind and get rid of obsession. It finally achieves the internal serenity and harmony. Mind can make differences on matters, optimizing the whole system and leading to sound circulation. The difference between human and animals lies in consciousness, which means that men can consciously adjust the psychology and mind for a sound circulation and better life.

As for the mutual relation between body and mind, Taoism and medicine hold the same belief. Based on beliefs such as "correspondence between man and the universe", and "unification of body and mind", with the guidance of *yin-yang* theory and theory of *Zang-fu*, they study and explain the rules of human origin and development. Thus, the principles guiding health preservation practices are determined.

4.2 *Understanding of health against the backdrop of global health*

As for global health, the nature of life should be explained not only by the mutual functions of the integral parts of a living thing, but also by the mutual relation between the living thing and the environment. Organisms are structured dynamically, who depends on the incessant changes of the integral parts. With hierarchy and the ability of self-adjustment, living things should be studied according to different hierarchy of the organism. Living things and nature form an integrated system, in which there are different hierarchies and parts whose characters, structures, functions and properties are determined by general structure and systematic laws. This notion is in line with our traditional understanding of health. Factors affecting health can be classified into different levels and aspects. The fusion of Confucianism, Taoism, Buddhism and medicine results in the premises to formulate a health system perfectly integrating physical, psychological and social factors.

Firstly, man is a part of the nature. As a species living in the nature, man should match the rhythm of nature, as well as its own *yin* and *yang* with that of the nature. Thus, health should be preserved in accordance with law of nature. "Man should adjust itself with the change of seasons so as to get adapted to summer and winter." "Spring and summer are good to grow *yang* while autumn and winter *yin*." This is interaction between living things and nature, which means that any changes of time and space would have an impact on human body. There is also a relation among different organs of man, and between man and the external environment. Man can stay in health when these relations are kept in balance. The periodical changes of climate, alternation of seasons, changes of days and nights, and geographical differences can all influence human body. Given the integration of human body, and the opposition and unification between man and nature, man should take different measures to preserve health in line with time, place and the condition of him. The relation between external environment and man, as well as the relation between regional disease and the general condition, should with careful analysis and consideration.

Secondly, man is a part of society. Living in the social network, man is inevitably influenced by the society. Personal relationship can result in different emotions, all of which have impact on human body. According to TCM, the seven emotions' resulting in diseases is a reflection of the interaction between *Zang* and the outside world, and a symbolization of the functions of the organs. Different changes of emotions can disturb the balance of *qi*, causing the malfunction of *jin-ye*. Too many emotions (joy, anger, worry, thought, grief, fear and surprise) would lead to the imbalance of *yin and yang*. "Anger damages liver, joy damages heart, worry damages lung and fear damages kidney." "Balance joy and anger, and enjoy the life." Man can only stay in fitness with calm and soothed emotions.

The body and mind form man as a whole, so do the systems. Systematically, the structure of a living thing is dynamic, which is preconditioned on the incessant changes of different parts. TCM believes that zang-fu is related with emotions, and that *zang* and *fu* are interacted with each other. The theory of yin-yang and five phases can be used to explain their mutual generation and mutual overcoming. The balance between yin and yang is crucial to keep healthy.

Health preservation is also a part of life. It includes components of life (including social environment, view of life, and personality, etc.), life style (daily activities, emotions, food and exercises, etc.). In this system, any oversight can cause failures of health maintenance and management. Only by "following the law of yin and yang, eating and drinking within a degree, rising and resting regularly, and being not too tired" can the fitness and spirit be kept in a healthy state.

4.3 *Impact of traditional health preservation on university students' self-management of health*

The fundamental goal of physical education in universities is to produce talents with all-round development of body and mind to meet the need of society. However, the current situation in universities is that university students, having come through a fiercely competitive examination, rarely reach the required physical standard, and are still declining in terms of physical quality. Due to the lack of previous education on health, unhealthy life style without self-management out of home and psychological problems of their ages, their physical and mental health worsens. Being physically matured yet psychologically immature, they tend to suffer loneliness of leaving home and home-sickness. Different from high school, university means a change of personal relations. The change of ways of study from being passive to being active, and the adjustment from one's own expected social role to the role expected by the society may cause psychological conflicts. With the rising of number of students come into worse studying conditions and less communication with teachers. Besides, they usually lack knowledge about disease prevention and control. All these problems haunt students day and night.

At the same time, they have enough knowledge to accept the traditional health preservation methods deeply rooted in traditional culture. The combination of traditional health preservation and their education of health is not only an important part of their quality-oriented education, but also an important method for them to develop comprehensive notion of health, as well as the basis for students to have a sound habit of self-management for all-around development and a premise for their success.

In traditional health preservation, there are plenty theories and method to adjust body and mind, including the establishment of the view of life, habits and daily life routine, and proper arrangement of food and drink. It is an entire system about health. It is of great significance for students' campus lives, as well as for their self-management of health in the future.

4.4 *Formulation of health management system*

In modern times, the definition of "being healthy" is more than being free form diseases. The World Health Organization (WHO) defined health in its broader sense in 1946 as "a state of complete physical, mental, and social well-being and not merely the absence of disease or infirmity." This is now a relatively complete and scientific definition about health. It is a multidimensional concept including 3 aspects, namely physiology, psychology and social adaptability, while social adaptability in the last analysis depends on the physical and psychological status. Good emotions can bring the best of physiology; otherwise, it would disturb or degrade some physical functions and cause sickness. The worsening of physical status may give birth to relevant psychological problems. Disability, diseases, and obstinate diseases in particular are usually the origin of worry, anxiety and depression, followed by some abnormal psychological status. Both as integral parts of a man, body and mind are closely related.

According to modern bio-psycho-social model, health is considered as a positive status with sound function. It is retained by meeting individual's biological, psychological and social need, which is by no means easy (WHO, 1948). This model holds the belief that both health and sickness result from the complex interaction of psychological and social factors. Macro-level processes and micro-level processes are equally important for being healthy or unhealthy. There is hierarchy among different levels of a living thing. Changing one of the levels will lead to changes in all other levels, which means that micro-level processes (like the change of cells) are embedded in the macro-level processes (like the change of social value). The change in micro-level has impact on the macro-level, and vice versa.

Man is both biological and social. Biologically, man should comply with the rules of life; while socially, man has great social adaptability. Any unhealthy factors, whether about life routine or social network, may damage health. To enjoy a happy life, one should take proactive measures, initiate self-defensive mechanism and build a self-management system of health on multiple levels.

The formation of this system comprises structure of life (views of life and spiritual quality, etc) and life style (life routine, food and drink, and exercises, etc).

Positive views of life and sound spiritual quality come first. Nowadays, due to grave pressure of employment, university students focus on their academic grades, neglecting the psychological health and social adaptability. Views of life refer to people's fundamental idea and attitude towards life, including value of life, goal of life, and meaning of life. With positive views of life and sound spiritual quality, students are more likely to deal with challenges in their lives with positive attitude. Traditional Chinese health preservation attaches great importance to the improvement of mind and morality, emphasizing adjustment of emotions and cultivation of sentiment. In this process, the most important is to keep a quiet and calm mind, with naturalness. Once given the birth, man has been occupied by various desires to improve the living standard and social status. This is necessary for social development and prone to distract men from paying attention to health, leading to disasters. Traditional Chinese health preservation reminds us to distinguish internal and external factors of life. Internal means physical fitness and life, while external means fame and wealth. Our chasing material wealth should be for the improvement of life, damaging life. Staying in naturalness does not mean being indifferent to everything, but to keep a quiet state of mind, restrain desires, keep in a proper degree and act carefully. Having desires but not greedy, progressive but without hurting himself, taking advantage of time and following the rules of nature, one can achieve the tranquil state of mind and stay in health. Those more likely to succeed do not have too many desires and pursue what is unrealistically high. This is what Laozi advocated to achieve anything with actions without intent. In traditional health preservation, there are many contents about training mind and cultivating sentiment in qigong of Taoism, and classics of Confucianism and Buddhism. They are our invaluable spiritual wealth.

Second is healthy life style. A considerable part of university students now know very little about what is a healthy life style, or cannot carry out even knowing it. Traditional Chinese health preservation believes that the right way is not detached from life, but in every daily activity. Health preservation actually is to choose a life style and keep balance. By paying attention to our daily activities, such as dinning and drinking, rise and rest and even every movement, one can be greatly benefited. Many experts of health preservation have reminded people to pay attention to minor things. It is not recommended to refrain yourself from doing something that you think is of little interests, nor to do something that you think is of little damage. Laozi once said "the tree which fills the arms grew from the tinniest sprout; the tower of nine stories rose from a heap of earth; the journey of a thousand li commenced with a single step." In his view, naturalness, action without intent and noninterference is the goal of one's life. Success grows from insignificant matters, so does failure. A trifle at the beginning can evolve into something matters. Thus, we should take measures before preventing disasters from happening. This principle serves very well as a principle for maintaining life and preventing diseases. When we neglect details in life and indulge ourselves in unhealthy lifestyles, we are approaching towards sickness. Such contents are plenty in traditional health preservation classics. Suwen, the first chapter of Huang Di's Classic of Internal Medicine, commences with talking about preservation health in accordance with different times of a day. All of it is about the relation between the routine of the day and health, and how to adjust oneself to match the change of nature. Traditional health preservation theory enlightens us on how to follow a healthy routine of the day.

The third is having a good command of certain ways to adjust emotion and getting into good habit of doing exercises. Due to the lack of health education in the colleges and universities, a certain number of students do not have proper ways to adjust emotion, neither are they used to doing exercises regularly. The traditional health preservation is rich in resources yet to be dug. The academic circles have made meaningful explorations both in Taoism and psychological modulation. Professor Lv Xichen, Yang Desen, et al. did applied research in the wisdom of Taoism psychological modulation, thinking that Taoism can relieve anxiety, depression, fickleness, self-abasement and so on in many ways, such as accepting and acknowledging the reality to relieve worry; complying with nature to ease the emotions; maximizing vacancy and guarding stillness with great vigor to calm down; being not affected by the outside world to hold the essence; remaining indifferent to favors and humiliations to widen the mind; accumulating richly and breaking forth vastly to accomplish great undertakings; benefiting all things to realize one's value; getting rid of the obsession on form to outstand; and grasping one's own understanding to keep a grand aspiration. Professor Yang Desen summarized the following philosophy from Laozi."If we could renounce our artful contrivances and discard our (scheming for) gain, there would be neither thieves nor robbers. Who is content Needs fear no shame. Who knows to stop Incurs no blame. From danger free Long live shall he. In harmony with other people and keep a moderate attitude. Conquering the unyielding with the yielding. Recover one's original simplicity and let nature take its course." By referring to the successful experience of foreign country and combining years of experience of mental treatment, he found the Taoism cognitive therapy. Via this therapy, with cognition and reaction transformed, plus meditation, shadowboxing, qigong and yoga, negative emotions can be adjusted, improper behaviors corrected and diseases prevented and cured. Chen Bing, a teacher, contends that Buddhism psychology not only have practical value on mental treatment, but also have enlightenment on scientific psychology and modern civilization. The mental adjustment of Buddhism remains to be exploited and integrated so as to be applied to college students' mental adjustment. Meanwhile, the favorable exercise habit is an important means to adjust emotion and keep healthy. In addition to the various sports program currently held in the colleges, the traditional health preservation abound with gymnastic breathing exercise balancing and integrating form, energy and spirit is quite effective to the healthy mind and body, such as the popularized gymnastic qigong recently.

5 CONCLUSION

1. The understanding of health from the perspective of global health is in accordance with that of the tradition. The theory of traditional health preservation contains profound wisdom, which requires being explained in modern language and annotated by science. It is of great

significance for the modern human beings to comprehend life and take an active part in the management of health.

2. The blending of Confucianism, Buddhism, Taoism and medicine in traditional health preservation provides theoretical and technological basis for the formation of self-management of health, which perfectly integrate physical, psychological and social factors.

3. Integrating traditional health preservation and constructing the college students' health self-management from the perspective of global Health is significant to their campus life and life-time health management as well as the influence on the society. The health self-management system includes: the life structure (social environment, view of life, character and so on; life style (daily routine, mood, diet, exercise and so on). In 21st century, China's global health strategy should focus on maintaining the national health, realizing the equally-shared health and medical service, guaranteeing the national economy and social benefits. They also take an important position in the political agenda around the world.

FIRST AUTHOR

Zhang Yan, Female, born in 1966 from Qingdao Shandong, Assistant Professor, Study Physical Education and Health Education.

REFERENCES

Chen Bing. Buddhist Psychology [M]. Guangzhou: Nanfang Daily Press, 2007.

Cheng Peng. Impact of Taoism on Current China's Sports Culture [J]. Journal of Hubei Sports Science, 2013 (6):480–481.

Global Health College of Tsinghua University [OL]. http://www.tsinghua.edu.cn/publish/jjh/6853/2011/20 110317171902334693373/2011031717190233469337 3_.html. http://baike.baidu.com/view/12048817.htm.

Lv Xichen, etc. Wisdom of Taoism in Healthy Psychology-Interactive Studies on Taoism and Western Psychological Treatment [M]. Beijing: Social Sciences Academic Press, 2008.

Zhu Xiongzhao, Yao Shuqiao, etc. Health Psychology (5th edition) [M]. Beijing: People's Medical Publishing House, 2006.

Zhang Zhuping. On the Popularization of *Shaolin Zen Music Ritual's* Operational Model and Impact on the Rejuvenation of Ethnic Sports Culture [J]. Journal of Beijing Sport University, 2010 (10):21–24.

Zhong Hui. Reconstruction of the Value of Sport From the Perspective of National Soft Power [J]. Sport Science Research, 2012(5):67–69.

Sports Engineering and Computer Science – Luo (Ed.)
© *2015 Taylor & Francis Group, London, ISBN 978-1-138-02650-6*

Things in sports training

Bao-Rong Chang
Department of Physical Education, Yan'an University, Yanan, Shanxi, China

Xue-Min Han
School of Applied Sciences, Hainan University, Danzhou, Hainan, China

ABSTRACT: Compared with the computer networking strengths and weaknesses, this paper discusses things in sports training application, which can provide some useful reference for researchers and practitioners who have concerned with IOT educational applications.

Through sports training based on fieldwork, it found that the current researchers detected at physiological indicators sports training, mainly depending on the researchers bringing the relevant instruments, to field testing, sampling data. In this way, spending a lot of manpower, material resources, we cannot accurately grasp the sampling time, and for some reason, it is difficult to keep abreast of emergency situations. The project to address the problem, the proposed construction of ZigBee-based wireless sensor network capable of physical training a continuous monitoring of the situation of things the system will be more scientific data for transferring directly to the laboratory; through the formation of ZigBee-based networking system right youth intended sports training environment and physiological status monitoring, researchers and coaches can provide scientific source for data analysis and decision support. This project not only provides the basis for the sports, sample teaching, monitoring of international competitions and support the volume and intensity control of sports training, it also can greatly promote the efficiency of Physical Education and sports competitions to achieve our initial experience in networking applications athlete management, sports equipment management, administration and other aspects of the sports industry.

Keywords: networking; sports training; application

1 THE MEANING OF THINGS

Things are important part of a new generation of information technology. Its English name is called "The Internet of things". As the name suggests, things are "connected to the Internet." Things are defined: by Radio Frequency Identification (RFID), infrared sensors, global positioning systems, laser scanners and other information sensing device, according to the agreed protocol, to any object connected to the Internet, information exchange and communication to achieve the objects of the intelligent identification, positioning, tracking, monitoring and management of a network. This has two meanings: first, the core and foundation of the Internet of Things is still, in the Internet based on the extension and expansion of the network; second, extended and expanded its client to any object and between objects, information exchange and communication. To sum up, the essence of things is mainly reflected in three aspects: first, the Internet features, namely the need for networked objects must be able to achieve interoperability of the Internet; second, identification and communication features that incorporate things "material" must to communicate with the automatic identification of objects and things (M2M) function; third, intelligent features, the network system should have automated, self- feedback control and intelligent features.

2 THINGS IN SPORTS TRAINING

Things in sports training application by sensing devices (radio frequency identification, infrared sensors, global positioning systems, laser scanners, etc.) and communication protocols sport factor (people, sports equipment, sports, sports, sports clothing a network sports environment, sports competitions, etc.) connected to the internal system to achieve sporting identify, locate, track, monitor and manage the exchange of information for the sport. Things in sports training application framework is a dynamic network, with the self-organizing capacity and public communication protocol standard, extended and expanded its client to factor between any sport, there is identity, virtual properties and intelligent interfaces, seamless integration and information networks.

2.1 The use of intelligent networking to build the training environment

Characteristics of the training environment directly affect the training performances. Training resources of the physical environment contain traditional training, which is limited, and the lack of a virtual learning environment to interact with the real world, and the integration of physical training environment and virtual training environment is also likely to stay at the surface. Through networking, it not only made real-world items watts connectivity, but also realized the real world (physical space) and virtual worlds (digital information space) interconnect, which can effectively support human-computer interaction, interaction of people and goods of the question, social interaction between people. The introduction of things of physical training has a digital environment, networked intelligent features, and virtual learning environments which can be integrated, especially sewing and timely capture, analyzing needs of teaching and coaching athletes, and accordingly, providing intelligent and resource training environment for coaches and athletes. For example, perceiving light sensor in the gym will always monitor a fluorescent UV light irradiation by light of this increased male hormones that enhance lung capacity, muscular load capacity enhancement, physical rapid recovery of UV light intensity control the stadium lights switch; athletes in the classroom using a computing device to read a local or remote call embedded sensor data object for the current training.

2.2 Things to better control the use of athletic training and competition

Changes in physiological and psychological state of master athletes in training and competition process are to achieve an important guarantee for a better level of training and competition results. Coaches should put wrist sensor on the athletes, which can automatically forecast temperature, blood pressure, heart rate and other data of athletes, when the data unforeseen circumstances, the data acquisition system will automatically monitor and alarm. Coach results through scientific data to provide for training and competition to ensure timely regulatory status of training and competition, to ensure the normal level of training or long play. In addition, data on physiological indicators of Things blood lactate, heart rate, hemoglobin, urine protein, serum testosterone and other monitoring exercise load conditions, to arrange a large

amount of exercise training to provide reliable scientific reference number. In this way, each athlete within their appropriate load limits were sufficiently large amount of exercise training to athletic performance is greatly improved.

2.3 Scientific training and competition facilities to improve athletic performance driven

Use sports networking technology, by implanting chips in training equipment, relying on the chip to monitor athletes' training, such as javelin athlete holding the gun position is incorrect, the height of the column of attack when hurdler enough angle shot during basketball player shooting low, diving into the water angle of these athletes by sports equipment and venues implanted chip to control, and then back to the chip via a communication protocol coaches or athletes. Feedback coaches or players according to an image mode may be, it can be voice, coaches or athletes training in a timely manner to adjust the co-operation of the feedback information according to the chip. While improving the level of training athletes, coaches workload and efficiency guidance will also be greatly improved, not only coaches to guide athletes training in the sports arena, but also to guide the athletes training at any place, so that it can guarantee good coach remote guidance of other athletes who train at the same time raise the level of reserves of the more outstanding reserve sports talents.

2.4 Things for mental training

High-level sports competition and psychological factors have become important weight balance of the game. Individual psychological characteristics analysis and diagnosis of athletes; using sports psychology knowledge to solve their psychological problems, eliminate their psychological barriers; using advanced psychological testing instruments, guiding psychological training in order to foster good psychological quality, stable in the game play its competitive level. By measuring the body temperature, ECG, EEG and other methods, or collect grasp consciousness variable athletes, physical variables, variable pay attention and focus, and then help athletes overcome psychological barriers through networking and the use of special methods to eliminate the fear before the race, stage fright and nervousness.

2.5 Use of things to play the role of monitoring instruments

Application referee, scoring the work is also very important. Now pay attention to a computer and other equipment misjudgment and projects have been resolved easily which are likely to cause controversy, such as boxing, gymnastics, taekwondo and some room problems. There is a football game electronic recorder that can record player shooting situations and locations, and timely reporting to the referee. Marathon runners installed on the soles with a widget, to prove athletes finishing the course from start to finish. Electronic sensing device set on the ground will receive signals emitted by athletes' foot, and so they can know the exact location and time. These applications are all based on science and technology, the development of modern computer technology produced.

2.6 Standardize the management of high-level athletes promote the overall quality of the athletes

Athletes management training and development is currently a difficult sport, because the success or failure affects the overall quality management of athletes, and even affects the athletes performance in domestic athletic arena. Therefore, the rational, efficient and scientific management of sports training and athletic teams have gained more and more individual attention. Things to supervise the use of sports athletes and management can not only improve work efficiency, but also can reduce the statistical work time; it is also able to quickly grasp, analyze and compare the level of training of athletes, changes in bodily functions, the results of increase in a timely manner summed up the experience, so it can be a complete athlete down to save training data, to provide a reference for future training and management, according to a variety of training data coaches and medical test results, combined with the current training purposes, the development of practical training and management programs. Some athletes drinking, fighting vices in sports, things global positioning technology can promptly grasp the whereabouts of athletes, sports teams to avoid the overall image detrimental of the event.

2.7 Things traceability for food quality and safety assurance and regulatory athlete nutrition

By integrating the use of electronic tags, bar code, sensor networks, mobile communications networks and computer networks to achieve quality and safety of agricultural products and food information seamlessly between different supply chains subject, the number of agricultural and food products to achieve logistics, to achieve the quality of agricultural products and foodstuffs tracking and tracing. With RFID as the main information carrier, relying on network communications, systems integration and database applications

and other technologies, the establishment of the "meat" Safety Information traceability platform, implemented by government regulation of breeding, slaughtering, marketing and other aspects of information technology in one platform ensure that each block meat "can be traced back to go to track, information can be stored, the responsibility to trace, product recall."

2.8 Things scientific selection for athletes

High-level athletes must start as soon as possible to impose specialized training, and it is possible to reach the world level. Things technology through sports body shape can be different in ages in different level of athletes training, and other indicators of physical fitness and physical function and exercise capacity for quantitative analysis, and evaluation of their athletic ability to predict its movement potential. Sports complex networking technology primarily through laser scanner detects indicators athlete body, and then through a certain protocol to the indicators back to the coaches, to provide an accurate basis for a reasonable selection of coaches to effectively improve further success rate of selection and reduce the athletes' attrition rate.

3 THINGS ADVANTAGES FOR SPORTS TRAINING

1. *Things can break the Internet, which is a virtual world. Its biggest drawback is that the user can only passively get some information, but not change the real world objects that storm, things can be achieved through the current network in the world exchange of information automation and remote control of mechanical equipment or related items between.*
2. *Some simple things operate data collection although advanced equipment, but no special operations personnel, coaches do not have to resort body Branch and researchers.*
3. *Networking involves multi-disciplinary knowledge, coaches do not have a subject expert for advice, training will be able to solve the problem.*
4. *The experimental training is important teaching tool to train athletes ability and creative thinking abilities, but the traditional experimental teaching has its limitations, for example, because of security issues or lack of experimental equipment, many experiments do not allow players hands. Things can provide a safe intervention for experimental teaching, sharing, intelligent experimental training environment, for example, each of the experimental equipment has helped with the use of digital property information, not*

the time when the experimental equipment will automatically activate the alarm system; experimenter can remotely control the remote into the IOT test equipment; experimental process data can be collected in real time and in an appropriate manner to the experimenter, digital, networked and intelligent experiment teaching.

4 THINGS INADEQUATE TRAINING FOR SPORTS

Although our networking technology has started, and gradually plays a role, the overall look of things sports technology research and development application is still in its infancy, low key products equipment and integrated system maturity, greater difficulty in large scale applications. In terms of information sensing, data collection for sports training, environmental monitoring sensor device types are not enough, full enough, the sensitivity is not high enough, the lack of domestically produced, low price and stable operation of the sensor, but also need to study further to small, precise, sensitive development. In the intelligent decision support, athletes scientific digital models generally do not establish that the lack of reference to the control computer analysis, the technical parameters of a single control system, integrated intelligent management level is not high enough. In the automation control equipment to support intelligent decision, the athletes cardiovascular, respiratory, muscular system automation equipment such as a remote control is not high enough, mainly sports-related monitoring things CNC equipment and technology development for a long time, invested more than, and it is difficult to obtain short-term economic benefits, which are expected to limit the promotion of speed.

5 CONCLUSION

Current research on things, mainly in logistics management, intelligent home and other areas to explore the field of sports training is still uncommon, but to build a digital sports training through advanced networking technologies, promoting sports team training to improve the level of education achieved management, training management, modern, fully meet the needs of the athletes in terms of learning, life, and sports training reform is undoubtedly a new attempt, hoping that this can provide some useful things for researchers and educational applications practitioners reference.

REFERENCES

[1] Yunhao Liu Things Introduction Science Press, 2010: Beijing.
[2] Sun Mingge, Zhu Xilin based software configuration SQL database technology under the king [J]. Micro Computer Information, 2006, 22 (3):109–111.
[3] U-Japan Policy [EB/OL]. <http:llwww.soumu.go.jp. 2009-10-22.
[4] Lijian Chen, ZHAO Guo-jie. Development of modern science and technology revolution and athletic sports [J]. Hebei Institute of Physical Education, 2007, 21 (2):12–14.
[5] Yang Zhiqiang, Xiao Qing China Mobile: all directions networking applications practice [J] Telecom World 2009 (11):40–42.
[6] American Sport Education Program Coaching Youth Basket-ball (Forth Edition) [M] Champaign, IL, USA:.. Human Kinetics, 2007.
[7] The Internet of Things [EB/OL]. <http:Nwww.itu. int, 2009–9-6. >
[8] Internet of Things in 2020: Roadmap for the future [EB/OL] <http://ec.europa.eu/. 2009.1 1–16. >

Sports Engineering and Computer Science – Luo (Ed.)
© *2015 Taylor & Francis Group, London, ISBN 978-1-138-02650-6*

Constructivism study of sports teaching from the perspective of ecological system theory

ChaoBing Yan
Physical Education College, Jiujiang University, Jiujiang, Jiangxi, China

ABSTRACT: In the face of sports teaching of information ecological imbalance problem, by using the literature material law, teaching practice law and logic analysis, based on the constructivism learning theory, this paper analyzes the structure and function of the sports information teaching ecological system. Besides, it also discusses the ecological situation. The construction of sports information teaching ecological system is to carry on from the following five aspects. Firstly, build the ecological extended information teaching system. Secondly, deepen the teaching ecological management. Thirdly, stimulate the students' initiative and creativity in teaching of information ecology. Fourthly, refine the learning situation design. The last one, strengthen the construction of teaching evaluation system. Sports' teaching of information ecology advocates situational teaching should be specific and real. However, this doesn't mean sports teaching of information ecology is against the abstraction and generalization. Therefore, the theory of situational teaching and learning in the sports teaching ecosystem is the direction of the development of physical education.

1 INTRODUCTION

Today, constructivism learning theory has become a hot theoretical application for domestic and foreign teachers. The theory is especially based on various teaching activities under the condition of modern education technology. The aim is to help establish a new method for the new teaching mode. According to the constructivist learning theory, about the creation of scenarios, no real teaching cooperation is not far from good, the conversation function is not strong, meaning construction is not obvious. All these are the problems of sports information teaching system at present. From the perspective of ecology, the perplexing relationships between teachers, students, administrators and information teaching system, the internal and external environment show all kinds of problems of the present sports teaching. That is the teaching ecological imbalance: imbalance between teachers, students and managers; the imbalance of the internal structure and teaching information system function; the imbalance between teaching subject and teaching of information system and the internal and external environment. Facing the problems of sports teaching of information ecosystem imbalance, through teaching practice, the structure and function of the ecological factors, this paper tries to analyse the informatization teaching ecosystem, evaluate the ecological situation and solve the problem of sports information teaching ecology. The situation of teaching ecological development is not good. To design and develop situation, collaboration, conversation and significance construction are the four big constructivist learning theories that attribute of sports teaching of information ecosystem.

2 ANALYSIS OF ECOLOGICAL SYSTEM OF SPORTS TEACHING

Ecological system of sports information teaching includes the sports information teaching system, teaching subject, teaching environment. Sports' teaching of information system is the core content and has its own unique structure and function.

2.1 *Structure of sports information teaching system*

According to the principle of ecology in Pyramid, the internal factors of sports information teaching system, from low to high, is divided into four levels of teaching, scene subsystem, answering subsystem, the examination and evaluation system, the teaching management subsystem.

Teaching management system includes teaching files, teaching management, and teaching platform access to information circular tracking. Examination and evaluation subsystem includes chapters self testing, self testing and self testing, the final stage of group collaborative learning self assessment, assessment of other people. Answering subsystem includes automatic question answering; answer the questions, video answering mail. Teaching scene subsystem includes courseware, virtual classroom, experiment, case upload and analysis.

Teaching system informatization in Pyramid shows the various subsystems and system status. System information teaching ecology in Pyramid shows more the more roles in guiding and restricting system, similar to the system of "nuclear". It provides teachers, students with the examination and evaluation of the target and the basis of teaching management subsystem in the syllabus. The teaching scene subsystem, the lowest, uses the subsystem most frequently when students are the most active.

2.2 Structure of the ecological system of sports information teaching

According to the teaching of information systems and all kinds of connections of information about inside and outside environment constitute the information education ecological system. Teaching subjects including teachers, students, administrators, school personnel. Environment refers to the information in the information environment of hardware and software, including the Internet, personal computer, operating system. Information environment refers to the multimedia courseware evaluation, excellent course construction project funded incentive policy.

The ecological system of sports information teaching has a perplexing relationship, the most basic contact structure "Teaching producer → The teaching of consumers" Link, such as "teaching video playback → video watching", "teaching materials → teaching materials to download or upload", "online homework → online assignment submission". A plurality of contact structure cross, basic and extended to form strip, tree, plexiform, mesh and other forms of "information technology to teach students state of chain". The general structure of it is "Information teaching resources → teaching of information production → teaching of information dissemination → teaching of information consumption → Teaching of information decomposition".

Teaching resources include the purchase of sports information from the computer and camera equipment. Information teaching production includes teaching information system development, design, information collection and case teaching. Teaching communication includes domain name and space Internet access, network purchase. Information teaching consumption includes video viewing, downloading and analysis, virtual classroom use case, operation, simulation experiment online homework and self testing, BBS use.

Information teaching decomposition includes teaching evaluation information management, network management, teaching and administrative staff management, since my account management.

To promote the development of ecological sports informatization courseware, the incentive policy is formed: "Courseware development → Teaching application → Courseware awards appraisal → Courseware development", "Construction of excellent courses → Teaching application → Excellent course awards appraisal → Curriculum construction".

"Sports teaching ecological node" is a variety of information, different levels and types, to teach students state of chain, belt, tree, cluster, network, ring joint formed by the interweaving of. All aspects of sports information teaching state of chain can be informative teaching ecological nodes, for example, sports teacher is teaching of information producers and may be informative teaching consumers. They are in the node information to teach students the ecological chain, becoming the growth point of science teaching. The growing point into the informative teaching information center is an important hub for maintenance information teaching environment. The teaching of each ecological chain organically unifies together.

Overall, the ecological system of sports information teaching consists of a plurality of informatization teaching students state of chain and information technology to teach students state ring to form a complex structure. In the exchange relationship of each link, in the chain through the material, information and energy, the linkage between the formations, reinforce each other, prosper together, decline. A biological growth is not in contact with the biological growth inhibition or promotion. Its biological effect is through the biological community space—"Biological field" interactions occur. The specific sports information teaching activities can also form a certain "information teaching ecological field". Here, teachers, students, management personnel are the main body. Through the information teaching system, both the inside and outside environment form and interact with each other, such as network the administrator through the management of teachers students, students through the management personnel supervisor. Information technology to teaching students state field refers to the unique one of educational informationization. It is distinguished from other information teaching space entity. It is the space carrier of information teaching hardware, information teaching software, teaching group and the information teaching system. Sports teaching has strong practicality and rich teaching resources to give information to the teaching ecology.

2.3 Functional analysis of teaching of information ecosystem

"Ecological niche" is the understanding of sports teaching of information ecosystem and information-based teaching ecological field function is extremely important concept. Niche is a species in a particular scale in a specific ecological environment of "function" status. It includes the environment "requirement" and "influence" and its two aspects of law. Niche includes the temporal niche, spatial niche, tropic niche. Multidimensional niche, can also be divided into resource niche and niche. Niche is the essence of organisms in the environment of resources. Niche requires physical teaching ecological construction to pay attention to the more advantageous position of sports teaching ecological system in teaching ecosystem. It also requires focusing on the development of information-based teaching by different factors in different functions. So the informationization teaching subsystem in the different functional positions have different division of labor, and coordinate and cooperate with each other.

It is required that physical teaching state of chain and each subsystem should be in own special position and play their respective roles. And the following goals should be achieved: first courses and after repair programmed are good. Each has its own unique features and mutual cooperation, such as interactive teaching and traditional teaching, teaching system, virtual classroom system simulation, online operation and test modules or subsystems. In the information of human body science based on experiment teaching, "design and development of information simulation experiment" will enrich the connotation of ecological niche. The informationization teaching experiment of "virtual" niche "become a real niche", for the traditional mode of experimental resources, and form a functional ecological chain complementary.

3 BASED ON THE CONSTRUCTIVISM LEARNING TEACHING OF INFORMATION ECOSYSTEM FROM THE PERSPECTIVE OF THE THEORY OF OPERATION OF THE GENERAL FORM

3.1 Based on the constructivism learning sports teaching mode and teaching methods under the guidance of the theory of

According to the school sports teaching rules of constructivist learning, to reflect the students' subject status at the same time, sports teaching mode fully meet the needs of students' Autonomous learning. In the whole process of physical education, PE teachers play a guiding, teaching, organization and incentive effect. He is teaching counselors. By creating a situation, collaboration, conversation and learning environment, the teacher improves the students' interest in learning, enhances the independent thinking and learning ability, develops their innovation ability. In such a model of physical education, the appearance of students' learning initiative can effectively broaden the students' thinking and learning. The teachers' organization of teaching has become more diverse.

In the constructivist teaching mode, exploring more mature teaching method are scaffolding instruction. Teaching ecological chain is: Scaffolding → Into the situation → Independent exploration → Collaborative learning → Effect evaluation; Anchored instruction, teaching ecological chain which is: Creating context → Defining the problem → Autonomous Learning → Collaborative learning → Effect evaluation; Random access instruction, teaching ecological chain is: Presents the basic situation → Random access instruction → The development of thinking training → The group collaborative learning → The assessment of learning effect.

3.2 The general form of sports teaching operation information ecosystem

About the ecological operation of sports information teaching, teachers, students and teaching management are the teaching subjects.

We can borrow "Ecological potential" concept in ecology to describe the status and trends of the survival and development of sports information teaching. Ecological potential refers to the potential, trend of niche. It indicates that the energy, biomass, individual number, biological resources units hare, ability to adapt to the environment, the influence and force, energy and material exchange rate and the level. Sports information teaching situation and trend are the informationization teaching factors in the information teaching. In the sports information to teach students the state field, characteristics of ecological potential depends on the factor of teaching factors (knowledge structure and sports such as teachers' knowledge and experience, students' experience, as well as the information teaching system modules or subsystems information reserves, the amount of information exchanged and rate, environmental influence and force etc.) and formation of teaching students to state field of environmental and resource level. Any teaching factor (such as students) of individuals always wants to have the fastest speed of information exchange. They share the maximum amount of information, along with the increase in factor density. Average individual use efficiency decreased. But if the density is very low, the comprehensive situation of mutual promotion between teachings cannot form factors, the factor of information exchange rate, the amount of information and the formation of ecological potential to become the largest. From the analysis of ecological niche, niche breadth of teaching factor on resource utilization ability, wide distribution, and other teaching factors, niche overlap between is the larger, and vice versa. The development of a variety of functional characteristics of the teaching module or subsystem has important significance on the formation of a network teaching system, which mutually promote comprehensive situation.

Sports teaching ecological benign condition should be unified balance and development. Teaching of information ecological balance refers to each subsystem and each factor and composition teaching existence and development, and a variety of internal and external environment information teaching ecological factors, keeping dynamic, harmonious proportions. Information teaching ecological status is reasonable. "Into the" balanced development, i.e... Teaching of information ecology is an important link of teaching innovation state information ecological best and final performance. A lot of information teaching form and teaching projects including network virtual classroom, video solutions, simulation experiments, constantly develop on the basis of the traditional, to meet the "need for collaboration, conversation", in line with the needs of the diversity of learning the increased diversity. The information-based teaching environment has a positive meaning for maintenance information teaching ecological balance and stability.

4 THE ECOLOGICAL SYSTEM OF SPORTS INFORMATION TEACHING OPERATION PROBLEMS AND SOLUTIONS

4.1 *On the problem of ecological system of sports information teaching operation*

Despite the need for independent learning, making sports teaching with other teaching form can't replace the existence value, reason for existence and development space. But in the actual teaching, to a certain extent, the sport teaching is the traditional teaching mode. Students' subject status and initiative are still not fully played, just "irrigation" as the "machine", "listen" to "the class", "everyone exchanges" as "the man-machine communication".

In the ecological system of sports information teaching in the presence of informationization teaching resources, inadequate investment lack creativity. The producer consumer participation is not positive. And weak management phenomenon is not rare. Sports information teaching system function module is not fully able to support learners' meaning construction: case base creation of scenarios, for example, students' independent analysis and problem solving skills with collaborative information tool are poor, especially the development and application of network online "simulation module" is almost zero.

4.2 *Improve the sports teaching of information ecosystem situation way out*

As an independent operation of the open system, the ecological system has certain stability. Stability of ecosystem refers to the ability to maintain or restore. The ecological system has its own structure. Its function is stable. The intrinsic reason for ecosystem stability is self regulating ecological system. Ecological system is in a steady state. It is called the ecological balance. Generally speaking, ecosystem diversity, composition of energy flow and material circulation way form self adjustment ability. On the contrary, ecosystem structure and composition of single self adjustment ability are relatively weak. Therefore, the structure and function of improving development initiative, the subject of teaching more beneficial to the constructivist teaching modules or subsystems can maintain or restore information teaching ecosystem stability. They are beneficial to the construction of meaning.

According to the principles of instructional design under Constructivism Learning Environment: take the student as the center, play situation, dialogue and coordination role, pay attention to the design of learning environment, use a variety of sports information education resources to support the "learning". The learning process of the ultimate goal is to realize the meaning construction, improve the sports teaching of information ecosystem. The existing situation can start from the following several aspects:

4.2.1 *The students play in the sports teaching ecology of initiative and creativity*

Expansion of student teaching niche, allowing them to the teaching environment has great influence and strong ability of resource utilization. The student is not a simple teaching of information consumers, but also can become the information-based teaching producer. Ecological construction should stimulate the students' initiative. Then the students will actively participate in the teaching of information, which reflects the students' pioneering spirits.

4.2.2 *Deepening PE teaching ecological management*

Ecological management of sports teaching guide, supports integrated function. The teaching management and network maintenance are to actively guide students to use information technology teaching resources, and the corresponding hardware facilities are provided at the same time (computer, etc.). Security, increases physical educational informationization investment; formulate corresponding policies and incentive system (such as quality course construction, multimedia teaching appraisal). By using the network to teachers, students, external personnel situation, historical records, frequency analysis and the efficiency of use of feedback, usage, Ecological management of sports teaching further promotes the construction and operation of sports teaching of information ecosystem.

4.2.3 *Niche expansion of sports information teaching system*

Expand the sports information teaching system in the teaching of information ecology in the ecological system. It is conducive to expand the occupation of teaching resources. There is more overlap in the classroom teaching and after-school self-study niche. So it is necessary to strengthen the sharing and use of teaching resources. Classroom teaching can use information teaching resources (The condition is that the classroom has Internet connection). The use of after-school learning and network teaching case (sports events such as from the network presented in the form of homework, pictures, video, text, etc.) should also be strengthened.

4.2.4 *Optimization design sports teaching learning situation*

Constructivist teaching design is different from the traditional PE teaching design. There is more emphasis on students' autonomous learning process. So the teaching design must be firstly designed learning environment to give the classroom to the students. If students can carry out learning activities according to the existing sports information resources for education or learning tools, they can feel the fun of fully autonomous learning and cooperation. In the process of teaching design, teachers should control the situation scale; achieve freedom but not messy the classroom atmosphere.

4.2.5 *To strengthen the construction of sports teaching evaluation information system*

Sports teaching evaluation is an important part of physical education teaching work. It is to realize the scientific management, an important means to ensure the quality of teaching. Evaluation system construction should follow the scientific laws, selecting evaluation means and method of science. Teaching evaluation information can include the students' self evaluation, group evaluation, teacher evaluation, peer evaluation, media evaluation. Evaluation index system construction can be formulated according to the teaching methods and teaching in environmental science. The evaluation should focus on training students' learning ability and innovation ability.

5 CONCLUSION

The ecological system of sports teaching is a teaching subject through information teaching system linked with surrounding environment culture ecosystem. In this system, various perplexing relations form a teaching ecological field. In the ecological field, each subsystem or module teaching subjects, teaching system is in different teaching niche last, but they cooperate with each other and influence each other, for a common goal: students to construct their knowledge. In the view of constructivism learning theory, situational creation requires enhancing the information teaching system "simulation (Teaching)", "case teaching" sub system development and application: "collaboration" requirements focus on the development of "virtual classroom" subsystems. "session" should give full play to "teaching evaluation" and other system functions. Sports teaching, advocates situational teaching, for the specific and real. But it is not against the abstraction and generalization. From the function of education, indirect experience form of learning is still a major. Students may not always start from direct experience. This process requires the informatization instructional design to pay attention to the original experience learning theory and practice of sports. And students are closely related, therefore, sports teaching of information ecosystem of situational teaching and learning of theory are the development direction of sports information teaching.

REFERENCES

[1] National Quality Curriculum Resource Network—Course [OL]. http://www.jingpinke.com/course.
[2] Ecological niche [BD/OL]. http://baike.baidu.com/view/103716.htm.
[3] Constructivism [BD/OL]. http://baike.baidu.com/view/79065.htm.
[4] Ji Dou Yong. Cultural ecology [M]. Lanzhou: Gansu people's publishing house, 2006:208.

Sports Engineering and Computer Science – Luo (Ed.)
© 2015 Taylor & Francis Group, London, ISBN 978-1-138-02650-6

The positive influence of online education technology on college PE teaching

HongSheng Zhao & Hong Zhang
College of Physical Education, Northeast Dianli University, Jilin, China

ABSTRACT: Online education is a new teaching model accompanied by the development of computer online technology. Online education technology in college PE Education can make up for the deficiencies in the classroom and stimulate students' interests in learning. Besides, the technology can also improve teaching effectiveness and develop students' sport awareness. As a result, it has a positive influence on establishing lifelong sports concept.

1 INTRODUCTION

Online education is a new education model that students get access to learning guidance and information by means of computer online technology under the guidance and collaboration of teachers. At last, students can achieve the learning objectives by adopting effective ways[1]. Traditional PE teaching cannot meet the requirements to fully develop people. Therefore, the application of online technology has been a necessity in the modernization of PE teaching.

2 STIMULATE STUDENTS' INTERESTS IN LEARNING, AND DEVELOP STUDENTS' LIFETIME SPORTS AWARENESS

Online education technology in PE teaching can provide a variety of sports material and make teaching more vivid and specific. In addition, it also pays more attention to the development of students' character, which plays very important roles in improving students' interests and passion for learning. Students can acquire a large number of knowledge about sports and health online. Then deepen their understanding of sport and understand the importance and significance of lifelong sports so as to lay a foundation for lifelong sports.

3 DEVELOP STUDENTS' ABILITY TO LEARN AND THINK INDEPENDENTLY

To solve problems depending on online searching has become a common means of students' after-school learning. Online education technology can better reflect the guiding role of the teacher and also the principal part of students. In the traditional teaching model, students mostly turn to teachers and students to solve problem. But the online teaching model makes a fundamental change about the problem-solving approaches, in which students need to settle the problems on their own. So it gives full play to the students' learning initiative and enthusiasm, which is helpful to develop students' ability of learning, thinking, and problem-solving independently.

4 ONLINE EDUCATION IS AN EXTENSION AND SUPPLEMENT TO THE SPORTS CLASSROOM TEACHING

In PE class, the knowledge that the teacher teaches is very limited. So to increase students' sports knowledge is a problem to be solved now. However, the Internet provides students with a lot of learning materials. Through the Internet, the boring teaching content becomes lively, and it can stimulate student's interests in acquiring sports knowledge. Since students can get relevant information in the whole process of teaching, fully use valuable resources and be properly guided and inspired by teachers. A virtuous cycle of PE classroom learning could be build to the utmost, the purpose of comprehensive education people can be achieved.

5 MAKE IT EASIER TO GRASP THE SPORTS TECHNIQUE

In traditional PE teaching, actions are often described by language. Clear action representation is an important basis for the formation of sports technique, which relies on the teaching process of explanation, demonstration, presentation, etc. However, some sports techniques are difficult to describe, and the effects of demonstration are not so good. At this point, teachers can use the computer to make the teaching content specific and visual, turn dynamic state to static state, turn fast speed to slow speed. Certainly, the purpose is to help students better learn the details of the action, better grasp the essentials of the technique and enhance the students' knowledge of the entire action. Multimedia courseware could use space design, such as two-dimensional and three-dimensional space, to make a three-dimensional analysis of the key point and difficulty. Moreover, the space design can also be used to make it easy and simple and make the complex and abstract action become vivid. So that the learning speed can be accelerated and the learning efficiency can be improved. The multimedia courseware can solve these difficulties more easily. Furthermore, it can help students understand the action, form concept, remember structures and build up a clear action representation in the brain.

Because of the diversity of problems that arise in specific teaching, it is difficult to collect all the teaching animation material. Then we can make sports teaching animation courseware by ourselves. In the teaching process, teachers can directly explain the demonstration actions through the courseware. Students who master the action can make practice. Those who do not master it can use the control buttons to repeat the demonstration.

At the same time, the key points and difficulties are very clear to find out. Meanwhile particular sounds and icons are used to remind students to pay special attention. In this way the key points and difficulties are solved effectively. Finally, it can make the students to grasp the essentials of the actions in a vivid, direct, proactive and three-dimensional way. Thus the teaching effect of PE class can be improved.

REFERENCES

[1] Du Wen. The P. E network (Tutorial) construction and application of [J]. 2007 Journal of Beijing University (1).

[2] Shi Tiefeng. Talk with the implementation of [J]. Education and the occupation of. 2007 design of remote network teaching platform (12).

[3] Zhang Tiexiong. The achievements in the research of university network sports curriculum based learning and management group. Sichuan Sports Science. 2004 (2).

[4] Chen Rong, Terry Chui. Current situation and Prospect of auxiliary teaching of college sports curriculum multimedia and network prospect [J]. Journal of Beijing Sport University (2). 2004.

[5] Hu Bin, Xiao Wei. Application of network technology in Distart physical education cutting. 2003 Journal of PLA Institute of Physical Education (4).

[6] Wu Wu, Hu Xiaoming, Tan Hua et al. Development China physical distance education [J]. Journal of Physical Education (6). 2001.

Sports Engineering and Computer Science – Luo (Ed.)
© 2015 Taylor & Francis Group, London, ISBN 978-1-138-02650-6

The role of Internet for the lifelong physical education system

Qiong Peng
P.E. Institute of Northwestern Polytechnical University, Shaanxi, China

ABSTRACT: As an integral part of lifelong education system, the lifelong physical education has been generally identified with a lot of academics from all over the world, and it has been strongly promoted in our country. Through research on Internet for its active and negative effects to lifelong physical education, this paper suggests to apply the positive influence of Internet to advance universal lifelong physical education, which are specifically reflected in the following aspects: 1. to inspire the exercise motivation; 2. to provide exercise guide; 3. to organize a communication platform; 4. to intensify the practical activities.

1 INTRODUCTION

In 1965, during an International Conference on adult education organized by UNESCO, Parl Lengrand, the Director of Adult Education Bureau officially proposed the term "lifelong education" which has later been spread all over the world. Also, he published *The Introduction of Lifelong Education* in 1970, which laid the theoretical foundation of lifelong education, and the concept of lifelong physical education was inspired by the theory of lifelong education and its impact. Lifelong physical education refers to a person's lifelong physical activity and physical education [1]. UNESCO prescribed precisely in the International Sports Charter that "P.E. is a necessary element of lifelong education in a comprehensive education system" and "physical activities and sport must be ensured by a global democratization of the lifelong education system to practice through everyone's life". Thus, physical education is an integral part of education, and the lifelong physical education is an integral part of lifelong education.

The concept of lifelong physical education has been immediately identified with lots of professional scholars, and vigorously promoted by national services. It gradually formed a new trend. Based on the body's inner development, combined with the effects of physical exercise on body, lifelong physical education take the initiative to meet the new demands of social development on human beings. Human activities show that, only to regularly engage in physical exercise can we live a qualified life.

Specifically, there are two major sections in lifelong physical education: first one is people constantly learn and participate in sports exercise and fitness activities to enhance physique and improve health level which gives lifelong physical education a clear purpose to makes sports an important

content of life; another one is the lifelong physical education can provide opportunities to practice and participate in sports activities for different stages or areas of life [2]. The establishment of lifelong physical education showed the importance for human life to improve the quality of it, which told us that sport should not just be an episode in life, but a theme through our entire life.

From the perspective of the process of lifelong physical education, preschool physical education, school physical education and community physical education consists of a hierarchy system of lifelong physical education. It is not a competition on the field or a challenge life limit, it is just an activity, but can promote our healthy, happily and joyfully. Because of sports activities, childhood has lots of fun; adolescents are filled with energy; the middle-aged are more colorful. In twilight sport keeps us away from problems of loneliness and pain.

Nowadays, people's learning, working and life are greatly changed by the rapid development of computer and the Internet. Internet connects up schools, institutes, libraries, presses, bookstores and other various information resources. On one hand, it constructs a mass knowledgebase; on the other hand, all of the world's excellent teachers or experts can provide same knowledge material and teaching guide from different angle. Anyone in any locations through Internet can access online question-and-answer and paper discussion. In this case, people have adequate alternatives, whether in time and space. So that self-learning becomes a necessary form of efficient study. Everyone at any time and place, can enjoy learning, living and entertaining freely through the Internet, and this is the real meaning of every learner. Therefore, lifelong education is available.

Whether or not the consciousness of lifelong physical education can be formed is directly related to a person's participation and willpower in physical exercise during his life. At different stages of growth, the body cannot maintain the same physiological and psychological characteristics, which needs adjustment to the real conditions. Knowledge and methods of school physical education is just the Foundation for lifelong physical education, only using different forms of physical exercise can achieve the best physical exercise effects. Networking makes it possible to meet the needs of various groups, so that the lifelong physical education to become an operable way to practice.

2 THE INTERNET EFFECTS ON LIFELONG PHYSICAL EDUCATION

2.1 *The positive effect of Internet on lifelong physical education*

2.1.1 *Internet provides the theoretical guidance for lifelong physical education*
Multimedia and Internet technology applied to physical education fields can break the limits of classroom teaching modes. Physical education teaching has changed from classroom-centered mode to independent type, from "stages sport" to lifelong physical education. Under the macro guidance, people at all ages learn and train within the atmosphere of science of harmony, so that the physical education process achieves universal lifelong goal.

2.1.2 *Internet provides a platform to people for lifelong physical education*
People love social communication very much. We tend to stay in groups: formal and informal groups, political groups, work groups, friends groups and so on. We can use various Internet tools like BBS forums, QQ Group, app group to set up a group of physical exercise, exchange the knowledge of fitness and training methods. In addition, many sports need teamwork, online communication makes frequent group activities as much as possible. People lack of willpower to exercise from time to time, the members or partners of groups can supervise, encourage and push each other to do something together.

2.2 *The negative effects of Internet on the lifelong physical education*

2.2.1 *Internet takes up people's time of exercise*
With the development of computer and multimedia technologies, the Internet has penetrated into all aspects of daily life, especially impact on the way people live. Nearly 80% students surveyed spend most of their spare time on Internet., veyed spend most of their spare time on Internet., a total of 40 hours online per week, while the time for physical exercise is negligible[3]. Overtimed Internet surfing occupies the time of reading, thinking, interpersonal contacting and even eating and sleeping. Internet is an integral part of modern life, but the negative influence of it to people's lives is also considerable. Since people overindulged in Internets, they don't have the time or energy to take part in exercises, which might cause serious impact on health.

2.2.2 *Internet provides people some incorrect knowledge of fitness*
Internet as a new mass media and contacts way, greatly meet peoples' need of sharing information resources, beyond spatio-temporal limit. But, it is rich in information which are imbalanced in quality. The truth is people lack of the capacity of professional identification and the Internet lack of effective management. Therefore, incorrect knowledge from Internet may lead to the serious physical and psychological effect on people.

Thus, how to guide people to exercise scientifically and reasonably also needs the attention and accurate guidance from the professionals, so as to form the habit of lifelong physical education.

3 SUGGESTIONS

3.1 *Through online media*

Advocate people to actively use the positive impact of Internet to promote lifelong physical education and stimulate motivation through online media. Tell people the importance of exercise in work, learning and life, constantly throw the issues about the sport people interested in, leading people to think positively and stimulate people's desire to exercise.

3.2 *Professional sports site*

We should provide exercise guidance through professional sports site; inform people of the knowledge of exercise. According to the difference between sports and individual physical characteristics, provide different types of targeted exercises to help people's proper physical exercise.

3.3 *Communication platform*

The government departments should organize activities through the Internet platform; sports experts are invited to discuss and answer people's questions online. People with various Internet communication means, like mobile blog, BBS Forum, QQ Group or Internet live can immediately obtain the professional guidance. Through synchronized online interaction or asynchronous discussion, it is convenient to communicate and exchange ideas on the exercise knowledge, skills, solution of difficulties and problems. In other words, Internet provides a new way of explaining counseling answering and discussion. Thus personal knowledge will be well reconstructed to provide effective methods for problem solving and scientific and rational help for exercise. Consequently, lifelong physical education is improved in the long run.

3.4 Strengthen the practice of physical education

From the perspective of physical education, narrow or broad, physical activities are always the main model [4]. Physical activity is an important part of lifelong physical education. In order to ensure the quality of exercises, we need to make capital of our knowledge to the content before practice.

The ultimate goal is to get people to learn sports and to practice physical exercise. Via the Internet platform, the importance and necessity of physical activity can be strengthened which in turn encourage people to exercise on a regular basis.

REFERENCES

[1] Zeshan Wang. 1994. Lifelong Physical Education. Beijing: Institute of P.E Press.
[2] Zhichao Lin & Jingyan Ren. 2001. The Review and Prospect of Reform on Physical Education in Colleges and Universities in China. Beijing: Beijing sport University Press.
[3] Biqiang Zheng & Yeyun Zhang. 2011. Reflection on the Students Craze for "Dwelling Style" Life. Journal of Fujian teachers University (Philosophy and Social Sciences Edition).
[4] Dictionary of Sports Science. 2000. Higher Education Press.

Sports Engineering and Computer Science – Luo (Ed.)
© 2015 Taylor & Francis Group, London, ISBN 978-1-138-02650-6

Continuing education curriculum construction for primary and secondary school teachers based on education informatization

Chao Yang
Institute of Higher Education and Regional Development of Yunnan Normal University, Kunming, Yunnan, China

Li Yi
International School of Yunnan University, Kunming, Yunnan, China

ABSTRACT: The research focused on the problems and causes in continuing education for primary and secondary school teachers under the circumstances of education informatization, and then posed countermeasures. With the respondents of teachers, curriculum makers and curriculum managers of primary and secondary school teachers of Yunnan province in China, it studied and analyzed the present situation of informatization degree, curriculum structure arrangement, curriculum practicability, curriculum management and evaluation in the form of questionnaire and interview, and then, analyzed cases with Factor Analysis Approach. Surveys show that we should enhance curriculum resource informatization degree, exploit diversified curriculum resources, perfect six curriculum modules, implement standard and scientific curriculum management, and establish plural curriculum evaluation system. Data and conclusion drawn from empirical survey and statistic analysis not only enrich theory for the continuing education curriculum construction, but also provide practical reference for measure making of education administrative authorities and teachers' training agencies.

1 PROBLEM DIAGNOSIS

1.1 *Single curriculum resource form and low informationization level*

Curriculum sources are closely related to region social and economic development level. Due to regional social, cultural differences in the country and dual economy structure, there are gaps between rural and urban areas, and cities; thus needs for primary and secondary school teachers are diversified in terms of content and modes. However, curriculum resources at present are mostly in textbook forms, and the multimedia resources are limited. Integration of sky-network, ground-network and human-network is not realized, sharable learning resources and informationized learning are limited.

1.2 *Deficient curriculum layers*

Curriculum layer means curriculum at national, provincial and county levels divided from curriculum management system and authorization aspect. In continuing curriculum education for primary and secondary school teachers, curriculum layers are not balanced and diversified with priority of external curriculum sources; national curriculums take the most part (curriculum recommended by

Ministry of Education and ministerial excellent curriculums dominate), and there are less province developed curriculums and no county developed curriculums. The single curriculum is far from enough for the need of different areas, various types of teachers and individuals.

1.3 *Incomplete curriculum structure*

Analysis of the existing continuing education curriculums shows that preliminary mode for curriculum module features has been formed, but there are also problems in terms of incomplete curriculum structure. First, in terms of content module, the curriculums cover six modules: teacher's ethics culturing, discipline teaching, modern education theory and methods, modern education technology and its application, class adviser's work, humanity and science knowledge, "humanity and science knowledge" modules only covers humanity curriculum and less science knowledge curriculum. Second, proportion of "modern education theory and methods" module is severely imbalanced. Most content of pedagogy curriculums is necessary for practical teaching; but theory learning can not be neglected. The fundamental curriculums, such as curriculum theory, pedagogy principle, education psychology, Chinese and foreign education theory have been included besides education laws, regulations and policies, pragmatic education scientific research methods. Public basic curriculums and education professional curriculum, which have been weak in pre-service education, are even weakened; therefore, real command of pedagogy and psychology curriculums is almost impossible. Third is separation between subject curriculum, comprehensive curriculum and inter-disciplinary curriculum. At present, continuing education curriculum for primary and secondary school teacher are organized in subject teaching form. With the deepening of teaching practices, branch division is more detailed and branch links are separated. Take physics as an example, physics in secondary school is not connected with chemistry, biology, geography and so on, and there is no inter disciplinary curriculums. Teachers' training is only the renewal or consolidation of the professional knowledge.

In conclusion, continuing education curriculum for primary and secondary school teachers lacks overall design and arrangement, layers between modules or curriculums are vague. Requirement for teachers' knowledge structure, especially for their capability is not included: curriculum structure is still "independent", disciplinary boundaries still exist; disciplinary communication and penetration are not enough for demonstrating the latest development trend and cutting edge scientific achievements. Disciplinary blending is not reflected, what teachers' teaches and learns is mostly knowledge curriculum and development curriculum is not enough.[1] In addition, over emphasis of knowledge systematicness and accumulation of knowledge leads to the neglect of knowledge structure optimization. All of these do not conform to the need of enhancing teachers' quality and modern teacher professional development theory.

1.4 *Focus on theory instead of practice, poor practicability*

Primary and secondary school teachers are experienced through years of teaching. And the experiences are the basis for structuring new knowledge. Proper use of the experiences and appropriate proportion of theory and practical curriculums can not only satisfy the need of teaching practice guidance and practicability, but also reflect features of teachers' continuing education curriculums setting. But our survey shows that continuing curriculum construction for primary and secondary school teachers bears distinct "disciplinization" feature with deficient focus on activity and practical curriculums, and imbalanced proportion of theory and practical curriculum. The survey is done with curriculum dichotomy (theory and practical curriculum) to compare and analyze the existing curriculums. Survey shows that, expectation

and reality of the continuing education curriculums disconfirms. Multiple or single expectations of teachers focus on teaching ability enhancing, actual problem solving, new idea and method obtain through continuing education curriculum; all of these indicate teachers' high expectation for the curriculum and their pragmatic pursuit. In addition, the existing teachers; continuing education curriculums put more emphasis on the theory which are not instructive for practice and actual teaching problems, and there are curriculums which are low-level repetition.

1.5 *Lack of scientific curriculum management*

Curriculum management includes planning of curriculum setting, guidance of curriculum development, and supervision of curriculum implementing, evaluation of curriculum content and so on. Completeness and feasibility of curriculum management system determines management and quality of training to a large extent. Surveys show that, major problems in teachers' continuing curriculum management are: first, lack of curriculum systematic planning. Curriculum planning is the plan or curriculum plan for teachers' continuing education in accordance with national, teachers' professional development within certain period. Surveys show that, lack of management, incomplete or unscientific curriculum plan will lead to the randomness of curriculum setting to satisfy teachers' need for education knowledge structure; lack of systematic planning in curriculum management will also lead to repetition of curriculum setting, which affects teachers' learning enthusiasm. As a result, learning resources are wasted. Second is low selectiveness of curriculum. In optional curriculums setting, training agencies select optional curriculum and trainees can only accept them; individualized selection of proper curriculums is not realized. Even trainees are free to select curriculum, they can only made the choice based on the name of the curriculum or the subject. In the selection system, teachers' selection of curriculums is blind to a certain extent. Third is lack of supervision of curriculum implementing. Surveys show that, teachers' training agencies are mainly in charge of the selection and realization of curriculum schedule, training experts and teachers and trainees. And there is lack of supervision for the dynamic curriculum implementing. Teachers' training agencies make uniform rules for curriculum content selection, syllabus and teaching plan writing, and they also learn trainees' satisfaction of the curriculums and trainers during and after the curriculum implementing in the form of questionnaire or interview to lay the foundation for future curriculum management. However, conformity of trainers' teaching plan and actual teaching situation and quality is rarely supervised and checked.

1.6 *Incomplete curriculum evaluation system*

Curriculum evaluation is a process to judge the curriculum effect with certain standard, information and professional knowledge, including the judgment of curriculum design, compilation, implementing and students' scores. One aspect is the evaluations of curriculum design, compilation and implementing. Curriculum evaluation means the basic check of curriculum plan, curriculum subject and teachers' planned curriculum subject and curriculum plan, connection between modules and so on. At present, continuing education curriculum for primary and secondary school teachers still focuses on curriculum exam scores, instead of the processes such as professional knowledge renewal, progress of teaching skills, teamwork spirit and ideological quality progress. The single evaluation form seriously affects direction and quality of teachers' continuing education.[2] Surveys show that curriculum evaluation for continuing education still puts emphasis on term end exam, assignment and attendance check, learning process evaluation, students' self-evaluation and team evaluation is neglected.

2 CAUSES FOR MAJOR PROBLEMS

Dialectically, problems arise as a result of internal and external factor. Surveys show the same in the field of continuing curriculum construction. The problem is a result of both internal and external factors, such as ideology, theory and practice interaction, curriculum concept and construction basis, curriculum requirement survey and construction guarantee and so on.

2.1 *Lack of the concept, insufficient emphasis on curriculum construction*

Of all problems in teachers' continuing education construction, the first is concept. It is the value orientation of active subject, as well as the subject's understanding and cognition of the object. Teachers' continuing education curriculum construction is implemented nationwide in different layers of teachers' groups; but its effects vary due to differences in subjects' concept and cognitive levels. Surveys show that causes for the problems in curriculum construction are: first is the lack of attention to the curriculum construction and lack of complete and systematic understanding. It is usually believed that curriculum construction is textbook construction. The key to curriculum construction is curriculum development and selection of textbooks. Curriculum construction guides textbook master to teach effectively, which is the attention to curriculum construction and neglect of the people, curriculum goal, curriculum organizing and implementing, and curriculum evaluation and so on. The second is teachers' understanding that curriculum construction is not related to them, it is the responsibility of theory researchers, curriculum managers, and training agencies. Teachers lack sufficient understanding the role of the reasonability of curriculum goal and curriculum structure optimization in teaching quality enhancing. Some teachers pay less attention to curriculum construction as they take the training as the way to get diploma or a task.

2.2 *Disconnection between theory and practice*

As the guidance of practice, construction or perfection of theory is an important direction and guidance. Teachers' continuing education curriculum construction at present is conducted vigorously with the guidance of curriculum theory. Experience curriculum theory, discipline curriculum theory and Taylor's curriculum theory are the guiding theory for curriculum setting and evaluation, and also the important standard for continuing education curriculum construction effect. But in practice, theory of teachers' continuing education curriculum construction is the same as that of students' education with no features. There is disconnection between theory and practice. Plan of curriculum construction, selection of textbooks and its basis are more determined by theory researchers' opinion instead of actual teaching and training of teachers. All of these lead to disconnection between the theory and practice of curriculum construction, such as in-adaptation of curriculum goal, textbooks selection and evaluation index to the requirement of teaching. In addition, the theory is not perfectly guiding the practice due to the imperfect theory system.

2.3 *Lack of the understanding of curriculum concept and its construction basis*

Curriculum concept in China is either translated or scholars' elaboration from concept of curriculum; there is no uniform definition for curriculum, and thus the concept of curriculum is vague in the process of curriculum construction. Understanding of the connotation of curriculum is not deep enough as the componental factors (curriculum target, content, curriculum organization and implementing, and curriculum evaluation) are not understood. Besides, in the process of curriculum construction, setting, goal setting of curriculum shall be done on certain basis. Continuing education construction for teachers shall be conducted in accordance with regional economic, scientific and technical, cultural development and the characters of trained teachers. Syllabus of the curriculum shall be characteristic with proper textbook, complete curriculum materials and scientific evaluation methods. Surveys show that, in continuing education for teachers, curriculum managers, implementers and learner lack sufficient understanding of the curriculum construction, thus leading to the old-fashioned concept, insufficient curriculum structure, single resource form, less scientific curriculum management and lack of practical value orientation in curriculum construction practice.

2.4 *Lack of field survey for the requirement of teachers' training curriculum*

Continuing education curriculum construction for primary and secondary school teachers flourishes in recent years, curriculum quality is also to be improved with the social need for talents. How to develop curriculum construction? What is the primary and secondary school teachers' need for curriculum training? What is the acceptance extent in primary and secondary schools? Most continuing education schools or agencies do not know the answers to the questions. Thus, the curriculum is hardly proper for the lack of understanding. Most continuing education curriculums focus on theory teaching, the skill training is far from enough for teachers' actual need, the curriculum setting is of little help in actual work. In addition, curriculum is not set based on requirement and the curriculum resources and training cost are wasted.

2.5 *Deficient guarantee for curriculum construction*

Guarantee service of most continuing education schools or agencies are not sufficient due to the imbalance of the circumstances of colleges, universities and training agencies, such as incomplete organization setting, different investment, different office facilities and staff, various faculty and training facilities, differences in teaching research, teaching schedule and class hours, imperfect training agencies construction and vague responsibilities, imperfect construction procedures. Less attention is paid to curriculum construction and development in some schools or agencies, there is no sufficient investment, specialized continuing education research office, network office or consulting office; in terms of faculty construction, there is no relevant professional training and further study policy, low teachers' degree of specialization reduces opportunities of practice.

Teachers engaged in continuing education for teachers do not understand the characters or need of the trainees; therefore, the effect of the curriculum is not ideal.

3 COUNTERMEASURES OF IMPROVEMENT

3.1 *Perfecting the six curriculum modules*

Strengthening teachers' ethics shall start from strengthening teachers' sense of responsibility and mission to enhance their psychological quality and adaptability; discipline teaching curriculum module shall help teachers' continual renewal and master of advanced teaching concept, supplementing of new knowledge, and boosting of problem solving and research abilities; modern teaching theory and method module shall help teachers in absorbing latest teaching research achievements, renewing teaching concepts, mastering new basic teaching methods, enhancing the ability of theory and practice combination ability; modern education technology module shall help in teachers' understanding and basic mastering of modern information technology, continual application of modern information technology in education; class advisor work module is mainly to help the advisor to understand and master the work standard, learn the rules and features of the work, improve the ability of problem solving and enhance class advisor's work creativity; human and scientific knowledge module training curriculums consist of human science and nature science, the former is mainly to help teachers broaden human and social science knowledge and the latter is to broaden nature science and technology for the necessary understanding and command of scientific methods.

3.2 *Diversified curriculum resource development*

First is the emphasis of curriculum resource material construction; teachers' continuing education agencies shall put emphasis on dynamic balance of curriculum construction terms and materials, as well as hardware, equipment and facilities construction; curriculum construction is to be treated as a long-term and fundamental work. Material resources include curriculum resources for perfecting primary and middle school teachers professional knowledge and skills, curriculum resources that renew teachers' knowledge and skill as a result of scientific and technical transformation, and resources that are useful in teachers' teaching and research ability enhancing; the second is strengthening network and media curriculum construction. On the one hand, it is to be made clear that education network and media resource construction shall be connected with the latest achievements of modern science and technical development; on the other hand, target is important. Targeting the actual need of the trained teachers, organizing form of the curriculum shall focus on the actual features of teachers' job. Apart from these, quality and quantity of the network and media resources are to meet the need of teachers.

3.3 Standard and scientific curriculum management

First, obligations and tasks at national, provincial and county levels shall be made clear and curriculum management permission is transferred. Curriculum plan suitable for local circumstances is made at provincial and county level, curriculum implementing plan is organized and checked, schools are instructed in terms of national and local curriculum implementing for school curriculum implementing and development; in addition, social curriculum resources are integrated in curriculum development; the second is systematic design of curriculum plan. At first, based on the guidance of national and provincial education authorities, teachers' need and advice, mid-term curriculum plan and annual training plan are made by experts, field teachers and administrative staff. [3] Next, curriculum plan is for all curriculums, involving curriculum development, implementing and evaluation process; third, curriculum development is important. Teachers are given sufficient time in preparing curriculum content, teaching methods and strategy. On the other hand, apart from the selection of textbooks recommended by Ministry of education and province, teaching materials communication with normal universities, teachers' continuing education is to be enhanced, characteristic textbooks, teaching materials, self-learning material and local textbook are compiled to provide rich resources for teachers' continuing education.

3.4 Diversified curriculum evaluation system

The first is the adoption of diversified evaluation forms. Learning and psychological features of primary and secondary school teachers shall be taken into consideration in the evaluation; based on the actual need of teaching activities, means and methods of different dimensions can be used with the emphasis on practicability. The second is evaluation with multiple subjects. Subjects of curriculum evaluation are teachers' continuing education agencies, teachers, educators and researchers, trainees (trained teachers) and other relevant people. In reality, the evaluation is mainly about curriculum implementing conditions, process and results; the combination of the multiple evaluations can help to realize the diversity of teachers' continuing education curriculum.

REFERENCES

[1] Deng Zejun. Problems Perspective in Teachers' Continuing Education Curriculum [J]. China Adult Education, 2007(10):41–44.
[2] Hao Weifu. Curriculum Setting and evaluation System of Primary and Secondary School Teachers' Continuing Education [J]. China Adult education, 2006(5):25–28.
[3] Wen Rong. Problem Analysis and Countermeasures for Teachers' Continuing Education Curriculum [J]. Continue Education Research, 2011(2):8–11.

Sports Engineering and Computer Science – Luo (Ed.)
© 2015 Taylor & Francis Group, London, ISBN 978-1-138-02650-6

The relation between anxieties and reading comprehension for students majoring in biological engineering and the enlightenment to teaching

Yushuang Zhu
School of Foreign Languages, Jianghan University, Wuhan, China

ABSTRACT: The study of a survey on students' anxiety, currently representing an important area in English learning. The survey is conducted among 120 second-year Biological Engineering majors. It analyses their anxieties and factors affecting their English learning, especially reading comprehension. It discusses the characteristics of distribution of the anxiety scores in students' English learning. Based on the above analysis, the author has come to the following conclusion: too much anxiety, the lacking of confidence, fear of negative evaluation and fear of communication do hinder their progress. Thus, some teaching methodologies and strategies to improve teaching proficiency are proposed accordingly.

Keywords: anxiety; anxiety scores; testee; correlation; questionnaire

1 INTRODUCTION

Anxiety is a kind of mental state of nervous or terrified. Language anxiety is a complex psychological phenomenon, which is specific to language learning. It happens only in the course of language learning. It's an apparent anxiety for self-consciousness, self-belief, self-affection and self-behavior, which is related with classroom language learning. (Move 1 stating current knowledge). Early in the 1960s, researchers had started to study the effects of anxiety on the achievements of language learning of students. Now this study is furthering. Some research findings indicate that language anxiety has a negative influence on students' achievements. However, some studies show no relation between them. E. Horwitz, M. Horwitz and J. cope claimed that anxiety may cause difficulty in students' reading comprehension. (Move 2 selective summary of previous researches) Recent study on anxiety in students' reading comprehension is not much. Less microexperiment was done on anxiety in reading comprehension. It deserves further studying. (Move 3 indicating a gap in previous research). This paper tries to throw a light in the relation between anxiety and achievements of students' reading comprehension. It aims to discuss the characteristics of distribution of the anxiety scores in students' English learning and analyze its cause and propose some effective teaching methodologies and strategies (Move 4 description of the present study).

2 LITERATURE REVIEW

Scovel (1978) argued that anxiety can be classified into facilitating anxiety and debilitating anxiety in terms of their characteristics. The former one can motivate language learners and help them establish confidence and accept new tasks. The latter one will lead students to escape their tasks and cause a feeling of wanting to escape anxiety. Williams (1991) has similar viewpoints. He pointed out that the classification of anxiety is affected by the degree of anxiety. He holds that a great anxiety may hinder students' progress. Afterwards, many studies were done on the relationship between anxiety and language learning and they had reached an agreement. Cardener, 1985, Young, 1992, Steinberg & Horwitz, 1986 found that anxieties in English learning had a negative influence on students' learning. The more anxiety a student has, the lower score he gets.

On the concrete demonstration of anxiety in English learning, E. Horwitz, M. Horwitz and J. Cope again pointed out that the anxiety is an apparent psychological phenomenon during reading and speaking and a reading comprehension obstacle caused by anxiety deserves further research. The research of the relationship between anxiety and reading comprehension should start with the nature of reading. Reading is a highly-complex intellectual activity. It does not exist in space, but only time. It is fast and irretrievable. You will not read it again unless you want to review it. In that case, readers should recognize, understand, process and store acoustic signals in such a short time so that signals can carry a meaning. Consequently, anxiety and tense caused by the incomprehension is a quite common psychological phenomenon which is likely to happen. Incomprehension and anxiety act as the cause and effect of each other. What is more, they influence each other and form a vicious circle which has a bad effect on reading activities and students' interest and enthusiasm.

3 METHODOLOGY

3.1 Subjects

120 second-year Biological Engineering majors of Luoyang University participated in this study. There are 54 boys and 66 girls. They are from 13 different provinces or areas with an average age of 20.

3.2 Instrument

1. FLCAS (Foreign Language Classroom Anxiety Scale) designed by E. Horwitz, M. Horwitz and J. Cope was used in this investigation. This scale form was widely accepted by most scholars in their research either at home or abroad. "So far the experimental result of the anxiety scale indicates that anxiety in foreign language learning can be measured effectively and reliably."(Horwitz, 1991, p39). FLCAS totally has 33 items which are designed to focus on learners' feelings in specific circumstances. 29 of them are concerned with the typical difficulties which are related with speaking, reading, writing, memorizing and speed of language processing.
2. Self-designed ELAS (English Reading Class Anxiety Scale) includes 20 questions which concentrate on some typical kinds of obstacles related with psychology, reading strategy, memory processing in reading class.
3. Based on the combination of FLCAS and ELAS, "Questionnaire On English Learning Anxiety" was designed. There are 53 items in the questionnaire. Every item has 5 choices: A. strongly agree B. agree C. neither agree nor disagree D. disagree E. completely disagree The choosing of A will get 5 points. B, 4 points; C, 3 points; D, 2 points; E, 1 point. The higher the average score is, the more anxious the students are.

3.3 Procedure

1. Use FLCAS to test students' anxiety degree in classroom learning.
2. Use ELCAS to test students' anxiety degree in the English reading class.
3. Use the examination paper of reading comprehension to test students' comprehensive ability and single ability. With the aid of descriptive statistics, especially M (mean) and SD (standard deviation), we described the results of the tests together with the overall situation of the reading test.
4. Acquire the Spearman-gradation parameter between the anxiety degree and the total score in reading test and investigate the correlation between them and analyze the data through SPSS (9.0).

4 RESULTS AND DISCUSSION

In order to know the relationship between anxiety in foreign language learning and scores of various parts of reading test, we used SPSS (9.0) to analyze our data and got the following results just as the following Table 1 indicate.

Table 1 shows that the anxiety degree in reading class is higher than in any other kinds of classes, that is, anxiety is more likely to happen during the course of reading comprehension.

It is probably related with the learners' physical and psychological constitutions. During the course of reading training, readers are in a passive position which is quite different from where they are in reading, writing, translation or other study of foreign language. The course of reading is also a course of receiving and decoding signals. In the natural communication, linguistic signals are conveyed quickly, continuously, rectilinearly and irretrievably. The speed of receiving and decoding signals is determined by the stored experience which already exists in learners' head. For example, some responsive words, some useful grammars, phonetic rules, and so on. The more experience you store, the more quickly you response. In a consequence, students often have heavy psychological burden, especially in attention, during the course of reading training. Low-level students are constantly very nervous and tense. When they meet new words or difficult sentences during reading,

Table 1. Descriptive statistics of anxiety questionnaire.

BT	N	FLCAS		ELCAS	
		M	SD	M	SD
M	54	2.831	0.528	2.914	0.566
F	66	2.802	0.535	2.971	0.561
T	120	2.821	0.526	2.932	0.558

Note: BT: group; N: number; FLCAS: Foreign Language Classroom Anxiety Scale; M: male; F: female; T: total.

the more difficult the reading comprehension is, the more time they spend on. They even remember that when they move on to the next part. In that case, they seem to be reading although actually they have stopped reading or got nothing at all. Then it runs in a vicious circle. Non-concentration is common to be seen in reading. Thus it will cause incomprehension and furthermore, anxiety and horror. In the questionnaire, 71% of the testees chose A or B when answering the question "I feel more tense and nervous in the English reading class than in other class." And 74% testees chose A or B when answering the question "During the English reading class, I concentrated on nothing except the words pouring into my head." The score of anxiety reaches 2.85 and 2.91 respectively.

No matter in reading class or foreign language class, the anxiety of the testees is mainly demonstrated in the following 2 aspects: One is from outside world or negative self-evaluation; the other is from the communication activities.

First, the score of negative evaluation is often high with an average 3.0. The sense of anxiety is mainly demonstrated in the following 3 ways:

1. Have difficult for using English;
2. Thinks that others are better than him in learning English. 73% of the testees confess this.
3. Worry about a low think of him from others. 80% of the testees have such worries. This result is consistent with Horwitz's view on "fear of negative evaluation."

Watson & Friend's explanation of negative evaluation is like the following: a preliminary fear of other people' s evaluations or a depression by others' negative evaluations or worry that others will give a negative evaluation. No matter "fear" or "depression" or "bad preliminary feelings" will probably hurt a student's self-esteem and self-confidence. A person who has a low expectation of future achievements will think "He is inferior to the others." Furthermore, this constant reminding will badly hurt his self-esteem and self-confidence and cause an apparent anxiety before start. Student who is sensitive to other students' evaluations will avoid answering the teacher's questions. When the class is discussing or answering questions, he will not start a topic or propose any questions. He will not interrupt until he has to. Keeping silent, smiling only, answering questions merely with "yes" or "yah" or nodding are some typical behaviors among those students.

Second, the lower the he thinks of himself in learning English, the higher degree of anxiety he possesses. The highest score of fear of communication in the

questionnaire comes from the question that "It embarrasses me to volunteer answers in the English reading comprehension class."

5 CONCLUSIONS

1. Anxiety has a negative correlation with the score of English reading comprehension and a correlation with various aspects of ability.
2. A fear of communication activities and a negative evaluation from others or him are the two major reasons for anxiety.

Based on the above research, we think that it's every teacher's responsibility and duty to help students to reduce anxiety as possibly as he can. Create a relaxed classroom atmosphere, promote characters and increase students' confidence in improving their reading comprehension ability. Through more fundamental training, we should make students master a large number of responsive words, applicable rules of phonetics and grammar in order to overcome the reading comprehension difficulty and master some reading comprehension techniques and methods in order to improve the speed and ability of receiving and decoding linguistic signals.

There are still some limitations in either theories and actual operations in this research which need to be improved. The development of foreign language reform and theories may learn from the findings of this study in some aspects.

REFERENCES

Gui, S. (1977). Methodology in Linguistics. Beijing: Foreign Language Teaching and Research Press.
Hao, M. (2001). Research on the relationship between English Achievements and achievement motivation as well as anxiety state. Foreign Language Teaching and Research, 2, 111–115.
Liu, R. (1991). Language Testing and its Methodology. Beijing: Foreign Language Teaching and Research Press.
Wang, C. (1990). Applied Psycholinguistics. Changsha: Hunan Educational Press.
Wang, M. (1999). Of Foreign Language Teaching. Hefei: Anhui Educational Press.
Wang, Y. (2001). Anxiety and its Influence on Foreign Language learning. Foreign Language Teaching and Research, 2, 122–26.
Woods, A., Fletcher, P. and Hughes, a. (2000) Statistics in Language Studies Cambridge University Press.
Young, D. (1991). Creating a Low Anxiety Classroom Enviroment: What Does Language Anxiety Research Suggest. The Modern Language Journal, 75, 283–94.

Sports Engineering and Computer Science – Luo (Ed.)
© 2015 Taylor & Francis Group, London, ISBN 978-1-138-02650-6

To practice and explore the reformation of ethnic preparatory Chinese course in applied undergraduate college—taking Baise University as an example

Xiaozhen Lu
Baise University, Guangxi Zhuang, China

ABSTRACT: Along with the development and reformation of national higher education, applied undergraduate college will become an important part of higher education. Facing with these new opportunities and challenges, how to find a suitable teaching mode for development is an important subject for ethnic preparatory Chinese course in applied undergraduate college.

Keywords: applied undergraduate college; preparatory Chinese; course reformation

1 INTRODUCTION

In recent years, along with the development and reformation of national higher education, ethnic preparatory education has entered the development stage of educative reform. The quality of preparatory students has improved greatly. There are A Minority preppy and also B Ordinary National preppy. What's more, preparatory school level becomes more rich and varied. These changes to the Preparatory Education have brought new opportunities and also new challenges. Due to complex ethnic composition, different geography and different cultural background, Preparatory Students' basic knowledge and study ability have great difference. Therefore, contrary to the particularity of students in different basics, levels and professional directions, it becomes very important to carefully study of building curriculum system, effectively improve the teaching quality and provide protection for the entire talented person training system works. Combined with my school Chinese curriculum reform, this paper further clarifies the new ideas of applied language matriculation colleges curriculum reform and development.

2 DEVELOPMENT STATUS OF APPLIED UNDERGRADUATE COLLEGES

Since 1999 college enrollment, "upgrading to university" of local colleges are more than 640 (include independent colleges, private colleges), accounted for about 55% of the National Undergraduate Colleges and universities. Now, this kind of colleges' development encounter a serious bottleneck: on one hand, the graduates' employment pressure increases; on the other hand, the enterprises there are a structural shortage of talents, local colleges are difficult to support the local upgrading of industry. In addition, some important vocational colleges' school level, school strength and the employment rate have exceeded some local colleges in the same region. Development of Local Undergraduate Colleges appeared more and more narrow dilemmas. Therefore, the state decides to learn the experience of building applied technology university from Europe especially Germanic and make a policy to transform and develop.

From the policy guidance of the current national education Ministry for higher education, the future university pattern is: the first level is part of research university; the second level is the research teaching university; the third level is the application type undergraduate colleges and universities; the fourth level is the vocation college. Since the colleges enroll, "upgrading to university" of local colleges are faced with a new situation, which is transferring to the application technology university. Macroscopic situation development of higher education forms "reverse" trend to the transforming development of new undergraduate institutions. It becomes an inevitable chose to train applied talents for the survival and development of newly established colleges.

The target of training applied talents is to cultivate the undergraduates, who have one specific knowledge and skill. According to the application talent training mode, in discipline construction, research direction is mainly engaged in the applied research and developing research; specialty construction and talent training are focusing on technology system; talent training is faced with industry (enterprise) demand, which highlights practice, application and utility. The curriculum is arranged to strengthen the practice teaching; increase the elective courses and small classes; the curriculum resources development corresponds to the talent training objectives. To achieve the goal of training applied talents, the teaching methods, teaching methods, examination and evaluation ways should be accordingly adjusted.

3 ORIENTATION OF APPLIED UNDERGRADUATE COLLEGE PREPARATORY CHINESE CURRICULUM

In 1992, the former State Education Commission promulgated the higher schools of minority preparatory "education program" and clearly put forward the ethnic preparatory education belongs to the higher education category and ethnic preparatory education is a special level of higher education, "Pre", "fill" binding. The guiding ideology of preparatory Chinese course syllabus is to cultivate the students' reading ability and education ability. In other word, localization of curriculum is "fill": "fill" the senior high school students with relatively weak foundation of knowledge and ability.

Compared with high school students, matriculation students' abstract thinking and rational thinking ability are stronger, their life experience is richer, their knowledge is more varied, and they are more self understudied and more comprehensive social cognition. The law of physical and mental development and cognitive features decide that the preparatory Chinese teaching of minority nationality is different from the middle school Chinese teaching, which belongs to the basis of education nature. The author thinks that preparatory Chinese teaching syllabus ignored the important characteristics of "pre" as a transition to university education. Preparatory education is not education after middle school and also not the simple repetition and extension of high school teaching.[1] Pre "is the core of it. In other word, preparatory education is adaptive learning stage before undergraduate education. Along with the gradual enlargement of school scale, to help preparatory students, whose basic knowledge is weak, develop into qualified college students as soon as possible, we must relocate training target of the preparatory Chinese course.

According to the training target of applied undergraduate colleges and combined with the reform of the current ethnic preparatory Chinese course, applied undergraduate college preparatory position of Chinese course is focusing on the link up with undergraduate education. Therefore, every aspect of preparatory curriculum teaching objectives, preparatory curriculum setting, classroom teaching, and evaluation methods should be closer to the university.

4 THE REFORM MODEL OF APPLIED UNDERGRADUATE COLLEGE PREPARATORY CHINESE CURRICULUM

There's only one year for preparatory school. How to let the preparatory students from different ethnic groups, cultural background, mode of thinking, and language habits of preppy soon develop into qualified college students in such a short period of one year? As one of the three major preparatory courses, Chinese course is different of the normal education teaching. Chinese course must grasp the core of Chinese education and adopt ultra conventional measures/method to let students' get great progress. For applied undergraduate college preparatory students, we should try hard to train them as applied undergraduate college students. Thus, we should reform and perfect Chinese curriculum mode of the preparatory undergraduate students to meet the requirements of times.

4.1 Re-orientation of teaching course goal, refine curriculum setting

Curriculum is the basic way to realize the training goal and the comprehensive carrier of knowledge, ability, quality system and value system.[2] Therefore, the curriculum setting should be reflected in how to best improve the students' individuality; how to highly train the comprehensive ability of students. The curriculum teaching should be located at curriculum function and students' reality. The Chinese curriculum teaching goal should not only use knowledge standard as the standard. And the Chinese curriculum teaching should reflect the tool character and human character of Chinese. And the goal of Chinese curriculum teaching should focus on improving students' thinking ability; appreciate level; oral express ability and writing ability. In this way, Chinese curriculum teaching can realize the real value of Chinese.

According to the curriculum structure of Applied Undergraduate colleges and combined with the characteristics and capabilities of preparatory college students, our school divides the comprehensive, highly Literary Courses "reading and writing" is into three courses: "Mandarin and communication", "Writing", and "Reading and appreciation". Each course is divided into two semesters. The three courses are all provided for arts classes. Practical writing "course" and

"Mandarin and communication" course are provided for science classes. These three subjects have their own teaching goals: Mandarin skills and communication classes should focus on improving students' "speak" ability; writing classes should focus on improving students' "writing" ability and reading and appreciation course is focus on improving students' "reading" ability. In guarantee of the respective independent teaching courses, the three courses should cooperate and coordinate in the teaching content, teaching schedule, extracurricular activities, homework and examination.

4.2 *Reform teaching content, reflect the times, targeted and open*

Textbook is an important part of the course. At present, preparatory course has no one practical and authoritative textbook. The teaching materials used in preparatory course of Chinese colleges are not uniform. Some use the national college preparatory textbook "Reading and writing" (Tianjin University Press, 1993), which is slightly old-fashioned in the arrangement of the contents; some use the "University Chinese" as a textbook from different publishing house, whose pertinence is not too strong.

For the use of preparatory course materials, applied undergraduate colleges and universities should focus on actual ability of preparatory college students and make the accurate position (delivering qualified undergraduate talents to undergraduate college) to flexibly choose textbooks. First, we should choose some essays, which can reflect time characteristics, praise traditional Chinese national culture to arouse students' imagination and creative ability; train students' ability of appreciation of the beauty and improve students' Chinese accomplishment. Second, we should select some practical teaching materials to guide students strengthen practical training and better adapt to the rapid development of society. Correspond to the three courses described above, our college selects materials are Chen Xingyan editor of the "Mandarin spoken English Course" (Tsinghua University press), edited by Xue Ying "The new case of Applied Writing Course" (Beijing Institute of Technology press), College of education of Guangxi University For Nationalities compiled "Reading and appreciation course". We think the three textbooks are practical and full of target for the preparatory undergraduate students.

4.3 *Reform traditional single class teaching model, build varied efficient class teaching model*

Classroom teaching is the main position of personnel training. Classroom teaching mode is the main means to realize transformation from knowledge to ability. Training applied talents need dynamic classroom and varied teaching mode. The traditional classroom takes old teaching method, such as "chalk and talk", "cramming", "What I say goes". Classroom teaching for applied talents should be liberated from the traditional teaching method. In the other word, we should transform from "teacher driven" to "task driven"; transform from the "one-way output" to "exchange"; transform from the "program" to "focus"; transform from "closed" to "open"; transform from "The common" to "generality, individuality and" and transform from "Body type" to "The dominant type".

The core idea of preparatory Chinese course in applied undergraduate universities is to cultivate the students' language application ability; let the students learn how to learn; promote the development of students' inquiry ability; develop good language learning methods and habits and form necessary language literacy for lifelong learning. Therefore, the implementation of classroom teaching should be around these cores. In the "reading and appreciation" course, we mainly take research and open teaching method; highlight the importance of reading and practice; at the same time teach the theory and method of literature criticism; help students consciously use knowledge of literary theory to instruct reading and improve the ability of reading and appreciation. For the "Writing" course, students have been sufficiently trained to write basic article in the middle school and lack of practical writing. In order to help students build new writing knowledge system to meet undergraduate course of all majors, we focus on the practical style of writing. "Mandarin and communication" course is a skill lesson. We adopt scientific, effective and flexible teaching methods to focus on teach the knowledge of mandarin communication and express, such as teach knowledge, train skill and train thinking ability. And we mostly focus on strengthening the Mandarin expression and communication ability of students. Ensuring the independence of their teaching courses, three courses should cooperate and coordinate in the teaching content, teaching schedule, extracurricular activities. This is very helpful to improve students' comprehensive quality of Chinese ability.

4.4 *Reform curriculum evaluation system; promote the improve of the student's Chinese accomplishment*

Curriculum evaluation is an important part of the cultivation system of innovative talents. The scientific and effective examination should be guided by the innovative education and quality education and should be based on the development of students' knowledge, ability, quality and psychology, etc. and fully give play to the regulating, evaluation and feedback effect. The traditional mode of education evaluation is single. In many cases, traditional education evaluation is just a result of exam paper. Examination contents "value knowledge, despise ability; value memory, despise innovation; value written, despise application". This is opposite with the innovative talents training target. Therefore, we must reform the traditional evaluation way and grasp the connotation of understanding, good memory and applying, etc.

How to establish scientific and reasonable evaluation system for applied undergraduate college preparatory education, making the best use of the teaching and learning "two-way test" function and better meeting the needs of the reform of higher education are important problems for deepening the reform of teaching, which cannot be ignored. In the Chinese curriculum evaluation, according to the target of training applied talents, our school takes the flexible examination method. For example, for "reading and appreciation" course, we take the open book examination and examination contents are all subjective question. "Writing I" course takes the close book examination. The form of examination paper is increasing subjective component and appropriately reducing objective questions, like the fill in the blank, choice, judgment, which have the standard answer. And the target is checking the students' critical thinking ability. On the other hand, "writing II" requires students write papers; "Mandarin and communication I" is organizing all the students to participate in the mandarin level test and the test scores of students will be used as the final exam scores; the evaluation of "Mandarin and communication II" is carried out in the concrete practice teaching. The final score consists of test score and peacetime score. Test scores take 60% and peacetime score takes 40%. The peacetime score includes: homework, attendance, classroom practice, reading notes, etc. This approach can reduce the proportion of test scores in curriculum examination and make the constituents of scores and evaluation varied. And also, this approach help teachers effectively control the students' learning process and also help students develop the correct study method and the habits and enhance the ability of students' self evaluation.

In a word, with the development of higher education, preparatory Chinese courses in application oriented institutions must find a foothold and find a breakthrough in the curriculum setting, teaching goals, appraisal of curriculum resources, class teaching, and checking assessment and so on. In this way, we can find our own position in the wave of education reform.

ABOUT THE AUTHOR

Lu Xiaozhen (1969–), female, Zhuang, Guangxi Baise, master of Baise University, research direction Chinese lecturer, linguistics and applied linguistics. Address: School of education Baise University foundation No. 21 Guangxi Baise Zhongshan two road (533000).
Tel: 13607766769 e-mail: ywxz609@163.com

REFERENCES

[1] Dong Yinghong. Ethnic Preparatory Chinese textbooks compiling principles of minority education research [J], In 2007 third period.
[2] Zhang Junzhen. Study on the setting principle of Shaanxi province undergraduate courses. Shanxi Education (Higher Education) [J], In 2012 ninth period.
[3] Deng Yunchuan. Look to the pre university education from basic education reform—the reform of University Preparatory Chinese course as an example. Study of Ethnic Education [J], In 2005 fifth period.

Sports Engineering and Computer Science – Luo (Ed.)
© 2015 Taylor & Francis Group, London, ISBN 978-1-138-02650-6

The consideration of environmental factors—an empirical study of students' oral English output from school of life science and technology

Dan Zhou

School of Foreign Languages, Wuhan Textile University, Wuhan, China

ABSTRACT: The present study intends to investigate the factors affecting life science and technology majors' oral English output from the intrinsic and extrinsic aspects, in other words, the environmental factors and affective factors influencing college English learners' oral English output in college English teaching and learning settings. Based on the findings, recommendations for more productive and humanistic teaching and educational design and practice were provided for EFL instructors and learners in China.

Keywords: oral English output; life science and technology; affective factors; input

1 INTRODUCTION

Learners' oral English proficiency is a subject that drew attention to itself as early as the beginning of the 19th century. Generally speaking, there are two factors to affect students' oral English in class. One is they fail to find suitable words to express themselves, and the other is they are afraid of making mistakes and being laughed at by classmates. But these studies are either in a general sense (e.g. Rivers and Temperley, 1978, etc.) or take as subjects ESL children (e.g. the Wong-Fillmore study and that of Joan Rubin's in Skehan, 1989:73–79) or high school students (Chen, 1997:15–19). Few studies have been done with college and university undergraduates, and in the practical English teaching in China. The study of environmental and affective factors may be one of the solutions to the problems of Chinese EFL learners' poor oral English proficiency, so that instructors may create a more ideal language learning atmosphere and students may develop a more positive attitude toward the target language in order to lower students' affective filter and facilitate great language acquisition by way of more voluntary participation in class and more practice after class.

2 LITERATURE REVIEW

Foreign language learning is a very complex process involving a lot of factors, which influence language learning from various aspects. Generally speaking, these factors belong to the following aspects: (1) learning environment; (2) personal differences; (3) cognitive process; (4) learning results. Learning environment includes social-cultural environment and language environment. Personal differences include many different personality traits, such as age, intelligence, personal affective factors, the existing knowledge (including native language), personality, motivation, attitudes, and so on. Cognitive process involves two kinds, one is the process during learning, and the other is the process at the output of the target language. Learning results, according to some theories, are divided into linguistic and non-linguistic, while some other theories divide it into inter-language system (Wang Zong-yan, 1988: 190) and the practical target language production.

3 METHODOLOGY

3.1 Subjects

A total of 305 second-year students of five English learning classes from school of life science and technology at Huazhong University of Science and Technology participated in the study. All the subjects, who enrolled in college in the year 2012, had studied English formally for six years in junior and senior high school and one year in college. The subjects ranged in age from 17 to 20. The following table shows the number of the students of various departments.

Students' participating in study

Number	Class description	Number of subjects
1	Biochemistry and Molecular Biology	60
2	Bioinformatics	60
3	Biological Science	60
4	Biotechnology	62
5	Biomedical Engineering	63

3.2 Survey instruments

The instruments used in this descriptive study were classroom observation, questionnaire, interview and SPSS software. It tried to find out which was the most significant factors influencing oral English output. Classroom observation and Interview also used in this study.

3.3 Procedure

The whole data collection lasted for one semester, about five months.

Questionnaire: The survey was handed out at the end of their third semester in the academic year of 2012–2013. The survey was finished in class and collected immediately.

Classroom observation: During the whole semester, the author observed the five classes, once every two weeks, sitting at the back of the classroom, and recorded what had been observed: teachers' teaching, students, performances, classroom communications, and so on.

Interview: At the end of the semester and after the questionnaire was collected, and some data were analyzed, the author interviewed 10 students randomly chosen. The interview was then sorted into two parts: students' opinions about the learning environment, including their ideas on their teachers, the textbook, syllabus, and their opinions on the relationship between their speaking ability and affective—personality factors. The author also interviewed the five teachers, who all cooperated quite well. They told the author their teaching methods, their opinions about their students, such as their personality, about teaching materials and syllabus according to their teaching experiences.

4 RESULTS

Table 1. Teacher/student activities in classroom.

Teacher	Class	Speak English in class	Teacher-student centered class	Feedback to answers
Mr. Wang	1	70%	Teacher-centered	Corrected students' mistake immediately
Ms. Li	2	40%	Teacher-centered	Repeated the correct answers
Ms. Liu	3	40%	Teacher-centered	Paid much attention to students' grammatical mistakes

(Continued)

Table 1. (Continued)

Teacher	Class	Speak English in class	Teacher-student centered class	Feedback to answers
Ms. Wei	4	50%	Teacher-centered	Corrected students' mistakes immediately
Ms. Zhu	5	20%	Teacher-centered	Paid much attention to correctness rather than fluency

Table 2. Enter regression (oral English output): model summary.

Model	r	R square	Adjusted r squares	Std. error of the estimate
1	1.000a	1.000	1.000	

Predictors: (constant) Personality scale, teaching methods, attitude and motivation, environmental factors, teaching materials, social cultural factors, language anxiety, self-esteem.

Table 3. ANOVA of the nine factors/ten levels.

Source	Squared total	Degree of freedom	Average total	F	Significance
SS between	24	8	3	4.54	P < 0.05
SS within	27	41	0.66		
Total	51	49			

According to F distribution, suppose P = 0.05, F (8, 41) = 2.18. Since F = 4.54, which is greater than 2.18, there is the significant difference among these factors affecting oral English output.

5 DISCUSSION

5.1 The relationship between affective factors and students' oral English output

5.1.1 Attitude and motivation

Generally speaking, the results of this investigation showed that the students like to and even long to learn oral English. They attached great importance to their oral English proficiency, and more students have instrumental motivation than students have integrated motivation, but they lack practice and guidance, and are prevented from practicing by psychological or affective factors (e.g. fear of making errors, afraid of being laughed at by others). It is a sad reality of our colleges and universities that the students have spent ten years, or even longer, learning only a kind of "mute" English. It would be almost impossible for the learners to really gain a command of the foreign language by learning it in this way.

5.1.2 *Anxiety and self-esteem*

Anxiety arises from communication comprehension, poor performance, fear of negative evaluation and bad learning experience. Anxiety is likely to have a debilitating effect on EFL learning, especially on oral English learning. Students' anxiety in class is connected with teachers' behavior. Accordingly, anxiety may well be caused by the traditional teaching method in China. Text-centered teaching separates teachers and students, and teachers' absolute authority still pervades much of the class, which hinders understanding between teachers and students. There has always been a tendency, on the part of the teacher, to claim superiority over his or her students and, consequently, to lose sight of his or her role in class. The teacher who evinces these characteristics keeps on blaming the students for their aberrant behavior and "unsatisfactory" performance; he hardly ever bothers to make a probe of the students' cognitive, emotional and psychological background. In this case, students are more anxious and thus reluctant to participate in class.

5.1.3 *Personality*

Personality has fairly low correlations with oral English proficiency. But from the observation and interviews with students, it seems that extroverted students are better in speaking, the reason of which maybe that extroverted students have more chance to practice both in class and outside class. All in all, it is revealed that oral English output is significantly influenced by the level of language anxiety they experienced in language class and their language anxiety in turn is influenced by a series of other environmental and affective factors, such as classroom atmosphere, the proper arrangement of classroom teaching, and so on.

5.2 *The relationship between environmental factors and students' oral English output*

5.2.1 *Teaching methods and teacher-student/ student-student interaction*

From the observation and interview, we see that only one teacher uses the communicative teaching approach, which emphasizes students' participation and communication in class. The pattern of college English teaching is still teacher-centered. Most of the teachers spend most of the class time on the explanation of grammar rules and usage of vocabulary, and on the analysis and translation of the text sentence by sentence. They think that students can grasp and use what teachers have taught them, yet to grasp a language mainly depends on students' input and intake of the materials and large mount of practice repeated a lot of times. According to modem teaching methodology, language learning process falls into three stages: input (reading and audio/visual), intake (processing and memorizing), and output (speaking, writing and translating). Without the stage of intake, there will be no output. One can hardly envisage a language learning situation in the absence of an interaction of the student with his/her fellow students, the teacher and the textbook. Each time the student interacts with any of theses sources, she makes various hypotheses about what she is learning, and accepts or rejects them, trying out new ones. What should be pointed out is the importance of human interaction in the classroom as a condition for successful oral English learning.

5.2.2 *Teaching materials, syllabus, and methods*

Teaching materials, syllabus and methods are important component in any education. Teaching syllabus is the guide to materials selection and methods application. The course books for students are designed according to the teaching syllabus. Some teachers take a one-sided approach to the teaching goals of the College English Teaching Syllabus. But they neglect the requirement of "communicating information by means of English". They lay the importance to reading materials input, seldom giving consideration to intake and output. This symptom determines their materials selection and teaching methods in class. Teachers only explain the text in great detail, do the exercises connected tightly with the comprehension of the text and the usage of words and phrases, but neglect the exercise for oral English practice.

5.2.3 *Social cultural factors and students' oral English output*

Seen from the questionnaire, about half of the students have the social cultural problems in learning oral English. They do not want to take risks to give the wrong answers. They think it is more secure to keep silent in class, so as to obey the traditional Chinese cultural values and customs, or not to lose face. The influence is rooted deeply in their minds, even if they do not realize it or they do not admit this point. He/she fears whether he/she can use the cultural norms accepted by both cultures, these factors, therefore, are strongly interrelated and they all influence students' participation in class and practice outside class.

6 CONCLUSION

The classroom environment provided a lot of knowledge about the target language, but it provided little time and opportunity for each subject to practice his/her oral English. Besides, the features of most of classroom exercises required the subjects to pay much attention to language components teachers taught, and teachers' questions in class were mostly aimed at the language components and the understanding of the learning materials, all of which made the subjects have no opportunity to express their ideas coherently. Most of the subjects had positive attitude toward oral English learning, but due to their poor English basis, teachers' bias on oral English, as well as the influence of the textbook, they did not have high motivation in learning and practice oral English. Because of the influence of Chinese traditional cultural and moral values, they usually kept silent in class, which hindered their practice of oral English. Other affective factors such as self-esteem, anxiety, and personality all prevented them from opening their mouths in class.

REFERENCES

Arnold, J. (2000). Affect in Language Learning. Beijing: Foreign Language Teaching and Research Press.

Betas, M. (1990). Contexts of competence: Social and cultural considerations in communicative language teaching. New York: Plenum Press.

Claire, k. (1999). Context and Culture in Language Teaching. Shanghai: Shanghai Foreign Language Education Press.

Cook, J.B. (2000). The Communicative Approach to Language Teaching. Shanghai: Shanghai Foreign Language Teaching Education Press.

Section 2: Social sports and sports economy

Sports Engineering and Computer Science – Luo (Ed.)
© 2015 Taylor & Francis Group, London, ISBN 978-1-138-02650-6

The compared research of Adapted Physical Education between China and the USA

LinJie Wei & Zhe Li
Xian University of Architecture and Technology, Xian, Shaanxi, China

ABSTRACT: Adapted Physical Education (APE) is originated from China, but it has a full development in the United States. The wholesome legal system provides a good condition for developing APE. This article describes the origin of APE and compares the relative policy and laws of APE in China and the United States. Compared with America, Chinese adapted physical education exists the following problems: defective relevant laws and regulations, lack of public infrastructures, inadequacy of attention from the whole society.

Keywords: Adapted Physical Education; origin; development; law; China; the United States; compare research

1 INTRODUCTION

Adapted physical education is an individualized program including physical and motor fitness, fundamental motor skills and patterns, skills in aquatics and dance, and individual and group games and sports designed to meet the unique needs of individuals. The word "adapt" means "to adjust" or "to fit". Adapted physical education is viewed as a sub discipline of physical education that provides for safety, personal satisfaction, and successful experiences for students of differing abilities. Adapted physical education may include students who are not having disability but might have unique needs that call for a specially designed program. This group might include students who are restricted because of injuries or other medical conditions, the low fitness (including exceptional leanness or obesity), inadequate motor development, or low skill, or those individuals with poor functional posture. These individuals might require individually designed program to meet unique goals and objectives[1].

2 ORIGIN

Although significant progress concerning educational service for individuals with disabilities has been relatively new, the use of physical activity or exercise for medical treatment and therapy is not new. Therapeutic exercise can be traced in 3000 B.C in China. As early as written in the Qin and Han Dynasty in 《Huangdineijing》 which is recorded through guiding, massage, it restores the physiological function, the practice of rehabilitation exercise for the disabilities. The ancient Chinese famous doctor used Deep Breathing, Wuqinxi, Eithtkam, and all kinds of recreational activities for treating other disorders such as paralysis, mutilation of muscle atrophy, to restore physical and mental function in patients with examples [2].

As it can be seen from the course of human history, intellectual and physical education of disabled people indicates that social material, politics, the progress of science and technology civilization, the disabled physical education have become important symbols of social civilization standard in a country.

However, it is a recent phenomenon that physical education or physical activity is to meet the unique needs of individuals with disabilities. Efforts to serve the population through physical education and sport have been given significant attention only during the 20th century, although efforts began in the United States in the 19th century.

3 DEVELOPMENT

3.1 *China*

In China, the disabled sports and education have the development of small scale during the period of the late Qing dynasty. In 1874, Beijing established the first Chinese special education school which named "Be bright eye academy". Shandong Penglai County has set up the first Chinese school for the deaf in 1887. Children in these schools do some physical exercise like kicking shuttlecock and shadowboxing, but it was the early time of disabilities physical exercise. After the founding of new China in 1949, the Chinese government had highly focused on disabilities sports and established a series of laws to guarantee the development of disabilities' sport.

Under the lead of government, some welfare enterprises and institutions, community, special-education schools have widely carried out various forms of mass sports activities, holding nearly 300 disabled physical education training classes. Nearly 10,000 people have attended it. The number of amateur athletes who take part in the disabled games has reached more than 200,000. There are hundreds of high level coaches and referees in sports training, competition organization working for the disabled people.

From 1982, Chinese disabled delegates attended several Paralympic games, including "South" game for the disabled people, Special Olympics, the deaf Olympic Game, Bland Games and the World Championship and got more than 2,000 medals. Chinese disabled delegates, for the first time, entered the world of the top10th Olympic Game is in Atlanta, USA in 1996. In 11th Paralympics held in 2000, the Chinese sports delegates finished sixth. The Chinese delegates in the fifth, sixth, seventh sessions "south" continuously kept gold MEDALS and total MEDALS for the first games for the disabled. In the 2008 Beijing Paralympics and the 2012 London Paralympics, the numbers of Chinese delegates' total MEDALS is in a row for the first two times. Especially in 2012, the Chinese delegates got 95 gold medals, 71 silver ones, and 65 bronze ones. The total number was 231, and it was in the first place [3].

3.2 *The United States*

In 1838, physical activity began receiving special attention at the Perkins School for pupils with visual disabilities in Boston. This was the first physical education program in the United States for students who were blind. According to Charles. E. Buell's (1983) account, it was far ahead of most of the physical education in public schools.

Most students in the history with adapted physical education recognized that medically oriented gymnastics and drills developed as the forerunner of modern adapted physical education in the United States in the latter years of the century. Sherrill (2004)[4] states that the physical education prior to 1900 was medically oriented, preventive, developmental, and corrective in nature.

From the end of the 19th century to the 1930s, programs began to shift from medical education, and concerns for the whole child emerged. Compulsory physical education in public schools increased dramatically, and physical education teacher training (rather than medical training) developed for the promotion of physical education.

This transition resulted in broad mandatory programs, which consist of games, sports, rhythmic activities, and calisthenics designed to meet the needs of the whole people. Individuals unable to participate in regular activities were provided corrective or remedial physical education.

4 POLICY AND LAW

4.1 *China*

The Chinese government has given great importance to the disabled sports since the state established. In order to gradually make the disabled sports more standardized, institutionalized and legalized, the relevant state departments successively formulated and promulgated a series of laws and regulations for the development of undertakings of physical culture and sports for the disabled in China. These Acts and Laws are as follows:

4.1.1 *People with Disabilities Act of the People's Republic of China (1990)*

The 36th rule says: "Nation and society encourage and help disabled people to join in all kinds of culture, sports and recreation activities to enrich the mental needs of disabled people." The 37th rule says:"The culture, sports, recreation activities should face to the basic level and suit for different needs of disabled people."

4.1.2 *The Sports Law of the People's Republic of China (2000)*

The 16th rule says: "The whole society should care about and support the elder, the disabled sports activities. The governments shall take measures for the elder and the disabled to get into sports activities." "Schools shall create conditions for organizing sports activities suitable to the characteristics of the disabled students." The 46th rule says: "Public sports installations and facilities shall open to the public. Students, the elder and disabled people have the preferential measures. At the same time, government shall improve the utilization rate of sport facilities."

4.1.3 *The National Fitness Program Outline (1995)*

The 15th rule says: "Extensive disabled sports and fitness activities improve the physical quality of the disabled people and provide the equal opportunity for them to participate in social activities, rich the fitness method of the disabled sports, cultivate sports backbones, and raise the sports level of the disabled people."

4.1.4 *2001–2010 Sports Reformation and Development Outline (2000)*

The 11th rule says: "Pay close attention to the elder and disabled sports. The sports organization shall pay attention to the elder and disabled's characteristics and help, and guide them to take part in the sports activities."

4.2 The United States

Four laws or parts of laws and their amendments have had significant impact on adapted physical education or adapted sport: IDEA, section 504 of the Rehabilitation Act of 1973, the Olympic and Amateur Sports Act, and the Americans with Disabilities Act (ADA).

4.2.1 Individuals with Disabilities Education Act (IDEA)

This act was designed to ensure that all children with disabilities have chance to have a free appropriate public education (which emphasizes special education) and related services designed to meet their unique needs and preparing them for employment and independent living. In this legislation, the term "special education" is defined to mean specially designed instruction at cost to parent and guardians to meet the unique needs of a child with disability, including instruction conducted in classroom, at home, in hospitals and institutions, in other settings, and in physical education (OSE/RS, 2002). IDEA specifies that the term "related services" means transportation and any development corrective, or other supportive services required to assist a child with a disability to benefit from special education.

4.2.2 Section 504 of the Rehabilitation Act of 1973

The right of equal opportunity also emerges from another legislative milestone that has affected adapted physical education and sport. Section 504 of the Rehabilitation Act provides that no other qualified individual with disability or by reason to that disability can be excluded from participation, or be denied the benefits, or be subjected to discrimination under any program or activity receiving federal financial assistance (Workforce Investment Act of 1998). An important intent of section 504 is to ensure that individuals with disability receive intended benefits of all educational programs and extracurricular activities. To be equally effective, a program must offer students with disabilities equal opportunity to attain the same results, to gain the same benefits, or to reach the same levels of achievement as peers without disabilities.

4.2.3 Olympic and Amateur Sports Act

The Olympic and Amateur Sports Act of 1998, has contributed significantly to the provision of amateur athletic activity in the United States, including competition for athletes with disabilities. This legislation has led to the establishment of the United States Olympic Committee (USOC) and has given it exclusive jurisdiction over matters pertaining to U.S. participation and organization of the United States in the games. The USOC will encourage and provide assistance to amateur athletic programs and competition for amateur athletes with disabilities, including the feasible, the expansion of opportunities for meaningful participation in programs of athletic competition for athletes without disabilities.

4.2.4 Americans with Disabilities Act (ADA)

The Americans with Disabilities Act has been passed in 1990, whereas section 504 focused on educational right. This legislation extended civil rights protection for individuals with disabilities to all areas of American life. Provision includes employment, public accommodation and services, public transportation, and telecommunications. Related to adapted physical education and sport, this legislation requires that community recreational facilities, including health and fitness facilities, should be accessible, and appropriate reasonable accommodations should be made for individuals with disabilities. Physical educators must develop and offer programs for individuals with disabilities to give them the ability to participate in physical activity and sport experiences within the community.

5 THE QUESTIONS

Although historical research has shown that adaptive physical education was originated from ancient China, but it was developed slowly in modern China. Chinese vigorous development of competitive sports undertakings for the disabled is not bringing the correspondingly development of Chinese adapted physical education. Compared with American, Chinese adapted physical education has following problems.

5.1 The relevant laws and regulation system is not sound

The relevant laws and regulation of USA specific lists indicate that disabled individuals should enjoy the priorities at school. The education department has considered the diplomas, which disabled individuals will face when doing physical education teaching plan and scheme, which fully embodies the equality of human rights. In contrast, the Chinese government tended to adapt the policy and regulation of sports education, but it didn't take action. In fact, the physical education class of special school in China has no syllabus, no teaching scheme, no textbook, no aim in the 1980s. It was until 2007 that the ministry of education issued 'the blind, deaf school, intelligence school compulsory education curriculum experiment plan', which just simply explained the physical education class for blind, deaf, and intelligence school such as to make the students know some knowledge about sports health, and ensured that students have one hour each day for physical exercise to strengthen students' physique, improve the students' sports ability, etc. It doesn't include how to reform and carry out the criteria of physical class and the physical class administrative mechanism.

5.2 Lake of public infrastructure investment

The laws and regulations related to the adaptive physical education set by the U.S. government are not only just limited to the education, but also a combination of public transport, electricity, communications, public infrastructure, such as the laws and regulations of the relevant departments which made more comprehensive protection for rights for the disabled from all aspects. In American schools, hospitals, shopping malls, public transportation department and sports sites and gyms and other public service organizations, almost everywhere has dedicated channel, seat, parking and other services for the disabled. These facilities fully embodied the first thought of American government of human rights. To some great extent, Chinese disabled individuals have difficult in transportation, school and work, due to the inadequacy of infrastructure investment for the disabled.

5.3 Adaptive physical education doesn't get attention of the whole society

The understanding of adaptive physical education depends on the society and the progress of the times. The main body of adaptive physical education is disabled, and the physical education for the disabled cannot be simply interpreted as physical training which is not only limited to meet the physiological needs of disabled persons, but to help the disabled through sports activities in the mastery of sports skills, improve the physical quality at the same time, and improve their quality of life. Under the system, although Chinese competitive sports undertakes for disabled people to get a good development, but adaptive physical education has not received extensive attention of the whole society. The disabled people still belongs to the edge of the society. Their social welfare and the right to education have not been truly respected. Most disabled people just study in special school. The opportunity of employment is small. In this case, it lets the adapted physical education alone.

The United States is the country of best development the adapted physical education in the world.

The perfect legal system ensures that the disabled individuals can fully enjoy equal education power. The developed public infrastructure and hardware input for the disabled individual provides convenient conditions for various social activities. The whole society respects people with disabilities, and gives the full affirmation to their citizens' basic rights. These made the good condition to develop the adapted physical education. China should improve the related policies and regulations of the adapted physical education as soon as possible, and strengthen the construction of infrastructure for the disabled individual, and enhance the whole society to attach importance to disabled individuals in education, employment and health care consciousness, and develop Chinese adapted physical education in an all-around way.

BRIEF INTRODUCTION OF THE AUTHOR

Linjie Wei, female, (1980.7–), Master, the lecturer of Physical Education Department of Xian University of Architecture & Technology, her main research direction is Physical Education & Sports Human Resource Management.

Zhe Li, male, (1974.4–), Master, the lecturer of College of Information and Control Engineering of Xian University of Architecture & Technology, his main research direction is Education & Smart Building.

REFERENCES

[1] Joseph P. Winnick, EdD. University of New York, College of Brockport. [M] Adapted Physical Education and Sport (Fifth Edition), P4.
[2] Disabled person. Baidu encyclopedia.
[3] Medal count—2012 Paralympics. [J] Palaestra, No. 3 2012, 56–57.
[4] Sherrill, C. (2004). Adapted physical activity, recreation, and sport: Cross disciplinary across the lifespan (6th ed.). Dubuque, IA: Wm. C. Brown.
[5] The second national handicapped person sampling survey in 2006 main data bulletin [N] Chinese Times, 2007–5–2 (4).
[6] Yi Guo, The research of Chinese disabled employment discrimination, [D] Northwestern University, 2008.
[7] Terry L. Rizzo, Top 10 Issues in Adapted Physical Education A Pilot Study—How Far Have We Come, Palaestra, (2013), 27, 21–26.
[8] Lee, S.M., Burgeson, C.R., Fulton, J.E., & Spain, C.G. (2007). Physical education and physical activity: Results from the School Health Policies and Programs Study 2006. Journal of School Health, 77, 435–463.
[9] Kelly, L. (2011). Connecting GPE and APE curricula for students with mild and moderate disabilities. Journal of Physical Education, Recreation, and Dance, 82(9), 34–40.

Sports Engineering and Computer Science – Luo (Ed.)
© 2015 Taylor & Francis Group, London, ISBN 978-1-138-02650-6

Social status research on Anhui Province disabled competitive sports

DaWei Cao
Assets Management Department, Huaibei Normal University, Huaibei, Anhui Province, China

Chao Liu
Sports Department, Huaibei Normal University, Huaibei, Anhui Province, China

ABSTRACT: Research approaches such as on-the-spot investigation, interview and literature review are used to analyze the social status of disabled competitive sports in Anhui Province, involving social environment, organization management, and economic situation. Disabled sports policy laws and regulations implements were pointed out to intensify with great effort, as well as strengthening the social attention. Management system should be improved, which finally formulate systematization, institutionalization, legal management. Policy input funds should be increased. All social forces should be mobilized. More channels for the funds should be raised, as well as doing well with disabled sports training center as the guidance role. Construction and operation training bases should be make great efforts to coordinate with other public stadiums.

Keywords: Anhui Province; social status; competitive sports; disabled sports

1 INTRODUCTION

It has been long for the development of disabled sports in Anhui Province. The disabled person's life, study and health got the concern of the government since new China been founded. From the 1950s, the disabled were organized to join in physical training and different kinds of sport competition in areas such as Fuyang, Tongcheng and Wuhu[1]. By the end of the establishment of the Federation of the Disabled in Anhui Province in 1988, the disabled person sports had received great attention in the whole society, more and more organized physical training and sports competitions appeared, at the same time, some young athletes were discovered. Some of them got regular training, and achieved good results, for national disabled sports athletics level which was very low at that time[2].

From the Federation of the disabled in Anhui province established in 2007 to the Seventh National Paralympic Games, through long-term training and the accumulation, young disabled athletes had made new achievements in sports competitions in local and abroad. Among them, in March 1992, Anhui Province delegated to participate in the National Paralympic Games for the first time, and won 5 silver medals and 7 bronze medals. In May 1996, in the Fourth National Paralympic, they gained gold medal for the first time, they won 6 gold, 5 silver and 7 bronze medals in total. From then on, Anhui Province began the exploration of local road for disabled sports, and the biggest feature of that time was occasional. Because of lack of funding, top athletes and young athletes for long time, they became more and more serious since Seventh National Paralympic Games. The main characteristic of this stage was that, where there was competition there was sport training, and where there was no competition it stopped. Although there were some domestic elite disabled athletes in some individual events, it could not cover the middle or lower level of Anhui disabled sports at that time. To sum up, in today's new situation of the developing disabled sports, how can Anhui province's competitive disabled sports take advantage of their own advantage to get new development, reinforce their deficiencies, and shorten the distance between the provinces in strong competitive disabled sports to achieve the goal being one member of the second class in China's competitive disabled sports and promote the National Paralympic Games birthplace is the main focus. Therefore, it is necessary to study the present situation of competitive disabled sports in Anhui Province. It not only focuses on the present, but also for the future, in order to achieve further development. Research methods such as literature material, survey, expert interviews, mathematical statistics and logical analysis are used in this paper. Status quo of Anhui Province disabled person competitive sports is regarded as the research object, the leaders and coaches who engage Anhui Province disabled person athletics sports and athletes are worked as survey object. Through investigating and analyzing the basic status quo of Anhui Province disabled person competitive sports, the author finds the existing problems and puts forward the corresponding proposal.

2 STUDY ON SOCIAL ENVIRONMENTS FOR COMPETITIVE DISABLED SPORTS IN ANHUI PROVINCE

2.1 Policies and regulations of competitive sports-related of Anhui province

In order to provide a good social environment for the development of disabled sports, Anhui provincial government has always attached great importance, according to laws and regulations of the state and relevant departments, they put forward a series of specific requirements and tasks combining with the their own circumstances.

According to Anhui province national economic and social development of the Twelfth Five Year Plan and the Chinese disabled person enterprise Twelfth Five-Year Plan development compendium, the Disabled Sports are required to further concern and support to enhance the handicapped sports cause[3]. In October 2006, the first award document in recognition of outstanding disabled athletes' policy was promulgated in Anhui Province, which has get positive effect to promote and facilitate the development of disabled sports in Anhui Province[4]. In October 2007, with some suggestions promulgated by the State Council about strengthening our province disabled person sports work, some new ideas and effective measures were put forward, also combining with the characteristics of Anhui Province[5]. These laws and regulations promulgated not only reflect the attention from the Party and government to the disabled sports, the equality of human rights protection, but also embody the humanistic care for the vulnerable groups. They guarantee the development of the competitive disabled sports. The enforcement of these policies and regulations and some specific measures is still needed to be increased further to the intensity of Anhui Province, also the disabled sports social and physical environment needed further improvement.

2.2 Social concern degree for disabled competitive sports

Disabled sports are a part of the enterprise for the disabled, and have been proud to become a unique part in sports. The disabled athletes can take sports as a display of self-ability and ways to improve inner spiritual world, show the society their ability instead of disabled[6]. Once realizing self-value and social acceptance, they may expand their communicating field[7]. Therefore, the socialization degree is directly related to the development of sports for the disabled. But the research finds that the degree of social concern is very low in Anhui province, which involves few competitions held, few media reported, few watching audience and social sponsors, small range participating people. It restricts the popularity and socialization for the disabled sports in Anhui Province.

3 STUDIES ON MANAGEMENT OF COMPETITIVE DISABLED SPORTS IN ANHUI PROVINCE

3.1 Management of organizational structure on Anhui province disabled competitive sports

Through the investigation, the disabled sports management subject in Anhui province had been temporary situation for long time. On one hand, it was under the direct administrative management of the Anhui Disabled Persons Federation. On the other hand, it had to accept business guidance from Anhui province sports bureau. The province Disabled Persons Federation had the function of management and operational guidance to the Disabled Persons Federation of different cities and training bases. National Paralympic Committee of Anhui province delegation is assembled by the provincial government, the provincial Party Committee Propaganda Department, provincial Disabled Persons Federation, the provincial Civil Affairs Department and the Provincial Sports Bureau, which is responsible for the entries task, coaches' transferring and training, athletes' selection and training, the delegation' s equipment procurement. The provincial delegation was directly managed by provincial Disabled Persons Federation. Anhui Province Sports Association for the disabled was established in April 1984 in Hefei, and renamed to Anhui Province Disabled Sports Association in 1993. The majority of city Disabled Persons Federation has disabled sports association, which is responsible for the management and coordination of the disabled sports and related matters. In fact, it is mostly useless, without due role. In 2012 Anhui province disabled sports training center was set up, it has a small amount of staffing and training venues, from then on, Anhui province disabled person sports developed towards stability and effective direction.

3.2 Sports event project setting up of competitive disabled sports in Anhui province

Sports event project setting and layout of Anhui Province competitive disabled sports must be on the basis of social economy and achievements of Anhui Province. Some major match such as Paralympic Games, Asia Paralympic Games and National Olympic Games are participated, which need a comprehensive planning.

The Anhui Disabled Persons Federation has a clear planning. Track and field, swimming will be two infrastructure projects. Advantage projects, such as table tennis and weightlifting will be strengthened, which are also the national key projects. New individual projects should be break through, especially the project heavy athletics and blind judo, which are suitable for our province. Training work has been started in swimming, track and field, table tennis, weightlifting, shooting, blind goalball, sitting volleyball, and deaf basketball.

4 STUDIES ON ECONOMIC CLASS OF COMPETITIVE DISABLED SPORTS IN ANHUI PROVINCE

4.1 Sources of funding and investment of competitive disabled sports in Anhui province

Anhui province is an underdeveloped economically areas, the development of disabled sports funds has been not fixed for a long time, with narrow funding source channels. 100% of the funds are from the public undertaking expenditures, whose amount is not fixed, from 200 thousands yuan in 2008 to 1800 thousands yuan in 2012. The government regulates an annual extraction funds from 8% welfare lottery, but the actual for the disabled sports is also very limited. Enterprises from social donation and the other way are very small and random. At present, Anhui disabled sport is actively striving for the provincial government and financial departments' attention and devotion. From 2009, funds from the sports lottery has been extracted, in addition, Anhui Disabled Persons Federation has launched initiatives to ensure the disabled sports first-line and second-tier athletes training funds paying from the provincial employment security funds for the disabled, to make every effort to support training work of each project. It will also gradually realize the three class linkage and jointly to do training security work for province, city and county.

4.2 Sites and equipments of competitive disabled sports in Anhui province

With the establishment of Anhui province disabled person sports training center and provincial disabled

sports training bases founded, such as Anqing blind gateball base, Tongling sitting volleyball base, Wuhu deaf basketball base. The problem for elite disabled athletes' training has been greatly solved.

5 CONCLUSIONS

1. *Disabled sports policy and funds need increase, at the same time, all forces of society are mobilized to raise more funds by different channels, using a variety of opportunities to seek cooperation from companies, striving for material and financial support.*
2. *Anhui province disabled sports training center must consummate to improve the management efficiency, and in a long-term, perfect the disabled sports development strategy management institution. In order to make the disabled sports systemized, institutionalized and legalized, then relative policies, laws and regulations must be established and improved.*

ACKNOWLEDGMENT

This work was financially supported by the Anhui Provincial Bureau of sports (ASS2013312).

REFERENCES

[1] Anhui province disabled person sports activities. Anhui Province [EB|OL] http://www.61.191.16.234:8080/was40/index_sz.jsp?Rootid = 3292&channelid = 36620.
[2] Zhao Jingyan. The status quo of Yunnan province disabled sports development survey [D]. Master of education graduate degree thesis of Yunnan Normal University, 2006:24.
[3] The people's Republic of China Law on the protection of disabled persons [EB|OL]. http://www.cdpf.org.cn/zcfg/content/2007–11/29/content_50523.html.
[4] Interim about athletes and coaches who have made outstanding contributions in Anhui Province [EB|OL].http://www.ahzwgk.gov.cn/XxgkNews-Html/OA077/200802/OA077050600200802002.Html.
[5] Views of the general office of Anhui Provincial People's Government on Further Strengthening the work of disabled sports [EB|OL].http://www.ah.gov.cn/zfgb/gbCONTENT.asp? Id = 4252.
[6] Linstorm, H. Integration of sport for athletes with disabilities into sport programs for able-bodied athletes [J]. Palaestra, 1992, 8(3):28.
[7] Sherrill, C. Social and psychological dimensions of sports for disabled athletes [J]. Sorts Science Review, 1986, 5(1).42–64.

Sports Engineering and Computer Science – Luo (Ed.)
© 2015 Taylor & Francis Group, London, ISBN 978-1-138-02650-6

Study on the investment situation in leisure sports of electric power enterprise staff in Hebei Province

ShuPing Xu
North China Electric Power University, Beijing, China

ChengBin Ji
Beijing College of Politics and Law, Beijing, China

ABSTRACT: Taking the electric power enterprises in Hebei Province as an example, this paper analyzes the investment situation of power company workers' leisure sports. Meanwhile it also discusses the significance and role of electric power enterprise workers' leisure sports. Therefore, the author hopes the leadership of the power enterprises will increase investment in workers' leisure sports. In return, it will enhance the physical fitness of employees, improve labor productivity, and promote better development of electric power companies.

1 INTRODUCTION

In this world, besides light, air and water, two things, leisure and games, must possess in all life. Without leisure, all lives can not be sustained. Without games, all lives are difficult to evolve. The higher the degree of social civilization is, the more we should pay attention to leisure and games. Famous American leisure expert Jeffrey at the annual meeting of the American Society for Leisure noted that in the 21st century the leisure economy prevails. Sport is an important part of people's leisure activities, and also widely accepted as a way of life. Leisure sports refer to the general term for a variety of sports activities in daily life that people use for entertainment and leisure. It is the result that modern people keep seeking and inevitable outcome of the social development to a certain phase.

While leisure sports in power companies generally refer to the mass leisure sports activities that are performed or carried out inside the companies. Its main target is the entire staff, and the main purpose is to promote the overall health of employees through physical exercise. Only in this way, they can provide enterprise employees with easily accessible sports opportunities, stimulate their spontaneous exercise motivation, cultivate their conscious exercise habits, and facilitate integrating sports into life.

As the promulgation and implementation of China's National Fitness Program, starting from social practice, many scholars collected a lot of practice survey data, and conducted fairly systematic researches on leisure sports. However, researches on electric power enterprise workers' leisure sports are still relatively weak. Through the investigation on the investment situation in staff leisure sports of dozens of power companies, the thesis seeks to provide a theoretical reference for the implementation of investment in workers' leisure sports of Hebei Electric Power enterprises and overall fitness program.

2 SUBJECTS AND METHODS

2.1 Subjects

This thesis takes the investment situation in workers' leisure sports in 10 power companies of Hebei Province.

2.2 Research methods

2.2.1 Literature
In CNKI, the author read a large number of documents relating to the research, and fully retrieved, analyzed and utilized the literature related to this study.

2.2.2 Questionnaire
The author developed a questionnaire based on the research purposes. 100 questionnaires were distributed and 98 were returned. The effective rate is 98%, and it accords with the requirements of this research.

2.2.3 *Mathematical Statistics Act*

All data was entered into computer, and then processed by using Excel software.

3 RESULTS AND ANALYSIS

3.1 *Findings*

3.1.1 *Investment status in leisure sports management of power enterprise workers*

The cultural qualities of the person in charge sports are an important factor in the development of power companies' leisure sports. This directly or indirectly affects the development of the electricity enterprise workers' leisure sports.

The author surveyed the cultural structure of the person in charge of Hebei electric power enterprise workers' leisure sports. Then the survey data shows that the men at the wheel in electricity workers' leisure sports are showing a trend of higher education: bachelor degree or above account for 45%, three-year college degree 40%, and secondary education accounts for 15%.

As can be seen from the above findings, in Hebei Province the large and medium-sized electricity enterprise workers' leisure sports management is affected by the planned economy. The management modes are lagged behind, and this is hindering the implementation of the current electric companies' leisure sports.

Most sports management teams are lack of full-time managers. Besides, most management posts are held concurrently by workers union members. In the mean time, these officers don't have a fully understanding on sports-related knowledge and public health projects because of their own reasons and companies. They are unable to resolve some thorny issues when developing enterprises' recreational sports activities. Furthermore, they often feel powerless, thereby affecting the electricity enterprise workers to carry out leisure sports.

Table 1. The questionnaire results of the cultural structure of the people in charge of Hebei electric power enterprise workers' leisure sports.

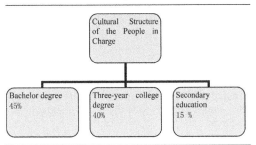

3.1.2 *Investment status in power enterprise workers' leisure sports facilities*

In most power companies in Hebei Province, there exists the situation that the actual expenditure is less than planned investment in leisure sports. The reason for this phenomenon is that the sports value of the leaders and organizers is relatively backward. It results in the funding to leisure sports becomes a mere formality and the effectiveness is poor. Most companies invest less in the planning and construction of sports venues and facilities.

According to Table 2, survey result displays that the existing sports facilities are mostly concentrated on basketball, badminton, table tennis and several other projects. Facilities are old and inadequate. In 10 medium-sized power enterprises, there are a total of 16 basketball stadiums, 22 badminton stadiums, 50 Ping-Pong tables, 8 gyms, 2 swimming complexes, 2 soccer fields, 1 ground track field, and 3 activity centers. As for 10 such enterprises the above mentioned stadiums are too insufficient and simply cannot meet the needs of employees' sport. Apparently, it has become a constraint bottleneck for the electricity employees' casual sports development.

3.1.3 *Investment status in power enterprise workers' leisure sports expenditure*

Sports consumption reflects sports funding directly. Sports consumption refers to that people buy or use sports tangible products and intangible services to meet the needs of their sport. Power Enterprise leisure sports funding is the economic foundation for powering enterprise workers to carry out leisure sports. Surely, the funding is also an important

Table 2. Table of sports projects carried out by enterprises.

Category	Basketball	Badminton	Work-break exercise
%	100	100	99
Category	Table tennis	Tug-of-war	Cards and chess
%	99	95	95
Category	Shuttlecock and rope skipping	Football	Taijiquan
%	80	80	70
Category	Tennis	Long walk	Swimming
%	60	60	30
Category	Others		
%	20		

factor for the smooth development of leisure sports of power enterprise workers. Electricity enterprise workers' sports consumption status is manifested in the following areas (see Table 3).

From Table 3, we can see that electricity workers hold a quite profound understanding of "spend money on health" and have a relatively strong sports consciousness. Only with the concept of recreational sports and consumption awareness can make people participate in the exercise more actively. Middle-young people are the backbone of the future society and the backbone of the family. It is even more needed for them to deepen their understanding of the physical and mental health, health conscious-ness, and the concept of lifelong physical exercise. Therefore, it is necessary and urgent for the power companies to increase investment in sports funding.

The survey on leisure sports funding of electricity workers primarily was implemented from two aspects of funding amount and funding sources: First, in sports funding amount, the survey results show that most power enterprises invest 30,000 and 60,000 Yuan every year. Compared to other funds the proportion of sports investment funds is very low. It is not difficult to explain that the company's values and understanding of sports have impact on business investment, and hinder the healthy development of electric power enterprise workers' leisure sports.

In terms of per capita consumption on funding for sports, the amount is smaller. The survey also found that there is a trend that the larger the enter-prise is, the smaller amount of per capita expenditure is. It also reflects the greater enterprise is, the less the management, organization and care of the electric-ity enterprise workers' leisure sports is. Secondly, in the enterprise in order to achieve "overall fitness", support from the unit alone is not enough, work-ers must bear a part. While whether the enterprise workers are willing to invest their own money to sports, it is up to their investment awareness.

In terms of sports funding sources, the survey showed that the main sources of funding are: union funds, administrative funds, self-raised by the trade unions and workers themselves, trade unions and administrative funding, commercial sponsorship,

Table 3. Electricity workers' amount of consumption (per year) in sports.

Sum of consumption	Male workers (%)	Female workers (%)	Total (%)
50 or lower	10.5	12.1	22.6
50–100	5.1	7.8	12.9
100–150	9.9	10.9	20.8
Above 150	29.7	11.8	41.5

and so on. Among them, corporate executives and union funding are the main source of electric power enterprises to carry out leisure sports. Moreover, it is conducive to the planning the resources alloca-tion from an overall perspective of the power com-panies. And in this way companies can form clear rules of responsibility and obligation of workers sports, and remove all possible interference, so that enterprises leisure sports revolves around the power. But this enterprise investment system leads workers leisure sports activities to be noneconomic, lose self-survival and development capabilities. At last it makes power companies depend more and more on such investment.

3.1.4 Planning and organization functioning of electricity workers leisure sports investment

Power companies' investment planning in lei-sure sports activities is basically designed around contests. Taking ball contests as the main content is the basic characteristics of electric power enter-prises to carry out organized recreational sports activities. Most of the contests are conducted inside the power enterprises organized by enter-prises or employees themselves.

The planning of electricity workers' leisure sports activities reflects various power companies' number of sports activities and funding condi-tions. But the actual effect is often far away from the operation expected. Furthermore its guiding ideology is not associated with the company's cul-ture building and long-term development. This also shows that the power company's leisure sports are lack of coherence and systematic work in organi-zation and operational methods.

3.1.5 Investment status building leisure sports associations and clubs of power enterprise workers

Compared with the electric power companies of developed countries, sports associations and clubs organizations within the power enterprises in Hebei not only have a large gap in the number. Actually, the existing associations or clubs have not yet estab-lished a complete rules and regulations. Normally, let alone formally register in the power department. Private voluntary organizations of sports associa-tions and clubs have been strongly supported and advocated in the Western developed countries. They have also played a positive role for the realiza-tion of the countries' overall development strategy. Electricity companies should mobilize the enthusi-asm of employees to develop their creativity in the enterprise, set up their own associations belonging to the staff and the club organization. The result shows that it is a beneficial measure to promote the healthy development of leisure sports activities of electricity enterprise workers.

3.2 Result analysis

3.2.1 The problems of the investment status of Hebei electric power enterprise workers' leisure sports

In the management for enterprise employees' leisure sports working rights and responsibilities may not be implemented is a prevailing the problem. Now power enterprises' understanding toward physical work has a certain degree of bias, and most of the units and functional departments do not take full advantage of their rights. They do not accurately grasp the goals and tasks of electricity enterprise workers' leisure sports. The division of responsibilities is not clear so that the organization of enterprise workers leisure sports becomes a mere formality.

It can be seen from field surveys, every power company works out sports programming annually. In the variety of investments, the proportion of sports work is clearly insufficient in the aspect of either time or investment funds. A considerable part of the power companies regards the sports work as a way to improve employees' health and productivity. They do not corporate culture and business strategy and long-term development of power companies with leisure sports together. Functions and goals of leisure sports in modern power companies are diversified. Certainly, the long-term development planning of leisure sports in power companies should unify with the overall development plan of the companies. Companies should increase funding in sports facilities according to their abilities, and meet the needs of employees' leisure sports as far as possible.

3.2.2 Raise awareness of the importance of power enterprise workers leisure sports

Currently, some problems and deviations in the funding in power companies' leisure sports and sources of funding are too narrow. Our society is in transition, not only society needs to recognize the value and function of sports, but also the work thinking of electricity workers' leisure sports must also keep pace with the times. If they cannot change the work ideas from the planned economy era, walk out of the original mode of thinking, and avoid the mere formality of physical work, it is impossible to break new ground for power enterprise employees' leisure sports and play special effects of leisure sports in the power companies' culture.

4 CONCLUSIONS AND RECOMMENDATIONS

1. *Strengthen the construction of sports organizations and emphasize the actual results of employees' leisure sports in electricity enterprise*

2. *Strengthen effective funding and support the establishment of the electricity workers sports associations and clubs*

   The factual use of funds may be used for technical training, funding grassroots sports associations, clubs and venue facilities construction, etc. Sources of funding can be multiple channels, such as the absorption of corporate sponsors, the grass-roots units and staff remuneration models and so on. At the same time, the companies should also be adapted to local conditions, combined with enterprise economy. It also useful to establish a more complete enterprise workers power sports associations and sports clubs.

3. *Strengthen the construction and management of sports facilities investment*

   Sports facilities within enterprises are necessary for employees to participate in sports activities, and reflect the companies' image and appearance as well. Currently corporate decision-making leaders have been fully aware of the business of this situation that sports facilities can not meet the needs of employees' sports. This situation has also become an important factor restricting the development of leisure sports of power enterprise workers. Eventually, it is worth learning many foreign companies plan construction of sports facilities in the initial phase.

4. *Strengthen the building of electricity corporate culture*

   On the one hand, enterprise workers' leisure sports should take serving each employee's health as the purpose. On the other hand, in order to realize self-worth in leisure sports, we should give full play to the wisdom and intelligence of employees, and make employees involve, organize, plan and manage the activities.

5. *Strengthen the connection of electric enterprise leisure sports with the sports work of society and communities*

   Electric enterprises should communicate with the local sports bureaus and sports associations to exchange technology business. In order to develop and utilize a wider range of social sports resources, they'd better participate in sports activities organized together by enterprises and society.

REFERENCES

[1] Ge Ning. Survey of College Teachers in Fitness Exercises Analysis [J]. Journal of Nanjing Sport Institute (Social Science Edition), 2002 (2).

[2] Zhao Yujuan. Investigation and Analysis on Large and Medium Enterprises Sport in Henan Province [J]. Journal of Hebei Institute of Physical Education 2009 (11).

[3] Hao Xiaogang. Study on the Current Status Deficiency and Development Trend of Corporate Sports [J]. Journal of Hebei Institute of Physical Education 2003 (4).

[4] Guo Qin, Zheng Xiao-ping Study on Status and Countermeasures of the Enterprise Sports in Pearl River Delta [J]. Journal of Sports Adult Education 2008 (11).

[5] Zhang Haibo. Analysis on Enterprise Sports from the Deep Structure of Chinese Culture [J] Modern Economic Information, 2009 (4).

[6] Xu Dongming. Operating Mechanism and Prospect of Chinese Urban Community Sports Management [J]. Sports Adult Education, 2004 (4).

[7] Wu Luosheng. Study on the Opportunities of Xi'an to Develop Sports Fitness and Entertainment [J]. Sports Adult Education 2004 (5).

[8] Zhang Xuelian, Cao Jihong, Shi Dawei. Problem Faced by Sports Development of Chinese workers in Social Transition [J]. Journal of Shenyang Institute of Physical Education, 2004 (2).

[9] Tian Ping. How Embarrassed the Employee Sports Is [J]. Foreign and Chinese Corporate Culture, 2001 (2).

[10] Wu Fei, Wu Xinyu. Establish a New Pattern of Physical Work Adapted to the Market Economy [J]. Journal of Hebei Institute of Physical Education, 2003 (6).

[11] Lan Jiufu. Values in Social Transitional Period [J]. Beijing Normal University Press, 1999.

Sports Engineering and Computer Science – Luo (Ed.)
© 2015 Taylor & Francis Group, London, ISBN 978-1-138-02650-6

The research about dilemma and operation law of the local government's supply of sports public goods in W.Y.K. township in China

LuoJing Zhu

The College of P.E., Central South University of Forestry and Technology, Changsha, China

ABSTRACT: The paper stated the dilemma and the operation law of the local government's supply of sports public goods in W.Y.K. town in China through the field investigation. The study found that the local government was facing with some difficulties in process of the supply of sports public goods in W.Y.K. town as follows: 1) the supply of sports public goods' marketization and socialization didn't make town cadres more likely to work hard. 2) The supply of sports public goods' marketability and socialization has been constrained by under-developed economy. 3) The function of sports public goods' supply can't run effectively because TV broadcasting services, cultural undertakings and sports undertakings cannot be organically integrated in the unified administrative department's functions. Besides, we found there are some operation laws in process of the development of sports culture in W.Y.K. town as follows: 1) Achievements of one's official career is the most important factor in process of the development of sports culture in W.Y.K. town. 2) Sports staff can't focus on the development of sports public affairs because they are faced with the complicated affairs in W.Y.K. town. 3) It is the hidden rule that the village committees obtain the fund of rural public services from abnormal channels.

Keywords: sports public goods; operation law; supply

1 INTRODUCTION

The development of sports culture in China is still in a marginal place. Many sports scholars tend to view the development of sports culture standing in their position. So we often read many related article were written by sports scholars suggest that leaders pay attention to sports culture. There are some factors leading to the marginalization of sports culture's development in W.Y.K. town such as the aim of overall development of town, the supply system of rural public goods, achievements in one's official career and so on. The study tried to reveal the deeper causes of the marginalization of rural sports culture's development so that we can see the true situation of the development of rural sports culture.

2 GENERAL SITUATION OF W.Y.K. TOWN

W.Y.K. town locate in the middle and lower reaches of the Yangtze River. Y.X. county government is the superior government of W.Y.K. town. And Y.X. County is a state-level poverty-stricken county [1]. Chinese government set a standard for confirming qualifications of state-level poverty-stricken county. It is economic index that annual per capita income is less than 1300 RMB ($ 208). All poverty-stricken counties will obtain the central government funding for poverty alleviation. There are 18 administrative villages in W.Y.K. town. The total populations are approximately 40,000 (See Table1). Although Y.X. County is a state-level poverty-stricken county, W.Y.K. town is the most vigorous economic region in Y.X. County (See Table 2).

Table 1. Human geography environment of W.Y.K. town.

Index	Data
1. Total population	40,000
2. Civil servant (except policeman, tax official, industrial and commercial official)	36
3. Total area (km²)	77.31
4. Cultivated area (km²)	14.54
5. Area of forest land (km²)	14.63
6. Water area (km²)	21.28

Table 2. Economic environment of W.Y.K. town.

Index	Data
1. Fiscal revenue (million RMB)	8,000
2. GDP (billion RMB)	16.8
3. Value of industrial output (billion RMB)	14.2
4. Annual net income of farmer (RMB)	4674

3 REVIEWING OF THE SUPPLY MECHANISM OF RURAL SPORTS PUBLIC GOODS

3.1 Generalization about the supply mechanism of rural sports public goods

The supply mechanism of rural sports public goods comes from "the experiences of Xian'an". Hubei Provincial Party Committee made policy about comprehensive supplementary reforms of villages and towns according to "the experiences of Xian'an" [2]. It is the core content of the policy that the governments paid for social organizations, and social organizations provided rural public goods. There are two goals of policy: improving work efficiency and saving the human cost. It can also be called "socialization of public services" in this paper.

3.2 Controversy about socialization of public services

The policy maker, Yaping Song, the dean of Hubei Academy of Social Sciences, is a strong supporter of socialization of public services [3]. He said that the governments above the county level once had agencies in town. The population of employee of these agencies reached up to more than 200,000 people. In its early days, the agencies can basically meet the needs of rural public services. These agencies worked inefficiently with the development of the rural society. And it resulted in wasting of resource that the agencies employed so many extra people. So socialization of public services is sustainable way of development.

There are other scholars supporting Yaping Song's viewpoint. Xiaoping Qin pointed out that socialization of public services is a new idea. And the supply mechanism have many advantages such as improving the local government transformation, reducing cost, increasing public resource utilization, motivating civil servant's passion and so on [4]. Zhangbin has the similar concept in his article [5]. From an academic point of view, the studies lack empirical research.

However, many people have different opinions on this matter. Professor Xuefeng He proposed the ever-popular suspects. He holds the idea that the marketplace can't provide enough public services which originally came from the agencies in villages and towns [6]. On one hand, Professor He admired reform spirit. On the other hand, the actual effect about socialization of public services is not positive in practice. There are two important reasons: 1) Rural public services can't be standardized. And it can't be fully market-oriented.

2) Rural public services can hardly be put into effect without financial support in reality. In other words, the local government takes incumbent responsibility and obligation of providing public services.

It is the bone of contention that whether the supply of rural public services can be marketized through analyzing the concepts carefully of the two sides.

4 THE LOCAL GOVERNMENT WAS FACED WITH SOME DIFFICULTIES IN PROCESS OF THE SUPPLY OF SPORTS PUBLIC GOODS IN W.Y.K. TOWN

4.1 The supply of sports public goods' marketability and socialization didn't make town cadres work harder

It's the using principle of county financial funds that 30% of financial funds are paid for public services, and 70% of financial funds are paid for employee's salary. On the surface, most of financial funds are paid for employee's salary. In fact, however, annual financial wages of sports administrative department's staff are 15,000 RMB ($ 2401.5) each. Annual financial wages of temporary workers are 6,000 RMB ($ 960.6) each. Obviously, the salary level can not motivate employee's passion.

4.2 The supply of sports public goods' marketization and socialization has been constrained by under-developed economy

Different from developed areas, the government of town is not attractive to social organization. Social organization would be unprofitable in economically backwardness countryside areas because the town government can not afford to supplying public services. The supply of sports public goods has been in a dilemma. On one hand, the town government forces the sports public service to marketize; On the other hand, Social management organizations are loath to invest supply of sports public goods because of the pursuit of maximum profit of market.

4.3 The function of sports public goods' supply can't run effectively

TV broadcasting services, cultural undertakings and sports undertakings cannot be integrated in the unified administrative department's functions organically, because they are set in the unified administrative department in W.Y.K. Town. The administrative department is called for center of broadcasting, culture and sports. The most profitable public affair is TV broadcasting services. The center of broadcasting, culture and sports earned about 300,000 RMB ($ 48030) per annum through charge Cable TV fee and maintenance fee. Hence, public affairs about culture and sports hardly obtain attention of staff because the two public affairs may not be profitable. This is one of causes of the marginalization of rural sports public affairs.

5 THE OPERATION LAWS OF THE DEVELOPMENT OF SPORTS CULTURE IN W.Y.K. TOWN

5.1 *Achievements of one's official career is the most important factor in process of the development of sports culture in W.Y.K. town*

Organization Department of the CPC Central Committee repeatedly stressed that GDP isn't an important achievement indicator of the local party and government leadership. But achievement view focusing on GDP still profoundly influenced the development direction of W.Y.K. town. For instance, we found that 8 of all indicators involved economic according to task list of economic and social development of W.Y.K. Town in 2012 (See Table 3). And none of indicators involved sports public services. The interviewed officials said to me, "At present, absorbing investment is the core task. We think that we can

afford to supply of public goods only by developing economic."

5.2 *Sports staff can't focus on the development of sports public affairs because they are faced with the complicated affairs in W.Y.K. town*

Theoretically, the duties of center of broadcasting, culture and sports should include Cable TV service, developing culture and sports. In fact, staff center faced with the complicated affairs in W.Y.K. town. For example, the town government asked all civil servants to work for land takeover, because W.Y.K. town will emerge into development zone in nearby city in the future. And the work already had lasted for more than a year.

5.3 *It is the hidden rule that the village committees obtain the fund of rural public services from abnormal channels*

The village committees tried to obtain support of funding on account of the limited funding. And then it resulted in competition. To obtain funding, the village committees find in different ways. For example, some village committees held ancestor worshipping ceremony. The village committee will contact with clansman throughout the country. Village head frankly said, "Ancestor worshipping is one of aims. Expanding relationships is the core aim of the ceremony." The village head contacted with influential clansman through ceremony so that the village committee will get more support.

Table 3. Task list of economic and social development of W.Y.K. town in 2012.

Assessing items	Target values	Score
1. Introduce investment	100 Million RMB	25
2. Tax revenue coming from trading enterprise	6 Million RMB	10
3. Fixed-Asset Investment (FAI)	1.08 Billion RMB	10
4. Industrial added value	152.31 Million RMB	8
5. Introduce small and medium-sized enterprises	Snnual value of production of enterprise reached up to 20 million RMB	7
6. The construction of small towns	All the quotations are subject to confirmation of Urban Construction Bureau	12
7. Agriculture base	2–5	10
8. Village renovation	Set up 2–3 rural demonstration sites	8
9. Building of administrative efficiency	Refer to index of Y.X. County government	3
10. Create credit	Create credit goals of town	2
11. Social assistance and insurance	Accomplished task of people club bureau	2
12. Safety of food and pharmaceuticals	Strengthen supervision	1
13. Safety production	Zero defects	Deduction

6 CONCLUDING REMARK

In China, the township government is the most grass-roots level of government, farmers in all kinds of relations with the countries. The W.Y.K. town government was faced with the urgent affairs such as land takeover, livelihood, economic transition and so on, because the W.Y.K. town has been in economically backwardness countryside areas. It is also the real reason that township and village leaders couldn't treat sports public services as focus. Therefore, urging leader attaches great importance to the sports public services being divorced from reality is not at all worthwhile.

ACKNOWLEDGEMENTS

1. The research is supported by Social Science Fund of Hunan Province. (Grant No. 13YBB228).
2. The paper is supported by Research Fund of Central South University of Forestry & Technology. (Grant No. 2013YB07).
3. The research is supported by Open Fund of the Research Centre of Leisure Sports of Hubei Province in 2013.

REFERENCES

[1] State-level, poverty-stricken county.http://baike.baidu.com/view/1597474.htm?fr = aladdin, Baidu Encyclopedia.
[2] Hubei Provincial Party Committee's suggestion about comprehensive supplementary reforms of villages and towns. Government Document of Hubei Provincial, [2003] No. 17.
[3] "socialization of public services" benefits famers. http://news.xinhuanet.com/politics/2009-05/25/content_11430151.htm, 2009-05-25.
[4] Xiaoping Qin, Zhigang Wang, Wangjian, Qingshan Hu."To keep things with money" New ideas for reform of rural sport public service provision mechanism. Journal of Shanghai university of sport, 2012-1:32.
[5] Zhangbin, Xiaoping Qin."Use money to buy service": new ideas for sports basic public service provision mechanism. Journal of Shandong institute of physical education and sports, 2012-10:6.
[6] Xuefeng He. Why is the reform of "socialization of rural public services" wrong? http://www.infzm.com/content/7511. 2008-01-23.

Sports Engineering and Computer Science – Luo (Ed.)
© *2015 Taylor & Francis Group, London, ISBN 978-1-138-02650-6*

A strategy study on synergetic development between college and mass sport resources in Shandong province

Jing Zhou

Department of Physical Education, China University of Petroleum, Huadong, P.R. China

ABSTRACT: Sport resource development strategy is always a front subject in the field of sport research in our country, and it plays a decisive role for the booming of Shandong province college sport and mass sport. This paper uses the method of field survey, questionnaires and mathematical statistics for investigation in 44 universities and 10 cities in Shandong province. The results show that: 1) Sport stadiums of Shandong province mainly belong to the education system and their facilities are great, which coexist in paid and free opening forms. It can effectively ease the situation of the lack of stadiums resource of mass sport in Shandong province. 2) The development of human resource of the mass sport in Shandong province is rapid, but the overall level is low. It is difficult to meet the increasing public demand. The only way to solve the problem better is to blend college teacher resource in mass sport system. 3) The operation of college sport stadium resource system needs the investment fund of mass sport. So making full use of the college financial support and resource can solve the needing problem of high efficient and sustainable development. 4) The synergetic and complementary development of the mass sport and the college sport resource of Shandong province in the facility equipment, manpower, financial resource and other aspects will be the effective way of promoting the rapid development of sport resource in Shandong province.

1 INTRODUCTION

As the most extensive social and cultural phenomena, sport is highly respected in the world today and as the way of life, education, mode of production, art form, it reflects the social function. The basic function of starting spending, spurring economic growth and keeping fit is more and more obvious[1]. Shandong is a populous, economic and sport province, and with the gradual realization on comfortable life, the lifestyle of people's health and physical activity has changed dramatically. Physical exercises and consumption gradually become an indispensable part of life. The National Games has held in Shandong in 2009, which not only stimulated the enthusiasm of people for physical exercise, but also increased investment for sport in the stadium resource, human resource and government financial investment. In the newly build 42 stadiums for this National Games, some stadiums are completed by colleges or government respectively, the others are joint completed by universities and local government. In promotion of such a policy, we can see the important position of college sport resource in mass sport and competitive sport. Due to a huge sport population of the mass sport and various needs, this article wants to find the influence factors of college sport resource in solving the deficient phenomenon of the mass

sport resource, and seeks the development strategy to promote the synergetic development of college sport resource and mass sport resource in Shandong, and to provide decision-making reference for rational allocation of sport resource in Shandong.

2 METHODOLOGIES

2.1 *Objects*

The college sport resource of 44 colleges (4 institutions under the ministry of education, 1 physical education school, 10 comprehensive colleges, and 29 higher undergraduate institutions) and the mass sport resource, including the stadiums, teachers, social instructors and financial investment are this paper's objects.

2.2 *Methodology*

2.2.1 *Field investigation method*
Through visiting 10 main cities (Jinan, Qingdao, Weifang, Dongying, Zibo, Jining, Rizhao, Binzhou, Yantai, Weihai) in Shandong province, we understand the distribution of local sport population, the supply, demand of physical resource, and firsthand material.

Table 1.   The distribution and recovery of questionnaires.

Objects	Distribution	Recovery	%	Available	%
Teacher	80	70	88%	68	97%
Social instructors	40	35	88%	30	86%
Official	20	20	100%	18	90%
Exercise groups	50	46	92%	42	91%

2.2.2   *Questionnaire method*

1. Questionnaires were mainly designed around the sport stadiums' equipment, human resources, financial investment, utilization and opening of stadiums, management mode and other aspects.
2. The validity of the questionnaire (See Table1). In order to ensure the validity of the questionnaires, some experts have involved checking the draft. The questionnaire will be recognized after the experts' review and comment. The 90% experts believe that the questionnaire meets the need of study purpose, clear language, appropriate technical term, and the 10% of experts believe that the survey questionnaire meets the basic need. No experts believe that the questionnaires do fail.
3. The reliability of the questionnaire. Because there are amounts of single choice, the kind of questions is in line with out-half reliability test methods, so we carry out the semi-statistical analysis of single choice. We get "half test" product-moment correlation coefficient-the rhh is 0.86. After we use "Spearman–Brown" formula rtt = 2r/(1 + rhh) to be corrected, we get out half reliability coefficient-the rhh is 0.79.

2.2.3   *Mathematical statistics method*

By the EXCEL and SPSS statistics software, we conduct the comprehensive analysis of all data.

3   RESULTS

History and reality of mass sport and college sport development around the world indicate that resource in the process of developing sports is by no means a simple type of resource. Instead, it requires that different types of resources play different roles in different levels. Each subsystem in the system of mass sport and college sport cannot exist in an isolate situation. There is a constant exchange of materials, technology, human resources and information between projects, which has a great influence on the establishment and development of all the projects. The interaction may strengthen relationship among projects, open up the developmental space and bring about more opportunities for improvement. Complementary effect is basically a kind of increase in quantity. It is a way to mine physical resources that were previously unprofitable by assembly and improve integrated efficiency by making the most of it. Complementary effect emphasizes more on integrated efficiency than individual efficiency. On one hand, combination between different individuals offers an opportunity for underutilized resources to obtain the full utilization. That is to say, integrated efficiency is strengthened by enlarging the scope of using resources.

3.1   *The status and problems of stadiums*

By the end of December 31, 2003, there are 850,080 sport stadiums in line with the requirement of the fifth national physical census in the various systems, trades and forms of ownership of our country. There are 547,178 standard and 302,902 non-standard sport stadiums, which cover an area of 2250 million square meters. The construction area is 75.272 million square meters, and the site area is 1330 million square meters.

Shandong Province has 47,379 sport stadiums until now. Sport stadiums cover an area of 109 million square meters, and the total construction area is 5.48 million square meters. The total area is 101.97 million square meters and the cumulative investment is 8.91 billion RMB.

The education system shares the largest percentage in a total of 31,968 stadiums in Shandong Province, accounting for 67.5% of the total number of sport stadiums in the province. Sport system has 1140, accounting for 2.4% of the total number of these stadiums; the rest are other systems, accounting for 30.1% of all stadia[2]. In the education system, the university's sport facilities are the highest level of equipment. According to the requirement of the Ministry of Education, colleges should have much more 3 square meters of the stadium area for each student, and the stadium requires good facilities, while the national average site area is only 1.03 square meters each person, the specific requirement of stadiums, facilities and other aspects of the site are difficult to meet.

According to the result of the fifth national sport census, nationwide colleges totally have 549,654 stadiums, accounting for more than 60% of the national stadiums. Many colleges not only have a high level of training stadiums and training teams, but also have a higher level of stadium operation and management system. 42 new stadiums have been built for the Eleventh National Games Held in 2009 in Shandong province and many of them are accomplished by universities. For example, Shandong Sport Institute, Binzhou campus and Binzhou sports school have a total construction area of about 300,000 square meters, which holds the contractor 11 Games judo and Chinese boxing competitions; Weifang city University Stadium has a construction area of about 20,000 square meters, which holds the contractor 11 Games women basketball, volleyball and team competition; China University of Petroleum (East China) stadium has a construction area of 18,767 square meters, which has the contractor 11 Games men volleyball team match; Ocean University of China covers an area of 13,570 square meters Stadium, including gymnasium, swimming pool and the student activity center, with a total construction area of 37,779 square meters. Stadium construction area is 21,520 square meters. Swimming pool area is 2438 square meters. The activity center area is 11,290 square meters. There are 4145 stadium seats, which hosts the 11th National Games women basketball group match (See Table 2).

Unfortunately, however, such abundant resources of university sports facilities, except regular use in teaching training, are not fully and effectively opened and utilized. Some facilities are basically left underused in weekends and holidays. Even in class, sports facilities in some universities are not made full use of, quite a contrast to shortage of public sports facilities in community. Therefore, there is

no doubt that the best course of promoting community sports in all rounds is to fully develop and utilize the function of sports facilities by opening them to the community and improving the utilization rate so as to serve the public. As an important part of sports market in China, university sports serve as the turning point, leading to socialization of campus physical education and connection where learning and application join together.

Besides, it is the key link to carry out the national fitness program to develop fitness awareness, lifelong exercising habit and sports ability. It is an important approach for these schools to configure and develop sports resources reasonably and to improve resource utilization efficiency, bringing the economic and social benefits of university sports resources into play. This is an important way for the sustainable development of university sports.

The services of college sport stadium mainly arm at high-level team training in universities, teaching physical education of school students, as well as facilitating extra-curricular physical activity places for faculty and students.

According to the result of the survey, college sport stadiums in general meet these three aspects on time and space requirement. In the time of morning, evening, weekend and holiday, more than 90% of the outdoor stadiums are open to people (See Table 3). But only a few outdoor stadiums have to be paid, including some of the outdoor swimming pools and plastic tennis courts and so on with higher operating cost. The outdoor track and field, outdoor artificial turf soccer field, basketball field, volleyball field, roller skating facilities are basically free open. The opening time requirement of college indoor stadiums is very strict, and site maintenance and management is also similar to the community's fitness club. The 96% of indoor stadiums have to be paid, which in some degree alleviate the cost of university stadium maintenance, enhancing the quality of service and improving the stadium's equipment and facilities. And many colleges open this window for the community to establish a good reputation of the

Table 2. The stadium status of 10 major cities in Shandong province.

City	Number	%
Jinan	4018	8.4
Qingdao	4551	9.6
Yantai	3656	7.7
Weifang	2971	6.2
Zibo	4986	10.5
Dongying	2281	4.8
Jining	3365	7.1
Weihai	1617	3.4
Sunshine	1474	3.1
Binzhou	2388	5
Total	31307	65.8

Table 3. The opening stadia of 44 colleges in Shandong province.

	Yes		No	
If outdoor stadia are open	40	91%	4	9%
If indoor stadia are open	30	69%	14	31%
If indoor stadia are open by paid	29	96%	1	2%
If outdoor stadia are open by paid	6	15%	34	86%

stadium operators and to enhance their own social status, which is good for the college admission, student employment and the close cooperation of government and enterprises (See Table 4).

3.2 *The status and problem of human resource*

By the end of 2009, there is a total population of 93.09 million people in Shandong Province. The social sport instructors of Shandong province are 82,050, with an average of 1135 people. This is higher than the 595:1 ratio in Beijing, which shows that Shandong province is a lack of social instructor in comparison with the developed city.

However, the number of instructors is not the only constrain of social development of mass sport. According to the result of survey, the comprehensive guide ability of social instructors of Shandong Province is not strong. Proportion of the national social instructors is only 0.2%, and the lowest level of the national levels instructors accounts for 74.5% percent of total. In conclusion, we know that besides improving instructor social groups, we should strengthen their own quality and ability at the same time (See Table 5).

By the end of 2009, there are a total of 159.3 million students in 128 high education schools of Shandong Province. According to the survey result in 44 colleges of Shandong Province, there are 2759 full-time physical education teachers.

Table 4. The status of Shandong social sport instructors[3].

Level	Quantity	Proportion
National	164	0.2%
National one	4587	5.59%
National two	15852	19.32%
National three	61127	74.5%

Table 5. The basic situation of higher education in Shandong province.

		Graduate students	Undergraduate students
Number of schools	Quantity	31	128
	changing	0	2
Number of enrollment students (million)	No. of people	2.2	50.1
	Increasing (%)	16.9	–2.5
Number of student total (million)	No. of people	5.7	159.3
	Increasing (%)	12.8	3.8

The title of professor ratio is 35%; the rates of associate professors, lecturers and teaching assistants are 36%, 53% and 4% respectively. Compared with the year 2000, we can see that the proportion of high-level PE teachers is larger and lower-level ones are smaller. There are more and more teachers whose title is professor and as the qualifications of college recruiting for the full-time sport teachers have become increasingly demanding, the result is that the number of teaching assistants is in declining year by year. Social sport instructors generally have many other practical problems. For example, there are the backward teaching methods, the narrow ways to further studies, the overall low levels of education. Full-time college PE teachers are higher than the social instructors in all aspects of ability. In the environment of college, the teachers learn by themselves and motivate constantly, which virtually promote the speed and level of the education industry (See Table 6).

"The Promotion Law of Japanese Sports" puts forward "national and local public groups can not interfere with the facilities of the school education circumstances and should strive to make the school sports facilities for physical activities". In China's public culture and sport facilities 6th, the State encourages the schools and other units within the culture and sport facilities open to the public. In the personnel of human resources of university sports, Zhu Lian-Zhi points that "college physical education system in our country has a large number of high level scientific research personnel, including teachers, coaches and team management, which can cultivate a large number of talents for the mass sports". In the college sports information resources, Yu Ping believes that "the university sport department is the latest sport information access to comprehensive utilization of transmission, and it also is important place to have excellent library facilities and information.

3.3 *The status and problem of financial resource*

According to the survey result, Shandong Province's mass sport fund shows good momentum. In recent

Table 6. The status of full-time PE teacher in 44 colleges of Shandong province.

	Number	Percentage
Total number of teachers	2759	100%
The average of 44 schools	62.7	
Professors	193	7%
Associate professors	993	36%
Instructors	1462	53%
Assistants	110	4%

years, the financial allocation of three sport sectors (provincial, municipal and county) of our country increases year by year. In 1995, public fund for sports is the 7.6 million, which accounts for 5.03% of all; in 1999, public fund for sports is the 13 million, which accounts for 5.67%. In 2003, public fund for sport is the 21.25 million, which accounts for 6.65%. The consciousness of sports consumption of the residents directly affects the sport behavior and levels. Statistic results show that 64.56% of community residents agree that the investment of physical fitness is worth. More and more people prefer to high-quality, entertaining and modern physical training mode. The running found of College stadium system is usually anchored in fund of school finance. Because their running needs to raise fund by themselves to support maintenance facility, equipment improvement, the department of welfare and other aspects, the main ways are to absorb the social capital investment by positive social integration, competitions, opening stadiums, and holding in long-term and short-term exercise classes without affecting the classes of teachers, students and leisure exercise.

Under the condition that university sports works well; the present sports resources should be tapped to establish a relation of exchange, where university sports and masses sports share resources with each other to highlight the overall advantage of university resources. We should create condition for carrying out all kinds of services in the course of intensive development, positively explore operating pattern and access of commercial sports resources, develop the potential of university sports resources and arouse the working enthusiasm of teachers through compensable service.

To improve utilization of university sports, the existent organization in single, intersected and closed state must be reformed to establish a new form that affiliates optimization of university resources to enhance the harmony between university sports and masses sports and deal with the relationship between social benefits and economic benefits to realize their maximum in accordance with law of market economy as well as relevant policies and regulations.

4 CONCLUSION

1. Sport stadia of Shandong province mainly belong to the education system and their facilities are great, which coexist in paid and free opening forms. It can effectively ease the situation of the lack of stadiums resource of mass sport in Shandong province.
2. The development of human resource of the mass sport in Shandong province is rapid, but the overall level is low. It is difficult to meet the increasing public demand. The only way to solve the problem better is to blend college teacher resource in mass sport system.
3. The operation of college sport stadium resource system needs the investment fund of mass sport. So making full use of the college financial support and resource can solve the need of the high efficient and sustainable development.
4. The synergetic and complementary development of the mass sport and the college sport resource of Shandong province in the facility equipment, manpower, financial resource and other aspects will be the effective way of promoting the rapid development of sport resource in Shandong province.

ACKNOWLEDGEMENT

Supported by "the Fundamental Research Funds for the Central Universities", Project NO. 09CX04053B, Subject NO. R091402B.

REFERENCES

[1] Ming Li. 2006. Status study development and utilization of resource in Hubei province. *China Outstanding Doctor/Master Dissertations Full-text Database.*
[2] Zhen Li. 2009. Sport stadium's present situation investigation and the idea of the use of integrated reformation in Shandong province. *Master thesis, Shandong Normal University.*
[3] Xin Hong Wang. 2009. Integration implementation mechanism research of Shandong province sport human resource. *Master thesis, Shandong Normal University.*
[4] Kai Gao. 2007. Interactive development research of Shandong province community sport and college sport. Master thesis, Nanjing Normal University.
[5] Jing G. 2009. The university as the center for regional sport development research. *Master thesis, Southwest Jiao Tong University.*
[6] Jie Dong & Yong Jun. Zhang. 2006. The public investigation and analysis of sport resource in Shandong province. *Journal of Shanghai Institute of Physical Education*, (5): 20–23.
[7] Bao Zhen W. 2010. The Shandong public sport resource present situation research in national constitution. *Journal of Shandong Sport Institute*, (7): 17–21.
[8] Tie Min S. & Chang Zheng Z. 2005. Paid service present situation and feasibility study of college sport stadiums in Shandong province. *Journal of Shandong Sport Institute*, (6): 51–52.
[9] Ling Hua R. & Yu-Shan T. 2007. Synergetic outlook of social sports resources under pan-resource background. *Journal of Shanghai University of Sports*, (2): 1–5.

[10] Zhen Li. 2009. Investigation on status quo of Shandong sports facility resources and assumption on the integrated reform of the operations. *Master Thesis, Beijing Normal University.*

[11] Feng Yan L. 2006. Study of the "Eleventh Five-Year Plan" of sports business in Shandong province. *Master Thesis, Beijing Normal University.*

[12] Shui Sheng P. et al. 2005. Research on the status quo of sports industry in Shandong province and its strategic objective for development and its countermeasures in 2010. *Shandong Institute of Physical Education and Sports,* 21(3); 19–21.

[13] Guang Feng Yuan. 2003. The study of the status and development and the configuration countermeasures of Shandong urban area sport resource. *Capital Institute of Physical Education,* 15(4): 25–27.

[14] Tong Yan Li & Yun Hong Z. 2009. Tentative research on the development of competitive sports resource. *Journal of Chengdu Sport University,* 35(6); 14–15.

[15] Yan Fang Geng. 2009. The study of the impact on the sustainable development of competitive sports in Shandong province laid by the eleventh national games. *Master Thesis, Qufu Normal University.*

[16] Yang Li & Yong Gen J. & Yu Ying S. 2000. Investigation on rational distribution of P.E. resources in Shanxi colleges and universities. *Journal of Physical Education,* (2): 76–79.

[17] Jie Yuan. 2011. College sports resources service city community sports development countermeasures. *Journal of Nanjing Institute of Physical Education (Natural Science),* 10(3): 100–111.

[18] Ping Zhou & Shao-Bo X. & Zheng-Qian Z. 2011. Research on the current situation and countermeasures of college sports resources serves of rural community sports. *Journal of Guangzhou Sport University,* 31(1): 34–37.

[19] Lian Ming Zang. 2001. Appraisal about the utilization efficiency of sports resource in colleges and universities. *Journal of Physical Education,* 8(4): 76–78.

[20] Li Hong Yu. 2011. The strategic choice of the status quo and optimization of allocation of sports resources in higher education institutes—A case study of higher education park in Hangzhou, Zhejiang Province. *Journal of Beijing Sport University,* 32(2): 92–96.

[21] Jian Zhang. 2011. Investigation and research on college sports resources opening to the surrounding communities. *Journal of Harbin Institute of Physical Education,* 29(3): 30–33.

[22] Jian Qing Zhu & Hu-Ping C. 2006. The structure of sport resource and the choice of optimal collocation manner of common colleges in east area. *China Sport Science,* 42(6): 104–110.

[23] Jin Ping Xiong & Shun Yi Li & Dan Chen. 2011. A study on the sports resources' allocation status and operation countermeasures in Jiangxi colleges and universities. *Liaoning Sport Science and Technology,* 33(1): 23–26.

[24] Yi Lao. 2005. Cultivation of the primary social sport instructors with P.E. resources of colleges and universities. *Journal of Shenyang Physical Education Institute,* 24(2): 40–41.

[25] Rong Li & Zi Xing Hang. 2001. Discussion on development and disposition of resource of university P.E. education [J]. *Journal of Nanjing Institute of Physical Education,* 15(6): 29–30.

[26] Yi Xiang Liu & Xin Feng. Social allocation of college sports resources in China [J]. *Journal of Wuhan Institute of Physical Education,* 2009, 43(1): 70–72.

[27] Li Cheng Tang & Li Hui Tang. 2010. On the status of university sports resources service the fitness for all. *Sichuan Sport Science,* (2): 82–85.

[28] Rui Xia Huang. 2004. Discussion on construction of the common service system of P.E. resource in colleges and universities. *Fujian Sports Science and Technology February,* 23(1): 57–58.

[29] Qiao Ling Li & Hong Ling Hu. 2008. Simply analysis the concept and definition of university sports resources. *Journal of Nanjing Institute of Physical Education,* 22(1): 84–86.

[31] Cheng Liu & Guo Gui Wu. 2009. The study of using colleges and universities sports resources to develop community sports activities. *Journal of Harbin Institute of Physical Education,* 27(6): 35–37.

[32] Xiao Ping Liu & Yu Liu Tao. 2007. Research on the universities sports resources and their social sharing in China on the basis of resource deployment theory. *Journal of Sports and Science,* 28(3): 94–96.

Sports Engineering and Computer Science – Luo (Ed.)
© 2015 Taylor & Francis Group, London, ISBN 978-1-138-02650-6

On functions of sports in building a well-off society in an all-round way in China

Chuan Zhou
P.E. Department of UESTC, Chengdu, Sichuan, China

ABSTRACT: This paper demonstrates that sports play an irreplaceable role in building a well-off society in an all-round way from the perspective of sports itself and sports industry. Taking competitive sports and mass sports for example, this paper performs profound analysis on the functions of sports. Besides, it further discusses the development of sports industry and its significant functions in building a well-off society in an all-round way. The study shows competitive sports contribute to creating a new diplomatic method. They also enhance the sense of national identity, strengthen the national consciousness, and orient the way of building a well-off society in an all-round way. The functions of mass sports in building a well-off society in an all-round way lie in the positive role of school sports, social recreation sports, and family sports. Sports industry promotes the growth of the Gross National Product (GDP) and stimulates domestic demand. It is beneficial to the adjustment of national market economy structure and effective integration of social idle funds.

Keywords: sports; well-off society building; positive role; China

1 INTRODUCTION

Born in the company of human production and living, sports have played a positive role in human development and progress. They are not only the product of civilization and a cultural form, but also an organic part of socialist culture. Meanwhile, sports, as an educational means as well as a physical activity, play an important role in shaping healthy bodies, cultivating sound spirits and improving the comprehensive quality of people. A well-off society in an all-round way is a stage of the development process of socialism modernization with Chinese characteristics. It means that the life quality will be improved further on the basis of subsistence and reaching the level of comprehensively affluent. In the process of building a well-off society in an all-round way, sports, with its unique social functions, will play a great and irreplaceable role in improving people's health quality and promoting all-round development of human beings.

From two main parts of sports'—competitive sports and mass sports, this paper elaborates the functions of sports in promoting the development of China's socialist culture and the construction of a well-off society in an all-round way. And it further analyzes the development of sports industry and its vital functions in building a well-off society in an all-round way [1].

2 FUNCTIONS OF COMPETITIVE SPORTS IN BUILDING A WELL-OFF SOCIETY IN AN ALL-ROUND WAY IN CHINA

Competitive sports, also known as athletic sports, are a subsystem of big sports systems. Adopting the form of athletic contest, competitive sports aim at winning games medals and refreshing the record. Chinese competitive sports began in March, 1959. Guotuan Rong, a Chinese table tennis player, captured the first world championship for China at the 25th World Table Tennis Championships held in Germany. Thereafter Chinese competitive sports were on track and competitive sports career grew increasingly prosperous. By 2012, the total number of world champions won by Chinese athletes had reached 2326 [2] since the founding of the P.R.C. sixty years ago. The functions of competitive sports were mainly divided into four aspects: creating a new diplomatic means, enhancing the sense of national identity, strengthening the national consciousness and playing a guiding role in mass sports.

2.1 *Creating a new diplomatic means*

Former South African President Nelson Mandela said: "Sports has the power to change the world." Competitive sports have transcended the boundaries between nations and have become important communication channels between nations, attracting and influencing people from different countries with their own charming. In recent years, as the supplement of the traditional diplomacy, the status of public diplomacy has been rising in diplomatic system and soft power ascension. It has been valued and developed by various countries. Since the founding of the P.R.C., from world-wide famous "Ping-pang diplomacy" to Expo diplomacy, practical activities with public diplomatic characteristics has never been interrupted. With the deepening of globalization, our country also faces many challenges in building a well-off society in an all-round way. This makes it particularly important to develop public diplomacy. Competitive sports in our country have been developing rapidly in recent years, with rich soft resources including athlete resource, sports spirit resource, the whole-nation system and other system resource. At present, our country continues to strengthen resource utilization of elite athletes. Elite athletes and the excellent qualities they embodied are not only the representative fragments of national image and national conditions, but also play an important role in constructing and propagating national image and in promoting international identity. Giving full play to the role of non-government under the leadership of government, cooperating with each other, our country should take the advantage of soft resource superiority to allow the world to interpret China from different perspectives. In terms of public diplomacy, competitive sports play an important role in exporting national values and molding national images.

2.2 *Enhancing national identity and strengthening national consciousness*

Competitive sports are essential for enhancing the national cohesion, speeding up the process of building a well-off society, promoting the Olympic spirit, as well as facilitating development of economy. Recalling the course of Chinese sports in the half past century, it's clear that competitive sports played a significant role in enhancing national identity and strengthening national consciousness. Chinese competitive sports made remarkable achievements again and again in the international arena, which aroused strong self-confidence and pride of the Chinese nation, and formed a strong national cohesion. Especially when the Olympic Games were held in Beijing for the first time in 2008, each Chinese was filled with the sense of pride. As the main stadium of 2008 Beijing Olympics, the National Aquatics Center was the only Olympic stadium built by people from Hong Kong, Macao, Taiwan compatriots, and overseas Chinese. It contains the profound cultural connotation of "Olympics, the hundred

years' Chinese Dream", which fully embodies the patriotism of Chinese around the world with the same root and heart. About 1.3 billion Chinese people all participated in the Olympic Games in different ways. "Being together with the Olympic" became China's most popular slogan. In peacetime, sports reflect the existing value of the nation. Fighting for the country and nation becomes great spiritual power of every athlete. They are respected as a symbol of their own nation. Athletes who work hard to fight for their nation are the pride of the whole nation. All of these show the significant functions of competitive sports on cultivating national cohesion. Its positive role lies in encouraging national spirit, condensing patriotic feeling, improving community solidarity and cohesion, awakening people's sense of belonging, and stimulating people's sense of national pride.

2.3 *Positive guiding role of competitive sports*

Competitive sports provide a role model for the society by leading young people to love sports with stars and inspiring the people to actively participate in sports. During the 1950s and 1960s, China's economic development has not yet started; accordingly, China barely had sports status in the international arena. In 1959, Rong Guotuan won the first world championship for China, which greatly encouraged the whole country to learn to play ping-pong and set off an unprecedented ping-pong heat in the whole country. At that time, no matter in urban or rural areas, in front of residential buildings or in elementary schools, you can see ping-pong tables everywhere, in different materials, including cement, stone, and wood. You might see students playing ping-pong on their desk, or workers and peasants playing ping-pong on the ground with a rectangle drawing. During the "three-year economic difficulties" period in China, the number of people who played ping-pong still reached 90 million nationwide. In the 1950s, there were few world championships and record of Chinese athletes. But mass sports were in full swing, with the public enthusiasm for sports ignited by champion title and record. Sports became the career of the whole country. The ping-pong champion, especially, led the public into participation, which contributed to the enduring of Chinese ping-pong sports.

3 FUNCTIONS OF MASS SPORTS IN BUILDING A WELL-OFF SOCIETY IN AN ALL-ROUND WAY IN CHINA

Mass sports are another subsystem of sports undertakings, which is the soil for cultivating sports talents. It plays indispensable role in building a well-off society. Mass sports can be divided into three categories, including school sports [3], recreation sports, [4] and family sports [5]. Their functions in building a well-off society are elaborated as follows:

3.1 The role of school sports

School sports are a purposed and planned educational process, which purposefully guide the students to learn more based on their hobbies. The ultimate aim is to develop students' lifelong sports ideas and prepare good exercise habit for their future in the community. It is the middle part that connects family sports with social sports. Therefore, school sports are sure to play a positive promoting role in building a well-off society. First, school sports have great impact on the implementation of lifelong sports. School sports are the cradle for training talents, and the base for developing a lifelong physical exercise habit. School sports have an important impact on overall development. Many facts show that all the interests, hobbies, and habits of adults in the community for lifelong sports are developed at school. Second, schools are equipped with the hardware facilities and personnel resources. With the close relation with social recreation sports, school sports have great potential to guide and promote the development of social sports. There are plenty of sports fields, perfect sports facilities, and professionals at school. Introducing all the sports knowledge and skills to the society will make great difference in promoting social sports activities, improving the national constitution, and carrying out nationwide fitness programs.

3.2 Enhancing national identity and strengthening national consciousness

Competitive sports is essential for enhancing the national cohesion, speeding up the process of building a well-off society, promoting the Olympic spirit, as well as facilitating development of economy. Recalling the course of Chinese sports in the past half a century, it's clear that competitive sports played a significant role in enhancing national identity and strengthening national consciousness. Chinese competitive sports made remarkable achievements again and again in the international arena, which aroused strong self-confidence and pride of the Chinese nation, and formed a strong national cohesion. Especially when the Olympic Games were held in Beijing for the first time in 2008, each Chinese was filled with the sense of pride. As the main stadium of 2008 Beijing Olympics and the only Olympic stadium built by people from Hong Kong, Macao, Taiwan compatriots, and overseas Chinese, National Aquatics Center contains the profound cultural connotation of "Olympics, the hundred years' Chinese Dream", which fully embodies the patriotism of Chinese around the world with the same root and heart. About 1.3 billion Chinese people all participated in the Olympic Games in different ways, with "Being together with the Olympic" becoming China's most popular slogan. In peacetime, sports reflect the existing value of the nation. Fighting for the country and nation becomes great spiritual power of every athlete. They are respected as a symbol of their own nation. Athletes who work hard to fight for their nation are the pride of the whole nation. All of these show the significant functions of competitive sports on cultivating national cohesion. Its positive role lies in encouraging national spirit, condensing patriotic feeling, improving community solidarity and cohesion, awakening people's sense of belonging, and stimulating people's sense of national pride.

3.3 Positive guiding role of competitive sports

Competitive sports provide a role model for the society by leading young people to love sports with stars and inspiring the people to actively participate in sports. In the 1950s and 1960s, China's economic development has not yet started; accordingly, China barely has sports status in the international arena. In 1959, Rong Guotuan won the first world championship for China, which greatly encouraged the whole country to learn to play ping-pong and set off an unprecedented ping-pong heat in the whole country. At that time, no matter in urban or rural areas, in front of residential buildings or in elementary schools, you can see ping-pong tables everywhere, in different materials, including cement, stone, and wood. You might see students playing ping-pong on their desk, or workers and peasants playing ping-pong on the ground with a rectangle drawing. During the "three-year economic difficulties" period in China, the number of people who play ping-pong still reached 90 million nationwide. In the 1950s, world championship and record of Chinese athletes were little, but mass sports were in full swing, with the public enthusiasm for sports ignited by champion title and record. Sports become the career of the whole country. The ping-pong champion, especially, led the public into participation, which contributed to the enduring of Chinese ping-pong sports.

4 FUNCTIONS OF MASS SPORTS IN BUILDING A WELL-OFF SOCIETY IN AN ALL-ROUND WAY IN CHINA

Mass sports are another subsystem of sports undertakings, which is the soil for cultivating sports talents. It plays indispensable role in building a well-off society. Mass sports can be divided into three categories, including school sports [3], recreation sports [4] and family sports [5]. Their functions in building a well-off society are elaborated as follows:

4.1 *The role of school sports*

School sports are a purposed and planned educational process, which purposefully guide the students to learn more based on their hobbies. The ultimate aim is to develop students' lifelong sports ideas and prepare good exercise habit for their future in the community. It is the middle part that connects family sports with social sports. Therefore, school sports are sure to play a positive promoting role in building a well-off society. First, school sports have great impact on the implementation of lifelong sports. School sports are the cradle for training talents, and the base for developing a lifelong physical exercise habit. School sports have an important impact on overall development. Many facts show that all the interests, hobbies, and habits of adults in the community for lifelong sports are developed at school. Second, schools are equipped with the hardware facilities and personnel resources. With the close relation with social recreation sports, school sports have great potential to guide and promote the development of social sports. There are plenty of sports fields, perfect sports facilities, and professionals at school. Introducing all the sports knowledge and skills to the society will make a great difference in promoting social sports activities, improving the national constitution, and carrying out nationwide fitness programs.

4.2 *The role of social recreation sports*

With the improvement of economic development level and life quality, recreation sports, which are based on sports, have become a symbol of economic power in a well-off society. Recreation sports refer to the positive and free sports. Recreation pleasure, rest, and relaxation are the main targets. Under increasing competitive pressures in society, people tend to improve health, relieve stress, and spend recreation time through sport. At present, China is in a critical period of building a well-off society, with social function growing, and recreation way increasing. Recreation sports abandon the ranking pressure of sports competition and return to the real function of sports, which is actively participating in various recreational sports activities in the case of relaxation. In this way, recreation sports promote tremendously to accelerate the process of building a well-off society. First, recreation sports promote spiritual civilization in building a well-off society. Recreation sports activities can not only meet people's recreational needs, get rid of the daily pressure, but also enhance friendship, promote emotional communication and ease the contradiction between people. Also, they can safeguard social stability, promote people's psychological and mental health, and accelerate the construction of spiritual well-off society. Secondly, recreation sports guide people to right health concept, help promote healthy and civilized lifestyles and social values, helps people pull themselves together, and release the pressure. Consequently, advanced culture can be spread, and the grand goal of building a comprehensive well-off society can be better achieved. Lastly, recreation sports promote economy development of a well-off society. The development of recreation sports promote that of tertiary industry through the optimization and upgrading of industrial structure. A great amount of high-relevancy service industries stimulate further improvement of the social security system. What's more, a lot of service staffs are needed for recreation sports activities, which adds employment opportunities and eases the employment pressure problems to a certain degree.

4.3 *The role of family sports*

Family is the cell of society, the basic unit of social life, the cradle of the growth of a person, and the only place where the earliest and extensive life-long education takes place. Sports consciousness and habits among family members not only promote the vigorous development of mass sports. Also they can promote the process of building a well-off society and facilitate the positive impact on families and society. Ultimately, it helps to enhance the overall national constitution, improve the quality of the whole nation, and promote the economic prosperity and life stability. It is essential to improve people's health quality to promote overall improvement in a well-off society. Health quality affects people's life. Family sports have the following functions. First, they enhance the physical quality of family members and help achieves higher satisfaction in spiritual life. Second, they can promote emotional communication and relationship among family members and improve internal unity of the family. Third, they can strengthen the relationship among family members, promote family stability force, and eventually achieves warmth and happiness of the whole family. All of these help to build a healthy and civilized lifestyle, and better promote spiritual civilization in a well-off society.

5 FUNCTIONS OF SPORTS INDUSTRY IN BUILDING A WELL-OFF SOCIETY IN AN ALL-ROUND WAY IN CHINA

With the acceleration of China's market economy development process, people's living standards are increasing continuously, which provides material conditions for the formation and development of the sports industry. [6] China's sports market is booming, with enormous market potential and business opportunities. Sports industry is an important part of building a well-off society, and another way of existence of sports cause under the condition of market economy. Sports industry is not only in line with the demands of socialist market economic restructuring, but also plays a significant role in stimulating economic growth, promoting employment, and expanding domestic demands. Sports industry plays an increasingly important role in building a well-off society in an all-round way in China.

5.1 Development of sports industry promotes the growth of GDP and stimulates domestic demand

Sports industry is developing rapidly worldwide, becoming one of the pillar industries of the national economy in many countries. Sports industry in Europe, America, and other developed countries accounted for 1% to 5% of the GDP, with an average growth rate of 20% per year. Despite the relatively late start, Chinese sports industry develops in a rapid speed and is beginning to take shape. According to data of the State Sports General Administration, sports industry value reaches 222 billion in 2010, accounting for 0.55% of GDP, with an increase of 13.44% over the previous year. The growth rate is higher than the GDP growth rate. Currently most countries are in the era of excess economy. In this circumstance, measures should be taken to stimulate and encourage consumption, to expand domestic demand by creating new consumption hotspot. Due to continuous economic and income growth, optimization of urban and rural consumption structure, enhancement of awareness of health and life quality, an increase of recreation time, sports consumption is rising in our country with strong growth momentum.

5.2 Development of sports industry facilitates adjustment of market economy structure

At present, the proportions of first and second industry are decreasing while the proportion of tertiary industry is increasing with an upward trend. Sports industry belongs to the category of tertiary industry. Therefore, the booming of sports industry will promote the development of tertiary industry and adjustment of market economic structure.

5.3 Sports industry enables effective integration of social idle funds

Sports industry, such as sports lottery is intended to raise funds for the development of sports. For a long time, our government establishes free allocation for sports industry, which cannot guarantee the long-term development of sports. Sports lotteries help to raise social idle funds, which expands the sources of funds, increases job opportunities, and provides more power for the development of sports cause.

In brief, sports industry and other industries are closely related. It has broad influences, including stimulating economic growth, vigorously exploring sports consumption, and contributing to the expansion of domestic demand. In this way, people's sports consumption ability is raised when they love and care more about sports.

6 CONCLUSION

This paper, based on the main line of sports and sports industry, elaborates the significant role that sports plays in building a well-off society from different levels and aspects. The functions of sports in building a well-off society in China include physique fitness function, political function, and economic function. With the continuous development of social economy, people's living standards continue to improve. The spiritual demands will increase correspondingly. As an important part, sports will play an increasingly significant role in building a well-off society.

REFERENCES

[1] C. Li. 2010. Strategic Studies on the Coordinated Development of Mass Sports and Competitive Sports. *Physical Institute of Shanxi Normal University* (5):43–47.
[2] X.G. Ma. 2011. Functions of Competitive Sports on Strengthening National Consciousness. *Academic Journal of Pingdingshan College* (2):10–13.
[3] Y.Q. Zhang. 2006. Shao Xiaojun. Study on the Relationship between the Well-Off Sports, National Fitness, and School Sports. *Education and Occupation* (4):48–49.
[4] F. Lu.2005. Recreation Trend of Sports in Well-Off Society. *Academic Journal of Wuhan Institute of Physical Education* (3):7–10.
[5] Y. Guan. 2000. Family Education in Sociological Perspective. *Tianjin Social Sciences Academy Press.*
[6] J.D. Jiang. 2007. Situation and Countermeasure Study of Chinese Sports Industry. *Market Modernization* (49):376–377.

Sports Engineering and Computer Science – Luo (Ed.)
© 2015 Taylor & Francis Group, London, ISBN 978-1-138-02650-6

The nature and characteristics of rhythmic gymnastics

LiLi Zhu

Department of Physical Education, Wuhan University of Science and Technology, Hubei, Wuhan, China

ABSTRACT: Based on the natural and rhythmic movements, the rhythmic gymnastics is a women's sports event played bare-handed or with light apparatus in the accompaniment of music. It can be divided into two groups—the mass rhythmic gymnastics and the athletic rhythmic gymnastics—including both collective and individual events. The collective events require the five athletes to complete two sets of different movements together. In one set, the same apparatus is used, while in the other set different apparatus are used. Individual events are played with rope, ring, ball, rod, gauze kerchief, colored ribbons, sword, fan and hydrangea, which can be practiced both collectively and individually. Nowadays with the rapid development of rhythmic gymnastics, we are required to be more aware of the developing trend of the rhythmic gymnastics and understand its connotation well. So it is very important for us to explore the essence and characteristics of art.

1 THE ESSENCE OF RHYTHMIC GYMNASTICS

The rhythmic gymnastics is a women's sports event played bare-handed or with light apparatus with music and on the basis of the natural and rhythmic movements. There are various movements practice in it, such as the bare-handed movements practice. It combines ballet dance with all kinds of walking, running, jumping, turning, balancing together, with the practice of movements like swinging, encircling, rolling, throwing and catching with.

There are different kinds of light apparatus. Through these practices, all parts of the body muscle are evenly developed, improving the physical qualities of coordination, flexibility, suppleness and bouncing. More importantly, rhythmic gymnastics training with music can bring the practitioners nice body shape and graceful posture. This also fully reflects the natural beauty of the human body and the vigor of youth, letting them to feel the charm of the art and cultivate elegant temperament. That is why so many people are in favor of it, especially the young girls.

2 CHARACTERISTICS OF RHYTHMIC GYMNASTICS

2.1 *Because the rhythmic gymnastics is divided into two groups—mass rhythmic gymnastics and athletic rhythmic gymnastics, they have characteristics both in common and respectively. Let us talk about the features in common first*

2.1.1 *The rich content and a great variety in rhythmic gymnastics*
There are two kinds of movements in rhythmic gymnastics, one is bare-handed, and the other is with

light apparatus. Based on ballet, bare-handed movements is a integration of the essence of ethnic dance, gymnastics, martial arts, acrobatics and drama skills, plus a variety of body movements like walking, running, jumping, balancing, turning etc. The movements with an easy-going carriage and beautiful music make people enjoy the beauty so much.

The movements with light apparatus are mainly conducted by the coordination of the body movements with the apparatus. While playing, the apparatus and the body movements should act as one unit. In another word, the apparatus should act as the extension of each part of the body, achieving the perfect combination of the body movements with the apparatus. With different characteristics, each apparatus goes with main movements. For example, the main movements for rings are throwing, catching, rolling on the und, crossing the ring, swinging and looping in the shape of word 8 etc. The main movements for rope are swinging forward, backward and sideward, jumping over rope or rope jump, throwing and catching, rotating, swinging, looping in the shape of word 8 etc. The main movements of ball group include throwing and catching, bouncing the ball, ball rolling, wobbling, 8 movements on the ground, looping around arms or not. The movements of stick group are based on looping in small circles, knocking while throwing and catching, "pre-swing" of arms, swinging, looping, with apparatus, "pre-swing" with apparatus, looping in 8 word shape etc.. The movements of ribbon group are the serpentine (4–5 waves), spiral (4–5 rings), the use of "wind", circle, 8 word, throwing, throwing in small range, whole body or part of the body passing through or over the patterns formed by the ribbon.

2.1.2 *Rhythmic gymnastics have a highly artistic and ornamental value, mainly embodied in its healthy beauty of the body, the fantastic beauty of the music and its modeling beauty in difficult movements*

2.1.2.1 The healthy and physical beauty of the rhythmic gymnastics

The athletes who do rhythmic gymnastics training regularly for a long time have very soft body line, graceful posture, and elegant temperament. It gives out the information of health and young. This reflects the artistic characteristics of rhythmic gymnastics. Therefore, athletes are required to blend their own emotions and feelings in the movements of rhythmic gymnastics so as to enhance body and facial expression. As a result, they can give it a strong ornamental value.

2.1.2.2 The beauty of music in rhythmic gymnastics

The music used in the rhythmic gymnastics not only foils the atmosphere, but also the means to present various forms and techniques in it. So music is said to be the soul of the rhythmic gymnastics. The music in rhythmic gymnastics can be a piece of piano music or a symphony. And a variety of folk music is also widely used. The music for rhythmic gymnastics should go well with the rhythm and the style of the movements. It should be chosen according to the temperament of the athletes and the apparatus to fully convey the artistic conception of the music and the movements. The music with a graceful melody and a distinctive style can not only stimulate the athlete's passion and make their personality prominent, but also enhance the artistic appeal and exaggerate the atmosphere of the performance. All these will surely impress the audience with fantastic beauty.

2.1.2.3 The beauty of difficult molding in rhythmic gymnastics

Some of the difficult moldings in rhythmic gymnastics can be completed only on the basis of a certain quality of flexibility and strength. The difficult movements are mainly composed of the combination of some running, jumping, balancing, turning and apparatus juggling, rolling. To complete the difficult movements, the athletes are required to have better quality in flexibility, more skilled techniques in apparatus handling, which makes the event more competitive. The difficult molding, composed of the combination of apparatus and the body, can create a vivid picture for the audiences. It fully demonstrates the beauty of human body and exaggerates the warm atmosphere of rhythmic gymnastics, thus enhancing the artistic charm of rhythmic gymnastics.

2.1.3 *The innovation of rhythmic gymnastics movement*

Artistic gymnastics is traced back to the late nineteenth or early twentieth in Europe. Having developed for nearly 100 years, it has already formed a complete system of movements. But after becoming one of events in the Olympic Games, rhythmic gymnastics has got a higher requirement in its creation and innovation. After the new rules of rhythmic gymnastics were promulgated in 2005, new requirements were put forward to the degree of the difficulty, the innovation and special artistic effects. Therefore, these have promoted a new development of the rhythmic gymnastics movements. The development and the innovation of the rhythmic gymnastics movements will keep going, making it more charming and fantastic.

2.1.4 *Rhythmic gymnastics requires practitioner a good ability in performance*

Rhythmic gymnastics is a piece of art. The practitioners should not only bring the audiences nice beauties, but also can stimulate and experience the inner feelings during the process of their presentation. They should make themselves well understood with the expressions of their face and body movements, from every part of the body and every detail of the movements. If they fail in the cultivation of expressing ability, rhythmic gymnastics movements may become boring and unattractive.

In addition, practitioner's music accomplishment is also a very important factor. Music should be unique in rhythm, beautiful in melody, various in style. With the different kinds of music, rhythmic gymnastics movements can put more color to the art through rhythm degree, the speed of the amplitude of the movements, and well-proportioned movement of the whole set. Music brings positive effects on human muscle and nervous system. These positive effects are embodied through both the internal psychological changes and the external morphological changes of the practitioners. Therefore, if they are, in making a movement, inspirited by the music, and their passion of creation is stimulated, practitioners will achieve a perfect combination of emotion, movements and music.

2.2 *The respective characteristics of mass rhythmic gymnastics and athletic rhythmic gymnastics*

2.2.1 *The characteristics of mass rhythmic gymnastics*

2.2.1.1 Diversity in content and apparatus

Mass Rhythmic Gymnastics transplanted a lot from rhythmic gymnastics, modern dance and folk dance movements. Fans, gauze kerchief and swords are used as apparatus, adding new elements to the rhythmic gymnastics, reflecting the diversity of dance styles and artistic forms.

2.2.1.2 Easy to learn

Mass rhythmic gymnastics movements are usually basic movements, with lower requirements in completing, so they are easy to learn and practice.

2.2.1.3 Low requirement in practice site and easy to carry out

The mass rhythmic gymnastics courses need only a piece of empty place and a tape recorder to start with. Of course, it will be much better if there is a carpet, a mirror and some Barres.

2.2.1.4 The higher popularity

Due to the characteristics that mass rhythmic gymnastics is easy to carry out, it has become more and more popular. With the nationwide fitness programs started, people's sense of health has been strengthened greatly, especially their pursuit of beauty. Many women are willing to do the gymnastics physical training to keep fit with beautiful body shape. With the rapid development of the sports industry and the increase of the fitness clubs, gymnastics fitness course has become integral, gaining great favor from a large number of people. Many schools, especially the universities have set up the course of gymnastics, making the gymnastics the main means for girls to do exercise. That greatly promotes the cultivation of a healthy body, graceful posture and elegant temperament in the new generation of female students.

2.2.2 The unique characteristics of athletic rhythmic gymnastics

2.2.2.1 The most unique characteristic— competitive challenge

The main purpose to improve the competitive level in athletic rhythmic gymnastics is to participate in competitions and win a good place. With the promulgation of the new international rhythmic gymnastics rules in 2013, the new code of points for the difficulty levels increased continuously. Therefore, more time and space are allowed for the arrangement of the artistic value to meet the selection and application of the apparatus. The new regulations changed the difficulty levels and the amount of difficulty in set movements. These of course will promote the rhythmic gymnastics to develop toward the direction where the artistic style is highlighted. At the same time, it also means that the outstanding athletic ability will be brought into full play, promoting a worldwide integral improvement in rhythmic gymnastics. Finally, it can go with higher techniques and more fierce competition.

2.2.2.2 The improvement of art in athletic rhythmic gymnastics

In the international arenas, a high-level rhythmic gymnastics athlete presents series of vivid artistic pictures to the audience and the judges. Its artistic value is endowed by its arrangement and organization, and by the perfect combination of body with apparatus and the music with movements. The score for special art has also increased in the new rules, which enable the athletic rhythmic gymnastics to approach where the performances are. Their performances are described with the characters "difficult, new, skillful, excellent, fast and beautiful". Thus the arrangement innovation becomes the key point. Countries are making great efforts to study the new rules to increase artistic state of the set movements. They also try to restudy the principles of arrangement and set up a new arrangement pattern, which will promote the development of innovation in its movement system.

2.2.2.3 The technical characteristics of super-soft movement in athletic rhythmic gymnastics

In the new rules, a high-level flexibility is required in rhythmic gymnastics technical movements, especially in the super-difficult movements. In the international arena, many outstanding artistic gymnastics athletes can get good rankings in international competitions, mainly because of their good quality of flexibility. Some super-difficult movements can be accomplished only on the basis of good level in flexibility which enables the body movements to combine the apparatus perfectly. Therefore, the technology in flexibility has now become an important task in increasing the level of the athletic rhythmic gymnastics.

3 CONCLUSIONS AND REFLECTIONS

3.1 Conclusion

1. *The nature of rhythmic gymnastics is that the rhythmic gymnastics is a women's sports event played in the accompaniment of music, barehanded or with light apparatus. It is based on the natural and rhythmic movements.*
2. *The features of rhythmic gymnastics: rhythmic gymnastics is divided into two groups—the mass rhythmic gymnastics and the athletic rhythmic gymnastics. They have many mutual characteristics such as the rich content, great variety; with a highly artistic and ornamental value. Their innovation in the movements and their requirements for better ability in presentation are all the mutual features for them. At the same time, each of them carries some respective characteristics.*
3. *With the rapid development of rhythmic gymnastics around the world and the promulgation of the new rules, the rhythmic gymnastics level of our country needs to be further improved. To get fully prepared for the 2016 Olympic Games we should put our full strength into innovation and unique in the movements to become more competitive in the future.*

REFERENCES

[1] Tong Weizhen. Analysis and Prospect of Rhythmic Gymnastics Development in China. [J] Guangzhou. Journal of Guangzhou Sports University, 2005.

[2] Liu Lingchen the Prospect of Development Trend of Rhythmic Gymnastics through the New Version of Rules 2005 [J]. Hubei Sports Technology. 2005 (10).

[3] Zhu Ying. The Study on Aesthetic Value of Rhythmic Gymnastics [J]. Fujian Sports Science and Technology. 2006 (4).

[4] Zhang Qunhua. On The Improvement of Rhythmic Gymnastics Performance [M]. Teaching and Management. 2005 (8).

[5] Li Weidong. The Main Characteristics and Competitive Prospect of The Changes in The New Rules of Rhythmic Gymnastics [J]. Chinese Sports Science and Technology. 2003 (8): 13~15.

[6] Niu Lili. Diaozhaizhen. The Exploration of The Strategy and The Developing Trend of Rhythmic Gymnastics through The New Rules [J]. Hubei Sports Science and Technology. 2006 (3).

Sports Engineering and Computer Science – Luo (Ed.)

China-ASEAN community sports construction of cooperation

Zhao-Long Zhang & Ming-Ya Zhang
School of Tourism Culture, Yunnan University, Lijiang, Yunnan, China

NaiQiong Li
College of Physical Education, Qinzhou University, Qinzhou, Guangxi, China

ABSTRACT: Sports exchanges between China and ASEAN countries are beneficial to economic and social development in the region. Achieving the institutionalization, serialization and normalization of sports exchanges, in order to effectively reach the cultural identity of sports exchanges and cooperation, is a difficult problem. Through study on the sports cooperation development issues between China and ASEAN and how to build China-ASEAN sports cooperation community, it is believed that the construction of its mechanism should be based on three aspects—the organizational management, sustainable development and sports cooperation. It aims to coordinate and jointly build China-ASEAN sports cooperation for development.

1 CURRENT SITUATION OF CHINA-ASEAN SPORTS COOPERATION DEVELOPMENT

The economist Samuel Huntington has once stated that regional organization came up in getting of enough support from common cultures when talking about East Asia economic development[1]. The proceeding of integration of China-ASEAN regional economy should be based on some common cultures, such as sports. Through the implementation of *Guidelines to stimulate the development of sports industry* from the General Office of State Council, fields and contents of China-ASEAN sports cooperation has made great strides forward[2]. In order to bring sports culture cooperation into full play, we need to study China-ASEAN sports cooperation in order to realize win-win cooperation.

Keeping equality and mutual trust, strengthening regional cooperation, and realization of mutually beneficial development are the common aspiration between China and ASEAN. Further construction of CAFTA needs not only economic cooperation but also development in depth based on cultural communication. Sports culture cooperation plays an active role in promoting the establishment of political mutual trust. For example, in the early years of the new China, premier Zhou Enlai built close relationship between China and Burma after spending the Water Splashing Day with the Prime Minister of Burma Wu Barui under the severe diplomacy environment, making people deeply feel the special relationship carried by sports.

The football match of China-Vietnam Festival in 1955 is said to be the earliest sports cooperation activity organized by two governments. From then on, including sports, China made many cultural cooperation agreements with Indonesia, Cambodia and some other ASEAN nations in succession. Nowadays, the subscription of *China-ASEAN Cultural Cooperation Memorandum of Understanding*, *Nanning Declaration* and *China-ASEAN Cultural Industry Interactive Program* require the continual enhancement of sports cooperation and bring sports in promoting mutual political trust and economic development into full play.

2 DILEMMA OF CHINA-ASEAN SPORTS CULTURE COOPERATION

Deficiency in coordination and management mechanism constrains further development of sports cooperation. Sports culture cooperation cannot do without the governments' positive promotion, for example, how to exploit the advantages of regional resources to the full and coordinate relevant policies of different countries? How to avoid the phenomenon of much communication less cooperation and how to formalize chronic sport communication? Moreover, problems such as protection of relevant intellectual property rights, preferential policies and financial supports would be difficult to be solved without unified management organizations. The imperfection of regional sports industry cooperation mechanism constraints the expansion of sports cooperation.

Currently, China-ASEAN sports culture resources haven't reach the best exploitation and sharing effect. Sports cooperation between two places is still mainly mutual folk sports culture visits and small-scale sports products manufacturing.

The imperfection of regional sports industry cooperation mechanism makes it difficult to give play to each positive sports policy and integrate rich sports culture resources. Thus, it is urgent to exploit perfect sports culture cooperation mechanism. The deficiency in training of sports talents and cultural protection mechanism are barriers to expand sports cooperation. For example, Guangxi Sports International Center, which is at the frontier of China-ASEAN sports culture communication, has only several professionals knowing foreign sports cooperation well, making a great contrast with fast-paced development needs. Moreover, a feasible mechanism is needed to ensure the fostering of extrovert and compound talents, structuring of effective cooperation methods in manufacturing, studying, learning and politics, exploitation of effects of sports cooperation, expansion of research in traditional sports.

3 FRAMEWORK OF CHINA-ASEAN SPORTS COOPERATION COMMUNITY

3.1 Organization and management mechanism of sports cooperation community

The first step is to build an organization to coordinate sports culture cooperation. Intensification of communication and contacts of China-ASEAN sports culture cooperation, as well as strengthening the organizational leadership of sports cooperation communication, can concentrate disperse traditional sports program association or organizations and form a comprehensive, systematic sports cooperation system. Currently, none of China-ASEAN senior sports management sections establish special organizations for relative sports culture integration, and none of these sports culture cooperation has such relevant projects. Before this time, many universities in China have established sports cooperation programs with ASEAN nations. However, most of these programs are planned according to each unit's purposes without getting effects. So it is urgent to establish a special and powerful coordinating organization to organize and manage those relevant programs into one target and ensure the working of cross-regional cooperation. The establishment of coordinating organization should be promoted by governments, integrating mutual help and cooperation of different departments and groups, in achieving sharing of information resources. For example, the purpose can be achieved through establishing China-ASEAN

sports communication cooperation base or rely on the already established China-ASEAN center, establish special apartments to develop organizational and communicational activities of sports culture. Resident offices can be established mutually; Guangxi or Yunnan province can bring their regional advantages into full play and try to establish local centers for sports communication or contacts. The establishment of sports culture communication management mechanism can guarantee the sound development of sports communication and cooperation systematically, and ensure China-ASEAN sports cooperation being fulfilled as planned.

The second step is to set up the sports cooperation community foundation. As many have slow economy progress, how to find more financial supports has become the basic problem in developing long-term sports cooperation. Under the circumstance of insufficient input into sports communication of ASEAN countries, the current model of sports culture cooperation between China and western countries can be used as a lesson, which means that relevant government departments play the leading role. The strengths of international organizations, non-governmental organizations and folk associations are brought into full play in order to promote multicultural communication in the form of sports development foundations. One of the financial resources is input from the government finance. Each country should set up special awards for sports communication[3]; the second resource is that governments should collect enough finance to develop China-ASEAN sports culture communication in the way of guiding the participation of social finance; the third resource is that sports industry programs can be supported in the way of striving for the backing of the current communication foundations and applying for China-ASEAN investment cooperation foundation and credit finance. Foundations should be in direct charge of the financial management of sports cooperation programs, and they should also be responsible for planning and designing sorts of regular communication activities, such as organizing mutual visiting activities between folk sports culture groups of China-ASEAN countries, financing and supporting some major sports cooperation programs.

The third step is to support folk sports cooperation programs. Communications between countries rely on the close relationship among their people. All in all, friendliness between nations is friendship among peoples of different countries. Folk sports communication is the original driving force in promoting China-ASEAN sports cooperation. Governments should actively build platforms for folk communication in order to enrich forms of sports cooperation. The first method is to bring people from cross borders into full play. They can pull the special strings in promoting regional sports cooperation. Nationalities of Southeast Asia have close relationship with many minorities of China, especially the fact that many minorities in Yunnan and Guangxi Province has kinship relationship with nationalities of South-east Asia countries. For example, the Zhuang nationality and Dai nationality, Nong nationality, Laji nationality, Bubiao nationality, Shanzhai nationality from Vietnam has certain kinship relations, so does Jingpo nationality in Yunnan and Keqin nationality in Burma[4]. The second method is to take full advantage of Chinese associations from overseas. These people should be not only watchmen of traditional Chinese sports culture, but also the best "marriage connection maker". The best example is their dissemination of Chinese martial arts[5]. The third method is that governments should take sweeping measures in policy supports. Action plans in implementation of the joint declaration of China-ASEAN strategic partnership facing peace and prosperity jointly signed by China and ASEAN clearly put forward the proposal of more encouragement and support for sports cooperation programs.

3.2 Continual development mechanism for sports cooperation community

Firstly, it is necessary to build data bases of sports culture resources together. Sports culture protection and communication are necessary. Development of programs aimed at international students who can promote sports culture, such as planning program of sports cooperation memorandum—ASEAN universities network programs, and academic communication planning program fulfilled by the Chinese Ministry of Education, need rich information resources in data base. The establishment of China-ASEAN data bases of sports culture resources can give full play to the advantages of network data platform. Then, some valuable traditional sports culture programs are in danger. As an important measure in protecting living cultural relics, building database platforms for ASEAN national sports culture resources can make a general investigation of traditional sports culture, grasp the real conditions of national sports resources in ASEAN countries, file, digitally save and protect traditional sports culture. The cooperation and development of creative sports industry are necessary. As a major driving force in developing culture industry, creativity is catalytic agent in the evolution of cultural patterns[6]. Traditional ethnic sports are important parts of national culture and also "a rich mine" in cultural industry exploitation of modern countries. Building databases of national sports resources of ASEAN

countries is convenient for researching national sports resource conditions, serving for cooperation of national sports industry. Obviously, protecting the diversity of traditional sports gene base has important significance to creative development of cultural industry[7]. It is worthwhile to note that the establishment of databases doesn't end after inputting the data. A series of operation management systems must be established to guarantee the quality and continual development of database construction. Individual domain names should be applied after finishing every part in order to open up retrievals and provide sports cooperation development service in the form of WEB.

Secondly, it is necessary to build bases for talents training and scientific research of sports communication together. As talents is a major element in culture communication and cooperation, China and ASEAN countries should build bases for talents training of sports communication and cooperation together to ensure the needs for talents in sustainable cooperation. Building this base need to finish three tasks—talents training of sports communication, research for sports cooperation, development for industry of sports cooperation. Give full play to the roles of bases in universities and their teaching programs in ASEAN countries. Some sports teachers of relevant majors should get centralized training to understand the current needs and serving forms of sports culture integration based on universities and their programming resources; training courses should be regularly provided to social personnel and staff of companies related to ASEAN sports activities based on the social service function of bases. The problem of cultural transmission could be solved by experts' participation of research based on universities' rich human resource and research strength[8]. Enterprises can exploit sports markets with bases, market products with traditional China-ASEAN sports culture characteristics, and make bases into cradle of culture industry. Many universities in Nanning of Guangxi Province has established China-ASEAN training base for sports talents, China-ASEAN sports information center, China-ASEAN sports communication cooperation center, China-ASEAN national sports research center, China-ASEAN collaborative innovation center of culture communication and development. We should allow full play to the existing bases and research centers and build more and more efficient platforms for the big picture of China-ASEAN sports communication development.

3.3 Mechanism for markets development of sports cooperation community

First of all, policy, ensuring mechanism of sports industry cooperation, should be established. The development of sports industry rely on governments' policy supports to a large extent apart from its own abilities, such as the conduction of football match competition held on Lantern Festivals between Dongxing of China and Mangjie of Vietnam. Under the guidance of the policy "get enterprises into sports culture and get sports culture into markets" from the government of Guangxi District, above 70% of sponsor fees are provided by more than 50 local enterprises. Secondly, policy ensuring system is the booster in promoting the development of sports industry communication and cooperation, for example, the implementation of the law of foreign investment from Vietnam government provide more loose investment environment and convenience for foreign businessmen to invest in Vietnam. After the implementation of the new law, some sports companies from Guangxi collaborated with Vietnam enterprises to exploit the sports products market together, and the majority ownership belongs to Vietnam enterprises according to rules. Some other ASEAN countries, e.g., Singapore has established plans of "Sports Singapore" to strive to be a powerful nation of sports industry in Asia. Malaysia has established "the state's sports policy" and radicated the development of sports industry into a major field of social development and national construction[9]. Guangxi Autonomous Region has also established *Action plans of China-ASEAN sports communication and cooperation* in 2011. Deficiency in policy ensuring mechanism of sports industry and collaborating organizations of China-ASEAN sports industry makes it difficult in promoting the implementation of many collaborating programs.

Secondly, it is necessary to establish a mechanism for resource distribution. China and ASEAN countries have rich national sports resources. According to incomplete statistics, more than 300 national sports activities such as athletics, games, dancing, performance, festivals, nourishing of life and so on are owned by China and ASEAN countries together. The key to transfer these special resource advantages of two places into real industry advantages is to establish corporate mechanism for resource distribution. Theory of industrial finance also shows that industries with regional characteristics are easy to be formed when having the advantage of centralized resources. For example, although some recently popular sports events are frequently carried out, they are still mainly small-scale individual events held in China, making unnoticeable communication values purposed on sports culture communication. Under the framework of China-ASEAN sports cooperation community, a number of key competition projects should be carried out in order to exploit the gathering advantage of these resources to the full and effectively avoid the phenomenon of industrial isomorphism in this region.

3.4 *Conclusions*

In the long term sports cooperation community is an important pillar in promoting the stable relationship between China and ASEAN countries. We should take it seriously from a nationally strategic height, and enlarge the contents of sports cooperation fields in succession in order to establish more comprehensive and deeper friendship and promote mutual trust and cooperation between China and ASEAN countries because of sports. The construction of sports cooperation community will be beneficial in promoting better communication and coordination on both sides, strengthening regional sports culture participation constantly, taking each other's traditional sports culture seriously, and enhancing the competitiveness of regional sports culture industry.

REFERENCES

[1] Zhong Jingwen. How to make western development without cultural identity [J]. *National Solidarity*, 2000(4):20.
[2] (US) Huntington, Zhou Qi. *The Clash of Civilization and the Remaking of World Order*, Beijing: Xinhua Press, 2010:136.
[3] Xu Fenfen. Study on sports culture communication and cooperation between Fujian and Taiwan in the development of "Haixi" [J]. *Journal of Chengdu Sport University*, 2008(1):37.
[4] Li Naiqiong, Wang Jinghao. Study on China-ASEAN sports communication and development of cross-borders ethnic groups [J]. *Journal of Shenyang Sport University*, 2012, 31(6):129.
[5] Zhang Zhaolong, Qin Weifu, Li Naiqiong. Study on the integration development between Sepak takraw and chinese sports culture [J]. *Sports Culture Guide*, 2013,(3):116.
[6] Yin Bo, Feng Xia. Study on creative industry of sports culture in Beijing [J]. *Journal of Physical Education*, 2010, 17(6):22.
[7] Li Xiaochun, Zhangzhaolong, Study on the commercial promotion of sports activities of minorities' traditional festivals in Guangxi [J]. *Journal of Xi'an Physical Education University*, 2010, 27(4):42.
[8] He Chuansheng, Zhangzhaolong. Study on the current situation and measures in sports culture integration and development in China and ASEAN countries [J]. *Journal of Xi'an Physical Education University*, 2014, 28(1):27.
[9] Jiang Minglang. On China-ASEAN professional sports communication and cooperation[J]. *Around Southeast Asia*, 2008(4):52.

Sports Engineering and Computer Science – Luo (Ed.)
© 2015 Taylor & Francis Group, London, ISBN 978-1-138-02650-6

Music in sports activities—take the World Cup theme song for example

TingTing Zhou
Art College, Shandong University, Weihai, China

ABSTRACT: Large-scale sports activities have a sonorous theme song, the theme song is not only the symbol of sports activities, but also the best propaganda for the events. This article uses literature method to comb the evolution of World Cup theme song, analyze the theme song from the World Cup, and discusses the perfect combination of music and sports.

1 INTRODUCTION

FIFA World Cup is the football game of the highest honor, level, value, and profile. World Cup and Olympic games are two world class events. The influence and the broadcast coverage of the former are larger the letter. The World Cup is the most coveted divine glory for all countries in the field of football. Whoever becomes the champion of the World Cup is regarded the strongest in the world. The whole world will be crazy about them and the player would be seen as the national hero forever, therefore World Cup is the ultimate dream of every football player. The World Cup is held for every four years, any FIFA member countries (regions) can sign up for the event. The message expressed by the World Cup has long been appreciated: various mascots and colorful classic goals. Of course, the theme song cannot be ignored.

Theme song is the main symbol of the movie, TV, animation, drama and so on. The theme song for World Cup is created for football only. The opening ceremony of the World Cup usually last for only a few hours, and officials must be brief and efficient, theatrical performances must be compact, so the highlight is singing the theme song. It is a way to publicize the World Cup, is also a medium to spread the human civilization and establish a communication platform. Blatter, FIFA President reiterated "There is only two languages in the universe, one is the football, the other is music". In the world of football, music serves as a prop, but the quality of the song influence the World Cup greatly. A successful theme song is enough to draw the outline of the aesthetic characteristics of the World Cup and even leads people's aesthetic tendency of the tournament. This article will mainly analyze the World Cup theme song from 1990 to 2014, and it will study the perfect combination of sports and music.

2 INTRODUCTION OF THE PREVIOUS WORLD CUP THEME SONGS

2.1 *The 1990 World Cup in Italy: Un'estate Italiana*

Some people think this song is by far the most beautiful World Cup theme song, it is melodious and inspiring. Italians combine the sea breeze in the peninsula of the Mediterranean with their understanding about football. And then form a charming song. The sonorous voice of the male singer Moroder and the femal singer Nannini broke the snow of the mountain; fly across five continents, so everyone heard their heart beating with football. It is not just the expression of the joy for the events, but also reflects a nation's deep affection for football. The passionate singing and the brilliant Roman architecture are the most intimate chorus of music and football.

2.2 *The 1994 World Cup in the United States: Glory Land*

This theme song is closer to the combination of folk and pop rock on the musical style, full of vast wilderness of the American West. The song expresses neither heroism nor "peace, friendship, passion" or any other classic traditional theme. Although the song is entitled Glory Land, it doesn't have much to do with football. In other words, whether the style or the image of the song doesn't conform with the grand Olympics. So this song is very alternative in the 90s.

2.3 *The 1998 World Cup in France: The cup of life*

Ricky Martin is a world-class idol singer and leads the Latin music. This theme song is of strong interactivity, singers and fans can sing together, share the joy. As the song's drum rhythms and horns blare with a strong sentimental meaning, the audience can easily resonate into the emotional climax. The song was very popular after the World Cup and many football programs use this song to heighten the atmosphere. In addition, this song ranks top among all the sings from 30 countries around the world with enormous influence. Another theme song of this World Cup is DO You Mind if I Play. In this song, the white and the black take turns to sing, the tune is lighthearted and full of tropical temptation. The two songs are of different styles, but the French style is not obvious, and the singers are not Frenchman. This reflects a certain integration of race, nation culture of the time.

2.4 The 2002 World Cup in Japan and South Korea: Boom

The World Cup is held in Asia for the first time. The style of music reflected the difference between Asian and European pop music. So it appears the three World Cup theme songs. One is Boom with simple melodies, strong rhythms, popular genre and exotic feelings. The second is the Anthem. This theme song has two versions; the electronic music version is combination of the Oriental style and electronic style. Symphony version is classic and grand. It is the best ending song for the album. The third is Let us come together with fresh melody. It is unique among all the World Cup theme song. For it is more like a happy love songs or inspirational song.

2.5 The 2006 World Cup in Germany: Time of our lives

This song was composed by the famous composer Jurgen • Ailuofu song of the the Sony BMG record company, produced by Super producer Steve • Mike's and sung On June 9 in Munich by world famous Il Divo together with the R&B Queen Toni Braxton. This is the most lyric song in the history of World Cup. It is not as explosive as the song by Ricky Martin in 1998 nor as classic as Un'estate Italiana, but it is also a good song.

2.6 The 2010 World Cup in South Africa: Waka Waka

Recently, FIFA (FIFA) and Sony Music Entertainment jointly announced that the song wrote and partly produced by the famous Latin singer Shakira "Waka Waka (This Time For Africa) was chosen as the official theme song for the 2010 World Cup in South Africa. Shakira with South African musical group Freshlyground sung this song together at the closing ceremony. Blatter said the World Cup fans pay much attention to the theme song as they do to the mascot and the logo. It is also an important part of this exciting sports event. This song is full of the African rhythm character. I'm looking forward to hearing it in the World Cup, and see Shakira's and Freshlyground's performances at the closing ceremony."

2.7 2014 World Cup in Brazil: Todo Mundo

This is official theme song released. The top artist of Brazil MTV Gaby Amarantos with samba bands Monobloco cheerfully performed Todo Mundo. Soccer and samba are both symbols of Brazil. People also use the passionate Brazilian samba to welcome football fans around the world.

3 THE PERFECT COMBINATION OF MUSIC AND SPORTS

We need different music style to express the same theme. Music creation should have enough freedom, we should insist on the tenet of "independence and freedom". However, we shouldn't create art for its own sake. After all, art is the product of society and reflect the social life. Innovation is difficult, but it is also the magic of successive World Cup theme song. Every World Cup theme song has tried to break through stereotype and make innovation, leaving deep impressions to the world. These impressions are exactly the innovation of the music. The theme song as an important means of propaganda and expressing the World Cup, the creation level and style are directly related to the host country's understanding for the World Cup. At present, in big sporting events such as Olympic Games and the football World Cup, dogmatism thinking has gradually retreat and the addition of popular element has become the trend. Because pop music is easier to capture throb law of the human mind and emotion and are more likely to resonate with the human instinct feeling.

The combination of music and sports inherits the spirit of the World Cup and shows the music culture. Theme song as an important sign of the World Cup, has become a football culture, and even an important part of sports culture. Whether in the content or in form, the theme songs have a common characteristic. That is reflecting the unique national culture of the host country, and show their love for football. In the form of expression, the previous World Cup host nations are mostly European countries before 90s, their musical expressions mainly focused on Western classical music. With the development of football games, especially since the 90s, more and more Asian,

American and other countries host the World Cup, World Cup performance of music was full of a strong ethnic flavor and at the same time adopt the popular pop music. With the development of modern media such as radio and television, the theme song after 90s is more widely spread. The popularity of Un'estate Italiana and The cup of life were the best proof. It is this that makes music the best mass medium that brings the World Cup to the public.

Music serves as a carrier for transmitting and expressing people's thoughts and emotions. The theme song for big events take the responsibility for cultural heritage. The combination of sports and music is very necessary for the evolution of the history and social civilization.

Using the large-scale sports event as a platform and using music as a connotation will become the theme of music creation. Let us all be in the appeal of the previous World Cup theme song, enjoy world peace and share the culture communion.

REFERENCES

Cui Dongye. The World Cup football culture market, Journal of business culture, the academic version, 2007.8.

Li Yan-Transmition and culture, Hang Zhou, Zhe Jiang university press, 2009.10.

Liu Shuang, Yu Wenxiu-Cross-cultural communication: Dismantling cultural walls, Harbin, Hei Longjiang people's publishing house, 2000.1.

Tian Shiping, Li Kunxian. The World Cup and football culture, Journal of Shan Dong sports institute, 2007.4.

Wang Xiaodong, Wang Liancong, Ye Wei- Comparison of TV viewing between Athens Olympic Games and the Japan and South Korea World Cup, Journal of Shanghai sports institute, 2005.5.

Wu Jiani. The audio-visual feast in Media convergence times: Reports from the 2010 World Cup, The press, 2010.4.

Sports Engineering and Computer Science – Luo (Ed.)
© 2015 Taylor & Francis Group, London, ISBN 978-1-138-02650-6

Investigation of the incorporation of sports and art

Xiao Gong, Yanqiu Guan, Haoyuan Xu & Xiaoying Zhuang
Beijing Film Academy Modern Creative Media College, Beijing, China

ABSTRACT: In order to study the relationship between sports and art better, and guide the development of modern sports and art, combining with theory and historical analysis, this paper considers that sports and art have the same origin and aesthetic effect, which can bring people spiritual pleasure. This paper especially makes an analysis of the incorporation of sports and art from the historical view, and sees that the ancient social sports and art shows us with a state of combination and separation coexisting from their derivation, sports and art shows an irreversible trend to integrate in the process of development. The incorporation of sports and art has become a universal phenomenon up to today, which not only enhances the endless interests and charm to the sports life, but also brings infinite pleasure to people. The incorporation of sports and art might create fine art works.

1 INTRODUCTION

Sports and art respectively belong to two different cultural categories, but sports and art have the same derivation. They are both created in the process of people's social labor, so there is a common point. Both of them have aesthetic functions, and they can bring spiritual enjoyment to people. However, sports are different from art. After humans enter into civilized society, the boundaries of sports and art are more obvious. The contact among sports, art and beauty is less than before, but people's aesthetic quality enhances unceasingly with the further development of society, and sports and art gradually incorporated. Now the incorporation of sports and art becomes more tightly, and sports are continuously improving aesthetic taste and value through the incorporation with art. The incorporation of sports and art has become a necessity of historical development. Therefore, the relationship between sports and art is changing along with the historical development, and they are gradually incorporated in dynamic.

2 GENERAL INSTRUCTIONS FROM THE POINT OF VIEW OF THE ORIGIN OF SPORTS, THE ANCIENT SOCIAL SPORTS AND ART SHOWS A STATUS OF COMBINATION AND SEPARATION

Sports and art have a common derivation, but they are not strictly called "sports" and "art" at the beginning. They are the form of labor and entertainment created by original people in order to meet life demands, just because they have some elements of modern sports and art, they are known as sports and art. For example, the original dance is just a worship means to pray for good or a means of celebration for celebrating harvest created by the primitive tribe. According to records of "Art lecture outline", there is a kind of Gymnastics Dance in the primitive society, a tribal dance. There are hundreds of people dancing. People dance to celebrate the harvest when collecting fruit or getting plenty of animals. From Plekhanov's records, we can see that the original sports and art showing a state of incorporation. Not only that, sports and art appear in ancient time. For example, in the ancient Greek Olympics, there were both sports items and art competition with rich musical sense. People listened to music after games, and watched art programs to get physical pleasure and relax, so that sports and art were incorporated perfectly in the ancient Greece Olympic Games. In addition, sports and art in the ancient Greek culture were integrated through other forms, such as sculpture, painting, architecture and other items, which not only showed the beauty of art but also expressed people's pursuit of physical health. At that time, the most famous statue was "Discobolus", the typical incorporation of sports and art. It always played an important role on historical development of human civilization. People could see persistence beauty and strength through this statue. They could see a history and a culture relating to Hellene's aesthetics psychology and Hellene people's respect and love for life.

But the ancient sports and art were not always incorporated together. Sports were originated from social life, serving for the social service. Its direct utilitarian purpose was far greater than the aesthetic purpose, which means there must be separation between sports and art. For example, the ancient sports were served for military to a large extent.

Introducing sports to training could cultivate strong warrior. The ultimate objective of sports training was for fighting enemy and defending their territory. This kind of training was incorporated with kill and bloody. It was difficult to imagine the sports had any relationship with art, such activities were common in China history, such as Chinese Ansai waist drum. According to legend, military activities were often conducted in the Qin Dynasty Ansai city, the warrior tied their waist with drums, they would give an alarm in case of emergency, they could cheer when met the fighting, they could report news when meet reinforcements, and they could make celebration when meet victory, this kind of sport activities made huge effects on military exercises, and played positive role on Qin army killing the enemy. After a long history of evolution, the Ansai drum became popular with general population, and its form was continuously changing, and finally became the popular sports full of aesthetic sensibility at present, which shows people with new vitality. Ansai waist drum linked up with art, but the original Ansai waist drum has no association with beauty, things like that ancient sports were Inner Mongolia wrestling, horsemanship and archery, hunting and other activities in ancient times, which were originally developed for people's life and military, these sports highlighted the practical function of value, but the aesthetic value was not high.

3 FROM THE DEVELOPMENT PROCESS OF SPORTS, THE INCORPORATION OF SPORTS AND ART WAS AN IRREVERSIBLE TENDENCY

With the development of society, the aesthetic functions of sports were emerged increasingly day by day, sports had endless charm and great aesthetic value; but for art, there were less and less pure art works, art began to show a trend to close up to life and reality, the works with relatively independent aesthetic values such as literature, painting, music, and dance were gradually changed in the process of development, art began to incorporate with reality, but architecture, sculpture, and ceramics incorporated with reality more closely, there were less and less pure art works, art began to show a trend to close up to life and reality. Sports and art were developed in two-way, and incorporated in the reality, so the incorporation of sports and art was irreversible. We can confirm it through two specific examples.

We know that the Ansai waist drum is used for military activities originally, the artistic value is not high, but in the process of development, the sports have a very big change, and begin to enter from the military into people's life, while the aesthetic content has been enriched and the aesthetic value is to highlight. For example, at the beginning of the separation between Ansai waist drum and

military activities, through people's development, Ansai drum formed a set of more complex activity routines: run, jump, twist, twist, hide, step and other movement, all kinds of movement was changed from each other, which gave a great aesthetic experience to people. Later on, in Song Dynasty Ansai waist drum was incorporated with Yangko, and it became a common sport activity combining body building with aesthetic entertainment. Then this project got further development, gradually linked with music, dance, martial art, folk songs together, forming a unique artistic effect, which is very cool and comfortable and greatly inspired people's mind. Its aesthetic function was more outstanding.

There are different explanations about the origin of Chinese traditional martial art. People said that Chinese martial art was originated in the ancient war. The soldiers in the Warring States period summarized the fight and combat experience in the form of the initial martial art; someone said martial art was originated in the primitive society. It gradually came from the experience that the primitive men battled with wild animals. Anyway, Chinese martial art was in a very simple form originally, because it only had some simple actions, such as hit, beat, lift, throw, which was very ordinary and dull. While the purpose of the people who practiced martial art was simple, they just wanted to do physical fitness or for simple self-defense, competing force and war. It was closely related with our life. The original martial art lack of coherence beauty, harmonious beauty and complexity beauty. The aesthetic value is not high, but in Shang and Zhou dynasties, Chinese Traditional martial art had a new development where martial art began to become a training project. The governor in Shang and Zhou dynasties trained soldiers by "martial art dance", so martial art had a fixed pattern, and could show a kind of power. In Zhou Dynasty, "Xu" appeared in the society, which is equivalent to "school" in later time. Martial art became a program of education in "Xu". The popularization rate of martial art was greatly improved so that martial art skills can be fully developed. The "beauty" began to appear in martial art, and the ornamental value was gradually opening up.

Especially when the representative works "China martial art book" appeared, the Tai Chi theory of "Yin Yang Tao" not only made an important impact on the China health, but also laid the foundation on Chinese martial art system. It established a relatively completely theory of Chinese martial art, which played an important role in guiding the development and evolution of Chinese martial art. Since then, the incorporation of Chinese martial art and art has became more closely. Chinese martial art were inseparable with the art in patterns and ideas, and the aesthetic value of Chinese martial art is greatly improved.

In addition, there were many examples of incorporation of sports and art sports gradually in the development, such as archery, dancing, ball games and so on, having undergone from simple to complex, from low to high and from presented separately to presented incorporated with other art forms in the development process. The incorporation of sports and art is a necessity of historical development, and it is the objective requirement of human aesthetic improvement.

4 IN MODERN SOCIETY, THE INCORPORATION OF SPORTS AND ART HAS BECOME A COMMON PHENOMENON

In the long history of human being, sports and art are constantly changing. There were separation and incorporation, but in modern society sports and art have been incorporated together closely. Sports cannot leave art if sports want to be developed and play a more and more important role in people's life. With the development of modern science and technology, radio and television have become necessary products in life. People, watching sports programs just like watching movies and TV series, are always full of expectation, which also promotes the incorporation of sports and art in objectivity. Now, the field of pure sports and pure art become more and more small and the incorporation of sports and art become ubiquitous, which exists in many fields.

For example, we know sports architecture can show a king of art beauty in the construction field. Its practicability and ornamental value are equally important. A classic sports architecture can give people a sense of fashion, classical beauty, a kind of strength and a positive spiritual beauty. Therefore, the sports architectures in every city are unique. They are the classic buildings in the city, the embodiment of city spirit, especially the sports buildings in World large-scale athletic games, such as the Olympic Games, the Asian games and so on. It should be well known that Beijing Olympic Bird's nest costs huge. It is full of artistic beauty, and brings people with endless imagination. It has a smooth and beautiful shape which is grandeur and spacious, and it also give a feeling of warm, great and sublime aesthetic to people. It is Chinese pearl of wisdom, and it is also a perfect model of incorporation of sports and art. Moreover, the incorporation of sports and dance music is one way of exhibiting modern sports. The incorporation exhibits strength, beauty, vitality and passion from a higher level. Dance music plays a role of painting and regulating in sports, which improves ornamental value and aesthetic value for sport games. Therefore, it is widely used in synchronized swimming, bodybuilding exercises, artistic gymnastics and music, and it also becomes one of the competitive sports marking standards. In the modern Olympic Games, athletics projects not only have correct and standard requirements, but also require beautiful, relaxed and harmonious performances. It is inevitably linked to the injection of artistic elements, and the beautification o sports actions. Music always plays a role of sports action. The sports only with full flavor and beautiful can help athletes get high marks. After the strenuous exercises, there will be great help for adjusting body and restoring physical strength. If athletes can dance lightly, it will be the best choice for relaxing themselves. Of course, the training effect and competition records can be improved greatly.

The incorporation of the modern sports activities and art is very closely, which also produces large numbers of new sport projects. In the incorporation of sports and art people's various part of body are showing great vitality and charm. People are full of energy with unique temperament. In the process of continuing to create beauty and health enhancement, people also create sports art works as the acme of perfection, the incorporation of sports and art creates perfectness.

In a word, during the different historical stage of social development, there are both separation and incorporation of sports and art in the history, which all depends on the development degree of human civilization and the development of people's psychology. But on the whole, the incorporation of sports and art has become an irreversible trend, especially for the modern sports. The fields of modern sports keep expanding, and incorporated with art is extensively, which not only enhances the endless appeal and charm to the sports life, but also brings infinite pleasure to people's aesthetic life. The incorporation of sports and art is bound to create high-quality culture works.

REFERENCES

Anderson, J.R. and G.J. Julien, A Ikegami, R., D.G. Wilson, 1990. *Active Vibration Control Using NiTiNOL and Piezoelectric Ceramics,* J. Intell. Matls. Sys. & Struct, 20(2):189–206.

Hoffer, R. and D. Dean. 1996. Geomatics at Colorado State University, *presented at the 6th Forest Service Remote Sensing Applications Conference*, April 29–May 3, 1996.

Inman, D.J. 1998. Smart Structures Solutions to Vibration Problems, *in International Conference on Noise and Vibration Engineering,* C.W. Jefford, K.L. Reinhart, and L.S. Shield, eds. Amsterdam: Elsevier, pp. 79–83.

Margarit, K.L. and F.Y. Sanford. March 1993. Basic Technology of Intelligent Systems, *Fourth Progress Report,* Department of Smart Materials, Virginia Polytechnic Institute and State University, Blacksburg.

Mitsiti, M. 1996. *Wavelet Toolbox,* For Use with MALAB. The Math Works, Inc., pp. 111–117.

Sports Engineering and Computer Science – Luo (Ed.)
© 2015 Taylor & Francis Group, London, ISBN 978-1-138-02650-6

From the perspective of occupation of the contradiction between supply and demand at the present stage of our social sports instructor

HongGang Qu
School of Physical Education, Northeast Dianli University, Jilin City, Jilin Province, China

ABSTRACT: From the perspective of professional concept, this paper did research and analysis of China's professional type of social sports instructors at present stage, and the contradiction between supply and demand. Pointed out reasons lead to the problem, the types of professional characteristics and the social existence of the service object are the main factors that cause the lack of social sports instructors at the present stage in China, and based on these, this paper expounds the opinions and suggestions on the issue.

1 INTRODUCTION

With China's economic development and national income level rises ceaselessly, the broad masses of the people's pursuit of the spiritual life are increasingly prominent, physical and mental health attracting more and more attention. These requirements directly promote the development of mass sports in China and the growth of social sports industry. In 2011 the "comprehensive fitness program—2011 to 2015", referred as "with the increase of sports facilities, sports participation and sports personnel, sports education and counseling also increased". A large number of sports workers have entered this industry, there are macro guidance of national civil servants, direct practical fitness instructor, masses sports guidance that are warm-hearted public welfare undertakings of physical culture and sports and so on. We call them "social sports instructor". Since the first award ceremony of national social sports instructors, China has about 700000 people are various levels of social sports instructors, and the development of social sports in China has played a positive role, stimulating the development of society and economy. [1] However, for a long time, the quantity and quality of social sports instructors in China are not ideal, which is a recognized fact, restricting the development of social sports, affecting the quality of public sports service. From the perspective of occupation specific, through in-depth discussion and analysis, on the understanding of this phenomenon, this article provides suggestions and ideas for improvement and development of china's social sports instructor system.

2 OCCUPATION STANDARD OF SOCIAL SPORTS INSTRUCTOR

2.1 *The concept of occupation*

Occupation (occupation) has a lot of concept, summed up in the following definitions: the occupation is a means of subsistence that people engaged in the society; from the perspective of the society, occupation is social roles, a social obligation and the responsibility of workers for certain tasks, and workers can obtain the corresponding reward from the national economic activities; from the view of the need of human resources, occupation refers to the special labor positions of different nature, different content, different forms, and different operations.

2.2 *Occupation standard of social sports instructors in fitness*

In the national Ministry of labor and social security of the social sports instructors "national occupation standard", the social sports instructors of occupation orientation. Occupation: in the name of social sports instructors; occupation definition: engaged in sports skills teaching, scientific fitness guidance and management personnel in the mass sports activities; occupation grade is divided into four levels: primary, intermediate, advanced social sports instructors and social sports instructor.

3 TYPE OF OCCUPATION AND THE PRESENT SOCIAL DEMAND IN CHINA OF SOCIAL SPORTS INSTRUCTORS, THE VIEW OF ROUGH CLASSIFICATION OF SOCIAL SPORTS INSTRUCTORS

3.1 At all levels of civil servants engaged in social sports management

Civil servants at all levels of sports management departments in our country, are mainly responsible for social sports and mass sports activities guidance, organization and management, and the staff working in the department are some sports professional talents, some other professional personnel. In social sports work, according to their national policies and regulations, formulating the planning of regional social sports activities, organizing social sports professional learning and training, is mainly the macro management of social sports activities, not directly involved in social sports specific work. People in the sports department, job task decides they bear part of the responsibility of developing social sports, but also a part of social sports instructor. From the concept of occupation, they belong to the certain social roles, undertaking certain social responsibilities and obligations, and obtain the corresponding reward. [2] From the social point of view, this part of the social sports workers is makers and managers in the development of social undertakings of physical culture and sports in China. They broaden our social sports development scale, and adopt the development mode of macro guidance from the whole of the direction of development. They rarely involved in the specific social sports instructing practice, just in organization, implementation and management for social sports activities.

3.2 Work in various venues fitness sports professionals

This part is to obtain the most part of our social sports instructor qualification, basically it is just entering the work of sports professionals, and is mainly composed of social sports professional students, and sports professional athletes. There are a small number of fitness enthusiasts, their purpose of engaging in the activities of social sports guidance is very clear, is to society's survival. Because of its origin, it's engaged in professional sports, involved a certain level of sports skills, and sports related knowledge, so the industry can be active or passive. Occupation's basic idea is that people are engaged in the work as a way to make a living in the society. Starting from the concept, they are the typical sports occupation workers. They are mainly using the sale of their sports skills and knowledge, to provide services for fitness, in exchange for labor remuneration, in order to survive in the society. Looking from the social demand, this part of staff is the main force in the development of social sports occupation. Their work is the activities of sports guidance, obtaining the corresponding reward and occupation promotion, achieving social survival and manifesting the social value of their occupation. The degree of their work will be more and more high.

3.3 Guide the masses in various community sports activities to exercise guidance

Most of them don't get social sports instructor qualification: this part of social sports instructors is basically retired workers, cadres and workers. They basically belong to public welfare social activities, and they promote social activities mainly for the purpose of physical health and spiritual joy; there is also a little part is to obtain certain economic benefits through the promotion and guidance of the mass sports activities, but it belongs to a kind of incidental act, not the main purpose. In the view of occupation, their work is the area between occupation and non occupation, the main purpose is the sports activities, with some characteristics of the occupation, but with uncertainty. [3] Looking from the social demand, along with the development of society, the need for this part of the social sports instructors, is the main field of the current shortage of social sports instructors in China.

4 THE CONTRADICTION BETWEEN THE OCCUPATION NEEDS AND SOCIAL NEEDS OF SOCIAL SPORTS INSTRUCTORS IN CHINA

4.1 Occupation demand at the present stage of our social sports instructor

Cultivation of subject at the present stage of our social sports instructors is the students in the school sports colleges and retired professional sports athletes, because of social demand and employment system in China, they face the first choice of occupation, and the first requirement is to solve the problem of survival in the society. Therefore, their basic choice will be entering into the sports related industries, such as schools and sports related departments, but this is a small part, most of the other will enter into all kinds of fitness sites, becoming the instructor occupation sports fitness and practitioners. Because they are unable to choose social sports guide post welfare, social sports guidance also have no time and energy to participate in the public welfare. The direct reason is that we cultivate a lot of social sports instructors who cannot enter the social sports guide post. From the occupation demand, we develop the social sports instructors subject to occupation choice is to solve the problem of existence; this is the demand for their occupation. But our country at this stage, the post of social sports instructors provided cannot meet the basic needs. Therefore, they are unable to enter into social sports guide posts that are really needed. This is the supply and demand contradiction at both ends, and is also the main reason causing the relative lack of social sports instructors in China.

4.2 *Our social sports development trend and demand of social sports instructor*

Our social sports development has made considerable progress today, various fitness facilities and community physical exercise resorts increased, providing more choices for different fitness needs of the masses. At the same time, China's social sports development shows a clear trend, consumption scale of society sports fitness and the scale of spontaneous mass sports continue to expand, but the former is far from the latter in terms of speed and scale, development speed and influence, therefore, spontaneous mass sports becomes the main groups of social physical exercise in China. There are two main factors for the formation of this phenomenon: one is the continual improvement of economic level of our country, people's living level has lifted to a hitherto unknown height. Since the material demand is no longer a problem of survival, people's pursuit for their own health and the spirit is increasingly prominent. The other is the advent of the aging society, the level of medical and health conditions improved, people's life has been better and better, and the number of elderly population continues to rise. The proportion is increasing, the healthy demand of this part is more intense, and therefore, a large number of seniors participate in physical exercise. These exercises early are separate activities or small range of activities, from travel to simple Taijiquan, along with the community sports and activities. A large number of fitness dance abound of rich local characteristics and ethnic characteristics of the fitness way gradually develop, attracting a large number of physical exercisers. At present, under the social sports instructors training at all levels, various fitness facilities, sports guidance and related services are complete. Social middle class and above this level are the main target of service. Our country's economy continues to improve, thus this part of people increased, but the

increase in the number lags far behind in development of sports for the aged exercise group. But few young social sports instructor enter into this group to work. Elderly exercise group guidance is basically spontaneous commonweal guides. They come from the exercise groups, with high enthusiasm and instructive ability. But, equally obvious disadvantage, their number, age, physical health knowledge, teaching skills and guidance ability are poor; many problems such as the improper sports injury treatment also need to be noticed. [4] These phenomena and the large gap in demand of the social sports instructors are the main reason that leads to the lack.

5 SUGGESTIONS AND METHODS TO THE CONTRADICTION OF SUPPLY AND DEMAND OF SOCIAL SPORTS INSTRUCTORS IN CHINA

5.1 *Consider the occupation demand of the social sports instructors, focus on the training of occupation quality, broaden employment channels*

In addition to social sports guide post additional paid services, to expand a variety of jobs associated with social sports and occupation choice space to expand social sports guidance, solve the social existence and social value problem. From the analysis of the development trend of China's social and economic life, business social sports venues quantity will increase year by year, the economic proportion of sports industry will increase greatly. In background of vigorously promoting the development of tourism economy, the development direction of China's tourism industry has become the sports tourism industry. Outdoor sports in summer, and skiing in winter is the representative of this trend. Directly or indirectly, driven by the rapid development of social sports of the occupation, create the sports management, a lot of guidance and management positions. We are social sports instructor training, should focus on training students' occupation accomplishment, improving students' practical ability and the quality of their future occupation. Engaging them in various social sports job lays a solid foundation. [5] At present, many social sports professional students' occupation education colleges, only leads students into the posts but the students don't have the most basic occupation accomplishment. They are not very good to meet the requirement of posts, and this is a universal problem in our social sports instructor development process. Develop social sports instructors in the employment channels is the key to solve this problem.

5.2 The advantage of sports resources of the various regional sports college, the establishment of regional and the elderly in social sports instructor training system

Public mass sports activities is the main force of the development of social sports in China, and has become a main direction of the development of social sports. With the exercise group and the social sports guidance requirements increase, most of the existing social sports instructors can no longer satisfy the demand of exercise groups. In order to solve this problem, we can make use of educational resources and various sports colleges, and establish social sports instructors, elderly amateur training system. [6] Use school teaching time, the selection of the backbone of elderly exercise group as the early community physical exercise training and the intermediate social sports instructors. This will not only solve the commonweal social sports instructor shortages, but also can effectively improve guidance levels in elderly exercise instructor, guide the amateur sports training exercise for health. It's scientific, and it highlights the social function, saves the social cost.

5.3 Establish a social sports instructor evaluation system, there is a certain class must be included, a specified time accumulation of public social sports guidance, or reduce in rank or disqualified

The majority of the amateur sports training also needs to constantly improve the level of exercise and sports appreciation ability, which requires the sportsmen update the competitive level of learning. Amateur sports training groups of social sports instructor guidance's capability and competitive level is relatively limited, and can not meet the demand of long-term exercise group. The occupation of professional social sports instructors are mostly have the high level of competition and guidance ability, therefore, to establish commonweal social sports guidance system is reasonable in part of this group, can effectively solve some problems in amateur sports training. The social existence and social obligations two aspects embody the social sports instructors of social value.

5.4 The base construction of sports professional students practice, make commonweal social sports guidance by occupation blank area of students

Currently, social sports specialty occupation at a relatively low level of education in our country,
without adequate attention to the cultivation of students' practical ability. With China's economic and social development and the change of life, occupation and social sports jobs for college students are higher, occupation quality requirements will continue to improve, the traditional mode of social sports specialty education will no longer adapt to social sports jobs for practitioners of the occupation, therefore, the social sports professional occupation education will be the direction of development in the future. Improving the quality of students' occupation, we needs to grasp the rich theoretical knowledge and join high level of competition, on the other hand, also need to have practice and occupation training constantly. Through the establishment of experimental base practice for college students, so that students in the internship process can continuously improve their ability of social sports guidance. Training students actively participate in public welfare activities of social morality, and making perfect personality of citizens, it's in accordance with the essence of education. At the same time, a lot of students who step into the social sports guide posts can solve the quantity shortage of the commonweal social sports instructors, and also become the core members of amateur sports training group. Improve their level of social guidance can also promote our social sports activities.

REFERENCES

[1] Feng Huohong. Public social sports instructors policy adjustment research—the establishment of the system of social sports instructor public post (J). Journal of Beijing Sport University, 2012.7.

[2] Wang Huiwen. Social impact on the community sports instructors (J). Science Herald, 2011.1.

[3] Dai Jianhui. The social sports instructor occupation characteristic, function and development (J). Sports scientific literature communication, 2007.1.

[4] Guan Lijun, Tang Peng, et al. The basic point of sports social sports guidance team (J). Sports Expo, 2011.6.

[5] Zhang Xiaolin. Understanding and Thinking on the implementation of occupation standard system of social sports instructors in China [J]. Xi'an Physical Education University Institute, 2004.4.

[6] Yu Shanxu. Foundation and Countermeasures (J) development of commonweal social sport instructors in China. Journal of Shanghai Institute of Physical Education, 2001.1.

Sports Engineering and Computer Science – Luo (Ed.)
© 2015 Taylor & Francis Group, London, ISBN 978-1-138-02650-6

Risk management of major sports events

JiaBao Ye

Tianjin University, Tianjin, China

ABSTRACT: This paper analyzes the potential risk and reasons of major sports events, and puts forward the countermeasures to avoid the risk, constitutes the program of emergency of major sports events.

1 INTRODUCTION

In the year of 1985, riots made by hooligans in Brussels led to collapse, causing hundreds of deaths. In 1996, explosion and bloodshed happened in the Olympic Sports Park during the Atlanta Olympics. Recently, an increasing number of unexpected incidents happened in sports events and caused casualties and damages to property. Thus, risk management of major sports events is a very significant task.

Risk management is defined as the process which contains the management plans, identifies, estimates, evaluates, responds and supervises to the risks encountered by organizations, and it is the general term of those practices which achieve maximum security assurance with scientific management methods. An excellent risk manager must forecast possible risks in assess magnitude of the risk accurately at first, take effective measures to cope with the risk, and formulate normative procedures in the end so as to minimize the risk.

Because the scale of modern sports events is ever-growingly expanding nowadays and the complex and awkward factors exiting in sports events are increasing as well, which to some degree increase the probability of risk. In order to realize the maximum of efficiency, reduce risk and cut loss, risk management is very necessary in the operation of major sports events.

2 IDENTIFICATION OF RISK OF MAJOR SPORTS EVENTS

In the process of the major sports events management, managers should identify the various potential risks which may cause losses to events first, and they need to find the primary and secondary factors which cause these risks.

According to the general theory of risk management and characteristics of sports events management, we confirm and classify the risk management in the process of major sports events, mainly including: natural risks, management risks, human factor risks and infrastructure risks. The main reasons of those specific risks are in Table 1.

3 RISK ASSESSMENT IN THE PROCESS OF MAJOR SPORTS EVENTS MANAGEMENT

Generally, risks are defined by the relationship between the probability of accidents and the economic losses caused by the accidents. In order to

Table 1. Classification of risks and causes in sports stadium management.

Classification of risks	Causes
Natural risks	Caused by natural factors such as snowstorm, flood, hail, earthquake and typhoon
Management risks	Unreasonable management organization or mechanism, wrong and neglect decisions or judgments
Infrastructure risks	Inappropriate manufacture, procurement and supply of materials and equipments; untimely supply of materials and equipments; disqualification of materials and equipments
Human Factor risks	Dismission and forfeited of key personnel (teams); insufficient quality (ability, efficiency, health condition, responsibility and trait) of managers and technicians; crowded audiences

Table 2.   Classification matrix of risks in the process of major sports events.

	High loss	Moderate loss	Low loss
Frequently happening		Miscalculated money	Slip in the snack bar
Occasionally happening	Severe knee injury of players	Fall down around the sport stadium	Audiences without tickets
Rarely happening	Death of audiences	Body frostbite of audiences	Violence in the stadium

assess the risks quantitatively, the causes of accident must be considered. It is also needed to assess the probability in every accident and consequence of personal injury. The whole level of risks can be reduced by calculating the risks of every accident and adding them up according to the probability and the consequences of those accidents.

Once the risks or losses are confirmed, the risks need to be further distinguished, and the frequency and strength also need to be assessed. The frequency of risks draws on the possible number of risks and losses. Risk managers have to classify the frequency of confirmed risks and losses into "frequently", "occasionally" or "rarely". The severity can be determined by the loss degree into "high", "moderate" or "low". Severity and frequency are based on the past experience of the managers, and economic losses and personal injuries need be differentiated by the frequency and severity.

The matrix in Table 2 can be widely used in distinguishing risks. Once the risks are confirmed, it can be divided into nine types. By putting confirmed risks into this matrix, risk managers will assess risks successfully. However, it must be noticed that risk assessment is a dynamic process. For instance, if rainstorm happens frequently, it probably leads to more and more slide and injury. Sport stadium managers need realize these changes and make accordingly assessment.

4   RISK MANAGEMENT IN MAJOR SPORTS EVENTS OPERATION

4.1   Risk prevention

It is advised to avoid the risks which can cause great damage and occur frequently. A sports organization should consider not holding those events that have the possibility of resulting in financial loss. And it is not prudent to host sports events or activities which have already caused losses or lawsuits at other venues. All these types of events and activity programs should be avoided.

4.2   Risk transfer

Risk transfer means transferring risks to people or companies that are willing to take the risks. In

other words, it should be insured. As demonstrated in Tables 1 and 2, it often adopts the strategy of risk transfer for risks whose seriousness and frequency of occurrence are not that much, and it is not wise to host activities that may bring potential financial losses to the managing organization of the sports events. That is, purchasing different kinds of insurances are necessary.

Generally speaking, the types of insurances purchased by sports events organizers are: (1) public liability insurance. This is mainly aimed for all the organizers, coaches, referees, staff members, volunteers recruited by the government who participate in the events management. (2) accident insurance. It means the insurance that against all the personal injuries happening in the playing court, including all members as well as the spectators and it is a kind of common insurance nowadays. (3) property insurance. It is the insurance aimed at protecting sports equipment and facility. (4) insurance to cover the risk of sports events cancellation.

And the ways of purchasing insurances mainly are: (1) purchasing domestic sports insurance. (2) purchasing foreign sports insurance. (3) seeking sports insurance sponsor. (4) forming Chinese-foreign cooperative sports underwriter.

4.3   Maintain and reduce risk

Organizers of sports events should try to maintain and reduce the damage that may result risks. In the Table 2, the maintained and reduced risks are those who have less possibility of occurrence or smaller damage. Organizers could maintain these risks since they have tiny possibility of causing great losses. However, organizers should also be aware of avoiding financial loss and other incidents accompanied with the risks by taking effective preventative actions, which can be achieved by Standard Operating Procedure (SOP). SOP is a series of prescriptive documents, providing specific guidance and proper solutions to various situations. SOP is applicable to distinct kinds of risks. And risks can be effectively maintained and reduced through SOP.

As mentioned above, the processing model to cope with the risks happening in playing venue operation can be summarized as in Table 3.

Table 3. Processing model of risks in playing field.

	Great loss	Medium loss	Low loss
Usually occurred	Avoid	Transfer and reduce	Transfer or restore and reduce
Occasionally occurred	Avoid or transfer and reduce	Transfer and reduce	Maintain and reduce
Seldom occurred	Transfer and reduce	Transfer or maintain and reduce	Maintain and reduce

5 MAJOR STRATEGY OF REDUCING RISKS IN SPORTS EVENTS MANAGEMENT

5.1 Negligence and exemption treatment of facility

If the negligence has to do with the facility, then the equipment administrator should be responsible. Therefore, organizers of sports events needs to avoid risks in terms of facility maintenance and exemption treatment.

First, it is required to check the facility periodically to avoid the users suffering from damage and hazard; maintain the equipment and fix defection; alert the users, participants and spectators for unnoticeable damage and hazard; warn the users, participants and spectators for the inherent sports risks.

Second, using waiver and exemption. A waiver or exemption is like a contract, and it is right when service provider's negligence cause damage to the user and the user agrees to abstain the right to file a lawsuit. When a person signs a waiver or exemption right, it practically means abstaining the right to file a lawsuit against someone's negligence. A waiver or exemption means a person agrees not to file a lawsuit against a delinquent for the damage he may cause and is a way the service provider or professional seeking for self-protection by exempting ordinary negligence liability. Through waiver or service provider can ask injured costumers not to file lawsuits.

5.2 Court violence and effective control

The court violence caused by English hooligans in 1989 resulted in 96 casualties, and from then the court violence has been nearly unavoidable. In America, the most intensive violence concerning the spectators happened in professional matches. For example, some players of Denver Broncos were hit in the eyes by snowballs when leaving the court in 2001. During the two games held in Cleveland and New Orleans the team member were attacked by the furious spectators. Since then, it has become a huge task for the sports events organizers to prevent the players from the assaults of the overly passionate spectators. As the problems mentioned above and with the threatens from the terrorism, it has been a important component of risk management of sports events to foster safe and harmonious field atmosphere and strike hard against court violence.

First, it is necessary to have an emergency management preparation before the sports events. Principles to deal with the court violence are: control minor incidents and avoid casualties in major incidents. There must have precaution actions and specific emergency measures.

Second, tackling court group violence necessitates two aspects can work. The first one is to disperse the crowd reasonably and the other one is to persuade the crowd justifiably. The work of disperse should be carried out by the trained judicial officials and the security staff in uniforms. And the work of persuasion should be done by court administrators.

Last, employing cameras and high-resolution lenses to the game venues. Those cameras could send images back to the central observation room, enabling administrators to see the parking lot, ticket booth, the square and other areas. These facilities are conducive to preventing crimes and reducing violent incidents.

5.3 Medical treatment and emergency aid

For the common accidents such as slip or falling over in sports events, organizers should provide emergency medical aid and be able to cope with lethal or non-lethal medical accidents. So it can reduce the risk of sports events management to the minimum. The common-seen emergent medical aid model in sports events are as follows:

Medical treatment in the field. In-field medical treatment is provided for patients who have emergent medical accidents. It aims at offering the important site-disposal before being sent to medical station or other medical institutions. In many sports events, technical staff of emergent medical treatment or other paramedics offers emergent support to patients or perform basic life-sustaining medical work.

Emergent medical transport. Sending the injured or the patients to well-equipped medical institutions plays an important role in emergency medical treatment plan. It would be better to have ambulance to transport the patients immediately. If dealing with major accidents of the sports events, the local emergency medical service system should be prepared at any time and if the land transportation cannot fit the actual needs, then it is necessary to finish it by air transportation.

Emergency station. Emergency station should provide medical treatment levels which are considered appropriate by the sports events or the facility administrators. If serious and(or) complicated emergency medical accident happens, the emergency station should be able to provide emergency medical staff with proper environment and equipment to ensure the most timely medical treatment work.

REFERENCES

Allen Guttmann, The Olympics: A History of the Modern Games, Urbana and Chicago, University of Lllinois Press, 2002.

Holger Pruess, The Economics of Olympic Games. Walla Press, 2002.

Jeffrey G Owen, Estimating the cast and benefit of hosting Olympic Games: What can Beijing Expect from Its 2008 Games, Industrial Geographer, 2005.

Sports Engineering and Computer Science – Luo (Ed.)
© *2015 Taylor & Francis Group, London, ISBN 978-1-138-02650-6*

A study of analysis and countermeasures on constructing a good physical environment in school

Bin Xu

College of Foreign Languages, Northeast Dianli University, Jilin, China

ABSTRACT: The school sports environment refers to all the physical activities students participate in school, including objective physical facilities and subjective physical classes. This thesis will discuss how to construct a good physical environment in school, at first achieving the goal of physical teaching, and then helping students to develop a habit of lifelong physical training.

1 INTRODUCTION

According to a nationwide survey, Chinese students' speed, endurance, and strength quality have decreased, and the main reason for that is lacking of physical exercise. Building a good physical environment in school can be conducive to the cultivation of students' physical passion. As a result, sports environment in school will have something to do with students' health.

A good sports environment in school consists of physical facilities, different physical competitions, and physical classes. To build a good sports environment in school, the public ought to proceed from several different aspects, analyzing the current situation of the environment of college physical education, and coming up with reasonable suggestions.

2 OBJECT AND METHOD OF STUDY

2.1 *Subjects*

This paper takes Northeast Dianli University 2012 grade students as the research object. Questionnaires were distributed to 500 students randomly in the 2012 grade, and 492 pieces of paper were recovered; the effective rate of recovery was 98.4%.

2.2 *Research methods*

2.2.1 *Questionnaire*
The school is to offer PE elective courses. Students choose elective sports motivation and factors which affecting the students' passion to participate in extracurricular sports activities, sports entertainment games and physical activities in school. What's more, the survey will analyze the current social sports dynamic effects on students.

2.2.2 *Interview*
1. On the students of physical education elective courses and facilities, such as the view of the interview survey;
2. More detailed students' understanding of the school sports environment.

2.2.3 *Document*
1. Access to a large number of documents;
2. Related to the physical environment of literature collation, analysis and research.

3 RESULTS OF ANALYSIS

3.1 *The sports teaching is the basis to create a good environment of school sports*

When it comes to whether it's necessary to have PE elective courses under the circumstance of credit system, 65.2% of the students think that it's necessary to offer PE elective courses, while 15.6% of the students do not think that it's necessary, and 19.2% of students think that it does not matter. As to whether or not students are willing to take the choice of sports elective credits, 57.2% of the students are willing to, while 26.4% of the students are not, and 16.4% of the students said nothing. Factors that students are not willing to choose the sports elective courses are given in Table 1.

To create a good physical environment in school, the current mode of education should be changed. First of all, the technical requirements of teaching should be reduced. Then students' physical quality and interested hobbies should be emphasized. Finally, let students participate in sports as well as meeting credit needs.

The key to achieve good teaching goals relies on the good relationship between teachers and students. Teachers' words and deeds directly

Table 1. Students' attitudes toward physical education.

Factors	Don't like sports	Credit hard to get	Be afraid of tired	Sickness	Do not adapt to the teaching	Useless
Number	69	243	62	34	77	21
Percentage	13.4	48.6	12.4	6.8	15.4	4.2

affect students, so proper teaching methods can arouse the interest of students, fully mobilizing the enthusiasm of the students. The number of the students is organized according to the number of MBA. MBA program implements the lamination teaching, because reasonable placement of learning can make students learn better, and thus the program can achieve mutual promotional effect.

3.2 Facilities are the key to create a good sports environment in school

Investigation of students participating in extracurricular sports activities, says that those who often attend accounted for 46.2%, while those who sometimes occupy is 38.4%, and students who rarely participate in sports accounted for 15.4%. Factors affecting the students to participate in extracurricular sports activities, according to the literature index statistics as shown in Table 2. The analysis of the data can be observed from the table, so the good facilities are the key to create a good school sports environment.

First of all, the school sports facilities need to meet the students' need in quantity, focus on the situation, and gradually improve sports facilities, equipment requirements, providing students with a broad platform for movement and meeting the students' needs of sports participation.

Second, the school facilities need to meet the needs of students' movement in quality and environment. The school should periodically check the facilities, and clean the site. The beautiful and comfortable sports environment can be more effective to attract students to exercise.

3.3 Sports competition is an effective complement to create a good environment of school sports

Through the survey, more than 90% of the schools have carried out different forms of sports competitions, but the number of students who participate in those competitions is only 23.6%. Those who pay attention to the sports activities only account for 56.4%. Factors affecting students to participate in and pay attention to the competitive sports are shown in Table 3. Organized competition should

Table 2. Students participate in sports exercise frequency.

Factors	Often attend	Sometimes attend	Rarely participate in
The number	231	192	77
Percentage	46.2	38.4	15.4

Table 3. Factors affecting the students' participation in sports.

Factors	No suitable sports	Level is not enough	Matches with their	Own not to appreciate in value
The number	271	342	214	273
Percentage	54.2	68.4	42.8	54.6

be carefully prepared, and thus sports activities can make more students to participate in the game, pay more attention to the game, and focus more attention on the sense of collective honor. Schools should also take the open sports environment model, and at the same time students should participate in social and cultural activities, especially sports-cultural exchanges and social clubs in colleges and universities.

3.4 Film and television media publicity to create a good environment of school sports

Through the survey, 76.4% of the students pay attention to the sports news, and almost all the students have their own adoring stars. However, restricted by the school environment and conditions, the approach to obtain sports information is obviously limited.

In the survey, most of the students pay attention to the game or sports news, while schools offer little access to information. The school should make the class as a unit, setting up student in the activity room, arranging the teacher or counselor for management. In the activity room, TVs should be established for students to obtain information, making full use of electric media to influence students.

3.5 Use games to create a good environment in school

With the rapid development of sports, various sports have appeared. According to the survey, the number of students who contact with this type of game is 67.4%. And this kind of game mainly consists of 5 World Cup 6, 5 NBA 6, 5 speed 6, 5 Street 6. These games rely on the keyboard operation to move into another space. While students are in the game, they can get a lot of knowledge, and obtain the enjoyment and satisfaction at the same time. What's more, those games can improve their sports enthusiasm.

3.6 Art performing infection to create a good environment of school sports

Now art performance has become an important aspect in contracting a good school sports environment. Schools can offer the physical dance, aerobics, and other arts training on the basis of existing physical material. Organizations of professional cheerleading, rich art campus sports culture, sports art show held in some large-scale parties, and so on. All those forms can increase the artistic atmosphere, attracting more students to pay attention to the game.

4 CONCLUSION

Physical environment affects the people's participation in sports activities directly or indirectly. Students and school sports environment are linked closely, so building a good physical environment in school is directly related to the development of school physical education.

REFERENCES

[1] Xiong Maoxiang. On the systematic construction of sports environment [J]. Sports and science. In 2003 (06).
[2] Liu Yong. Study on the environmental factors of happiness of college students sports exercise. [J]. Guangzhou Sports University, 2005 (04).
[3] Jiang Junru. Analysis on physical exercise behavior of college students and the environment. The [J]. Sports scientific research, 2004 (03).
[4] highlight. In ordinary colleges and universities in China under the credit system to set up the sports elective course on [M]. 2003.
[5] Forest Zhicheng. To our ordinary colleges and universities elective investigation [J], Chinese sports science and technology, 2003 (06).

Sports Engineering and Computer Science – Luo (Ed.)
© 2015 Taylor & Francis Group, London, ISBN 978-1-138-02650-6

Influence of lifestyle on college students' physical health

Liang Li

Department of Physical Education, North China Electricity Power University, Beijing, China

ABSTRACT: The author surveyed psycho diagnosis scale of healthy lifestyle and students' physical health evaluation standard. Thus, this paper exposits the current situation about university students' way of life and physical health, and the correlations between them is included. Results show as follows: physical education plays an important role in strengthening students' physical health; healthy lifestyle has positive effects on physical health.

Keywords: health evaluation; psycho diagnosis; required courses of sports; university students

1 INTRODUCTION

Recently, there come the strategies of reinvigorating China through science and education and building socialist state with Chinese characteristics. Today the Party Central Committee puts forward again that school education must establish the guiding ideology—health first. Educators should regard education as the primary goal in the height of 'fostering' and improving the populace's cultivation. It comprehensively impels quality-oriented education. Health is the precondition for students to accomplish their learning tasks. What Chair Mao Zedong said in The Study of Physical Education and wrote in 1917 are: 'Physical education, the carrier of knowledge and the container of morality' and 'Physical education is indeed the most important for me. Being strong before learning, one can engage in far more efficient advanced studies in morality.' The physical condition of students has a direct influence on their input and outcome of learning [1].

The factors affecting university students' health include human biology, environment, life-style, hygiene facility and physical exercise ones. Based on scientific research, the author finds out that the law and the extent of influencing those factors above have on university students' health. Besides, the author also finds that having a clear recognition of physical education workers and principle makers helps them continuously correct and follow new working requirements. Then serve for advanced physical education. It also makes full preparation for university students before entering society positively and healthily.

Life-style means a sequence of living habits that people develop after having been influenced by certain national culture, social economy, and custom especially family for a long time. In modern society, it's more and more distinct that unhealthy behavior and life-style mainly affect human health. Human behavior is the generic terms of responses or activities that people have when they adapt to constantly changing environment for individual survival and breed continuation. It includes human's instinct activities, labor, interpersonal communication and other high-class social activities. Human's life-style is the hygiene criterion of groups and individuals that dominate in society and its constitution. It's also a sequence of living consciousness, living habits and living system. Naturally, all these are formed under a long-term influence of certain nationality, culture, economy, custom, criterion and family.

This research selected the psycho diagnosis scale of university students' healthy life-style in China's Education Online (http://www.cer.net) to classify the health level of graduates' way of life. As it is wellknown, university physical education dose not implement four-year's regular course. For a student, the end of university physical education means that the longest course he has taken in the career of study is over. The longest course is from the first grade in primary school, even earlier in kindergarten, to the second grade in university. There is a question that whether the final ending of this course means lack of physical education's essential place in students' life or not. The answer is definitely no. The students' physical health standard has been carried out up until now. Senior students will graduate under the influence of it. The variation tendency of students' physical health is the best explanation for their life-style.

Chart 1. Mark sheet on psycho diagnosis scale of university students' healthy life-style.

	1	2	3	4	5	6	7	8	9	10	11	12	13	14	15
A	3	1	2	0	0	3	0	1	3	3	2	0	3	0	0
B	2	3	3	3	1	2	0	2	0	0	3	3	3	2	3
C	0	0	0	2	3	0	3	3	0	0	1	0	3	3	1

2 RESEARCH OBJECT AND METHOD

2.1 Research object

Pick out students who participated in the testing of physical health standard in the year 2008–2009, 2009–2010, and 2010–2011 from 2112 senior graduates. There are 1977 eligible students.

2.2 Research method

2.2.1 Mathematical statistics method
All data root in school's physical health standard, statistically processed by SPSS10.0 software package.

2.2.2 Document literature method
Consult life-style-related information through books, periodicals and internet.

2.2.3 Questionnaire survey method
Adopt the psycho diagnosis scale of university students' healthy life-style. Totally 15 questions, score of which is in Chart 1.

Practically give out 1899 questionnaires, withdraw 1745, 1698 of which are effective.

3 RESEARCH ANALYSIS AND RESULT

3.1 Analysis on yearly variation tendency of students' physical health

The records of senior students' physical health standard started in 2008. It can be concluded from the chart that the number of excellent and well group shrinks. The higher the grade is, the passed and failed group expands.

In order to analyze the reason, we assume that the end of compulsory sport course leads to an obvious break point in the condition of physical health from sophomore to junior. Further researches on variation of physical health confirm the positive function of compulsory sport course.

The positive impact makes us laborious physical education workers feel double delighted and anxious. As a result PE class cannot be carried out through university. For the sake of adapting to the current education system and improving students' physical health, here comes the question.

The question shows whether physical educators teaching in the front line should consider implementing the thought of lifelong physical education and passing on healthy life-style to students rather than implement specific skill only.

3.2 Current situation and analysis on university students' lifestyle

Basically, the evaluation results of psycho diagnosis scale of university students' healthy life-style are divided into good (37–45), better (25–36), medium (13–24) and bad (below 12). Figure 1 shows the current life-style of senior students: good – 14.5%; better – 19.5%; medium – 34.7%; bad – 31.3%. If the total point is between 37–45, it indicates that one is good at studying, living and working and owns higher working and studying efficiency. Between 25~36, it indicates one has good life-style, the art of restoring energy in busy work and the potential of improving efficiency. Between 13~24, it shows that the health level of life-style is medium, one should try to improve his life-style. If below 12, it means bad life condition, one should determine to turnaround harmful living habits.

In order to further exposit the major factor affecting healthy life-style, the psycho diagnosis scale of medium (13–24) and bad (below 12) group has been analyzed. Sort the lowest-point choice of 15 topics as selection proportion, as in Chart 1 and Chart 2.

English Professor Brass pointed out in 1970s that human's health level is connected to 7 kinds of behavior, thereby affirms the relationship among behavior, life-style and health. He also brought

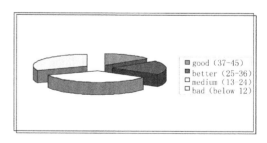

Figure 1. Component picture of university students' healthy life-style.

Chart 2. Analysis on yearly variation tendency of senior students' physical health.

	Excellent %	Well %	Passed %	Failed %	Passed rate
08–09	6.61%	43.78%	48.56%	1.05%	98.95%
09–10	5.76%	41.66%	48.73%	3.85%	96.15%
10–11	4.46%	39.26%	51.37%	4.91%	95.09%

Chart 3. The order of low-point choices in bad group's psycho diagnosis scale.

Sequence number	Topic number	Content	Option	Percentage
1	12	Towards sports, you generally	Dislike sports	100%
2	3	You generally eat in breakfast	Nothing	97%
3	14	How do spend your summer holiday	Negatively relax	94%
4	5	When having lunch, you are	In a rush	91%
5	15	In your opinion, esteem is expressed that	Achieving one's aim at any cost	90%
6	8	In extracurricular time, you usually	Go out	88%

Chart 4. The order of low-point choices in medium group's psycho diagnosis scale.

Sequence number	Topic number	Content	Options	Percentage
1	12	Towards sports, you generally	Dislike sports	100%
2	15	In your opinion, esteem is expressed that	Achieving one's aim at any cost	93%
3	3	You generally eat in breakfast	Nothing	91%
4	5	When having lunch, you are	In a rush	90%
5	8	In extracurricular time, you usually	Go out	88%
6	14	How do spend your summer holiday	Negatively relax	85%

forward 7 good behaviors and living habits, which includes exercising regularly and having breakfast every day. The international cardiac protection meeting in 1992 put forward that big four footstone of health are reasonable diet. They are defined amount of exercise, smoking cessation, restriction of drinking and psychological health.

3.3 The influence of lifestyle on university students' physical constitution

In light of the above research, compulsory sport class has positive impact on university students' physical health. This research chooses the impact phase in two yeas after finishing compulsory sport class, which is junior and senior. Through research we can see the variation tendency of physical health of good, better, medium, bad groups in two years. Under the interaction of weakening the effect of compulsory sport class, the effect of life-style on physical constitution can be seen vaguely. With the increase of health level in life-style, university students' physical health level maintains the original level. Meanwhile it tends to transform to a higher level. The

phenomenon illustrates the benefit of healthy life-style more sufficiently.

Many people know perfectly well that they have some bad ways of life, such as smoking, hypo motility, indiscipline diet, and so on. But they often have little change or give up halfway. It indicates that it's not enough to form a healthy life-style by knowing some information only. First the formation of healthy life-style needs the formation of healthy behaviors, which has great relationship with the formation of life-style step by step. The healthy life-style cannot be developed in one day, it needs a recirculation procedure.

People summarize it into 5 different phases:

1. Pre-consideration phase. The main thought is 'I don't want to change.'
2. Consideration phase. The main thought is 'I want some changes.'
3. Preparation phase. That is 'I have been ready to change my way of life.'
4. Action phase. 'My way of life has changed a little.'
5. Maintaining phase. 'I have a regular healthy life.'

3.4 Steps of developing a healthy way of life

Many health problems are related to individual behavior. If we want to solve these problems above, all we should change harmful life-style. The positive change can acquaint individual the best health condition. Self-control is the key factor to changing life-style and acquiring the best health condition.

There are steps to change harmful life-style:

1. The first step: definite the harmful life-style that needs changing, such as staying up, having no breakfast, smoking, and so on.
2. The second step: set targets. Break down long-term goal into periodicity small goals. Only formulate some goals once, plan action goals instead of result goals and objectify the goals.
3. The third step: formulate practicable and substantial goal plan. A thorough plan should include the definite time (number of days and time) of implement. For instance, if you set a plan that take relaxing exercise for 15 minutes every day, five days a week, your schedule should mark out which 5 days and the definite time to do it every day. In the meantime, prompting symbol helps you implement your plan. For example, your plan is taking a deep breath exercise every night before going to bed. You can tape a remindful scrap of paper on your table lamp.
4. The forth step: make records. Examine whether your plan has been implemented after one or two weeks. If most dates in your schedule have a tick during planned time, it indicates that you have been on the right track, and you can think about formulate next plan. If there are only a few of ticks, you need to find out the reason. Common reasons are: difficult goals, impractical. Too many goals cannot be realized in a short time. The schedule doesn't correspond with everyday life, and it needs properly adjusting. You cannot throw yourself into the set goal. In addition you need others' help and short of self-control ability.
5. The fifth step: assess and adjust. If you have accomplished most action in the schedule, go back to the second step and reformulate a short-term goal, then repeat the third and forth step. After achieving short-term goals many times, you may want to formulate a long-term goal, at this time, you are actually walking on the right track transforming into meaningful life-style. Once your new behaviors solidify to your good habits, they don't need ranking in schedule any longer. If failing to achieve, you should find out the reason why it fails and adjust your goal and make it practical. Then go back to the first step, and start again.

4 RESEARCH CONCLUSIONS AND SUGGESTION

1. *Compulsory sports class in university plays an active role in promoting students' physical health. PE teachers and relevant functional department should recognize the vital role of physical education and make students realize the positive effect that exercise brings. In addition, to strengthen students' awareness of life-long physical education is also needed. In practice aspect, they should develop and popularize rich and colorful group activity by facing the whole school. Besides, they should avoid competitive group activities so as to let more students take part in sports joyfully.*
2. *More actions could be followed. For example, adopting the strategies and measures of intervene behavior related to health; helping crowd change formed harmful behaviors and living habits; consciously adopting system procedure of promoting healthy behaviors and developing good life-style and so on. Surely, intervention method about healthy behaviors contains: regulation and organization; information spreading; environmental change; training, consulting, instructing, and so on.*

REFERENCES

[1] Wu Xin-yu; Fu Xiao-chun; Research on Relationship between University Students' Weight Index and Constitution Health Indicator [J]; Beijing Sport University Journal; Volume 08, 2006.
[2] Hu Li-jun, Yang Yuan-bo; Research on Relationship between Social Economy Development and National Constitution [J]; Sports Science; Volume 05, 2005.
[3] Yu Hao; Backward and Forward Research on Relationship between Social Class and Healthy Life-style [J]; Nanjing Social Science; Volume 05, 2003.
[4] Wu Xiu-qin, Xu Hong-feng, Chen Hua; Opinion Survey on University Students' Health Knowledge and Behaviors [J]; Sports Scientific Research; Volume 04, 2003.
[5] Chen Hai-chun; Discussion on Relationships of Sub-health, Life-style and Sports and Fitness [J]; Fujian Sports Science and Technology; Volume 02, 2003.
[6] Wu Bo; Questions on Psychological Health Standard [J]; Hebei University Journal (Philosophy and Social Science Edition); Volume 02, 2001.
[7] Ye Yi-duo; Re-recognition of Psychological Health Standard and Relevant Research [J]; Southeast Academic Research; Volume 06, 2001.
[8] Ye Yuan-hui; Reflection on Current Situation of Psychological Health Standard Research—and Balance Theory of Psychological Health [J]; Fujian Normal University Journal (Philosophy and Social Science Edition); Volume 02, 2001.

Sports Engineering and Computer Science – Luo (Ed.)
© *2015 Taylor & Francis Group, London, ISBN 978-1-138-02650-6*

Main problems in the development of leisure sports in China under the public service supply

JianPing Shu
Chengdu Sports University, Sichuan, China

Bin Yang
Sichuan University, Sichuan, China

XiaoYuan Wen
Chengdu Sports University, Sichuan, China

ABSTRACT: In recent years, our country sports made significant achievements in various aspects and great achievements in various aspects. However, in the sports system, our country provides a large number of inputs, but the leisure sports value degree is not high. Still there is a big problem. Mainly manifested in the sports, the growth of the total investment and finance are not synchronized, and the investment structure is unreasonable. Besides, the growth of the sports industry in economic income growth is not harmonious. Furthermore, the urban and rural sports development gap is big. Therefore, universal quality changes are needed to keep pace with the development of economy, and consumption patterns relative to these six aspects.

1 INTRODUCTION

As the growth of the national economy, state spending on public services is also included in the national important agenda of the 12th five-year plan. In the research on the leisure sports public service, we'll be the first to find out its main problem can be targeted to develop public service policy.

2 TOTAL SPORTS INVESTMENT AND FINANCIAL GROWTH

Along with the increasing investments of the national economy in sports. In 2011, the national fiscal revenue was 10.374 trillion Yuan, an increase of 24.8% over the previous year. The national fiscal expenditure of 10.893 trillion Yuan, there is an increase of 21.2% over the previous year. Among them, culture and sports and media show spending of 189 billion Yuan. It is a year-on-year increase of 34.8 billion Yuan and an increase of 22.5%. In 2011, culture and sports and media spending per person is RMB 140. Cultural and sports compared with last year increase speed is faster. But in general or less on the proportion of fiscal expenditure, and government financial investment to GDP and the proportion of fiscal expenditure is still in a very low level. So that the financial input in sports is very little per capita. Our country is not in conformity with the national fitness goals. According to statistical yearbook of the authority of the state and provincial governments, no government financial is input to the public sports statistics.

2.1 *The structure and the aim of the national fitness sports investment is not harmonious*

2.1.1 *Mass sports investment proportion of total spending less*

2011–2015, the national construction scheme is mentioned in the target task. The scheme covers urban and rural areas by 2015 and perfects the national fitness public service system. In our country, sports increases year by year from 2008, 20.529 billion Yuan to 25.417 billion Yuan in 2010. However, most of the sports funds are for administrative operation, sports management, sports competition, sports training and sports venues. At last, the above spending of 64% for several accounts shows less funding for recreational sports. For example, the funds of mass sports are used for fiscal spending to 2.131 billion Yuan in 2010. It accounts for only 8.38% of all sports spending 25.417 billion Yuan, accounting for 0.02% of the national public finance expenditure of 8.98742 trillion Yuan.

2.1.2 The development of sports service industry is still in its early stage

Statistical results show that: in 2007, the Chinese sports and related industries, "sports service" of the staff of 664500 people created the added value of about 32.25 billion, accounting for 25.49% of the total added value of sports and related industry. Besides, it also suggested that the Chinese sports and related industry in the tertiary industry accounted for a smaller proportion. In 2007, China's "sports service" staff of 512500 people, 18.06% of the total created the added value of 21.173 billion Yuan, accounting for 16.73% of the total. On the contrary, the sporting goods industry practitioners are for 2.292 million people, to create the added value of 100.887 billion Yuan, accounting for 80.78% of the total and 80.78%, respectively. According to the development of foreign sports industry, the developed degree is higher, the greater the proportion of sports services. At present the United States, Britain and other western developed countries sports provide service industry all over the sports goods industry. But seen from the above data, the added value of China's sports service industry is bigger than that of the sporting goods industry. The sports services in the sports and related industry are still for a low proportion.

2.2 The growth of the national fitness, sports industry and economic development and revenue growth

2.2.1 The development of china's sports and related industry begun to take shape

Statistics show that in 2008, accounting for 0.52% of GDP, China sports and related industry practitioners to about 3.1709 million people realized the added value of 155.497 billion Yuan. In addition, in 2007, the employees, "leasing and business services" are for the 2.472 million people in China. The electricity, gas and water production and supply industry" of the staff are of 3.034 million people. Added value in 2007, "Chinese furniture manufacturing (scale up)" created added value for 64.676 billion Yuan. "Pharmaceutical manufacturing (scale up)" created the added value of RMB 228.66 billion. Data analysis showed that overall, China's sports industry preliminary already have the prototype of the "new economic growth point". Especially the purpose is to absorb the positive role of employment. Even so, we should not be underestimated.

2.2.2 The Chinese sports and related industry grows fast

In 2007, the added value of China's sports and related industry was rising to 22.83% in 2006.

Afterwards, the sports service of added value growth rate reached 28.38%, year-on-year growth of 16.05% in 2008. Two years maintained sustained growth[1]. In 2007, China's Gross Domestic Product (GDP) is 26.58103 trillion Yuan, up 12.35% from 2006. In 2007, tertiary industry added value is of RMB 10.00535 trillion, the growth rate of 12.69%. From these data we can see clearly that the Chinese sports and related industrial added value rate are significantly faster than China's GDP growth rate. Besides, the growth of sports service is also significantly faster than the growth rate of the third industry in our country.

2.2.3 Chinese sports and related industry's contribution to the national economy is also small

It creates added value accounted for 0.49% of GDP, 0.97% of workers in urban professionals. In addition, the added value of "sports service" created by accounting for 0.32% of the value added of the tertiary industry. Meanwhile, professionals accounted for 0.27% of the workers in the tertiary industry[2].

Studies show that: in 2000, the British sports industry accounts for 1.8% of GDP, Canada (1.2%). Gap between China and foreign developed countries, by contrast, is more obvious.

2.3 Sports in the development of urban and rural development is not harmonious

2.3.1 A very big gap between the urban and rural sports facilities

According to the "fifth national sports ground survey statistics", Chinese township (town) village sports only accounts for the total number of 8.18%. And rural sports facilities are often being used, damaged. Besides, a lot of old and unsafe facilities mainly concentrated in the middle and primary school. In reality, combined with the current large removal at rural schools, most of the sports venues and equipment installations were seriously damaged and lost. The use of villagers exercise facilities would be much less. Thus, a disparity between urban and rural sports venue resources distribution shows a "city" and "rural disadvantage" phenomenon.

2.3.2 The government's emphasis on rural and urban sports gap significantly

Policy of mass sports in cities involves various aspects of construction, such as the capital, sites or organizations and so on. It has been a certain degree of recognition and security. However, the conditions of the rural mass sports development are relatively weak. According to the interview survey, it shows that the reform of rural sports organizations, sports department and other institutions resulted in the merger, reorganization, weakened sports institutions. Therefore, there is not enough staff and resources to grasp the sports work.

2.3.3 Sports human and financial resources of uneven distribution of resources between urban and rural areas

The government adjusted artificially low prices for agricultural products, and developed agriculture and rural at the expense of the interests of farmers. Therefore, at the beginning of the founding, the expense was provided city residents with a subsidy way to rapidly promote the industrialization of urban biased policies. As a result, leading to urban prosperity and rural poverty coexists in the dual social structure. Mass sports development directly affects the urban, have more and better sports resources than in rural areas. In 2000, the rural population accounted for 63.78% of the total population, but the government fiscal expenditure for rural accounts was for only 7.75% of current fiscal spending. 98% of the nation's cities were designed at the street office (and) job agencies or personnel. At the same time, 90% of the nation's social sports instructors were concentrated in cities.

2.4 Change of literacy and economic growth

2.4.1 Wealth and health

Under the market economy environment, market competition is intense. People's life rhythm is fast. In the face of social, people continue to strengthen individual competition, in order to not being eliminated in the competition. They have been being the social characteristics of the era. It is also easy to produce stress, fear and anxiety when people are at work, drinking in the dinner party, staying up late or eating with no regularity. People are often in pursuit of wealth at the expense of the health.

2.4.2 Heavy intellectual light energy

The innovation of technology, such as the high speed development of Internet, as well as the transportation is convenient, makes people liberated from manual labor. As decreasing physical activity at work, which "diseases of affluence" do people get significantly? Shown in the third national health test result, since 2000, adults are overweight and obesity rates continue to grow, 2010 adults are overweight rate reached 32.1% [3].

2.4.3 Heavy output light environment

The changing of social environment has also led to a sharp decline in the part of people's physical quality. Especially, some areas developed the improvement of economic output at in the expense of the environment. After introducing high pollution of

manufacturing enterprises, bad work environment can be seen everywhere. Therefore, the expense of the health for wealth has been increased.

2.4.4 Heavy light exercise to treat diseases

Although people's leisure time continues to increase, the idea of fitness is frivolous according to the people. Only after getting the disease and need to seek the aid of a doctor, they would pay attention to it. Lack of physical activity is an important cause of chronic illness. At the same time, a large number of experiments show that all the drugs have side effects. Using the right amount of exercise is beneficial to the health of the members of all ages and can reduce chronic disease.

3 CONSUMPTION IS RELATIVELY SINGLE, CONSUMPTION OF DEFORMITY

3.1 Per capita consumption and per capita growth out of proportion

With the constant increase of GDP, per capita consumption level is constantly improving. But the per capita consumption and per capita increases are out of proportion. The Engel coefficient of urban households in China in 2010 was 35.7%. Meanwhile, the Engel coefficient was 41.1% of rural households. Overall has entered a stage in our country, but the differences between urban and rural areas appear in many ways, such as regional differences, group differences and so on. Even general consumption, sex and developmental consumption are not identical.

3.2 Physical qualities to be improved

National physique monitoring results show that for the third time grown adults, older adults are overweight and obesity rates increase slightly, but power quality index decreases. Obviously, it falls between 0.6% ~ 6.2% to reduce men's, and women's lower amplitude was between 1.3% ~ 6.9%. Besides, the strength of the elderly in quality decline is not contained.

3.3 There is a big gap between the urban and rural consumption

Mainly from the per capita income, there is a big gap between the urban and rural. Rural consumer spending power is different, mainly on food. Rural residents pay attention to the basic survival needs, while urban residents are given priority to enjoy type and developmental consumption. The consumption structure of rural residents is slower than that of the urban residents [4]. The material basis is that the level of consumption will be increased by a function of consumer income.

3.4 *Lifestyle needs to be improved*

People attribute those to poor lifestyle. Current ratio is 37.3% [5]. The development of leisure sports aims to improve people's quality of life and make people realize the importance of correcting people's lifestyle.

3.5 *Consumption idea still needs to change*

Our country has shortage economy for a long time. Besides, it just enters the well-off stage and is also influenced by eastern traditional culture and consumption. The consumption idea is relatively conservative. There are principles about cash transactions: "tomorrow's money today can not spend"; "Ownership, use rights". Shop consumption, the pursuit of a high savings rate and consumption idea has been affecting the consumption of the Chinese people. In the case of the development of network information, the idea of store consumption has been changing. The new idea of consumption is to encourage online consumption. Furthermore, it has been shifting to pursue high consumption idea of reserve requirements.

4 SUMMARY

Provision of public services is a basic function of the government. The leisure sports in China lags behind economic development in the past period of time. Above of all, having accumulated for long-term is the main problem.

By looking for countermeasures, the author considers that we should gradually solve the problem. Thus, this article is for national social science fund project "leisure sports public services supply obstacles and paths for research" (11 bty013). Obviously, it turns out to be one of the achievements.

REFERENCES

[1] Chinese sports and related industrial statistics. Sports industry branch of China Sport Science Society. People's sports publishing house. 2011–12–1.
[2] 2008, China Statistical Yearbook. China Statistical offices. China statistics press, 2009.9.
[3] 2010 National physique monitoring bulletin. State General Administration of sport website. Group Secretary. 2011-09-02.
[4] Xiao Li. Comparative analysis of consumption structure of urban and rural residents. Research on Financial and Economic Issues. 2012.11: 138–144.
[5] Guo Jizhi. Analysis of the causes of unhealthy life style. Medical sociology. 2008 (29).5:56–58.

Sports Engineering and Computer Science – Luo (Ed.)
© *2015 Taylor & Francis Group, London, ISBN 978-1-138-02650-6*

Research on the developmental pattern of integration of badminton industry in Shandong Peninsula Urban Agglomeration

Lei Guo

Sports Department, Shandong Institute of Business and Technology, Yantai, China

ABSTRACT: The research of Shandong Peninsula Urban Agglomeration and the badminton industry is conducted. According to it, the thesis analyses the feasibility of peninsula badminton industrial integration development. Besides, it also puts forward the component of the integrative development pattern of badminton industry in Shandong Peninsula Urban Agglomeration. Then four parts were ascertained. They constitute the integrative development pattern. They are: a) ways of integrative development; b) coordinate mechanism; c) sequential option of industrial location and structure. Specifically, the unbalanced coordinating way of development and the coordinate mechanism of badminton association develop integrally under the lead of the government. Finally, both of them are acknowledged as the answers to the first two parts. As for the sequential option of industrial location and structure, we decided to proceed in two stages based on the point axis diffusion theory. In addition, we gave priority to the development of industry, which fit our special situation following the different function of urban structure. As a result, we want to build an integrally coordinate development pattern of badminton industry in Shandong Peninsula.

1 INTRODUCTION

The worldwide trend of urbanization made the metropolitan area become the basic unit constituting the regional economic development. From the view of economic development in China, GDP generated at the metropolitan areas in the Yangtze River Delta. The Pearl River Delta has occupied 28% of the whole. The key practice of the success lies in the integrative development of urban economy subgroups.

The rate of contribution Sports Industry has ever made to Chinese economic development increasing year by year. Attention has been paid to the P.E Integration as well in its academic circle. But there's no report on the research of Shandong Peninsula Urban Agglomeration. As a late starter in Shandong Peninsula, the badminton industry is obviously an emerging one. This thesis probes into the developmental pattern of integration of badminton industry by both qualitative and quantitative methods. It's based on the theory of industrial integration combined with Regional Economics in order to make contribution to the integration of the industry of P.E at large.

2 RESEARCH FIELD

2.1 *Definition of Shandong Peninsula Urban Agglomeration*

Zhou Yixing, who is from the Peking University thinks that it includes Yantai, Weihai, Tsingdao, Weifang and Rizhao those five cities according to geographic features. However, Jinan and Zibo can be also considered for their location in the city-intensive areas crossed by the Lanyan and Jiaoji railways. Dongying and Rizhao are added into the agglomeration as well. Because the former belongs to the peninsula and the latter is a seaside city near the peninsula. It is acknowledged that Jinan, Tsingdao, Yantai, Zibo, Weifang, Weihai, Dongying and Rizhao compose the Shandong Costal Economic Open Zone. Obviously, it is known as Shandong Peninsula Urban Agglomeration region[1], noted in <Notice about expanding the scope of the coastal economic open zones> by the State Council.

We can follow the release of <plan of Shandong Peninsula urban agglomeration regional development> in 2005 and <the comprehensive plan of Shandong Peninsula urban agglomeration>. Thus, the new star in regional economic plates— Shandong Peninsula urban agglomeration (here after referred to as Shandong Peninsula) grabbed Chinese' eyeballs once again. In the 2013's nationwide top10 most competitive urban agglomerations rankings, Shandong Peninsula ranked 4th after Yangtze River Delta, Pearl River Delta and the interrelated group of cities—Beijing-Tianjin-Tangshan Urban Agglomeration.

2.2 *Connotation of the integration of badminton cause in Shandong Peninsula*

The connotation of integration of badminton cause in Shandong Peninsula is the eight cities.

The government doesn't regard administrative division to make full use of industrial assemblage and complementation effect in terms of the inter-relationship of badminton cause and regional approach. Then the coordinative development in the eight cities and their badminton cause will be promoted via effective allocation and integration of resources.

3 THE FEASIBILITY ANALYSIS OF INTEGRATIVE DEVELOPMENT OF BADMINTON CAUSE IN SHANDONG PENINSULA

3.1 *Geographic location and administrative division lay the foundation for the integrative development of badminton cause*

In the world of standard economic theory, location is a very important resource. It affects the information, opportunities and space of economic development in this region directly or indirectly. Shandong Peninsula owns a really special location with the east extending into the sea. As can be recognized easily in the map, it's the key area serving as a bridge connecting the northeast of China, north China and provinces on the southeast coast of China. Traditionally though the Shandong Peninsula and the Yangtze River Delta are in one region—East China, the former is less impacted by the latter because of cultural difference. Further more, it cannot be greeted by the spring breeze from Beijing. Even there is the fact that they are sharing one geographical region—North China. There is something lying between the two. It's a vast rural area of Hebei Province. In that case, Shandong Peninsula must focus much more on the effective use of its own resources to realize better development in economic and cultural fields.

Another superiority of Shandong Peninsula is that the urban agglomeration belongs to provincial administrative region as well. Generally the government serves as the representative of public interest as well as the protector of its own interest. That is, when it comes to the construction of other integration course of the urban agglomeration, there are always some self-interest behaviors done by each province's government in resource-sharing and benefit distribution etc. In reality, the phenomenon becomes an obstacle to the development of integration. But it doesn't matter in Shandong Peninsula. Because the provincial government is pursuing the integrated planning on Shandong Peninsula to reduce the friction cost and the implementation one caused by various policies and system.

3.2 *The solid economic basis provides support for the development of integration of badminton cause*

The economic and social basis is the requirement for the integrative development of regional badminton cause. Shandong Peninsula also featured among the top5 urban agglomerations in China with its GDP of 6.22899 trillion RMB in 2001. Actually, it exceeds Liaoning Province's, the sum of Beijing, Tianjin and Tangshan and also the sum of Shanghai and Nanjing. GDP of Shandong has been ranking the 3rd steadily in China since 2011, to which the Shandong Peninsula has made great contribution. Take 2011 for example, the population of the eight cities in Shandong Peninsula (42.05 million) was about 40.7% of the total in Shandong Province. Its GDP was around 63.38% of the total. Though there's no extremely developed city in Shandong Peninsula compared with other urban agglomerations at home. The economic strength of each city was in relative equilibrium. In the national GDP ranking, the ranks of Tsingdao, Yantai, Jinan, Weifang, Zibo and Dongying were all in front of the 47th and stayed top50. Per capital GDPs of them also occupied the top6.

3.3 *The construction of a powerful province with abundant sports resources promotes the development of integration of badminton cause*

The aim to construct a powerful province with abundant sports resources was first put forward after the success of the 2009 National Games held in Shandong. The success was to realize the balanced and coordinated development of masses sports, competitive sports and P.E industry.

In 2010, the goals, developmental strategies and the safeguards were put forward in <The implementation opinions on accelerating the construction of a powerful province with abundant sports resources in Shandong>. In September 2012, the <Shandong government's implementation opinions on accelerating the development of P.E industry> came out with a series of superiority to the development of P.E industry. The opinions did good to promote the integrative development of that by integrating and allocating resources effectively in badminton cause.

3.4 *The current developing situation of badminton cause calls for the development of integration*

The badminton sports began rather later in Shandong Peninsula. That is, badminton sports there lag behind those in rich provinces in Southern China, except in Tsingdao. Tsingdao's badminton sport performance has developed to the top of the nation relying on its excellent industry environment. The region is crying for the support from Tsingdao. Also, Tsingdao needs to strengthen its lead in the badminton cause at home by building combination strength with neighbor cities. Meanwhile, integrative planning is needed as a result of different nature and varied steps of development so as to promote the coordinated and orderly development across all.

4 THE CONSTRUCTION OF DEVELOPMENTAL PATTERN OF INTEGRATION OF BADMINTON INDUSTRY IN SHANDONG PENINSULA

As is explained in the Oxford modern Cohen the double solution dictionary, 'integration' is 'the act of combining into an integral whole'. However, there's no mature consensus for the word 'integration' in the field of P.E. Shen ling cheng[2] thinks integration is a regional development pattern with work in cooperation with a due division of labor among interrelated regions. The purpose of them is to optimize resource allocation of social economy and then realize resource sharing, functional complementation, linked progress and interest sharing. Dongdong Wang[3] says that the development of P.E integration should follow the regular pattern of market-oriented economy so as to embody the principle on 'plan integrally complement each other's advantages and develop with original difference'. Obviously, the integrative development needs truly construction in the large as well as emphasis on the interregional cooperation and competition. It is great to promote coordinating development of the regions with lower internal friction. Generally, there exist the developmental pattern of badminton industry, the sequential option of location and structure of badminton industry and the coordinate mechanism of badminton industry. In conclusion, they are the basic constitution of the developmental pattern of integration of badminton cause in Shandong Peninsula according to the developmental connotation of integration of badminton industry in Shandong Peninsula. In addition, the constitution is combined with the definition of the developmental pattern of P.E industry in. On the construction of regional developmental pattern of P.E industry in the eastern provinces in China > by Cong hu ping.

4.1 The developmental pattern of integration of badminton cause In Shandong Peninsula

The unbalanced coordinate developmental pattern is considered as the right one of integration

in Shandong Peninsula in this thesis. The pattern is in terms of the basic ideas of theories such as 'balanced development' 'gradient advancement' 'growth pole' and 'spot-axis' etc. combined with the current situation of badminton cause in Shandong Peninsula. The essence of that unbalanced coordinate developmental pattern of integration is to incline appropriately. Besides, the essence is to invest the limited funds, technologies and resources intensively in the central part of badminton cause in Shandong Peninsula. It can be done on the premise of recognizing the existence of development difference. As a consequence, we can support and achieve the rapid and sustained development of the overall level of those regions. Meanwhile, the development differences among badminton industries caused by the different resources and nature of regions are recognized. We should hold a definite object in view. That is to give priority to the advantage industries and develop the potential industries with efforts. Finally, it can solve the bottleneck problem, which restricts the development of the whole industrial chain.

4.2 The establishment of the coordinative mechanism of integration of badminton industry in Shandong Peninsula

We take some coordinative patterns of integration of urban agglomerations in and outside China for reference. Especially we consider that of French urban agglomerations, combining the characteristic of badminton cause in Shandong Peninsula. Then put forward 'the coordinative patterns of integrative development association under the lead of the government'. That is:

First, establish the joint conference headed by the director of the Sports Bureau of Shandong Province and municipalities. As the highest decision-making mechanism in coordinating the integration of P.E industry in Shandong Peninsula, it's responsible for making a unified regulation system and policy in P.E industry. Meanwhile, setting forth the common development objects and planning is also needed. It's generally held once a year. Then, establish the integrative development association of badminton industry in Shandong Peninsula headed by the government authorities, business representatives and expert researchers. As the daily decision-making and leading institution, it's responsible for the further research of the market, fully soliciting public opinion and reaching an agreement. Finally make a joint commitment and sign the agreement. The association president is also a member from the joint conference and is held by non-government officials in principle.

4.3 Shandong peninsula badminton industry integration development location timing selection

Because of scarcity of resources and regional resources endowment difference between the objective realities, the author specifies the sequence of badminton industry development in different regions of the world successively. In determining the non-proportional coordinated development for Shandong peninsula badminton industry under the premise of the development mode, the first thing is to choose badminton industry advantage area. Then through the competitive area, it can lead to the development of radiative zone.

First stage: choose priority to the development of Qingdao, Jinan, Yantai, the formation of the competition and cooperation of badminton industry development pattern.

The development of area badminton industry not only has its own law of development and the inherent growth mechanism, but also with other factors affecting the coupling process. To choose badminton industry advantage area to research, we should combine two aspects of qualitative and quantitative from the perspective of industry and regional to analysis and judgment. Yan Wang[4] is in the doctoral dissertation research through the factor analysis method and pare to analysis. He selects the main factors affecting the development of sports industry sectors. Both included them cannot quantify sports cultural identity factors, there are hard to count the sports population and other factors. Badminton of three industries is all affected by urbanization level, per capita income, and badminton venue facilities three quantifiable factors. Check data and field investigation, data around the city (2012) as shown in Table 1.

The level of urbanization is to measure the urban development degree of quantitative index. Yu ling wu[5] considered many factors: a) analytic hierarchy process (ahp) and cluster analysis; b) screening of economy; c) population levels; d) lifestyle; e) environment four first-level indicators; f) calculation of 17 cities in Shandong province urbanization level comprehensive score.

In Shandong peninsula urban agglomerations, Jinan and Qingdao take higher scores and obvious advantages. Zibo, Dongying, Yantai, Weihai four cities were between 0.3 and 0.4, belonged to the second group. Per capita disposable income of Jinan, Qingdao, Dongying, Yantai ranked top four. Besides, badminton indoor venues are before the number three for Qingdao, Yantai, Jinan. He considered the population around the city, city influence and badminton athletics level. Qingdao, Jinan, Yantai have conditions obvious advantages in badminton service industry development, identified as the first focus on the development of three cities.

The second stage: take Qingdao, Yantai, and Jinan as the center, and point to area, then establish the radiation surrounding city badminton industry circle.

Qingdao, Yantai, Jinan three cities located in the east, north, west three bearing of Shandong peninsula, respectively. Considering the traditional geopolitical factors and current situation of the development around the city, the author searched eight cities, Qingdao—Rizhao, Jinan-Zibo-Dongying, Weifang-Yanta-Weihai. Finally, he planned three badminton industry economic circle, and radiation to the below 22 county-level cities. Firstly, use industrial circles to strengthen cooperation; Next, depend on the center city diffusion effects produced by the "under the spray effect" and "demonstration effect" to influence on the formation near region strongly and promote industry development; Then, make the competition with industry outside and stimulate interregional cooperation between structural badminton industry development; Finally, achieve the rational allocation of resources through the badminton competition between enterprises.

Table 1. Shandong peninsula each prefecture level three data comparison.

City	The level of urbanization	Ranking	Per capita disposable income of	Ranking	Badminton indoor venues	Ranking
Jinan	1.306	1	28892	1	310	3
Qingdao	0.901	2	28567	2	360	1
Yantai	0.334	5	26542	4	358	2
Zibo	0.335	4	24955	5	273	4
Dongying	0.340	3	27343	3	192	6
Weihai	0.303	6	25290	6	185	7
Weifang	−0.114	7	22508	7	201	5
Rizhao	−0.176	8	20098	8	152	8

4.4 Choose Shandong peninsula badminton's industry integration timing of industrial structure

4.4.1 In-depth development of Qingdao city badminton competition performance industry, improve the badminton competition regional core competitiveness, to further expand the radiation effect of growth pole

The development of Qingdao city badminton competition is taking shape. In 2009, it is undertaken firstly. The 11th National Game's all of badminton games, Qingdao undertook the 12th lift sudirman badminton mixed groups in 2011. Besides, it undertook the badminton championships, the national amateur badminton championship, and the national badminton super league in 2011. In addition, it is one of the domestic few hosted cities with so much competition. Qingdao badminton competitions are second to none, 2010, 2011 for two consecutive years by the national badminton super league title. But through our visit venues, the level of the game has very big impact on the income. Ticket sales still accounts for over eighty percent of the total incomes. In this situation, on the one hand, Qingdao should active play the role of the badminton mediation relying on the good natural environment, excellent performance and the hot market. It also depends on the improvement of business operations to improve business core competitiveness. At last, it can attract more international badminton competitions to hold. On the other hand, Qingdao must explore multi-channel industry development competition, such as media advertising revenue, badminton brokerage revenues, image of product development. Surely, the purpose is to get rid of the over-reliance on ticket sales awkward situation.

We would appreciate it if you make use of the enclosed Endnotes stylefile (Harvard.ens).

4.4.2 Efforts to develop the Jinan badminton competition performance industry, promote the further development of the badminton training industry

After the 11th National Games in 2009 successful held, Jinan's sports industry development environment was further improved. Jinan should seize this rare development opportunity, and actively strive for holding large national badminton competitions. Also to take advantage of political center, it organized all levels of the badminton match. Thus with Qingdao together, the entire peninsula badminton competition performance market is on the rise. At the same time, Jinan badminton mass base is strong and competitive level is higher. Besides, it has formed a good badminton culture atmosphere.

Jinan should standardize the badminton training market further. Finally, it should improve the quality of training, in order to take more responsibility for the provincial badminton athletics reserve personnel training.

4.4.3 Develop badminton training industry of Yantai (and other five cities)

Badminton training is badminton fitness entertainment industry forward related industry. And the badminton training of all kinds of personnel tends to become the consumer of badminton fitness entertainment industry team. At the same time, it can also be for badminton athletics team reserve talented teenagers. Badminton of Yantai city is in a steadily rising stage. It is full of enthusiasm of the masses to participate in the badminton fitness. The badminton filed is also adequate. But Yantai overall competitive level is not high. It is far to meet the demand of the market. So the badminton technology talent shortage becomes the bottleneck for the development of badminton industry in Yantai. Industry Association should provide badminton coaches job training opportunities, and apply amateur coach hierarchy. The Association also should break the monopoly competitive talents, resource sharing. Certainly, it should advance retire the free flow of high level athletes or coaches area.

5 CONCLUSION

The research of Shandong peninsula urban agglomeration and the badminton industry is conducted. According to it, the thesis analyses the feasibility of peninsula badminton industrial integration development. Besides, it also puts forward the component of the integrative development pattern of badminton industry in Shandong peninsula urban agglomeration. Then four parts were ascertained. They constitute the integrative development pattern. They are: a) ways of integrative development; b) coordinate mechanism; c) sequential option of industrial location and structure.

Specifically the unbalanced coordinating way of development and the coordinate mechanism of badminton association develop integrally under the lead of the government. Finally, both of them are acknowledged as the answers to the first two parts. As for the sequential option of industrial location and structure, we decided to proceed in two stages based on the point axis diffusion theory. In addition, we gave priority to the development of industry, which fit our special situation following the different function of urban structure. As a result, we want to build an integrally coordinate development pattern of badminton industry in Shandong Peninsula.

REFERENCES

[1] Yi Xing Zhou, Shandong peninsula urban agglomerations development strategy research, China building industry press, 2004.

[2] Shen Ling Cheng. Sports leisure market in the Yangtze river delta integration and coordinated development research. Journal of Beijing sport university vol 33, pp30–31, 2010.

[3] Dong Dong Wang, Changzhutan integration of sports development research. Hebei Physical Educational Institute. vol 23, pp36–37, 2009.

[4] Yan Wang, Our country sports industry choose regional advantages and its development development research, Doctor's thesis, 2011.

[5] Yu Ling Wu, Shandong province urbanization research and comparison, Bachelor's thesis, 2011.

Sports Engineering and Computer Science – Luo (Ed.)
© 2015 Taylor & Francis Group, London, ISBN 978-1-138-02650-6

"Satisfaction" research on stable development of high-end fitness clubs in China

YangCheng Tang
Hunan Vocational College of Foreign Studies, Changsha, Hunan, China

ABSTRACT: *Objective*: In view of the outstanding problems about stable development in China's high-end fitness clubs, some experts discuss the application value and effect of steady development.

Method: a) Above all, taking the customers (216 persons) and staff (64 persons) in five high-grade health clubs as the experimental object, we mainly use interviews, questionnaires, mathematical statistics and logical reasoning methods to analysis related data etc; b) Then we get the main conclusions and suggestions that the overall customer satisfaction is not high during the stability development in high-end fitness clubs. Obviously, the man' is higher than the woman's in general. With the longer membership, the greater age, the higher monthly income and lower education, customers have the higher satisfactions. Career satisfaction in turn for individual customers is individual households, others, managements, authorities, company employees, teachers or scientific researchers. Employee satisfaction in service operation is satisfactory. Under the experience economy in Changsha, male staff satisfaction is higher than female in general. Besides, the employee working nature of satisfaction in turn is professional managers, coaches, front desk service, logistics service, and membership consultants. Thus, high-end fitness clubs should do as follows: a) strengthen the management of project construction, position membership card with better price; b) focus on customer membership requirements among the authority of personnel, company employees, teachers and researchers; c) improve their satisfactions; d) develop more customers; e) strengthen the study; f) improve their academic and professional technology; g) improve their working enthusiasm and creativity; h) focus on work tedious and difficult among the front desk service, logistics service, the membership consultant; i) increase their pay; j) try to improve their satisfaction to retain them.

Keywords: satisfaction; high-grade; fitness club; stable development; research

1 PROBLEM AND THE SIGNIFICANCE

Since reform and opening up from the 1980's, China's economy has grown rapidly. China's economy today ranked second in the world. However, people's living standards and economic income level has got unceasing enhancement. Individual demand and sports consumption show diversified development trend.

In today's China sports consumption project, the earliest marketization is the sports fitness. After 30 years of development, the national fitness clubs get fast development. They have formed more than 3400, which have large scale, large number of investment, participants, expanding managements and employees. Besides, they benefit the people, widely are involved cities all over the country and have huge development space.

Compared with the United States, there is data show that an average of 8 persons has one in sports consumption. Moreover, an average of 1 million people in China have less than 1. Visibly, China has a huge potential fitness market, leading to open a large number of fitness club, and displays diverse characteristics.

Having asked for more advice of experts and teachers, the current high-end fitness club is defined as: a) overall investment scale is more than 10 million and equipment class is high; b) The site is 3000 square meters and decorates luxuriously; c) Staff service level is high and service concept is advanced and characteristic; d) Ordinary membership card price is above 3000 Yuan; e) And can fully satisfy consumers' physical exercise, physical rehabilitation care, psychological care, integration needs of leisure and entertainment venues.

The number of Fitness club increase sharply, but few can have their own characteristics. More than 80% of the club are homogeneous competition and price wars and don't form their own unique business characteristics. High-end fitness club is fewer in number, which leads to increasingly fierce

Table 1. Stable development statistics of China's high-end fitness club.

Indicators	Group	Number	Percentage (%)
Changsha New York Fitness Club	Employees	12	18.75
	Customer	39	18.06
Shanghai China Bally Club	Employees	13	20.31
	Customer	47	21.76
Beijing Jade Bird Fitness Club	Employees	11	17.19
	Customer	33	15.28
Shanghai Yizhao Wade Fitness Club	Employees	13	20.31
	Customer	42	19.44
Shenzhen Catic Fitness Club	Employees	15	23.44
	Customer	55	25.46

competition. Member of the customer is the god of operator. They bring wealth and profit to the club directly. Furthermore, the employee contacts with customers. Employees are the key to the fitness club management and development. Finally, how to improve their satisfaction rate is the core problem in the development of fitness club.

According to the characteristics of China's high-end fitness club, take out five of them as example. They are Changsha New York Fitness Club, Shanghai China Bally Club, Beijing Jade Bird Fitness Club, Shanghai Yizhao Wade Fitness Club and Shenzhen Catic Fitness Club. Carry out questionnaire survey among five high-end fitness club customers S (n = 216), employees (n = 64) and confirm the above groups as the research object at the same time.

2 RESEARCH METHODS

2.1 The literature material law

A large number of related books, documents and materials on service operations and fitness club service operation are referred during paper writing.

According to the Internet (China National Knowledge Internet, Wanfang database, Baidu and Google), the author retrieved to collect relevant papers, journals, process. After analysing the above data, the author studied the current situation, countermeasure and the development of China's high-end fitness club, which provides the theory basis of data analysis. These premise adequate theoretical preparations for paper writing.

2.2 Interviewing method

Interview a few domestic experts and experts in Beijing Sports University on telephone; Directly

interview a few managers, coaches, staff and customers in the high-end fitness club all over the country; Listen to and record their valuable advice on the fitness club. All these play a key role for the study.

2.3 Questionnaire survey

Referring to relevant literature, the author listened to the guidance of the important suggestion of relevant experts and teachers carefully. Then modified again and again, argue and designed 《China high-end fitness club customer satisfaction questionnaire》. Five grade evaluation methods are adopted in this questionnaire. The scores show for 1–5 points: very dissatisfied for 1 point; not satisfied for 2 points; general for three points; satisfaction for four points; very pleased for 5 points. A total of 12 questions appear total score of 60 points.

The higher the score is, the greater the satisfaction is:

1. Average score is 3 points;
2. Score below 2 points said the satisfaction rate is very low;
3. 2–3 points said satisfaction rate is low;
4. 3 to 4 points said satisfaction is common;
5. 4 points above said it's very good.

Total average score is 36 points:

1. Below 24 points said satisfaction rate is very low;
2. 24 to 36 points expressed satisfaction rate is low;
3. General satisfaction is 36 to 48 points;
4. 48 points above said high satisfaction.

This data helps explore the present situation of Changsha high-grade fitness club service operation performance, problems and effective measures.

《China's high-end fitness club customer and employee satisfaction questionnaire》 adopted the method of random sampling, Table 2 shows:

1. 260 pieces are given out;
2. 230 are taken back, and the rate is 88.46%;
3. 16 invalid questionnaires are removed, effective questionnaire are 216, and the overall effective rate is 93.91%;
4. 70 questionnaires are given out to employees, and questionnaire recovery rate is 94.29%;
5. Effective questionnaire is 64, and questionnaire effective rate is 96.97%.

2.4 Mathematical statistics

Questionnaire data is input SPSS16.0 statistical software, and related data is processed:

1. T test, $P > 0.05$ said there is no significant difference between two groups of data;

Table 2. Distribution statistics on high-end fitness club customer and employee satisfaction survey in China.

Types	Issue number	Recycling number	Recovery rate %	Valid questionnaires	Effective recovery rate %
Customer	260	230	88.46	216	93.91
Employee	70	66	94.29%	64	96.97%

2. $0.01 < P < 0.05$, said two sets of data have significant difference;
3. $P < 0.01$, said data has a very significant difference in both groups, in order to carry out further systematic research;
4. In the mean comparison T test, $p < 0.05$ in the table, due to the large table and data analysis;
5. In the mean comparison T test, $p > 0.05$ indicators in the table removed.

2.5 Logic reasoning method

By using comparison, deduction, induction, analysis and reasoning methods, corresponding conclusions are achieved. It is based on the analysis of the logic judgment on survey data, customers' interview and the content of the management personnel and related expert interview.

3 RESULTS AND ANALYSIS

3.1 Customer satisfaction indicators research in high-end fitness clubs

3.1.1 Overall satisfaction rate for the customer is not satisfied

After statistics SPSS16.0 software processing, Table 3 shows:

1. The customer in the "employee and the manager is very easy to contact", "traffic and parking convenience", "Strongly professional staff and friendly to people", "coaches' image and temperament" and "make friends with other members", the satisfaction rates are 3.19, 3.10, 3.01, 3.1, 3.00 > or = 3, visibly, customers in the above aspects is satisfactory in general;
2. Customers in the "courses in fitness", "open time", "reasonable service operation procedure", "reasonable membership card price", "help me to improve the function of body shape", "environment and trust", "well site planning", rates are 2.50, 2.47, 2.60, 2.48, 2.98, 2.86, 2.98 < 3, and visibly, customers in the above aspects in general is less satisfactory, especially in the "courses in fitness" and "reasonable membership card price";

3. The "total" rate is 34.19 < 36, and customer is the god of operator.

Their satisfaction directly affects the business profit and service operation. Thus, overall satisfaction rate of high-end fitness club service operation in our country for the customer is not satisfied and optimistic.

3.1.2 Overall satisfaction of men membership is higher than women', and there are significant differences

After statistics SPSS16.0 software processing, Table 4 shows:

China's high-end fitness club customer gender ($n = 58$ male, $n = 158$) mean differences in "total" value is 2.56 and ($F = 6.01$, $P = 0.02 < 0.05$).

Statistically significant shows that male and female customers have significant difference in "total". Finally, this shows that in China's high-end fitness club customer satisfaction men are high than women, and there are significant differences.

3.1.3 The longer membership, the higher satisfaction

After statistics SPSS16.0 software processing, Table 5 shows the following. Member of the customer time (less than 1 year $n = 46$; $n = 68$; 1 year 2–3 years $n = 53$; More than 4 years and $n = 49$) on the "total" mean differences are 1.27, 2.80, 2.91, and ($F = 10.93$, $P = 0.00 < 0.01$).

Statistically significant state member of time on the "customer satisfaction" has a very significant difference. This is because the shorter the member of time, the less understanding they have of the overall situation of fitness club. Therefore, they need some time to adapt to the fitness club, obviously, members. The longer membership, the higher satisfaction and has a very significant difference.

3.1.4 Customer's monthly income is higher, accordingly, overall satisfaction is higher

After statistics SPSS16.0 software processing, Table 6 shows:

1. China's high-end fitness club customer' monthly income ($n = 3000$ Yuan and below 22; 3001–6000 Yuan, $n = 88$;

Table 3. Customers' overall satisfaction survey analysis (n = 216).

Indicators	N	Mean	Std. deviation
Courses in fitness	216	2.50	0.75
Open time	216	2.47	0.70
Reasonable service operation procedure	216	2.60	0.64
The employee and the manager easy to contact	216	3.19	0.79
Traffic and parking convenience	216	3.10	0.51
Reasonable membership card price	216	2.48	0.72
Strongly professional staff and friendly to people	216	3.01	0.71
Coaches' image and temperament	216	3.17	0.80
Make friends with other members	216	3.00	0.81
Help me to improve the function of body shape	216	2.98	0.52
Environment and trust	216	2.86	0.49
Well site planning	216	2.76	0.61
Total score	216	34.19	4.79

Table 4. Gender satisfaction survey and analysis (n = 216).

Indicators	Gender	n	M ± SD	Difference value	F value	P value
Total score	Male	58	35.92 ± 3.23	2.56	6.01	0.02
	Female	158	33.36 ± 5.06			

Table 5. Investigation and analysis on member of the customer time (n = 216).

Indicators	Member of the customer time	n	M ± SD	Difference value	F value	P value
Total score	Less than 1 year	46	31.73 ± 3.95	−1.27	10.93	0.00
	1 year	68	33.00 ± 4.33	−2.80		
	Two-three years	53	35.80 ± 3.39	−2.91		
	More than 4 years	49	38.71 ± 4.39			

Table 6. Investigation and analysis on customer's monthly income.

Indicators	Monthly income	n	M ± SD	Difference value	F value	P value
Total score	Less than 3000 Yuan	22	31.50 ± 4.18	−0.18		
	3001–6000 Yuan	88	31.68 ± 5.13	−2.60	8.06	0.00
	6001–9000 Yuan	42	34.28 ± 4.37	−2.47		
	Less than 9001	64	36.75 ± 4.87			

2. 6001–9000 Yuan, n = 42; below 9001 Yuan, and n = 64) on the "total" mean differences, followed by 0.18, 2.60, 2.47, and (F = 8.06, P = 0.000 < 0.001).

Statistically significant is that the customer's income has a very significant difference on the "total". This is because the customer monthly income is lower, and their general requirements to the club are higher. Obviously, the higher the monthly income, the higher customer satisfaction on the "total score" is. It has a very significant difference.

3.1.5 *Under the experience economy in Changsha, the highest degree is higher, the satisfaction rate is lower*

After statistics SPSS16.0 software processing, Table 7 shows:

1. China's high-end fitness club customers' highest degree (postgraduate n = 35 degree or under-graduate n = 87;
2. n = 55 technical secondary school or high school, junior high school and the following n = 39) on the "total" mean differences, followed

Table 7. Investigation and analysis on customers' highest degree under the experience economy (n = 216).

Indicators	The highest degree	n	M ± SD	Difference value	F value	P value
Total score	Postgraduate	35	30.3750 ± 5.08429	−3.8917	7.567	0.000
	Undergraduate	87	34.2667 ± 3.75943	−0.2333		
	Technical secondary school or high school	55	34.5000 ± 4.37384	−2.9375		
	Junior high school	39	37.4375 ± 3.84654			

Table 8. Investigation and analysis on customer's career (n = 216).

Indicators	Career	n	M ± SD	Difference value	F value
Total score	Officer	47	32.82 ± 4.13	4.06	
	Teachers or researchers	28	31.38 ± 4.80		0.002
	Individual	55	36.66 ± 4.76		
	Management	48	35.16 ± 3.36		
	Employees	28	31.70 ± 4.87		
	Other	10	36.25 ± 3.67		

Table 9. Investigation and analysis on customer's overall satisfaction (n = 64).

Indicators	n	Mean	Std. deviation
Monthly gross income satisfaction	64	3.36	1.04
Getting along with customer satisfaction	64	3.08	0.95
Getting along with employee satisfaction	64	3.41	1.05
Reflecting value in job satisfaction	64	3.59	1.04
Reflecting professional competence in job satisfaction	64	3.61	0.98
Constantly improving the comprehensive quality total	64	3.84	0.96
	64	20.93	4.71

by 3.8917, 0.2333, 2.9375, and (F = 7.567, P = 0.000 < 0.01).

Statistically significant that the customer's career was very significant differences in "total score". Visibly, higher the customers' highest degree, lower the satisfaction. It has a very significant difference.

3.1.6 *Career satisfaction in turn is individual customers, other personnel, management personnel, authority, company employees, teachers and researchers*

After statistics SPSS16.0 software processing, Table 8 results show:

1. China's high-end fitness club customers professional (officer n = 47 teachers or researchers n = 28; individual n = 55; management;
2. n = 48 employees n = 28;
3. Other n = 10) on the "total" value of 32.82, 32.82, 32.82, 35.16, 31.70, 36.25, and (F = 4.06, P = 0.002 < 0.01).

Statistically significant is that the customer's career has significant differences on "total score". Visible, it includes career satisfaction in turn for individual customers, other personnel, management personnel, authority, company employees, teachers and researchers. There is a clear significant difference.

3.2 *Staff satisfaction indicators in service operation*

3.2.1 *Employee satisfaction rate is higher, overall situation is more optimistic in service operation*

After statistics SPSS16.0 software processing, Table 9 results show:

1. Employees in the "monthly gross income satisfaction", "getting along with customer satisfaction", "getting along with employee satisfaction", "reflecting value in job satisfaction" and "reflecting professional competence

Table 10. Investigation and analysis on employee's gender satisfaction (n = 64).

Indicators	Gender	n	M ± SD	Difference value	F value	Indicators
Total score	Male	33	22.63 ± 4.35	3.46	12.43	0.001
	Female	31	19.17 ± 3.92			

Table 11. Investigation and analysis on employee's work nature satisfaction (n = 64).

Indicators	Worker nature	n	M ± SD	Difference value	F value
Total score	Professional managers	8	25.62 ± 4.06	79.313	0.0000
	Coaches instructor	22	25.22 ± 3.66		
	Front desk receptionist	10	19.20 ± 5.25		
	Logistics service	11	17.11 ± 5.38		
	The membership consultant	13	16.71 ± 3.92		

in job satisfaction" "constantly improving the comprehensive quality", "total" on the satisfaction of value are 3.36, 3.08, 3.41, 3.59, 3.61, 3.84 > 3;
2. Employees in the above aspects are generally satisfied;
3. The satisfaction of employees in the "total" value is 20.93 > 18.

Staff is the main part of the business. Their satisfaction and motivation directly affects the economic profits of the club and the service innovation concept. Under the experience economy, the overall satisfaction rate of employees in high-end fitness club will be satisfactory, the overall situation is optimistic.

3.2.2 *Overall satisfaction of men employees is higher than women employees', and there are significant differences*

After statistics SPSS16.0 software processing, Table 10 results show: The male t (n = 33) and female (n = 31) employees on a "total" mean difference of 3.46 and (F = 12.43, P = 0.001 < 0.01).

Statistically significant shows that male and female employees have very significant differences in "total". This is because the female employees in the overall service operation shows advantages in general. They do their own work and consider other less. The concept of innovation in their work is not enough. Their low income leads to lower satisfaction, while male employees in the service operation show the advantage. They are wagging the dog and have their own ideas in work. Their income is higher, so the satisfaction rate is higher also. Obviously, under the experience economy, overall satisfaction of male employees is higher than women. It has a clear significant difference.

3.2.3 *Under the experience economy, the work satisfaction of club staff in turn is professional managers, coaches' instructor, front desk receptionist, logistics service, the membership consultant*

After statistics SPSS16.0 software processing, Table 11 shows: Under the experience economy, the work nature of club staff (n = 8 professional management personnel, coach instructor n = 22, front desk service n = 10, logistics service, n = 11 membership consultant n = 13) on the "total" value were 25.62, 25.62, 25.62, 17.11, 16.71, and (F = 79.313, P = 0.000 < 0.01).

The work nature statistical significance shows a very significant difference in "constantly improving the comprehensive quality". Under the experience economy, the work satisfaction of club staff in turn is professional managers, coaches' instructor, front desk service, logistics service, and membership consultant.

4 CONCLUSIONS AND RECOMMENDATIONS

4.1 *Conclusion*

1. *In the stable development of high-end fitness clubs in our country, the overall satisfaction for the customer is not satisfied: a) Men's satisfaction rate is higher than women' in general; b) customers with longer member, greater age, higher monthly income and lower education have the higher satisfaction.*

Career satisfaction in turn is individual customers, other personnel, management personnel, authority, company employees, teachers and researchers.

2. *Empoyee satisfaction is satisfactory. Under the experience economy, male staff satisfaction is higher than women in general. The working nature of employee satisfaction in turn is professional managers, coaches, front desk receptionist, logistics service, and the membership consultants*

4.2 Recommendations

1. *High-end fitness club in our country should strengthen the management and construction of project, locate the membership card price reasonably, focus on membership requirements of the authority, company employees, teachers and researchers customer, improve their satisfaction to last their members and develop more customers*
2. *Staff in high-end fitness club in our country should do as follows: a) strengthen the study; b) improve their academic and professional technology to improve their working enthusiasm and creativity; c) focus on work tedious and difficult of the front desk service, logistics service, the membership consultant; d) try to increase their pay; e) try our best to improve their satisfaction to retain them.*

REFERENCES

[1] Zhu Jufang. from SheBin explore the growth of our country sports fitness industry policy choices [J]. Journal of nanjing xiaozhuang college, 2003 (9): 107–109.
[2] TaoXing. to establish customer service system to ensure that the service effectively for a long time [J]. Journal of marketing, 2002 (7): 14 to 15.
[3] Liu Ying. Establish a perfect customer service system [J]. Reform the BBS, 2004 (05): 32–34.
[4] Hu Hongquan, opportunities etc. Our business fitness club customer service system [J]. Shanghai industry, 2011 (02): 110–131.
[5] Guo Xiuwen, SPSS statistical software used in the sports [M]. Beijing: people's sport publishing house, 2007.
[6] Li, practical sports multivariate analysis [M]. Beijing: people's sport publishing house, 2007.
[7] Tian Maijiu. Sports training learning [M]. Beijing: people's sport publishing house, 2000.

Sports Engineering and Computer Science – Luo (Ed.)
© 2015 Taylor & Francis Group, London, ISBN 978-1-138-02650-6

Discussion on insurances selection of professional sports clubs

HaoSong Li
Tianjin University of Science and Technology, Ministry of Sport, China

ABSTRACT: The establishment and selection of sports insurances are discussed in this paper from a sociological and economic point of view, it is an indispensable part of the development process of professional sports clubs, it is the basis of security system of professional sports clubs, it is a concrete manifestation of professional sports clubs reflecting the healthy development of people-oriented. In this paper, property, scope, functions and classifications of various sports insurances are defined and described, and the insurance and economic functions of various types of insurances are analyzed. For professional sports clubs, how to choose the insurances and play their role as well as insurance functions are deeply researched and explored in this paper.

Keywords: sports insurances; professional sports clubs; insurances

1 INTRODUCTION

On July 10, 2006, in the section 13 rescheduling of Chinese Football Association Super League between Shenyang Ginde and Qinddao Jonoon, Ginde's Guinea foreign aid Bangoura was kicked at right eyes, resulting in eyeball rupture and lens rush away; December 7, 2006, in the 15th Asian Games in Doha, Qatar, South Korean equestrian athlete KIM Hyung Chil fell off the horse and was dead by accident when he was participating in the individual cross-country of equestrian three-day event; June 10, 2007, in the Women's Uneven Bars preliminary of National Gymnastics Championship and Olympic trials held in Shanghai, Zhejiang's young athlete Wang Yan, whose hind hit the high bar after completed the first somersault, then her head heavily fell toward the ground, resulting in fracture and dislocation on the second and third cervical vertebra. There are such serious sporting disability events around us in less than a year, it can be concluded that there is a great need to solve the sports insurances in the development process of professional clubs.

The insurances investigation of more than 30 professional sport clubs from Beijing, Tianjin and so on conclude that more than 80% of the professional sports clubs are applying and following medical insurance regulations and specifications for purchasing insurance, such type of insurance is too narrow to adapt to the characteristics of professional sports clubs, which damages the insurance coverage and protection functions of professional athletes and professional sports clubs to some extent, and cannot meet the requirements of the rapid development of professional sports clubs, to a certain extent, it also undermines the economic interests of the athletes and professional clubs.

2 SOCIAL FUNCTIONS OF INSURANCES OF PROFESSIONAL SPORTS CLUB

It is a social security mechanism for professional athletes and related sports organizations, which provides them with fundamental rights to survive and endow them with an ability to further production when occurred accidents. Its social functions lie in ensure that professional sports clubs can timely obtain compensation from insurance companies after suffering economic benefits loss and personal accidents of athletes, it can ensure the normal operation of professional sports clubs and the compensation of personal medical expenses, thus to keep the club and social stable, reflecting the level of economic development of a country and the people-oriented humanistic care. Therefore, under the conditions of market economy, it is an indispensable part of professional sports clubs and has a role to play in protecting the weak, maintaining social stability and promoting the general welfares.

3 ECONOMIC FUNCTIONS OF INSURANCES OF PROFESSIONAL SPORTS CLUBS

The insurer and the insured sign insurance contract, according to the insurance subject, insurance amount and insurance premiums, insurance companies conduct site survey, investigation and parameters proof-read on professional sports clubs and athletes who suffered economic loss, and then provide economic compensation in accordance with the requirements of insurance contact and the extent of loss, which is a compensation mechanism of market economy. It is a legal contract of insurance companies and professional sports clubs under the conditions of economy market, which makes the professional sports clubs suffered damages get a certain amount of economic compensation, let coaches and athletes get a certain amount of economic compensation when they are disabled or unemployed, let them back to life and new job as soon as possible instead of being in debt.

4 DEFINING OF INSURANCES OF PROFESSIONAL SPORTS CLUBS

The insurances of professional sports clubs in the world are varied, it is a microcosm of the insurance field, and sports insurance can be divided into property insurance and personal accident insurance from the macroscopic view. Insurance of professional sports clubs refers to the insurance contact between professional sports clubs and insurance companies, it is a risk management technique that transfer the sports personal accident risks and sports property risks faced by athletes, coaches, management staff and professional sports clubs to insurance companies; it is a risk management that makes social ricks and natural risks standardized and scientific through insurance subjects, insurance contracts, insurance premiums, odds and amount of insurance coverage, it is a security system for the operation and management of professional sports clubs.

5 INSURANCES SHOULD BE CHOSEN BY PROFESSIONAL SPORTS CLUBS

It can be concluded from the available statistics that the worldwide sports insurances are varied, from insurance for a finger to insurance for a physiological organ; from insurance according to personal requirements to insurance according to requirements of organizations; from insurance according to social demands to insurance of a country's law legislation; sports insurances cover all aspects, its amount reaches more than 100. Therefore, for our country, who operates professional sports clubs in terms of less than 20 years, researching and discussing the insurances in accordance with the current situation is an issue urgently be solved.

5.1 *Medical insurance of professional sports clubs*

Medical insurance, also known as medical expenses insurance, it refers to that the insurer provides economic compensation to the insured who needs to accept medical treatment due to illness or body injury, since the sports that professional athletes engaged in are very confrontational and competitive, resulting in the injured parts and diseases of athletes in professional sports clubs are complicated, diverse and multiple, this kind of insurance is the most important aspect of insurance system of professional sports clubs, it should be the first consideration to professional sports clubs.

5.2 *Life insurance of professional sports clubs*

Life insurance of professional sports clubs refers to insurance for professional athletes when their life is threatened by unpredictable accidents. For example, the South Korean equestrian athlete Kim Hyung Chil fell off the horse and was dead by accident in Asian Games in Doha. Insurance companies offered a certain amount of economic compensation. It is worth noting: when the professional sports clubs purchase life insurance, especially whether athletes participate in sports events and non-sports events or not, their death should be contained in the life insurance contact of professional sports clubs, life insurance should be the first consideration to professional sports clubs.

5.3 *Disability insurance of professional sports clubs*

It is an insurance contact that refers to when the insured is suffered unpredictable personal accident in training or competition, the insurance company is responsible for economy compensation for medical expenses and disability rating, the compensation for disability rating of athlete is the important issue to be solved, due to the sports trainings are cruel and the competitions are intense, athlete's different parts of the body are often hurt and disabled. Therefore, such insurance is extremely important to professional sports clubs, it should be the first consideration to professional sports clubs.

5.4 *Unemployment insurance of professional sports clubs*

Unemployment insurance refers to a social insurance when the insured is unemployed due to uncontrollable social or economic factors. Insurance companies pay the insurance payment to the insured according to the previously appointed insurance contract, thus to maintain their basic living standards. This insurance is generally listed into the national legal insurance areas, such as: professional sports clubs suddenly dissolute owing to economic reasons, for this reason, it is necessary for professional sports clubs to purchase unemployment insurance for athletes.

5.5 Pension insurance of professional sports clubs

Pension insurance of professional sports clubs is aimed at the professional athletes, the amount of pension is determined by athlete's age, length of sports service time, insurance premium and so on, the insurance companies pay pension payments regularly until they die. In Western countries, such insurance is listed into the national social insurance system. Such insurance should be listed into the insurance coverage of professional sports clubs according to their economic conditions.

5.6 Liability insurance of professional sports clubs

Liability insurances involved in insurances of professional sports clubs are: public liability insurance, employers' liability insurance, liability insurance of using sports equipment and sports professional liability insurance.

5.6.1 Public liability insurance

Public liability insurance in insurances of professional sports clubs is also known as liability insurance of stadium owner. As the owner of sports facilities, professional sports clubs purchase insurances for the security of people who enter the stadium to take part in sports, do training, watching games and other activities. Such insurance refers to various economic compensation liability, it covers when various sports facilities are used for competition, training and other activities, the third party's body injury or property damage caused by accidents should be assumed by the insured. Professional sports clubs must purchase this kind of insurance.

5.6.2 Employers liability insurance

Employers liability insurance of professional sports clubs is in accordance with the Labor law employment contact, the professional sports clubs purchase insurance for various personal injury accidents of professional athletes and employees in the course of employment, and the insurance companies assume economic compensation liability, this kind of insurance is another insurance or economic compensation in addition to personal accident insurance and sports disability insurance, as employment relationship, it is a necessary insurance in social insurance system.

5.6.3 Liability insurance of using sports equipment

It is a insurance and economic compensation liability when athletes are using sports equipment, in order to transfer the damages on athletes and coaches caused by shoddy products, clubs sign insurance contacts with insurance companies, insurance companies are responsible for the economic compensation of injuries and loss, this kind of insurance should be cautiously chosen based on economic strength.

5.6.4 Sports professional liability insurance

Sports professional liability insurance refers to when coaches and doctors in club are engaged in their own work, some coaches are eager to be success and some doctors make mistakes in the course of treatment for athletes, thus resulting in body injuries on athletes, professional sports clubs sign insurance contacts with insurance companies, the insurers are responsible for the economic compensation caused by mistakes of coaches and doctors, this kind of insurance should be cautiously chose according to economic strength.

5.7 Property insurance of professional sports clubs

5.7.1 Sports facilities insurance

Sports facilities insurance provides compensation for damages of sports facilities owned or rent by the insured (professional sports clubs), which are caused by fire, theft and other activities stipulated by the contract. The amount of compensation depends on the insurance premium of the damaged facilities. Such insurance should be seriously taken into account.

5.7.2 Sporting events insurance

Sporting event's successful holding as scheduled is an important means of obtaining economic benefits for professional sports clubs. If the events cannot be held as scheduled due to numerous reasons, professional sports club's profits will suffer losses. Therefore, for professional clubs, they can consider purchasing the sporting events insurance.

5.7.3 Financial insurance

Financial insurance refers to that due to the sudden outbreak of a regional financial crisis, the expected cost cannot satisfy the current needs of the game, resulting in losses and overdrafts of competitions held by professional sports clubs, then the insurance companies provide economic compensation to sports clubs, this kind of insurance should be cautiously chose according to economic strength.

6 SIGNIFICANCE OF INSURANCES OF PROFESSIONAL SPORTS CLUBS

6.1 *Guarantee the normal operation of professional sports clubs*

In the commercial economy society, professional sports clubs are relatively independent economic entities, the quality of their operating results is borne by themselves. In economic life, professional sports clubs often suffer natural disaster and accidents, resulting in increase in expense and decrease in income, thus to destruct their normal operations. It can be concluded from the process of industrial capital cycle, in order to achieve the proliferation of capital, operation should be kept running. Through purchasing insurance, professional sports clubs transfer the risks to insurance companies, making professional sports clubs continue normal operation and increase profits, thus to improve the safety factor in the operation of professional sports clubs.

6.2 *Guarantee the economic interests of professional sports clubs*

In the management process of professional sports clubs, the loss caused by risks cannot be estimated, insurance can fix this kind of unpredictable loss in the form of insurance premiums, and professional sports clubs can put insurance premiums as a fixed expense to relieve the worries about risks. The insured of insurances of professional sports clubs includes professional sports clubs, professional athletes, coaches, support crews, managers and spectators. When they suffer risk loss within the insurance coverage, they can get timely and reliable financial compensation or insurance payment, thus to guarantee the insured's economic rights or interests and protect the club's economic interests from losses.

6.3 *Promote risk management of professional sports clubs*

As economic enterprises who specialize in dealing with risks, insurance companies have accumulated plenty of experience of risks handling and disaster prevention. They advocate disaster and risk prevention, reduction of insurance subjects to increase their own economic efficiency. So they can give a lot of opinions and suggestions about risk management to professional clubs and urge them to conduct risk management, thereby improving their abilities of disaster prevention and enhancing their risk management awareness and abilities.

6.4 *Promote the improvement of insurances of professional sports clubs*

At present, China's professional sports clubs purchase medical insurance, accident insurance, stadiums insurance. However, it is not enough to ensure the rapid and healthy development of professional sports clubs. In order to reduce the cost, some professional sports clubs even do not purchase insurance, which is extremely unusual. With the perfections of various laws and regulations as well as the rapid development of sports insurance industry, the insurances of professional sports clubs will be improved to ensure harmony and stability in clubs as well as their economic interests.

6.5 *Promote the athlete to improve awareness of insurance*

With the further development of China's sports market economy, the competitions of sporting events are increasingly intensifying, for professional sports clubs, the probability of the occurrence of danger and gaps between various insurances will appear increasingly. Therefore, urging the professional sports clubs to strengthen the awareness of the risks helps to enhance the insurance awareness of the management personnel and promote the development of insurance of professional sports clubs, guaranteeing their political and economic interests.

ABOUT THE AUTHOR

Li Haosong (1962–), male, Tianjin, associate professor of Tianjin University of Science & Technology.

REFERENCES

[1] Qiu Xiaode, Sports Insurance [M]. Beijing Sport University Press, Beijing, 2006. 5.
[2] Jia Jing, The Influence of WTO Entry on China's Insurance Industry and Development Forecast [J]. Modern Finance and Economics, 2000 (1):16–19.
[3] Wan Guohua, The Influence of WTO Entry on China's Security Market and Countermeasures, [J]. Modern Finance and Economics, 2000 (5):31–33.
[4] Liu Yifan, Notion on Development Strategy of China's Banking Industry on the Background of the Approaching Accession to the WTO [J]. Modern Finance and Economics, 2000 (5):23–26.
[5] Qiu Xiaode, Exploration and Discussion on Social Insurance of Athletes [J], China Sport Science, 1999 (3):36.
[6] Qiu Xiaode, Feasibility Study on Self-preservation and Self-help of Athletes [J]. Journal of Beijing Sport University, 1999 (3):8–10.

[7] Qiu Xiaode, Analysis on the Status of China's Athletic Sports Insurance [J]. Zhejiang Sport Science, 1999 (3):57–60.

[8] Qiu Xiaode, Thoughts on the Development of Policy Financial Institutions of China's Sports Industry [J]. Journal of Shanghai University of Sport, 2000 (2):5–8.

[9] Qiu Xiaode, Thoughts on the Development of China's Sports Bond [J]. Journal of Beijing Sport University, 2000 (4):444–446.

[10] Qiu Xiaode, How to Immediately Develop China's Sports Gambling after Joining the WTO [J]. Journal of Beijing Sport University, 2003 (6):734–736.

[11] Qiu Xiaode, How to Confront the Anti-dumping of International Market for Sports Industry after Joining the WTO [J]. Zhejiang Sport Science, 2002 (5):59–61.

[12] Qiu Xiaode, Research on Changes of Market Mechanism of Chinese Sports Insurance after Joining the WTO [J]. Chengdu Sport University, 2004 (6):15–18.

[13] Qiu Xiaode, Research on Improve the Overall Competitiveness of China's Sports Insurance after Joining the WTO [J]. Tianjin University of Sport, 2006 (6):214–216.

[14] Liang Xiuling, International Insurance Practice [M] Tianjin: Tianjin Science & Technology Translation & Publishing Co., LTD, 1998:133–273.

Sports Engineering and Computer Science – Luo (Ed.)
© *2015 Taylor & Francis Group, London, ISBN 978-1-138-02650-6*

On the capital operation of sports events

XiaoDe Qiu

Physical Training Department, Tianjin University of Finance and Economics, Tianjin, China

ABSTRACT: This paper studies the basic elements to explore the construction, methods and some limitations of the capital operation of sports events. The results indicate that the capital operation determines the failure and success of the sports events. Thus, it is necessary to operate the sports capital in a scientific way. This paper seeks to explore a theory to contribute to the development of sporting events in China.

Keywords: sports events; sports events capital; capital operation of sports events

1 INTRODUCTION

Background: With the deepening of China's economic reform and opening up, Chinese economic strength is growing with the flourishing of sports industry. Various world-class championships, World Cup, Champions Cup and comprehensive competitions are held in China, which have greatly promoted the development of international and domestic sporting events in China. Nowadays, keeping higher social prestige at lower cost by holding competitions with international influence is a principle, governments at all levels should adhere to achieve the Scientific Outlook on Development and coordinate the overall development. The capital loss of major international sports events is reflected in several aspects, as shown in Table 1.

Take the Athens Olympic Games in 2004 as an example, whose total expenditure is over 80 billion Euros. It ranks only second to the Moscow Olympics in 1980. The huge cost overruns affect the economic growth of Greek after the Olympics which fell to the lowest point in 2005. Greek Deputy Finance Minister Pei QuSidu Castro made it clear that it is impossible to make back the money

of hosting the Olympic Games in a "short-term", which is worth deep reflection and vigilance. Judging from past experience, for any country or region, the rapid development of sports events is an important sign at the beginning of its rising. Such as: Barcelona Olympics of Spain, Seoul Olympics, Los Angeles Olympics, and so on. It is widely acknowledged that the development of sports events is closely related to the growth of national economy, which indicates the development of the national economy, the rejuvenation and strength of the Chinese nation. To create a level playing field in the socialist marketing economy requires us to carry out "fair, just, harmonious and open" sporting events. It will undoubtedly play an important role in the development of the national economy. The data shows that the direct revenue of Los Angeles Olympics in 1984 is 3.29 billion $; the revenue of Atlanta Olympics in 1996 is 5.1 billion $; the revenue of Salt Lake City Winter Olympics in 2002 is 2 billion $ despite its small size.

Therefore, the study of capital operation of sports events will play a positive role in promoting the development of various sports events, saving cost for the country and creating economic benefits.

Table 1.

Overdraft	Defective item			
	Construction of stadiums	Improvement of urban traffic	City renovation	Improvement of communications equipment
Olympic Games in Mexico 620 million	48%	21%	19%	12%
Olympic Games in Montreal 1 billion	61%	17%	15%	7%
Olympic Games in Moscow 9 billion	52%	21%	14%	13%
Olympic Games in Greece 8.1 billion	44%	23%	22%	11%

2 DEFINITION OF SPORTS EVENTS CAPITAL

The capital of sports events includes the public investment of the government, social and charitable organizations, the funds of the investors and sponsors, the resources of partners, goods, services, or facilities of suppliers which are applied in the operation of sporting events and so on. After the reasonable permutations and combinations of the mental and physical capital by the organizing committee, the sporting events could be regarded as a carrier to utilize the joint-stock system, authorization system, commission system, chain management system by raising, building and operation of the sports events capital, and obtain the maximum economic benefits under the participation of relevant personnel, the manipulation of the operators, as well as the propaganda of media. To sum up, it is a process of sports events capital appreciation, an inevitable phenomenon when the market economy develops to a certain stage, a high-level sports business activity. It helps to promote the development of sports and related industries. Sports events capital includes funds, material conditions (such as the venue, equipment, and so on) and human resources salaries, training costs and various expenditures.

Without these, it is impossible to organize a sports event, even the large-scale sporting events. Therefore, sporting events capital is the economic basis to determine various sporting events.

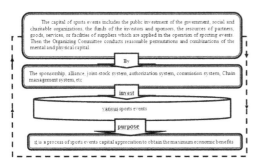

3 CONSTRUCTION OF SPORTS EVENTS CAPITAL

The construction of sports events capital is to excavate the growth potential of the sports events to the maximum degree and raise funds to serve the sports events. The research results indicate that the construction of sports events capital can be divided into several parts: 1. The financial system of sports capital funds: 1) sports funds; 2) sports bonds; 3) sports stock; 2. Sponsorship system of sports events capital: 1) sponsorship; 2) advertising and so on; 3. Operation system of sports capital market: 1) television; 2) tickets; 3) derivatives of sports events; 4. Cooperative arrangements of sports events capital: 1) cooperative partners; 2) franchise suppliers; 3) chain operations of sports events.

4 CAPITAL OPERATION OF SPORTS EVENTS

Generally speaking, the operation of sporting events capital includes three parts: first, serve for the operation of the sports events and rationalize its operating structure and industrial structure; second, serve for the operation of the sports events and produce more sports products suited to various audiences by the expansion of capital, meanwhile, be alerted to the risks in the production and management; third, serve for the sports groups and the profit growth of the organizing committee and obtain profits by the purchases of assets and fixed assets. Thereby the capital operation of sports group has two basic functions: one is industrial management; the other is capital operation.

From the perspective of current capital operation of sports events, the operation of sports events mainly focuses on the industrial capital operation which based on the core content, the financial capital operation includes the sports events, transaction of property rights, exchange of property operations, intangible capital of sports events, sports brand operations and so on.

4.1 Securitization operation of sports events capital

This capital operation is a concept relative to the production and operation of sports events. Its main idea is to regard the operation of sports capital as an economic activity, which is independent of the sports events capital (development of additional sports events products) and based on the valuated and securitized capital or materialized capital. It could improve the operation efficiency by the flow, fission, combination and optimal configuration, and so on. Such as issue sport stocks, sports lottery, sports fund, sports bonds in the sporting events.

4.2 Practical operations of sporting events capital

The industrial capital operation of sports events is carried out based on the core content of sports event and its target consumer group. The industrial capital operation of sports events involves the sports stadiums, sports venues, sports equipment, sports consumables and other hardware facilities as well as sporting events teams, media, propagandistic facilities. It could obtain profits by the integration and the necessary operation. The industrial capital operation of sports events mainly relies on good sports facilities, stadiums and excellent teams. It could make profits from various ways: ticket, television, development of additional sports products and franchise suppliers. The Olympic Games, World Cup, American NBA and other sports events are models of industrial capital operations.

4.3 Property operation of sports events capital

The equity capital operation of sports events refers to the operational activities based on the property of sports events, from physical to monetary form or from monetary to physical form.

4.3.1 Naming rights

When holding the major international and domestic sports competitions, the title sponsorship of sport enterprises is used to expand the capital operations. In some developed countries, more than 70% of the large stadiums have title sponsorship that is an important revenue source of sports events.

4.3.2 Auction

Sports auction is an important way to operate the capital of sports events, for selling the proprietorship of sports events, the possession of sports substance, tangible and intangible assets to the capital possessors. This kind of exchange of interests and funds could achieve complementary interests, profit sharing and complement each other's advantages. The auction operation of the sports industry is promising and requires further improvement and development. Although it is still in the initial stage, there are so many areas can be explored.

4.3.3 Sale

The funds that obtained by selling sports products could be used to develop sports events: (1) by selling the mascots with the signature of the chairman of the organizing committee and the chairman of the NFs National Federations; (2) by selling the commemorative stamp and first day cover with the signature of athletes; (3) by selling the T-shirts with the signature of renowned athletes at home and abroad; (4) by selling the sports medals, commemorative plaque and commemorative coin which have been granted a certificate and have the signature of the officials in this competition; (5) by selling the race regulations, transcripts with the signature of renowned chief referees; (6) by selling parts of the sports equipment in major international sports competitions.

4.3.4 Transfer

The management rights of the stadiums after the sports events could be transferred to private companies or sports management agencies, which can effectively improve economic benefits, reduce costs and provide more employment opportunities, thus promoting the development of social sports industry.

For example: Chengdu Blades club is transferred to Hong Fu Enterprises Ltd. The controlling party of Chengdu Blades club, the Scarborough Group, held a press conference in Shenzhen; the chief operating officer announced the management right of the club has been transferred to the Hong Fu Enterprises Ltd, it is managed by the Chongqing Gele Investment Co. Ltd.

4.3.5 Leasing

Lease the stadium to certain sports organizations and international sports organizations after the sporting events, by charging fees every year to reduce the unnecessary expenditure and achieve boutique operations.

For example: after the Sydney Olympics in 2000, the original gymnasium was leased to a large exhibition organization in Sydney to earn generous rent by the construction exhibition, exhibition of sports equipment, apparel and automotive exhibition.

4.4 Capital operation of intangible assets of sports events

Capital operation of sporting events means the organizing committee utilizes intangible assets of sports event and carries out sports capital funds, such as the reputation, hosting right, management right, patents and promotional rights, thus achieving the capital appreciation. The intangible assets of sports events play an important role in capital operation and it is like an economic leverage to achieve capitals preservation and expansion.

Sports intangible assets are a product of long-term development of human culture and it could be rewarding for its owners and users as the distinctive intellectual property. According to statistics, the sports intangible assets account for 60% to 70% of the sports industry revenues in developed countries. Take the IOC as an example, IOC began to carry out TOP program in 1985 and the revenue of the first phase is $ 97 million during 1985~1998 and the revenue of the fifth phase is close to $ 600 million in 2001 in the marketing process of Olympics. In addition, the revenue of the sale of television rights has increased to the expected $ 1.67 billion of Beijing Olympic Games in 2008 which only $ 101 million of the 22th Olympic Games in 1980. Nowadays, the revenue of the intangible assets has reached 89% of the total revenue of IOC [2].

Capital is the currency used in sporting events or various social economic resources and a variety of economic instruments and marketing measures adopted to maximize the economic benefits. At the operational level, according to the characteristics and patterns of sports events, the management takes a comprehensive, three-dimensional and multi-channel business development mode, displaying its unique business and economic characteristics, which is the driving force behind the development of the current sporting events. It uses market rules to achieve the appreciation of values and growth in benefits through skilful management or scientific movement of capital. Sporting events can never start without previous operation and business development. Therefore, capital plays a vital role in the development and management of sports events.

Comparison of Olympic capital operation.

Nation content	TOP Program of OCOG	Business development
1984 Los Angeles	95 million $	887 million $
1992 Barcelona	350 million $	1.027 billion $
2008 Beijing	600 million $	2.132 billion $

5 KEY ISSUES OF CAPITAL OPERATION OF SPORTS EVENTS

5.1 *Sustainability*

The capital operation of sports events contributes to the sustainable development of the industry. With the deepening of Chinese market economic system, most sporting events have established a modern management system, protecting the interests of investors while promoting the transformation of the operating mechanism of sports industry and the enhancement of the management level. At present, most international and domestic enterprises seek to improve their social awareness and expand their brand through the platform, which requires the new initiatives and measures of the event operators to improve their international competitiveness.

5.2 *Optimization of sports resources*

The capital operation of sports events will promote the optimization of sports resources. Currently, sports industry in China mainly depends on the massive investment of the government in exchange for sports economic growth, thus leading to multiple drawbacks such as low capital efficiency, poor sports quality, decentralized investment and redundant construction. Through scientific sports capital operation, it will be conducive to the flow and restructuring of related assets to change the economic growth mode of sports events.

5.3 *Internationalization*

The capital operation sports events will further the international development of sports events. Capital operation will be allowed for the entry of more international capital in the sports industry to China, and create favorable conditions for more domestic enterprises to enter the international sports capital market. Besides, it will improve the international competitiveness of the capital of sports events in China. Also will make it closer to the capital development of global sports events through event elements of different countries and regions, and increase the participation in various kinds of international sports activities.

REFERENCES

[1] Huang Huaming. Risk and Insurance [M] China Legal Publishing House 2002.
[2] Xu Jinliang. Market risk management, China Financial Publishing House 1998.
[3] Liu Xinli. Market Risk Management. Economics textbook series of Peking University 2006.
[4] Sun Xing, Qiu Yuan. Research on the Market Risk Management Mode of Major Sporting Events 2005.
[5] Li Guosheng. Research on the Elements of Risk Management of Sports Market 2005.4.
[7] Qin ChunLin et al. Sports Management [M]. Higher Education Press, 2002.8.
[8] Zhan Jianguo. Modern Operating Mode of Sports Business Management. Beijing Sports University Press, 2003.5.
[9] Zhu Zhenhuan, Wang Heze. Real Olympics. Tsinghua University Press 2004.H.
[10] Xiao Feng, Shen Jianhua. On the Risk Characteristics of Major Sporting Events and Risk Management of the Market. Sports Science Research.

Sports Engineering and Computer Science – Luo (Ed.)
© 2015 Taylor & Francis Group, London, ISBN 978-1-138-02650-6

The universality and individuality between sports professional club insurance and sports disability insurance

Wei Mu
Tianjin University of Science and Technology, Ministry of Sport, China

ABSTRACT: This paper seeks to clarify the relation and difference between sports professional club insurance and sports disability insurance to some extent, and also seeks to define the nature, scope, classification and function of sports professional club insurance. This paper proposes that sports disability insurance is within the scope of sports professional club insurance, which is a branch and exception of sports professional club insurance, but it also has its own specific attributes. Therefore, scientific research and defining of sports disability insurance and sports professional club insurance are required to make sports insurance industry be more suitable for the inherent law of sports cause, to ensure the smooth development of Chinese sports cause, and then to promote the development of sports insurance industry.

Keywords: sports insurance; professional club insurance

1 INTRODUCTION

Currently, a vast majority colleges of sports technology, sports schools and sports professional clubs all purchase insurance by applying and following the regulations and specifications of sports disability insurance or sports medical insurance, which facilitates the time, energy and insurance programs of professional sports clubs' to some extent. But whether the sports insurance variety, insurance coverage or the amount of claims are involved in this insurance, to achieve is an issue worthy of study and reflection. The sports professional clubs insurance should be qualified with the security mechanism. Analyzed from the nature of insurance and the scope of covering insurance, sports professional clubs have sports human injury insurance, sports property insurance (fixed assets and mobile assets), sports economic benefit insurance (profits loss insurance and contract insurance), sports liability insurance (public liability insurance, product liability insurance, employer's liability insurance and professional liability insurance), sports event insurance and so on.

Sports disability insurance is one of athletes and coaches' personal accident insurances; analyzed from the range of sports insurance, it is far from the requirement of insurance that covering range of sports professional club, and one insurance loophole in certain aspects will bring irreparable economic loss to sports professional clubs. Therefore, a careful study on the relation and difference between sports disability insurance and sports professional club insurance is the first thing of sports insurance researchers should do, to tell sports professional club the coverage range of insurance and the minimum insurance varieties need to meet the basic requirements of sports professional club's insurance covering, which plays a positive role for us broadly grasping the areas and development trend involved in sports insurance.

2 DISTINCTION AND RELATION BETWEEN CONCEPTS

2.1 *Definition of sports professional club insurance*

Sports professional club insurance is the insurance that was bought by club in case that it will encounter property damage, interruption of economic interests, liability, casualty and other risks in its operation. Club is an applicant and is also the insured one, and the insurance company is the insurer. The main products of insurance are property insurance, public liability insurance, life accident insurance, unemployment insurance, event liability insurance, and so on.

2.2 *Definition of sports disability insurance*

Sports disability insurance refers to that the insurance company pays medical expenses and disability expenses for the accidental bodily injury of athletes and coaches and disability while they are at training or competition field or on the way to the training or competition. The applicants could be both professional sports club and athlete. If athletes and clubs are applicants and the insured, the insurance company is the insurer.

2.3 *Defining of sports professional club insurance*

Sports professional club insurance covers all aspects of sports insurance, which is a microcosm of sports insurance. And sports insurance can be generally divided into property insurance and human injury insurance. Sports professional club insurance is a risk management technology, the insurance contract is signed by both club and insurance company, and it transfers the risk of human sports injury and property to the insurance company, which the risk should be faced by athletes, coaches, managers and professional sports clubs themselves.

2.4 *Defining of sports disability insurance*

Sports disability insurance specifically refers to that a sports organization or professional club that signs an insurance contract with an insurance company for injury accidents athletes, coaches and managers may suffer in particular occasion, to make insurance company pay for the medical expenses and disability expenses if the insured person suffers from the injury accident.

3 SCOPE AND CLASSIFICATION OF INSURANCES

3.1 *Classification and scope of professional sports club insurance*

3.1.1 *Two categories of professional sports club insurance*

The first category is property insurance of professional sports club, and the other is accidental bodily injury insurance of professional sports club. Property insurance of professional sports club includes sports property insurance (fixed assets and mobile assets), sports economic benefit insurance (profits loss insurance and contract insurance), sports liability insurance (public liability insurance, product liability insurance, employer's liability insurance and professional liability insurance), and sports event insurance; while accidental bodily injury insurance of professional sports club includes death insurance, disability insurance and medical insurance.

3.1.2 *Scope of professional sports club insurance*

This kind of insurance covers sports property insurance, sports economic benefit insurance, sports

liability insurance, sports event insurance, human injury insurance and unemployment insurance.

3.2 *Classification and scope of sports disability insurance*

3.2.1 *Sports disability insurance is the most fundamental level of sports insurance system with a relatively narrow scope, and it is to insure and compensate the injury accidents and disability that happened on athletes and coaches in the sports arena.*

3.2.2 *Sports disability insurance is an insurance mechanism to compensate the injury accidents and disability of athletes and coaches happened in the sports arena, with no financial compensation for other properties, liability and third party liability insurance.*

4 FUNCTIONS OF INSURANCE

4.1 *Professional sports club insurance should be all-around, and its function is to ensure the economic benefits compensation of the clubs if it suffers from the loss of economic interests, intangible assets, liability accidents, event postponement and athlete disability.*

4.2 *Sports disability insurance is qualified with the function of compensating financially to disability and medical expenses if athletes and coaches are subjected to accidental bodily injury in the sports arena.*

5 RELATIONS AND DIFFERENCES BETWEEN THE TWO INSURANCES

First, professional sports club insurance should include property insurance and accidental bodily injury insurance, while accidental bodily injury insurance includes sports disability insurance. Sports disability insurance cannot be equated with accidental bodily injury insurance, it does not include personal death, but accidental bodily injury insurance does. Second, sports disability insurance faces the individuals in a group while professional club faces individuals and the club both. Third, generally sports disability insurance only compensates the expenses of medical treatment and disability for the victims, while sports professional club insurance compensates the loss of all aspects of professional club. Fourth, the product of sports disability insurance is single while the products of professional club insurance are diverse.

6 INSURANCE PRODUCTS THAT SPORTS PROFESSIONAL CLUB SHOULD CHOOSE

6.1 Medical insurance of sports professional club

Medical insurance, also known as medical expenses insurance is to compensate all costs caused by the insured person because of illness or accidental injury. As the sports which professional athletes are engaged in are strongly confrontational and fiercely competitive, the damage parts and diseases of the athletes in sports professional club are complex, diverse and repeated. Thus, this insurance is the most important link in sports professional club insurance system, and it is also should be bought by the insurance product firstly.

6.2 Accidental bodily insurance of sports professional club

Accidental bodily insurance of sports professional club refers to the insurers provide some compensations when professional athletes suffer from the personal injury directly caused by unpredictable accidents. It is worth noting that the accidental bodily insurance in sports professional club insurance is significantly different from that of sports injury insurance. Sports injury insurance refers to the insured person who can only obtain the compensation of medical treatment and disability if he/she is injured in sports activities. But accidental bodily insurance of sports professional club can offer compensations whenever the professional athlete is injured, and it is less specific than sports disability insurance that the sports professional club should consider firstly.

6.3 Disability insurance of sports professional club

It is an insurance contract that the insurance company compensates the expenses of the insured person's medical treatment and disability once he/she is injured accidentally in the training or competition field, and it focuses on the compensation according to the degree of athlete's disability. Sports training and competition are fierce and competitive, often resulting disabilities in different parts of athlete's body. Therefore, this insurance product is very important for professional club, and should also be their first consideration.

6.4 Unemployment insurance of sports professional club

Unemployment insurance refers to the insurer offers insurance benefits to the insured person in accordance with the prior agreement, to ensure his/her bottommost living standard, when the insured person is unemployed because of incontrollable social or economic factors. Therefore, it is necessary for sports professional club purchasing unemployment insurance for athletes within the club's capability.

6.5 Endowment insurance of sports professional club

Endowment insurance of sports professional club refers to the insurer offers pension to athletes at regular intervals until death, the professional athletes are the insured, and the amount of pension is determined by the age, length of sports service, membership fees and so on. The selection of this insurance can be determined according to the economic situation of sports professional club and the development situation of social material.

6.6 Liability insurance of sports professional club

The liability insurances involved in professional club insurance are public liability insurance, employer's liability insurance, liability insurance of sports equipment, and sports profession liability insurance.

6.6.1 Public liability insurance
Public liability insurance of sports professional club insurance is also known as liability insurance of sports field owner. As the owner of sports field's facilities, the professional club arranges insurance for the personal safety of those who come to the sports field to participate in sports activities, training, watching games and so on. This insurance mainly accepts various economic compensation liability of the insured person, when the facilities of sports field cause third party's personal injury or property loss in competition, training or other activities.

On the surface, the third party injured is the object of protection, but the real object is the insured side, that's the sports professional club. This insurance product is necessary for sports professional club.

6.6.2 Employer's liability insurance
Employer's liability insurance of sports professional club mainly accepts the various economic compensation liabilities, that's the professional club should undertake according to employment contract of labor law when professional athletes suffer from personal injury during employment period. This insurance product has some similarities with accidental bodily injury insurance and

sports disability insurance, and it should be chosen carefully based on economic strength.

6.6.3 *Liability insurance of sports equipment using*

The club signs the insurance contracts with insurance companies during the usage of the sports equipment, to transfer the economic compensation for athletes and coaches' injury caused by inferior products, so that the insurer would be responsible for the economic loss. And this insurance should be chosen carefully based on economic strength.

6.6.4 *Sports profession liability insurance*

When the coaches or team doctors in club engage in their own work, mistakes sometimes caused by coach's anxiety for success or the treatment of doctor for athletes personal injury occurred in team. The insurer would be in charge of compensating the economic loss that caused by the mistakes of coaches or team doctors club should have signed sports profession liability insurance with the insurance company before that. And such insurance should also be chosen carefully based on economic strength.

6.7 *Property insurance of sports professional club*

6.7.1 *Sports facilities insurance*

Sports facilities insurance is to provide compensation when the sports facilities owned or leased by the insured sports professional club, they are damaged due to fire, theft and other acts stipulated in the policy. The amount of compensation is determined by the replacement cost of damaged facilities. And such insurance should be considered seriously.

6.7.2 *Sports event insurance*

The successful holding of sports event as scheduled is meaningful for professional clubs to gain economic benefits. If the sports event cannot be held as scheduled successfully, the sports professional clubs will suffer profit loss. Therefore, it is necessary for professional clubs to consider sports event insurance.

6.7.3 *Finance insurance*

When a sudden outbreak of financial crisis occurs in a region, the perspective competition cost cannot meet the current requirement of competition, leading to losses and overdrafts of the competition held by sports professional club. And in this situation, finance insurance will work, and the insurance company will compensate for the loss of sports club. And such insurance should be chosen carefully based on economic strength.

7 SIGNIFICANCE OF SPORTS PROFESSIONAL CLUB INSURANCE

7.1 *Guaranteeing the smooth operation of sports professional club*

In commercial economy society, sports professional club is a relatively independent economic entity, and it undertakes all the responsibilities resulted from the quality of its operating results. In economic life, professional sports clubs tend to suffer from natural disasters and accidents, resulting in increasing costs, decreasing income and destruction of normal operation. Analyzed from the process of industrial capital circulation, capital must move continually to realize the increment of value. Through insuring by sports professional clubs, the risks are transferred to the insurance company, so that sports professional clubs can operate smoothly and gain profits.

7.2 *Guaranteeing the economic benefits of sports professional club*

The loss caused by risks is inestimable, but this inestimable loss can be fixed in the form of paying insurance expenses through insurance. And sports professional clubs can treat insurance as a fixed expense to relieve the worries on risks. The insured of sports professional club insurance has professional athletes, coaches, support staff, administrators and spectators. When they suffer from the risk loss under the coverage of insurance liability, they can get timely and reliable financial compensation or insurance payment, and the economic rights or interests that the insured person should enjoy would be guaranteed.

7.3 *Promoting the risk management of sports professional club*

As an economic enterprise that specifically deals with risks in society, insurance company has accumulated a wealth of experience in handling risk and preventing disaster. They will propagandize the prevention of disasters and risks, to reduce the loss of the insured subject, and to improve their own economic benefit. Therefore, the company will give lots of advice to the professional club for risk management and supervise and urge the implementation of risk management, to enhance the disaster prevention capabilities of the club.

7.4 *Promoting the products perfection of sports professional club*

Currently, the major insurances purchased by sports professional clubs in China are medical insurance, accident insurance and sports field insurance, which are not enough for ensuring sports professional club because of rapid and healthy development. Even some sports professional clubs often do not purchase insurance to reduce costs, which is extremely unusual. With the completion of various laws and regulations and the rapid development of sports insurance industry, the products of sports professional club insurance will be perfected.

7.5 *Enhancing the insurance awareness of professional athletes*

With the further development of sports market economy and the intensification of the level of sports competition in China, the gap between the probability of the risks arisen in sports professional clubs and the insurance products will become evident, thus forcing the increasing enhancement of any sports professional clubs' risk awareness, which is conducive to the enhancement of the managers' insurance awareness and the development of sports professional club insurance.

REFERENCES

[1] Qiu Xiaode. Sports Insurance [M]. Beijing Sports University Press, Beijing, 2006.5.
[2] Jia Jing. Impact of Joining WTO on Chinese Insurance Industry and the Development Forecast [J]. Modern Finance, 2000 (1):16–19.
[3] Wan Guohua. Impact of Joining WTO on Chinese Securities Market and Its Countermeasures [J]. Modern Finance, 2001 (5):31–33.
[4] Liu Yifan. Development Strategic Conception of China Banking after Joining WTO [J]. Modern Finance, 2001 (5):23–26.
[5] Qiu Xiaode. Discussion on the Social Insurance for Athletes [J]. China Sport Science, 1999 (3): 36.
[7] Qiu Xiaode. Status Analysis on Competitive Sports Insurance of China [J]. Zhejiang Sport Science, 1999 (3):57–60.
[8] Qiu Xiaode. Reflection on Developing Policy-based Financial Institutions in China's sports Industry [J]. Journal of Shanghai University of Sport, 2000 (2):5–8.
[9] Qiu Xiaode. Thoughts on Developing China's Sports Bond [J]. Journal of Beijing Sport University, 2000 (4):444–446.
[10] Qiu Xiaode. How to Quickly Develop China's Sports Gaming Industry after China's Accession to WTO [J]. Journal of Beijing Sport University, 2003 (6):734–736.

Sports Engineering and Computer Science – Luo (Ed.)
© 2015 Taylor & Francis Group, London, ISBN 978-1-138-02650-6

Strategic researches of complementary development in resources of mass and competitive sports in Shandong Province

JiaoYang Xia

The Department of Physical Education, China University of Petroleum, Huadong, P.R. China

ABSTRACT: Realizing coordinated, complementary and win-win development in resources of mass sports and competitive sports in Shandong Province is important. Because this strategy is important to promote Shandong's great stride from large province to strong province of sports, and also the effective approach to realize high-efficient development of resources. This thesis makes a research on the current situation, complementary development theory and application and strategy of mass sports and competitive sports in Shandong Province, with the method of interview, questionnaire investigation, comparative analysis and mathematical statistics. The main conclusions are as follows: there are significant discrepancies in goals, policy, managing, finances, human resources and mass foundation between mass sports and competitive sports, which provides favorable conditions for complementary development of the two systems. And this kind of complementation may result in mutual promotion and development of the two systems, form a win-win situation, make both or at least one system upgrade the efficiency, which is called complementary effect and finally reach a new strategic level.

1 INTRODUCTION

Sport resources are the fundamental factors to determine competitiveness of sports and its sustained development. Sports competition is actually the competition of resources on human resources, stadiums, managing and fund investment. Shandong is a large province of its population, economy and sports, and the successful holding of Olympic Games in 2008 in China and the 11th National Games of the People's Republic of China in Shandong brought unprecedented development opportunities for physical education in Shandong Province. Shandong has made unprecedented achievements in mass sport and competitive sport. The national sports movement has developed in all areas. The reform of sports is being deepened successively and the service system of mass sports has being perfected in Shandong Province, which the aim is to strengthen our people's physiques promptly. On the other hand, the level of competitive sport has been raised and unprecedented achievements have been made in the process of improving the management mechanism of competitive sport, strengthening the team construction for competitive sport and improving logistics support of top-level athletes. In the 11th National Games of the People's Republic of China, Shandong had 63 gold medals, 44 silver medals and 46 bronze medals, the total number reached 153 and ranked first in China.

However, Shandong has a large population, the development of mass sports is delayed, sport infrastructure is still weak and the good atmosphere that all society show concern for and give support to the development of physical education has not been shaped. Therefore, comparing with the developed province there is still big disparity in sports economic development. All these problem affect and limit the development of competitive sport in Shandong Province. This thesis focuses on the coordinate development of mass sport and competitive sport in Shandong and the targets of common improvement, probes deeply into strategy of coordinated and complementary development in both mass sport and competitive sport and provides theoretical support for realizing of Shandong's great stride from large province to strong province of sports. Supported by "the Fundamental Research Funds for the Central Universities".

2 METHODOLOGIES

2.1 *Objects*

The main research objects are human resources, stadium resources and financial investment of mass sport and competitive sport in Shandong Province.

2.2 *Method*

2.2.1 *Interview*

Interviews by the unstructured method with 30 experts and scholars have been taken and the content of the interviews mainly focuses on the following three aspects: 1. Fundamentals and feasibility of the complementary development of mass sport and competitive sport; 2. Approaches and breakthroughs to the sustainable development of resources of mass sport; 3. Developing model, problems and breakthroughs in competitive sport.

2.2.2 *Questionnaire investigation*

1. More than 100 experts and scholars such as managers, coaches, and instructors of social physical education accepted the questionnaires, which mainly concerns with the present situation of resources and strategy of coordinated development. The results are shown in Table 1.
2. Validity of the questionnaire. In order to ensure the effectiveness of the questionnaire, varieties of experts and scholars examined and verified the questionnaire. The initial draft was sent to three experts respectively and was modified according to experts' examination and comments. After modification, the questionnaire would get the acceptability among experts. 90% of the experts think it conforms to the research goal, the expression of the language is quite clear and terms are quite appropriate. And 10% of the experts claim that the questionnaire basically meets the needs of the investigation. All experts agree that the questionnaire conforms to the research goal.
3. Questionnaire adopted the verification methods of test-retest reliability. During one month of investigation, 50 respondents were retested and the test-retest reliability was 0.81, which indicated the reliability of the questionnaire was very high.

2.2.3 *Comparative analysis*

Comparative analysis with actual data was made and it mainly focuses on internal track, extrinsic contradiction and inherent relations of complementary development in mass sport and competitive sport.

2.2.4 *Mathematical statistics*

This thesis uses statistics statistical on computer with statistical software such as EXCEL and SPS

Table 1. Questionnaire and retrieved qualified questionnaire.

Research object	Managers	Social instructors	Coaches
Number of given questionnaire	20	40	40
Retrieved number	20	35	36
Response rate	100%	88%	90%
Qualified number	18	30	32
Effective rate	90%	86%	89%

and makes a synthetic analysis with the result of investigation and questionnaire.

3 RESULTS

3.1 *Complementary effect*

History and reality of mass sport and competitive sport development around the world indicate that resource in the developing of sports is not a simple type of resource. Instead, it requires that different types of resources play different roles in different levels. Each subsystem in the system of mass sport and competitive sport cannot exist in an isolate situation. There is a constant exchange of materials, technology, human resources and information between projects, which has a great influence on the establishment and development of all the projects.

The interaction may strengthen relationship among projects, open up the developmental space and bring more opportunities for improvement. Complementary effect is basically a kind of increase in quantity. It is a way to make the physical resources that were previously unprofitable by assembly and improve integrated efficiency by making the most of it. Complementary effect emphasizes on integrated efficiency more than individual efficiency. On one hand, combination between different individuals offers an opportunity for underutilized resources to obtain the full utilization, which means integrated efficiency is strengthened by enlarging the scope of using resources. On the other hand, combination between different individuals shares knowledge and information sufficiently, which means that the individual's ability of creating efficiency is improved by improving the utility of resources. We can see from Table 2 that

Table 2. Complementary basis of mass and competitive sport.

Differences	Type	
	Competitive sport	Mass sport
Goal	Gold medals and winning glory for their country	Fitness and recreation
Policy	National system	Social power
Managing	Independent management by project center	Union management by society and sports center
Finances	Sufficient	Insufficient
Human resources	Sufficient	Insufficient
Mass foundation	Bad	Good

there are significant discrepancies in goals, policy, managing, finances, human resources and mass foundation between mass sport and competitive sport. Whether these discrepancies can make the use of complementary principles maximize or not, the affection of resources and interest is the most economical and direct approach in the present developing process. Complementation can result in mutual promotion and development of the two systems and form a win-win situation. However, the final goal of complementation is to lay foundation for coordinated development, make two or at least one system upgrade the efficiency called complementary effect, and finally it can reach a new strategic level.

3.2 The current situation of competitive and mass sports in Shandong

Construction of sports talents will exert great impact on the development of sports undertakings in our province, and plays an important role in quickening it in all-around way, meeting people's common interests increase the sports productivity and strengthen that human resources are the first resource, and pay great attention to the impact of sports undertakings by science, education and construction of sports talents. Till 2009, there were 2462 sports coaches in our province, tending to gradually grow in quantity, compared with that of 2005, 2006, 2007, and 2008. There are only 49 national coaches, accounting for 2% of the total number; 17% are 419 advanced coaches; 37% are 915 intermediate coaches, and 56% are 1079 primary coaches. The percentage of the national coaches was much lower while the primary ones occupied over half in total. According to the fact, it is concluded that the coaches in our province are worse in education background, teaching experience and qualification and so on, therefore we need to adopt a variety of measures to strengthen the cultivation of sports talents, issue and carry out relevant policy measures. Till 2009, there were 3173 registered athletes in Shandong, which the proportion of the international outstanding athletes, national outstanding athletes, Level 1 athletes and Level 2 athlete was 1:7.6:11.5:55.5. On the whole, the number of high-level athletes was transparently more than the period from 2005 to 2008, while athletes of Level 2 were less than the last years. Comparing with other provinces, the merits of the sports talents on the reserve of the competitive sports in Shandong are the adequacy of basic sports talents and its large quantity.

However, we cannot neglect that probability of success of the basic sport talents is low and the lack of high-level athletes is also existed. According to Figure 3, the use ratio and success rate of the

reserved athletes for competitive sports are increasing. To realize the integration in sports talents by getting familiar with the current situation of the sports talents, and making scientific developing goals and strategies on the basis of that, is helpful. In order to share resources and great useful of the sports resources, matching, the idea of "large and strength" should reserved talents and competitive level in sports. At present, the total number of the people engaged in sports has reached 35,370, occupying more than 38% of the population in Shandong, showing another improvement on people's healthy quality. Besides, popular fitness organization is increasingly perfect. There are more than 3,000 sports associations of various levels, over 80,000 social sports instructors and 30,000 activity sections. We can get the conclusion that in the sports resources for the public are extraordinarily sufficient in Shandong. People show great recognition and passion to sport exercise. There will be a much stronger sports system if we make complementation of the two systems.

There are 47,379 gyms in Shandong Province, covering 109 million square meters with gross size as 5,480,000 square meters and total area as 101,970,000 square meters with the expense of 98.1 billion in total. Gyms for education take up a large part in all the gyms as 31,968 in total, accounting for 30.1%. But only in 6 cities the sports space per person reaches the national average number 1.03 square meters of the 17 cities, taking up 35.3%

Table 3. Current situations of the athletes in Shandong Province.

The level	2005	2006	2007	2008	2009
Sum	4286	4596	4073	3299	3173
International outstanding athletes	1	20	26	32	42
Outstanding athletes	80	39	45	99	318
Level 1 athletes	335	583	570	569	483
Level 2 athletes	3870	3954	3432	2639	2330

Table 4. Current situation of the coaches in Shandong Province.

The level	2005	2006	2007	2008	2009
Sum	1744	1298	1725	1820	2462
National coaches	18	30	32	41	49
Advanced coaches	272	224	187	213	419
Intermediate coaches	773	396	349	435	915
Primary coaches	681	648	1157	1131	1079

in all the cities across the province. While other 11 cities is less than 1 square meter, especially in western of Shandong, such as the four cites Jining, Linyi, Heze, and Liaocheng that have great population. The average sports space in the four cities is only between 0.06 and 0.16 square meters, only reaching the level in the 1970s, it shows that the infrastructure of gyms in our province is poor and weak. The lack of the gym resources is the major factor that limits the realization of the mass sports plans in our province. Only with the complementation of the sports resources, which needs to readjust the open gyms in sports, educational and mass systems, the equipments and infrastructure, we can meet the primary need in the most effective manner.

The sports lotteries in our country increase steady. During the period of "Eleventh Five-Year Plan", is 24.28 billion in total, getting 72.8 billion welfare funds. In order to increase the input of the mass sports, 60% of the welfare fund of the sports lotteries is applied to popular fitness business. Meanwhile, more money is raised from every part in society and financing channels are broadened, so that donation and sponsor are obtained, encouraging the private enterprise to invest the fitness business and promoting to socialize sports business for the form and diversified structure. The funds of the competitive sports in our province are mainly from the government, which satisfies the competitive teams in city and provincial levels, but the funds for the adolescent sports activities cannot be guaranteed, in spite the investment for the adolescent sports activities has been included as one of the annual budget in sports departments of all levels. However, with the development of the sports business, financing channels should be broadened and more investing power should be led to take part in the competitive sports, in order to get more funds for the cultivation of the reserved talents for competitive sports. Every part in society should be engaged in the development in the competitive sports through more channels and multiforms.

3.3 Strategic study of complementary development of the competitive sports

The competitive sports strategy, as a national policy in our country, supports and strengthens which have weak base of the people, should have higher competitive cost and less limit in the cooperation. However, these competitive activities, with its weak base for development and rely completely on the support of the government, will be difficult to get breakthrough. Once the support from the competitive policy of the government decreases, these projects will fall in parts.

With the development of the mass sports, such difficulty will be solved from various aspects by the improvement of people's recognition and knowledge of the various competitive sports, to attract more people engaged in the fitness team, enlarge the competitive effects and promote the development mode. Merits of the high speed of the development of the mass sports will complement to the competitive projects in order to match the pace of the form of the market economy, and strengthen the base for development. In addition, mass sport is the main resources of the reserved athletes.

In order to attract more competitive business, the base of the mass sports should be ensured because only with the increase of the number of people engaged in the competitive activities can use great development of the cause. In 21st century, popular fitness sports have been promoted throughout the country and the population of the sports activities is increasing. The sports fields have enlarged and its cooperation areas are broadened from large and medium-sized cities to towns and countryside. But some factors limit the development of the mass sports, such as systems, the costs, area and human resources, and so on, slowing down the speed of its development. Compared with the competitive sports, in order to complete the construction of the mass sports in our country, mass sport must combined with the competitive sports in any aspects, such as the scientific results, human resources of high level, perfect infrastructure, and organization of competition, and so on, so that mass sports will be served speedily by such mature and outstanding results and the sustainable, coordinated and win-win development will be realized.

Mutual complementary effect refers to the mutual increase in common benefit and overall effect of both sides through cooperation and combination. Mutual complementation and common get progress in terms of policy, manpower, material and finance of competitive sports and mass sports. By judging from the emphasis and support point of policies, competitive sports and mass sports cooperate and develop in complementation.

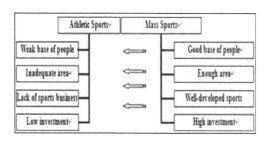

Figure 1. Analysis of the complementation from mass to competitive sports.

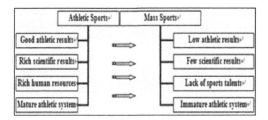

Figure 2. Analysis of the complementation from competitive to mass sports.

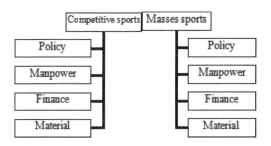

Figure 3. Analysis of mutual complementary effect between roller skating and skating.

The national institutes, social systems and public foundation could complement and adjust each other, increase consensus of different departments and play its due role, so that various sports resources may be fully integrated to make sports policies produce due effect.

Complementary system of human resources is divided into four parts—athletes, instructors, referees and scientific researchers. Exchange between different groups of human resources will promote the fast development of masses sports and competitive sports, and enhance sharing of human resources beneficial to develop and exploit limited resources to the largest extent, while improving their quantity of two events on all sides. Funding of competitive sports in our country mainly comes from investment of different levels of governments. But because masses sports rely on a part of sports, lottery revenue funded by the government and investment from society, enterprises and individuals.

Social involvement and investment have become an important resource for the development of mass sports in China. Except lottery revenue of governmental finance, the development of masses sports in our country is not funded by any special capital investment with most relying on social enterprises, groups, schools and individuals. The national investments of competitive sports mainly include establishment and operation of athlete groups, training of athletes, competitions and rewards.

The operating pattern of competitive sports is consistent with other governmental sections or public institutions, with its managers, logistical personnel, coaches and athletes managed and leaded by functional government departments as a part of national public institutions. As index of material resources includes sports courts and equipment two parts, competitive sports and masses sports could complement each other and develop synergistically in this aspect.

4 CONCLUSIONS

1. The significant discrepancies are existed between mass sports and competitive sports in goals, policy, managing, finances, human resources and mass foundation, which provide favorable conditions for complementary development of the two systems. And this kind of complementation may result in mutual promotion and development of the two systems, form a win-win situation, make both or at least one system upgrade the efficiency, which is called complementary effect and finally reach a new strategic level.
2. Human resources, stadiums and financial investment in resources of mass sports and competitive sports in Shandong have improved considerably. However, there are still many problems. Reserving of potential professional decreases, the investment for young athletes is limited, the facilities construction of sports fields is inadequate, and investment which from the society is insufficient. And at present all these problems need to be resolved.
3. Mass sports will complement competitive sports in four aspects of sports population, regional, investment of sports industry and sports cost. Meanwhile, competitive sport will also complement mass sport in four aspects of competitive achievement, scientific researches, human resources and competition. As a result, the win-win situation of resource sharing of the two systems will be formed, which speeds up the sustainable and effective development step of sports in Shandong.

REFERENCES

Bao Zhen Wang & Shui Jun Zhang. 2010. Actual example study of mass sports resource of Shandong province from the angle of national fitness constitution. *Journal of Shandong Institute of Physical Education and Sports*, (7): 17–21.

Feng Yan Luan. 2006. A study of the "Eleventh Five-Year Plan" of sports business in Shandong province. *Master Thesis, Beijing Normal University.*

Guo Qing Zhang & Peng Yu & Li Ping. 2011. Research on competitive sports resource conformity theory and practice of Changzhutan during the construction of Two-oriented society. *Bullentin of Sport Science and Technology*,19 (10): 4–5.

Guang Feng Yuan. 2003. The study of the status and development and the configuration countermeasures of Shandong urban area Sport resources. *Capital Institute of Physical Education*, 15(4): 25–27.

Jian Dai. 2007. Research on the integration model of competitive sports resources in Yangtze delta. *China Sport Science*, 27(6): 3–7.

Jie Dong & Shui Jun Zhang. 2006. Investigation and analysis of popular sports resources in Shandong province. *Journal of Shanghai University of Sports*, (5); 20–23.

Jun Qiu. 2012. "Strengthening the government" develops competitive sports "strengthening the society" develops mass sports. *Sport Science Research*, 33(5): 5–7.

Ling Hua Ran & Yu Shan Tian. 2007. Synergetic outlook of social sports resources under pan-resource background. *Journal of Shanghai University of Sports*, (2): 1–5.

Peng Zhang & Huang Lin. 2006. Metrological analysis of effective allocation of competitive sports resources. *Journal of Shanghai University of Sport*, 30(2): 25–29.

Shu Shan Zhang & Xu Jian & Liu Li qing & Han Fei. 2005. Study on optimization of sports resource distribution in transition society in Dalian. *Journal of Beijing Sport University*, 28(10): 1334–1336.

Shui-Sheng Pan & Dong Han et al. 2005. Research on the status quo of sports industry in Shandong province and its strategic objective for development and its countermeasures in 2010. *Shandong Institute of Physical Education and Sports*, 21(3): 19–21.

Tong Yan Li & Zhao Yun Hong. 2009. Tentative research on the development of competitive sports resources. *Journal of Chengdu Sport University*, 35(6): 14–15.

Wei Jun Liu & Ma Xiao Li & Shao Bin. 2007. Research on the inner impetus of interaction between competitive sports and mass media in the *process* of professionalization. *Journal of Beijing Sport University*, 30(7): 897–898.

Yan Fang Geng. 2009. The study of the impact on the sustainable development of competitive sports in Shandong province laid by the eleventh national games. *Master Thesis, Qufu Normal University*.

Zai Ning Zhang. 2009. Combination and optimization of the sports resource for high schools. *Journal of Sports and Science*, 30(6): 80–83.

Zhen Li. 2009. Investigation on status quo of Shandong sports facility resources and assumption on the integrated reform of the operations. *Master Thesis, Beijing Normal University*.

Sports Engineering and Computer Science – Luo (Ed.)
© 2015 Taylor & Francis Group, London, ISBN 978-1-138-02650-6

Theoretical construction of synergetic development between competitive and mass sports

JiaoYang Xia

The Department of Physical Education, China University of Petroleum, Huadong, P.R. China

ABSTRACT: "1+1 > 2" is the advanced subject in the field of China's physical theories and practice. According to it, the author introduces the harmonious development theory to these two systems. As a result, using it can make an overall plan and instruct the development of the physical system. Actually, the purpose is to make competitive sports and mass sports develop efficiency, optimization and sustainable. This construction will promote competition, corporation, distribution and complementation between the two systems, building the systematical, organized and high-efficient cooperating system by readjusting the limited resources in order to realize the promotion of the competitive and mass sports. There are three aspects stated in the paper: 1) synergy and synergy effects; 2) competitive and mass sports in the theoretical construction of resource allocation, management, information, organization and environment; 3) the four principles of the synergy development in competitive and mass sports: win-win principle, complementary principle, principle of shared resources and unity in diversity.

1 INTRODUCTION

Synergistic effect, as the result of synergy, refers to the overall or collective effect due to the interaction between subsystems in open and complex systems (Introduction to Synergetic). Under certain circumstances, it can serve as a power to organize, coordinate and create value and competitive advantage. These subsystems and related factors constitute the basic structures of competitive martial art, which are in close relation, collaborating with and influencing each other.[1]

With social sports resources being the basis, condition, content and means for the development of social sports, its quality, quantity, and efficiency become key factors affecting the sports function of society and, to a considerable extent, restrict the scale and speed of sports development in a well-off society in an all-round way. Therefore, it is necessary to pay attention to the synergistic effect of human brain, information and sense of identity, which is the only way to put the unordered sports ability into order.[3] With the development of science, the progress of science and technology and their continuous permeation in such fields as competitive sports, modern scientific and technological achievements are widely applied in competitive sports training, competition, decision-making, management, scientific research and other related ranges, making competitive sports a systematic event.[4]

2 METHODOLOGIES

2.1 Expert interview

In order to further understand the development of competitive and mass sports, the author has designed an outline for the interview about training, contests, coaches, athletes and other aspects in this study and had interviews by visiting many excellent experts and scholars in our country.

2.2 Questionnaire survey

In this subject, questionnaires survey has been designed for experts as well as for managers in sports management center and personnel in social sports guidance center of provincial and municipal levels.

2.3 Mathematical statistics method

The processing of data is mainly managed through Excel and SPSS statistical software for collection and analysis on the computer. By using SPSS16.0, expert opinions are chosen with Principal Component Analysis to have the R factor Analysis, in which factors whose given values are greater than 1 will be intercepted so as to facilitate analysis for the intercepted factor. After the factors loading with matrix have orthogonal rotation with the biggest variance, polarization will occur to each load of the factors when distances between the sizes of the absolute value become greater. In this way, the meaning of factors can be better explained. According to the calculating results, in the principle of maximum and similar properties, indexes are to be classified correspondingly and factors to be extracted and named for explanation.

3 RESULTS

Hermann Haken, with his profound insight, predicted that a new intersect discipline was to be born. In 1969, he declared the birth of "Synergetics", a new subject about synergy in his teaching in University of Stuttgart. He organized the international "Synergetics Conference", and experts from the world attended the conference on invitation. Attendees reached an agreement on fundamentals common in many phenomena in different fields and the same type of order parameter equation dominating different systems, which showed that Synergetics had got international recognition. Haken defined Synergetics as a learning about "interaction between different parts in various systems, forming new structural features that have never existed in the whole system on the original micro level".

The theory of Synergetics has been put into wide use in physical science, economies and other subjects. Since Igor Ansoff put forward the definition for Synergetics that "1+1 > 2" for the first time and established its economical connotation, the concept of Synergetics has served as the guiding principle in researching many issues of the theoretical circle and business community. The concept of Synergetics is a means to realize the overall effect of all kinds of resources. It regards the social sports activities as a whole or a system, form reasonable collocation and scientific use of various resources through synergy between resources including different organizations, different types and different time and space, thus making an organic combination of the tangible and intangible resources, ontology sports resources and available sports resources. This will turn the chaotic openness and utilization of a single resource into orderly utilization of multiple resources, breaking the traditional fragmented and independent resource management model.[2] Synergy in this essay means to make tangible and intangible resources of the event into an organic combination and rational collection, form their rational competition, utilization, allocation and complement and reduce wastes caused by independent operation and management. In this way, the limited resource, after integration, can be used to establish systematic, organized and efficient operating system so as to form a pattern where events can develop jointly.

Development is a philosophical terms, refers to the process where things change from the small to the large, from simple to complex, from junior level to senior level, during which old materials evolve into new materials. What causes things to develop is the general association of things while the origin of development can be attributed to their internal contradictions, namely internal cause. The aim of pursuing improvement lies mainly in improving the function of promoting harmonious development and the value of existence. Development in this paper mainly refers to the process where the overall level of the event develops from the weak to the strong, the operating systems improve from disorder to order and self-organization, and the existing operating and managing systems are replaced or optimized by better ones. By doing this, the competitive position of China's sports in the world can be further enhanced and form higher value of development.

Synergistic effect, as the result of synergy, refers to the overall or collective effect due to the interaction between subsystems in open and complex systems (Introduction to Synergetics). Under certain circumstances, it can serve as a power to organize, coordinate and create value and competitive advantage. The synergistic effect produced by synergy of competitive and mass sports consists of five aspects—management synergy, information synergy, resource synergy, organization synergy, and environmental synergy, while the synergistic principles supporting the five aspects are the win-win principle, complementary principle, the principle of resource sharing and the principle of seeking common ground while putting aside differences. Synergistic effect must conform to the support and guidance of the principle of synergy. Only on this basis, synergy, integration, competition and complementary effect developed among events can fully embody the advantageous benefits of 1+1 > 2 in the synergistic strategy in sports.

Synergistic principles are the basis and necessary requirements in restricting and orientating the development of synergy. Only by meeting the standard of synergistic principle, the synergistic effect of the sports event can be achieved. Synergistic principles mainly consist of the following aspects:

1. Principle of win-win. Only when the utilization of resources is to the benefit of all participants, synergistic relations of resources can be established. Although usually the revenue from that is not the maximum or the most optimal of the event, it is the best from the perspective of synergetic event group.
2. Principle of complementation. The synergistic parties have their own core competitiveness.

In the process of event management, their own weaknesses can be offset with partner's advantages so as to achieve the overall optimum to complement each other.

3. Principle of resource sharing. In the process of synergy, all of the participants communicate and share their own resources, information, organization and management with each other. Also they share the manpower, material resources, financial resources and information resources they have granted so that these resources can be made best use of.

4. Principle of seeking common ground while putting aside differences. There are common interests and conflicts among event principals or small events in the operating and managing system in the development of synergy. Under this circumstance, the paradoxical movement of seeking common ground while putting aside differences and putting individual conflicts toward synergy will be the most basic power in the development.

3.1 *Synergistic effect in resource allocation*

The allocation of resources mainly refers to the system constituted by people, property, goods and other elements of resources. Through reasonable choice of resources, they are rationally distributed to adapt to the requirement of systematic function of events, so as to realize the synergistic effect in resource allocation. The synergy of resource allocation refers to synergy in events and its configuration, mainly characterized by accommodation of elements and functions as well as reasonable combination and matching among elements; choosing

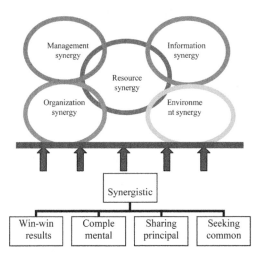

Figure 1. Synergistic effect and principles between competitive and mass sports.

the optimal portfolio to realize systematic functions and paying special attention to the rational use of human resources, so as to adapt to the requirement of development of event to achieve the synergistic effect.

Resource allocation consists of three levels— basal layer, application layer and target layer. Basal layer can be seen as the core of resource allocation. Only when this level of resource elements gets coordinated, complementary and optimized development, each content of the subsystem can achieve further optimization in resource and synergistic effect on operation of the system as a whole. The synergy of the application layer involves the synergetic development of many specific operating elements and implementing steps. It is the further development and application of the core elements, representing important practical significance. At the same time, the target layer is the ultimate goal of synergetic development for the system, so it also embodies the synergistic effect to get synergetic development in resource allocation of competitive and mass sports.

1. Basal layer. The basal layer of synergetic development of competitive and mass sports mainly includes four parts: human resource, material resource, financial resource, and information resource. Among these, human resource, material resource, financial resource belong to physical resource while information resource belongs to knowledge resource.

   Judging from the synergetic features of competitive and mass sports, the synergy of resource of basal layer is the most obvious and direct way and the core ability for the two events to develop. Meanwhile, the synergy of resource of basal layer plays a vital role in the development of the whole subsystems and even the synergetic system. It also restricts and confines the synergy of resource of application layer. Without the synergetic development of basal layer, real synergetic development of application layer will be impossible; neither is the synergistic effect for the whole system will. As a result, synergetic development will remain strategy on paper instead of in reality.

2. Application layer. Synergetic application layer of sports resources is, based on the competitive advantage of competitive sports, an organic combination of complementary synergy of resources such as various elements from different environment, aspects, subsystems, effective operating mechanism and so on. It is an organic combination of different organization system, technology system, management system and monitoring system. The synergetic application of competitive and mass sports should present a

dynamic disequilibrium state. With the dynamic change of external environment and the goal of internal members of the sports, the effect of application of resource synergy must also continue to accumulate, cultivate, develop and apply. In order to control and maintain influence of resource optimization and sharing of each event, some positive regulating measures in the organization, management and technical aspects of the event has to be taken, to ensure the order and harmony of internal resource utilization of competitive sports and to the ability of resource utilization.

3. Target layer. To form the optimizing effect of resource synergy is the ultimate goal of all levels of resource allocation synergy. When the operation and synergy of each system and aspect reaches the optimal state, the synergetic effect of competitive and mass sports will be achieved and the benefit will be increased.

3.2 Synergistic effect in management

Management synergy refers to that in the state of movement or crisis, the system, guided by the concept of synergy, promotes internal subsystems or elements to integrate synergistically with comprehensive means of management, such as interaction, cooperation and competition to achieve the optimization, complementation and synergy. So as to generate the order parameter for dominating the whole system, enabling it to realize self-organization, where the system goes to a new order state. The target of synergistic effect in management of competitive and mass sports is that the synergetic system of two events generates a new system that exceeds the original one in term of structure and function, making synergetic

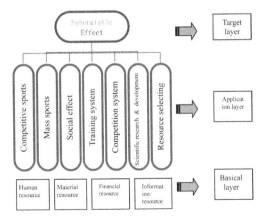

Figure 2. Synergetic development between competitive and mass sports.

development a highly adapting self-organization so that their individual functions and effects will be greater than the sum of the elemental forces as a whole. Synergistic effect in management is characterized with the following features.

1. Goal. The ultimate goal for management synergy is the synergistic effect of "1+1 > 2", whose essential requirement is to achieve the complementary advantages to aggregate and multiply functions. To reach this goal, the elements of the managing system of competitive and mass sports need to interact, coordinate and synchronize according to certain way of synergy and produce the order parameter dominating the development of the system. Then it will control the system to evolve in an orderly way, making the overall function grow to be the strongest to produce synergistic effect.

2. Nonlinearity. To achieve synergistic effect from management synergy, elements of the system should be strongly connected. It is not the simple sum of individual functions but the synergetic operation among elements and the optimal performance of the whole functions. The interaction and function organizing and operating the subsystems and elements are complicated and nonlinear, showing various hierarchy and causality of crossing.

3. Optimization. To achieve synergistic effect from management synergy of competitive and mass sports, all kinds of managing elements need to be integrated and different means be comprehensively used to promote the internal subsystems of the sports to interact, cooperate and coordinate with each other, achieving coherence and complementation. This embodies the concept of optimization. Besides, through the utilization of different ways and means, it can make full use of the advantages of each managing element to realize the ultimate goal of overall advantage and optimization.

4. Interaction. The purpose of management synergy in competitive and mass sports is to get the synergistic effect of management. However, it is difficult to realize with the single subsystem or element. Cooperation and interaction of various managing element are needed to enhance the effect of elements for the sake of overall development. This interaction can obviously produce nonlinear function thus achieving systematic effect.

5. Synchronism. The synchronism in management synergy stress that in the process of synergy, the cooperation of subject should be synchronized in term of time and space. Synchronism in term of time requires the decision-making system be closely connected and conform to the common

timing standard, while synchronism in term of space demands that the coordination among the system of decision-making subjects should make regular synergy and cooperation, just like the way the complicated traffic junctions direct the crossing vehicle to advance, halt or make a turn through the instruction of traffic lights and points men so that the traffic efficiency can be maximized and optimized. The synchronism in management synergy overcomes the discordance in the subsystems and elements, enabling the system to form a coordinated overall movement and turning it from disorder to order.[5]

Synergic effect in management, the organic integration of Synergetics and Management, is the synthesis of all activities and processes for the organization to achieve its goal in the changing environment by integrating various resources, coordinating the matching relationship among personnel, organization and links to produce synergistic effect and realizing the optimization.

Compared with the five functions, namely, programming, organization, personnel, direction and regulation in traditional management, synergetic management places more emphasis on planning and decision-making, organization and self-organization, human resources, guidance and prevention. With the updating and accumulation of technologies, projects and their management are growing widely used in all kinds of social organizations.

In the real world, synergy of project management is attempted to be made as the prior working means in more and more sports events. On account of competitive sports events, a complex, open and dynamic system, the success of projects must be built on the foundation of management, which means that when the management is accumulated much enough, it can be used, through the system of synergetic management, to settle problems encountered in events and coordinate relations among event members so as to make them work in synergy.[6]

3.3 Synergic effect in information

Information synergy is the foundation and technical support for sports system to make decision normally. To realize the high efficiency of two systems of synergetic development, the communication and exchange of information should be relied on much more.[7] In the process of management, organization and decision-making, information is changing greatly and there is enormous information in the world. The demand for spatial distance and agility calls for more timely information and greater synergy, especially in nowadays of information times. Information synergy is not only to realize the information sharing technically, but

to ensure each part of the operating system can pass the right information at the right time in the right place with the right means to the right systematic elements in terms of systematic operation. Information synergy of competitive and mass sports, with the means of information integration, realizes the organization and management of information to make related multi-information merged organically and used optimally.

3.4 Synergic effect in organization

The improvement of organizing ability is a long-term systematic project and an important supporting point for achieving the strategy of organization and development. Synergetic effect in organization puts emphasis on the establishment of complete and reasonable organizing regime, and structure on the basis of synergy in resource allocation, so that laterally the same level of sub-projects and activities of the organization. It can coordinate and cooperate with each other and vertically different levels of sub-projects and activities can combine organically. In this way, smoothness and harmony among different systems of the project and communication between both lateral and vertical information stream can be ensured.

3.5 Synergic effect in environment

3.5.1 Internal environmental synergy
Internal environmental synergy refers to the synergy of elements inside the system such as cultural atmosphere, internal policy, working attitude and operating model. It refers to the appeal and cohesion of the event to its members and stands for a greater representation of synergy. Internal environmental synergy is an invisible power, possessing the capability of promoting the sense and attitude of cooperation in members to complement each other and learn to cooperate. This synergy can make the

Figure 3. Diagram of organizing system of sports project.

whole project unite organically, cooperate with each other and fight together to effectively display the general power.

3.5.2 *External environmental synergy*

Internal environmental synergy refers to the adaption of the system to the outside world. Only when the event keeps harmony with the outside world then the outside prevention will be realized Internal environmental synergy puts emphasis on the strong adaptive mechanism of the system to maintain and improve the adaption brought by external stimulus or the change of external environment. It also includes the rationality of the external environment itself such as international political changes, the flexibility of domestic policies, competitive fairness, the integrity of finance and social organizing systems. Internal environmental synergy has certain effects on the production of synergetic effects of the system. The operation of the system of competitive sports events not only should achieve synergistic effect through their synergetic development, but also requires better adaption to external environment to give full play to synergetic effect and get greater benefit.

4 CONCLUSIONS

1. Synergistic development refers to the developing direction to regulate and instruct two kinds of sports. It is an organic combination and reasonable collection of tangible and intangible resource, forming the reasonable competition, application, allocation and complementation. It enables limited resource, after integration, to be built into systematic, organizing and efficient operating system so as to realize that the overall level of the two sports turn from the weak to the strong and the operating system from disorder to order and self-organization.
2. The synergistic effect produced by synergetic development between competitive and mass sports mainly consists of five aspects: management synergy, information synergy, resource synergy, organizing synergy and environment synergy, while the synergetic principles supporting the five aspects are principle of win-win results, principle of complementation, principle of resource sharing and principle of seeking common ground while putting aside difference.
3. Compared with personnel working on mass sports, personnel working on competitive sports are more familiar with the developing prospect of the event and pay more attention to the development of the event itself. Therefore, people working on mass sports can learn from professional skills in terms of training ways and techniques.

REFERENCES

[1] Guo Wang & Wei Hua Yao. 2004. On the development of athletics wushu from coordinate. *Wushu Science*, (04): 9–12.
[2] Hua Ran Ling & Yu Pu Tian. 2007. Synergetic outlook of social sports resources under pan-resource background. *Journal of Shanghai University of Sport*, 32(2): 1–5.
[3] Xiao Zhu Jing. 1989. 《Synergetics》 and modern training science. *Fujian Sports Science*, (11): 10–13.
[4] Yong Shang Wang. 1998. The synergetic effect of competitive sports. Journal of Tianjin University of Sport, 13(04): 79–80.
[5] Ming Guang Liu. 2006. Research on the method of collaborative optimization about the complex group decision-making system. *Tianjing University*, Ph.D thesis.
[6] Ling Li. 2004. A study of collaborative management in electronic commerce project. *Tianjing University*, Ph.D thesis.
[7] Zhi Yong Zhou. 2008. The research on synergetic capacities of corporate groups. *Dalian University of Technology*, Ph.D thesis.
[8] Chuan Min Wang. 2008. On the economic and industrial synergy development in the county territory. *Beijing Jiaotong University.*
[9] Run Bin Wang. 2009. Theoretical construction of sport operation mechanism. *Journal of Sports Adult Education*, 25(3): 23–25.
[10] Hui Ju Luo & Lu Xiao Cheng. 2010. Study on the building of the service chain theory of the sports industry in China. *Journal of Sports Adult Education*, 33(12): 30–33.

Sports Engineering and Computer Science – Luo (Ed.)
© 2015 Taylor & Francis Group, London, ISBN 978-1-138-02650-6

Strategies for the development of China's sports culture during the process of building a sports power

Hong-Chun Pu

Sports Department, Chengdu Sport University, Sichuan, China

ABSTRACT: Through the analysis of the deficiency and development potential of China's sports cultural strength, this paper shows that developing the sports cultural strength is the strategic choice to build the sports power, especially during China's progress from a sports power in size into a sports power in strength. From the view of the relationship between the sports cultural strength and the country's cultural soft power, it proposes that the strategies for the development of sports culture when building a sports power should follow the people-centered principles of scientific development, reforming and innovation and sustainable development. Besides, safeguard measures in aspects such as science and technology, human resources and laws and specific development strategies are also proposed for the strategies of the development of sports culture when building a sports power.

1 INTRODUCTION

It was proposed in the report of the 17th Party Congress to "improve the country's cultural soft power". And Premier Wen Jiabao pointed out in the government work report that "Culture changes the destiny of a nation" this year. All these are sufficient to show that the cultural soft power plays an important role in enhancing the overall national strength and developing the social economy, and China has put the development of cultural soft power as a new strategic "focus" for the great rejuvenation of China. At the same time, from the strategic view of national rejuvenation, President Hu Jintao has set the strategic objectives of sports development as "turning China into a sports power in strength from a sports power in size", specifying the direction and way for sports development in the new period. With its economization and globalization, sports have become a healthy lifestyle, an integral part of social culture, an important symbol of social progress and civilization and an important embodiment of a country's comprehensive strength and competitiveness; as it is involved in a country's economic and social development, thus affecting the country's political decision-making and policy development, sports culture has become a kind of national interest during the process of globalization.[1] In the process of strengthening national cultural soft power, choosing the sports culture development strategy with the development of sports cultural strength as the core has become the strategic, which need to build a sports power and enhance the national cultural soft power.

2 PROSPERITY OF SPORTS CULTURE AS AN IMPORTANT SYMBOL OF A SPORTS POWER IN STRENGTH

Sports culture is a broad concept that includes the dialectic unification of material culture, institutional culture, behavior cultural and spiritual culture at multiple levels, an organic integration of cultures of different levels and a cultural entity with core values of sports culture as the main content.

How to develop into a sports power in strength from a sports power in size? The answer to this question lies in the cultural strength of sports. The sports development strategy that "promotes sports by developing culture" will be the strategic choice to build the sports power in strength. Using sports as a way to promote Chinese culture in foreign countries can express the state will and demonstrate the international image of peace and progress while enhancing the international influence of China's sports, which specifically shows the ability of sports to improve the national cultural soft power. The sports development strategy that "promotes sports by developing culture" is necessary for sports development in China. Currently, a series of dissimilation phenomena of sports shows "the original cultural meaning of 'peaceful interaction' is lost for the influences of culture activity, of 'cups and medals mania' in the greater environment, such as the business and political environment of modern society". Thus, it is required that we should establish a competitive sports philosophy of "winning the medals and educating people" and national sports philosophy of "training physically healthy world citizens" from the view of the strategy of "promoting sports by developing culture", which is also the currently most important task of building the sports power in strength.[2]

3 CHINA'S LACK OF SPORTS CULTURAL STRENGTH

Mr. Samaranch once said before the Beijing Olympic Game that: "while China's strength in competitive sports has been enhanced rapidly, the impact of its sports culture is still relatively weak". Some scholars have also said that "as the soft power of national sports is not strong and the sports culture has little impacts, China lacks the voicing right in the global sports world".[3] The relatively weak influences and lack of "voicing right" are mainly manifested by China's disadvantageous status when participating in the international sports activities. For examples, the sports constitution and rules are developed by the Western sports powers, and China has no "right to define" the development of world sports culture.

China's lack of sports cultural strength is also manifested in the use of soft power resources of sports culture. Though we have advanced sports culture philosophy of "integrating the nature and human" and sports spirit that "as heaven revolves unceasingly, people should strive continuously for self-improvement", how to turn these soft power resources of sports culture into real soft power are the contents lacked in the current internal, and external publicity of sports culture. Similarly, China's lack of sports culture strength is also showed in the productivity of sports culture. The Chinese sports culture industry is still in a relatively backward state.

On the other hand, Beijing Olympic Games fully showed the cohesion strength contained in our nation's sports culture, making the national patriotic fervor and national pride enhanced. However, the internal service of sports culture still cannot meet the growing demands of public for sports culture. Facing the lack of sports culture strength, to meet the domestic demands for sports culture, develop the sports culture, have a long-term foothold in the world sports and "promote a more fair, healthy and prosperous sports world", we must improve our sports culture strength, rely on the strategy of "promoting sports by developing culture", and establish a new national sports education system in order to find a way out from the sports "cultural awareness".

4 DEVELOPMENT POTENTIAL OF CHINESE SPORTS CULTURE

Chinese sports culture, with profound and precious connotation and advanced development direction, owns the universal values, which can be integrated with the advanced sports culture in the world including the Olympic culture. For instance, the exercising philosophies represented by the Tai Chi Chuan, and its universe and world views of the harmony between man and nature as well as between man and man, is a useful supplement for the sports value of pursuing the extremity in Western cultures, and its extensive and profound theory and technical system fully demonstrates the unique creativity and outstanding achievements of China in the field of sports culture.

Similarly, the ideas of "bowing to competitors before competition, drinking together after competition and competing like gentlemen (The Analects of Confucius)" and "a perfect player owing benevolence, righteousness, propriety, wisdom, faith, goodwill towards others (Chart of Ancient Chinese Football)[4]" in Chinese traditional sports culture, containing rich sports culture ideologies, can be integrated with the Olympic spirit of "friendship, solidarity and fairness".

The diversified world development and globalization of culture provide opportunities for Chinese sports culture to show itself in the world. On the other hand, the world also calls Chinese sports culture to go out. To be a sports power in strength, China needs to develop its own sports culture. For the domestic aspect, citizens' physical health should be enhanced and the sports culture should be developed as the basis for building a sports power; for the external aspect, the soft power of sports should be strengthened and the international space for sports development should be created, which is the important tools to improve the national cultural soft power.

5 DEVELOPMENT STRATEGIES FOR SPORTS CULTURE DURING THE PROCESS OF BUILDING A SPORTS POWER

When developing the development strategies for sports culture to build a sports power, we should follow the principles that includes promoting the construction of sports cultural identity (the core values of sports culture) at spiritual level, motivating innovative development of sports systems at institutional level, and nurturing high-quality products of sports culture industry at material level. The specific content includes as follows.

5.1 *Constructing strategies centering on sports cultural values with Chinese characteristics*

The development strategies to build a sports power in strength should be established at the strategic level guaranteeing and expanding the national interest, including establishing development strategies and thinking mode with Chinese characteristics, and putting the strategy "promoting sports by developing culture" as the core. The core is to build universal values with "harmonious sports" as the main value goal and embody the national spirit centering on patriotism, and the spirit of era with reforming and innovation as the core while promoting the characteristic sports culture. During the integration of the traditional, modern, Chinese and foreign cultures, we should inherit the fine traditions, make use of our advantages and change the sports culture resource into the sports cultural strength to continuously improve the soft power of Chinese culture and enhance China's international influence and competitiveness.

5.2 Reforming the national sports management system to promote the development of sports culture

The rise of China's sports culture is the construction of the soft power of national sports and the promotion of development strategy, which for the construction of hard strength of national sports by establishing people-based sports culture and institutional culture. The construction of the soft power of national sports culture should be promoted through the innovation of sports management system, to lay the foundation for the construction of hard strength of national sports.

5.3 Developing public sports culture and cultivating a sports cultural atmosphere of universal exercising

A complete public sports service system is not only the basis to build a sports power in strength, but also an important criterion measuring the completeness of a country's social service system. We must strive to cultivate a sports cultural atmosphere of universal exercising, enhance the construction of public sports cultural infrastructure, coordinate the development of sports culture in urban and rural areas, support sports culture industry for public welfare, meet the growing public demands for sports culture and use the development of sports culture as a means to enhance the national cultural soft power.

5.4 Promoting traditional sports culture and creating national brands

China's national sports cultures constitute the diversified sports culture resources. The sports culture with ethnic characteristics is the main content and means of external publicity. It is the important measures for the development of sports culture to reasonably use, protect, inherit and study these cultural resources. Some non-material cultures are the living carrier and transmission of the traditional sports culture and an integral part of the splendid Chinese culture. It is the important strategic measures for the development of sports culture and the strengthening of national cultural soft power to effectively protect and inherit these cultural resources.

5.5 Promoting strategy of winning honors in the Olympic Games to enhance the international influence

Holding the major sports events represented by competitive sports can play a significant role in the promotion of national strength and the demonstration of external image. We should actively explore the potential, optimize the structure, improve the efficiency to promote the balanced development of all types of sports, and make use of the leading and radiation functions of competitive sports when building the sports power. At the same time, we should use the program "winning honors in the Olympic Games" to promote the implementation of universal exercising plan, make the promotion of the program "winning honors in the Olympic Games" as China's strategic objectives for the development of competitive sports and an important means to enhance the national image, improve the international influence and demonstrate the national cultural soft power.

5.6. Developing sports culture industries and exporting the sports culture

Sports industries are the material basis to enhance the whole sports cause and also the only way to change the mode of sports products. The development of sports industry is the core driving force to enhance sports cultural strength. Promote sports cultural industries by establishing sports culture development strategy of "promoting sports by developing culture". China's outstanding sports cultures represented by "Shaolin Temple" and "Tai Chi Chuan" have been influential abroad. The culture export needs diversified cultural symbols to enhance the country's cultural soft power. Therefore, it is another strategic measure for the construction of the sports power to carry out studies and promotion of sports culture, make domestic and foreign scholars understand and accept China's sports culture, and make good use of modern media technologies such as television, media and Internet to promote the spread and export of sports culture.

6 CONCLUSION

The strategy for the development of sports culture with the development of sports cultural strength as the core is the strategic choice for building a sports power in strength by promoting sports through culture development. It is also an important component of national cultural development strategy, which plays a great role in promoting the construction of national soft cultural strength while enhancing the national sports soft power. To become a sports power by promoting sports through culture development based on science and technology development, talent training and legal protection, it is required to adhere to the people-centered principles of scientific development, reforming and innovation and realize the sustainable development of sports culture as well as the sustainable development of people at the spiritual, institutional and material levels.

REFERENCES

[1] Tong Shao-gang. *Humanities and Sports-Sports and cultural interpretation.* China Customs Press, 2002.
[2] Research Group of Beijing University of Chinese soft power, *The practice of soft power in China—one of the concept of soft power*, People's Network, 2008.
[3] Sun Xi-lian, YU Xiao-hui, Mei Lin-qi, et al. *Taijiquan s international publicity and China s soft power promotion*, Journal of Wuhan Institute of Physical Education, 2008, 42(6): 72–75.
[4] Xie En-jie, Gao Song-shan. *The Modern Olympic Spirit and Chinese Sports Culture*, Journal of Capital Institute of Physical Education, 2003, 15(2): 11–13.

Sports Engineering and Computer Science – Luo (Ed.)
© 2015 Taylor & Francis Group, London, ISBN 978-1-138-02650-6

Functions of sports culture soft power in the transformation from a sports power in size to a sports power in strength

Hong-Chun Pu

Sports Department, Chengdu Sport University, Sichuan, China

ABSTRACT: Based on the literature review and logical analysis method, this thesis sets forth the source, connotation and relationship of sports power in size and sports power in strength, briefly discusses the connotation and structure of China's sports culture soft power, and summarizes its positive functions in transforming our country from a sports power in size to strength, so as to promote the pace of practical transformation into a sports power in strength.

1   INTRODUCTION

It has always been a goal of all Chinese people that our country transfers from a sports power in size to a sports power in strength. The success of Beijing Olympic Games makes Chinese people realize the dream we've had for a hundred years, and the unprecedented patriotic enthusiasm arouses the "power dream" of us again.

On September 29th, 2008, President Hu Jintao gave a speech in the Summing-up & Commendation Meeting of Beijing Olympic Games and Paralympic Games, pointing out that we should stick to the purpose of building up people's health and improving physical fitness and life quality of the whole nation, pay special attention and give full play to the important functions of sports in promoting people's all-round development and social and economic development, realize the coordinated development between competitive sports and amateur sports, and further promote our country's transformation from a sports power in size to a sports power in strength. [1] After the successful Beijing Olympic Games, the sports market of China is further expanded. More and more domestic and overseas sports industry experts start to consider about how to grasp the opportunity for development in post-Olympic economy and how to realize our dream of becoming a sports power in strength. Therefore, the sports culture soft power in China plays an important role on it.

2   CONNOTATION AND CONSTITUTION OF CHINA'S SPORTS CULTURE SOFT POWER

2.1   *Connotation of China's sports culture soft power*

Soft power refers to the integrated factors that affect development potential and international charisma of a country, such as culture, values, social system and so on, which collectively play a great role in improving and influencing of politics, economy, society and culture. Sports soft power refers to an attraction of thinking and culture reflected by physical activities or emerged in sports field of a country, a spiritual power reflected in code of conduct, values and political system, and a capacity that has an effect or influence on other countries.

Culture soft power refers to "the integrated power of a country formed on a certain production basis that influences the domestic citizens and other countries in the form of culture in a long period of time. It takes the value system as the core and mainly embodies itself in the form of attraction, cohesion, influence, selective and repulsive forces, and aims to achieve the intended goal by optional means." [2]

When peace, development and cooperation become the theme of the time, culture soft power takes an important part in comprehensive national strength and international competitiveness. According to the above analysis of three kinds of power, I think sports culture soft power is a culture type generated in specific environment, developing with the mutual influence, fusion, penetration and promotion of culture and sports. It is regarded as an important part of social sports culture.

2.2   *Constitution of China's sports culture soft power*

The constitution of sports culture is the inherent basis of keeping the sports culture system's integrity and great function in the development process, and it also determines the features of sports culture soft power.

After a comprehensive analysis of views pointed out by different researchers, I consider the constitution of sports culture soft power as a dissipative structure and open system composed of four main elements, which are sports material element, sports system element, sports action element and sports spirit element.

Sports material element is the basis of sports culture soft power and an important guarantee of sports practice. Sports system element is the key point, which plays a bond role among all elements. As the most authoritative element in this system, it formulates the whole character of sports culture soft power. Sports action element refers to the conventionalized action norms in sports practice, mainly reflected by the sports habits. Sports spirit element takes the initiative position and mainly includes sports thinking mode, sports aesthetic appreciation, sportsmanship, sports spirits and sports values, among which the sports values are the core of sports culture soft power.[3]

3 FUNCTIONS OF CHINA'S SPORTS CULTURE SOFT POWER IN THE TRANSFORMATION FROM A SPORTS POWER IN SIZE TO A SPORTS POWER IN STRENGTH

The core competitiveness of a country is reflected not only in the economy, but also in the soft power. It is quite important for us to improve the culture soft power to show our national image to other countries. In the same way, it is more important to develop our sports culture soft power on the basis of sports economy, so as to realize the purpose of transforming our country from a sports power in size to a sports power in strength. Our country was a sports power in size before, or more specifically, a competitive sports power in size, so it is necessary to make sports culture soft power play a part in order to turn China into a sports power in strength.

3.1 Economic function of China's sports culture soft power

In order to achieve the successful transformation from a sports power in size to a sports power in strength of China, economy is a necessary element that is regarded as a guarantee for improving national people's competitive sports level, physical fitness and athletic ability. As the superstructure, culture soft power promotes the development of social economy, raises working efficiency by improving labour force quality and mobilizing the initiative, promotes the development of enterprises by corporate culture, and directly turns them into economic effectiveness.[4]

From the experience of social and economic development, we can see that the national rise and fall and the transfer of world power's gravity centre are superficially determined by material matters, such as capital, resource and military power, but actually the fundamental factor is culture at spiritual level. Culture soft power can give a full play to its integrative function, carry out the people-oriented humanistic and spirit concern to satisfy people's spiritual demands, enrich their spiritual world, strengthen the spiritual force and promote their all-round development. With the help of improving every labour's comprehensive cultural quality and moral level, the level of civilization of the whole society will be improved. We should bring culture soft power's special functions of cohering, uniting, encouraging, guiding, educating and cheering into full play. We also need to mobilize labours' initiative of participating in economic production and stimulate their innovation energy to provide a strong spiritual driving force to economic development.

3.2 Political function of China's sports culture soft power

Political relations mainly include relations of class internal and class between, national and international relations and so on. Culture soft power can effectively promote the coordination of different political relations and the political stability.

A reassured state of people is the premise of political stability. If people are reassured, there is a political stability; if people are not reassured, there would not be the political stability. Therefore, both the formation and maintenance of the society's political stability cannot exist without the influence and guidance of culture. Facing a national crisis, culture soft power often can play a role of cohering and inspiring.

With the spirit of patriotism and common ideal, people can form up the cohesive force that unites them together with their cultural qualities and capability of settling problems improved, which make it possible to deal with the more complex democracy and function as a cultural guarantee of the development of democracy. Therefore, our country's sports culture soft power also plays an important role in the transformation from a sports power in size to a sports power in strength.

3.3 Diplomatic function of China's sports culture soft power

The main diplomatic function of culture soft power can be seen from the facts that a country's institutional innovation can attract imitators, a country's national culture can have emotional appeal to other countries and a country's diplomatic route, policy and operation mode have affinity. The correct diplomatic route and successful diplomatic operation of a country can promote the international prestige and enhance the international influence, so it will surely have an effect on the country's pursuit of transforming from a sports power in size into a sports power in strength.

3.4 Social function of China's sports culture soft power

The social function of culture soft power is mainly reflected in the promotion function of sustainable social development, which refers to the mutual adaptation of economic and social development with population, ecology, environment and resources. The view of sustainable development asks for sticking to the sustainability, coordination and fairness of development. It is required to correctly deal with the relationships among economic, social and ecological effectiveness, between immediate and long-term interests, and between the development of contemporary people and descendants.

Pay attention to the justice of development and distribution, and develop a way of civilized development with developed production, wealthy life and pleasant ecological environment. We definitely should not exchange for a temporary economic growth at the cost of violating environment and resources, or satisfy contemporary people's benefit at the cost of sacrificing that of descendants.[5]

Culture is a country's long-term accumulation and cohesion of national spirit and wisdom, and the precondition of the nation's existence, containing all genes of sustainable development in a country's process of development. This is the in-depth rationality of culture industry's promotion of social sustainable development. Besides, culture soft power can play a role in solving medical, employment problems and promoting people's all-round development.

REFERENCES

[1] Zhou Ai-guang. Exploration of the Connotations of "Big Country of Sports" and "Powerful Country of Sports". Journal of Physical Education. 2009, 16(11): 1–4.

[2] Liu Min-hang, Sun Qing-zhu, Fu Yu-kun. The Evolution from Sport Country to Sport Power Country. China Sport Science and Echnology. 2006, 42(3):33–36.

[3] Huang li. Investigation on the Compositions of Comprehensive Sports Power from the Connotation of a Strong Sports Country. Journal of Shanghai University of Sport. 2010, 34(4):15–20.

[4] Bao Ming-xiao. On Transition Strategy from A Major Sports Country to A Sports Giant Country. Journal of Nanjing Institute of Physical Education. 2009, 23(6):1–6.

[5] Zhang Hao. Research on the Counter measures in Composing the System of Sports Cultural Soft Power in Universities and its Development. Sport Science and Technology. 2006, 27(4):13–15.

Sports Engineering and Computer Science – Luo (Ed.)
© *2015 Taylor & Francis Group, London, ISBN 978-1-138-02650-6*

Research on the relationship between sports culture and the "new farmer" happiness in southern Jiangsu

BingShuai Wen

Military Sports Department, Suzhou Institute of Trade and Commerce, Jiangsu, China

ABSTRACT: Through the research on the relationship between sports culture and the "new farmer" happiness in southern Jiangsu, this paper aims to provide specific advice and suggestion about "new farmer" in agriculture modernization age. This paper defines the words "new farmer", the contents of "new farmer" happiness and the relationship of between sports culture and the happiness. The suggestions is as follows: construct the sports-culture project in southern Jiangsu from enhancing the sport culture contents, implementing the rural sports organization structure, founding farmers' sports associations, building rural sports-culture industry and establishing public sports sharing mechanisms and so on.

Keywords: "new farmer", sports culture, route, south Jiangsu

1 THE PRESENT SITUATION OF NEW RURAL SPORTS FITNESS IN SOUTHERN JIANGSU

1.1 *Insufficient sports culture connotation construction*

Overall, compared with the urban resident, the farmer in southern Jiangsu is backward at culture quality, thinking and living style. They often judge the new fashion and new things by conservative nationalism and geographical limitations. It's difficult for them to understand the theory of sports culture connotation, so the easier and effective way is making them understand the shallow theory by the intuitive outer expression. In the countryside, school sports have become the link of modern sports culture with "new farmer". The "new farmer" is trained with the sports skills in school education. But after graduation, they have few opportunities to be guided by experts, especially have no idea about the sports function, exercise and the science of health, so they is no clear requirements for the amount of exercise and exercise intensity. The propaganda of sports science and culture is almost blank in the countryside. Most farmers believe that agricultural working is sports activities, fitness can work in agricultural working. Which resulting in many farmers are not willing to participate in physical fitness activities.

1.2 *Villages and town's sports fitness institutions dummy or missing*

Each township in southern Jiangsu is the individual unit for the sports development. The ways of management are great differences between each town because of no laws and regulations specific norms.

The rural sports are managed by farmer associations or administrative. It is hard to guarantee at expenditure and the daily organizations of sport activities. There is no yearly plan and target for the activity, most activities are carried out by organizers' interesting. The topic problems are unclear responsibility ineffective organization. There is no dedicated organization and coordinators to manage the construction of farmer's sports fitness project. The farmers' sport fitness project assigned by the higher authorities is not implemented and managed, at the same time it is lack of supervision and tracking.

1.3 *Shortage of funds, lack of sports facilities*

The sports funds of southern Jiangsu rural are charged with the District, township (town) finances. Since there is no specific amount and uniform standards, the maintain funds was allocated by district, county and township, it has a lot of discretion. The key performance indicator of local government is economic development, political stable, farmer's sports activities are easy to be marginalized; it is hard to get the financial support. The constructions of rural sports culture often become the last priority. The development of sports culture is slow and stagnate. Sports culture in many towns is just a slogan, no positive actions.

2 EFFECT OF SPORT CULTURE TO PROMOTE THE "NEW FARMER" NATIONAL FITNESS CAMPAIGN

Sports culture that belongs to a branch of culture is inextricably linked with other cultures, is a part of culture and contains a unique identifier sports.

The other part is to integrate into the larger context of culture. Today most cited definition is: "sports culture, is a human created in physical activity and sports practice, and through the physical body shape, movement skills, sports equipment, material and intangible social attributes will, ideas, reflect the spirit of the times, revealing a distinctive mode of existence." [5]

In order to build a better relationship between sports culture and the "new farmer" fitness in south of Jiangsu, the sports culture is analyzed by the three layers which are very popular in academic research. So sports culture and happiness from the three levels of specific dock to implement, and thus enhance the new farmer's happiness in south of Jiangsu.

Start with outer layer of the sports culture analysis the effect of it enhances the new farmer's happiness in southern Jiangsu. Sports culture outer layer mainly refers to the physical level which refers to the material entity which is created and developed by human, including the sports structure only based the physical body, sports hardware facilities, sports products, sports events, sports intermediary, sports clubs and so on. It is the most directly physical media with the farmer. Farmer can directly involve in the sport by sports culture outer contents, let their body get a lot benefit by exercise, improve the opportunity to social communication and enhance mutual contact and team work spirit. At the same time make the farmer get a spiritual pleasure, a sense of accomplishment and the sensation of the mood. Sports also help training farmer brave and indomitable characters, surpass themselves, improving the ability of risk tolerance. It also will foster a healthy sense of competition, cooperation spirit and sense of fairness. Holding some amateur sports activities and sports events greatly enrich the farmers' cultural life.

The second level analysis is from the middle layer. The middle is a combination of spirit with physical, which is combination of natural with social development section, such as sports law, regulations, rules, sports organizations, sports organization system, public sports guidelines and so on. Sports culture's systematic protection, related sports regulations, especially releasing the legal documents about the farmer sports activity, the specific settings of administrative agencies for the rural sports, the mode of existence of rural sports association, the management rule system of the rural sports social community and so on. The contents of the middle layer of sports culture is the key policy and legal guarantee which ensure run rural sports activities orderly, configure reasonable physical hardware, ensure the professional sports funds allocation, develop rural sports organizations positive, legalization of sports institutions and so on.

The third is from inner layer. The inner layer is hard-core, is the spirit of sports culture, including sports values, sports ideology, aesthetic value of sports, sports philosophy, sports humanistic sprit, sports beliefs of different nation groups, sports ethics and so on, reflecting the human long-term production, life summary, the human pursuit of a better life humanities precipitation process of the sports culture of long-term accumulation. These are the crystallization of human excellence culture. Mainly by modern western pass into the modern sports local sports concept with traditional Chinese common components. In particular connotation of Chinese traditional sports culture and our national, religious and cultural traditions can be traced countryside, our traditional sports culture also influenced by China's traditional Confucianism, Taoism, and Buddhism. Therefore, the theory of the sports culture can enrich the cultural connotation of the new farmer; provide new supplies for the new rural culture, such as teamwork, self-reliance, good health, and Tai Chi effects. Including values, ways of thinking, aesthetic, moral, religious sentiment, and national character are the fundamental human existence and development. So pay attention to sports and cultural connotations of propaganda.

3 THE BUILDING OF THE "NEW FARMER" SPORTS CULTURE IN SOUTHERN JIANGSU

3.1 *Strengthen training and propaganda of sports culture in rural areas*

The government should vigorously promote the meaning of sports culture in rural areas. The implementation of sports culture project in south Jiangsu rural should be identified and accepted by the "new farmer". Let the "new farmer" to understand the connotation of sports culture and the implementation of the specific content. In accordance with the principle that "people-oriented, suit one's measures the conditions", closely integrated south of Jiangsu rural regional cultural characteristic. Through the township wind folk experience, local culture and other ways to enhance the education of rural youth sports culture and implementation of content recognition. Use the way of formal fresh, distinctive theme and content practical way of publicity, directly with farmer's ideas to reach a consensus effectively. Increase investment in media, such as through newspapers, television and radio to make the farmer sports culture's thematic posters, using mobile publicity vehicles, through the information media tools make farmer more easier to contact sports culture's details, the popularity of sports cultural science knowledge. At the same time, it also can be used the way the flow of propaganda, the National Physique Monitoring car into the countryside, experts for the physical condition of farmer's field pulse, the creation of sports and health lectures. And organize college sports majors to the rural as the sports instructors to popularize sports knowledge; develop the sports games of love to see and hear, rich village life, enhance the interaction; perfect social sports guidance function. For good practice models, establish long-term mechanism to improve science, adhere to the implementation of it.

3.2 Improve the physical construction of cultural organizations

National sports policy, sports authorities' provincial (municipal) level to deal with non-governmental organizations to strengthen the farmer's sports culture guide, to provide financial and material support as well as technical and policy. The administrative departments in Southern Jiangsu should start from the construction of a new socialist countryside, building harmonious and stable new rural districts to examine the great strategic significance of the construction of new countryside culture.

Clear the specialized cadres in charge of sports work and manage industry in the rural sports culture operation, annual distribution and management of the sports maintenance, management and supervision the village committee work on village peasant sport's operation. In the township level position setting, under the farmer sports culture management position, make its legalization, equipped with a full-time staff rational use the special sports funds which assigned to the village, carry out the physical activity and regular inspections, ensure the normal operation of rural sports and culture activities. Develop appraisal index, as the post assessment criteria. Strengthening village peasant sports organizations and backbone team construction, village Recreation Center for training and equipped with a part-time social sport instructors and managers, to ensure the normal development of the administrative villages in the National Physical Activity for village-level sports activities center.

3.3 Safeguard "new farmer's" Sports Association's independence and autonomy

Encourage farmer to the opening of a folk sports community in the village, township cultural departments to assist in the establishment of the articles of association, participation and management measures. The club has independent autonomy, membership in statute law, its nature is a non official, non-governmental organizations. [6]

Ensures that the farmer' cultural main body status in the construction of rural sports culture. The government plays a guiding and assisting role, do not involve in management. For example, set up by the farmer themselves village basketball team, old dance team, Tai Chi team association. The villagers choose suitable for the actual local sports concept, sports habits, and sports associations to participate in the sports cultural activities. The government culture department guidance, villagers to dominate, fully excavate the inherent characteristics of rural culture, sports culture as a way of life of the villagers to represent and promote the core values of farmer and rural life order, the sports culture is deeply rooted in the rural land.

3.4 Establishing and perfecting rural sports culture to a shared resource

To neighboring administrative villages to establish public sports venues, sports facilities and equipment hardware, such as establishing the natatorium, the public basketball court, the collective activities of the square, table tennis room, badminton hall and so on. Public resources format the town's administrative villages sharing mechanism. According to the villagers' actual sports activity, the adjacent village built public sports culture activity station. Make full use of the village primary school's sports venues, coordinate with the teaching time, and implement paid or unpaid open sharing mechanism. Meanwhile in the township build units which can get the planed welfare form the local government. The unit implements a membership system for their villagers. The villages can selective compensation for using the facilities, reference to Suzhou citizen card multi-function design patterns.

3.5 Building southern Jiangsu rural sports culture industry zone

Increasing investment in sports infrastructure such as the sports park, sports venues, the country activity room and other hardware facilities are the effective guarantee to carry out sports culture in rural. So the first problem we must consider is the construction of sports facilities. However there are a lot vivid examples we can find. Due to the lack of the long-term consideration for the design of the sports center, most of them are under deficit. The main focus issues we must plan before action are the construction of Sports Park, stadiums function of practical applications, the diversity of investment way, the scientific nature of the business model and the careful management. Building the facilities at rural area should be holistic design thinking. We should clearly distinguish the consumer type, the structure of consumer groups. To build cultural main content targeted, design the rationality of the commercial project settings, control the risk of operating costs, consider the acceptance of convenient paid service. So we can refer the Australia's successful model PPP at the investment and operations at the venues of our sports center. Government can take lower risk because enterprises take part in the construction and operation of the sports center. Especially it can effectively improve the venue operation management level after the completion construction, ensure the benefit of the management and avoid the waste of resources. According to the different natural conditions at different rural areas, we have to use different the PPP mode which can be divided into BOT (build—operate—hand over), BTO mode (built—hand over—operate), BT mode (build—transfer), TOT (hand over—operate—hand over), BOL model (built—operate—lease) and DBFO model (design—build—finance—operate) model. According to the economic conditions, the business environment and the participation of the government are different in the southern Jiangsu rural, we should choose different establishment model of Sports Park or venues. To build idiomatical, carrying capacity and operating infrastructure of rural sports culture service.

We need in depth native culture at rural areas of southern Jiangsu, such as working with culture affairs department to combine the Taihu culture, Wu culture, native food culture, historic human culture, sushi traditional architecture culture, folk arts and crafts culture and green rural tourism with the existing special resources, such as location, history and so on, of township, town, and village in southern Jiangsu.

To deep mining, processing, packing, combine the culture and sports facilities, sports and cultural products. By the promotion activities to the world, such as holding major sports events, sports cultural festival, sports tourism competition project and so on, which was led by the government, assisted by media, to build unique rural sports culture brands. For example the combination of the water resources of villages near Taihu Lake with activities like canoeing, dragon boating, swimming and so on; the combination of the space resource of mountain village with equestrian, mountaineering; the combination of ancient culture with kumara, Tai Chi, the traditional martial arts and so on. After broadcasting the brand worldwide, we could design and develop the relevant supporting products from the sports industry in the southern Jiangsu, develop the sports products industry, such as clothing and equipment, which has the local characteristics, then promote and scale up the products. Just like Jinjiang in Fujian province, to establish a sports clothing, shoes industry zone, many well-known independent brands and listed companies. Yangtze River Delta region is also the small and medium-sized enterprise intensive place; clothing, dyeing and finishing industry are very developed. These advantages greatly make our farmers can transfer to sports products. The key is in the guidance of our government, the scientific breakthrough, adapted to the development of the market economy rule. Government taken the led, mercerization of enterprise capital entry, independent operation, realize rural sports industry's open management mode, to make it to scale up, to achieve a win-win situation. To combine the traditional culture in the southern Jiangsu, physical elements and tourist industry together. Completely resolve the problems the lack resource and management of sports culture base, unique channel for promotion, neglected by sharing the resource. To put the sports culture rooted in the farmer's income generation industry.

ABOUT THE AUTHOR

Wen Bingshuai (1979–), male, Liaoning Xinmin people. A lecturer in Suzhou Institute of Trade & Commerce, master, mainly engaged in the study of physical education and sports culture

Contact: 18625262688, e-mail: 416860689@qq.com.

Mailing address: Military Department in Suzhou Institute of Trade & Commerce,

Suzhou Huqiu District xuefu road, No. 287

Recipients: WenBingShuai

Zip code: 215009.

Personalized sports information under the network environment the design and implementation of active service system.

REFERENCES

[1] Lu Yuanzhen, sports sociology. [M] Beijing. Beijing sports university press, 2002, 336.

[2] Wen Bingshuai. Chinese sports arbitration system analysis. [C]. Suzhou university. 2008, 13.

[3] Rishbana City. Austrilia Area Classifieation and Proposed Road Hierarehy to 2011. PIning Seheme MaP.

[4] http://su.people.com.cn/n/2012/1213/c347551-17859494.html.

Sports Engineering and Computer Science – Luo (Ed.)
© 2015 Taylor & Francis Group, London, ISBN 978-1-138-02650-6

Hung holdings analysis and study on the marketing strategy of limited competition

Lei Wang & Bing Li
College of Foreign Languages, Northeast Dianli University, Jilin, China

ABSTRACT: Event marketing strategy to enhance corporate visibility using events, plays an important role in the creation of good marketing environment. In this paper, through the holding of limited company in the Volvo Chinese open event marketing strategy to Hung Ming, we summed up the tournament selection strategy, the promotion strategy, and sponsorship strategy. At the same time related issues limited event marketing for Hung holdings and puts forward corresponding suggestions.

1   EVENT MARKETING

Event marketing has two meanings: one is the development of sports itself, the other the enterprise realizes its marketing purposes by events media. This paper mainly focuses on the second layers of meaning of event marketing, analysis and research of Hung Holdings Limited in the Volvo Chinese open marketing strategy.

2   HUNG HOLDINGS LIMITED EVENT MARKETING STRATEGY

Sponsors for the selection and recognition, there are certain standards, including: brand matching; matching products; matching market.

2.1   *Brand match*

Volvo China open is currently Chinese which is only a tour of Europe recognized by senior golf tournament; the tournament has a unique management model which is consistent from beginning to end, high-quality and high standard. The development has become a national champion golf occupation event, the event brings together a number of the world's top players, and it has an important influence in Chinese and throughout Asia, non negligible star Tournament Golf Tournament.

Hung holdings group specializing in golf course construction and management, so far with the world's top golf course designers, high-level stadium built 4 characteristics. The hardware foundation is for the selection of large scale events.

Hung company to build high-quality golf course, has a higher status in the China golf industry, the characteristics and China golf top competition Volvo China open a brand associated with a higher degree, so both a sponsorship agreement is feasible.

2.2   *Matching market*

The golf is a noble sport, but Hung company in the event the main push of the real estate products "coastal, Lake" is together in a seaside villa in combination with golf courses, each villa value is about 400, non ordinary consumers can bear, the target customer is the pursuit of quality excellence, life and love golf's high-end consumers.

2.3   *Products matching*

Hung company provides a venue for Volvo China open, products matching degree is very high.

3   STRATEGY SPONSORSHIP

3.1   *Sponsored plan*

Hung Holdings Limited in the open, with the playing field, bonus, as well as logistics. The playing field is Hung company "coastal lake," real estate projects in the golf course, the golf course to stop operation during the game, the entire game service. The game bonus of $3000000, a record high, becomes the 2012 Volvo China open a major bright spot.

3.2   *Sponsorship rights*

In 2012, held in Tianjin, the new Volvo China open, has a more wide marketing channels, the total value of the entire event media exceeded $30000000, covering more than 600000000 of the world's family.

Hung Holdings Limited access to the following events sponsorship rights: a giant advertising boards set inside the restaurant game volunteer position; VOLVO in Tianjin developers are involved in all activities. These activities include the kick-off ceremony, Crowne Plaza Hotel banquet activities and so on; into the enterprise information in the VOLVO film in the form of pictures; open match some of the volunteers to assist the company's work and activities. Hung Holdings Limited has a piece of publicity for the region in the ceremony of the exhibition.

3.3 Around the event promotion strategy

3.3.1 Customer maintenance and customer recommendation

Hung company in the customer maintenance reflected in available to the owners (i.e. purchased "coastal, lake villa." Products customers) distinguished viewing experience, the owners of the spectators privileges include: during the game, free in the coastal restaurant owners, exclusive private parking, carrying a family watching free. In addition, the owners will receive special identity sign, the import and the club which have identity special, convenient staff to provide better and more comprehensive services for the owners, the strategy to achieve the event marketing strategy to provide entertainment and recreational opportunities for key customers content.

Hung Holdings Limited, owners noble experience strategy, not only to the owners themselves feel their dignity, their attention, but also make the special value together with their relatives and friends to watch the race feel coastal lake villa products, which tend to buy the villa products. At the same time, the villa buyers daily circle of people, and the house of target customers a higher degree of coincidence, this also makes the villa purchase rate increased in the relatives and friends in.

3.3.2 Develop new customers

In the development of target customers, Hung company finds new customers through the "carpet" of information gathering mode. Hung company sets up champagne town in the only way which must be passed to enter the stadium, champagne town set up at the entrance of information collection, the audience voluntary in information collecting platform to leave your name, contact method, the court opinions and other information, to leave the audience can get the information of the memorial prizes and free to watch entertainment tickets and other small gifts. Sift through a large audience for the collected information, and then to meet the conditions of the population targeted marketing promotion, the maximum expansion of the target customer.

3.3.3 Products

Hung company other than the sponsors, the most special place is the game that is the company's products, which Hung company during the moment in the game on exhibition. In the open, most of the commercial value of the place is the venue, is each big media centralized regional reports, exposure rate is very high, for sponsorship, promotional products and services have the best effect. Tianjin Binhai Lake Golf Club and in the sale of the villa are weaved together, the audience in the process of watching the game, need according to the established route to follow the player through 48 holes, the total line length of 10 kilometers, takes about 4 hours to watch the match. The game process, the audience can see the game players, more intuitive to see the villa landscape, intuitive received the message villa, direct contact with consumers and the products produced in the process of.

3.3.4 Star promotion

Hung company to enhance the visibility, make full use of the star effect, the 2012 Volvo China open has a very powerful lineup of stars, athletes including the famous Spanish player Garcia, the former world number second; the famous golfer Montgomerie of Scotland, the previous industry ranked first; Ireland's famous players Harrington, three time grand Slam champion; American Hamilton, former Grand Slam champion; Casey of England, former Grand Slam title and famous Chinese player Liang Wenchong. The golf star player is a very important reason contest attracted many viewers, the star effect of these players, in order to gain more profit, is the core of sponsors sponsorship strategy.

4 CONCLUSIONS AND RECOMMENDATIONS

4.1 Conclusion

1. *In the sports business value being developed today, Hung Holdings Limited, according to the enterprise's own products and service advantages, seize the Volvo Chinese open this opportunity, acutely aware of the opportunities, saw the high overlap the audience and enterprise target customers of the event, the successful event marketing activities. In the limited rights, it makes full use of their own to provide site advantage, set the best stadium entrance, high density distribute publicity materials, holding the stars activity and effectively improve the enterprise and product visibility. In addition, it should pay attention to maintain old customers and develop new customers, collecting customer information, create a lot of potential marketing opportunities for the enterprise.*

2. *The whole event marketing activities also have details that need improvement: a commercial town has begun to take shape, but also did not reach the best effect, entertainment facilities, single, performance was not good enough, commodity is not complete, did not play the role of service station, can attract the audience is limited number; the acquisition of information to the visiting audience promised gift not completely fulfilled, damage the enterprise attempts to establish a high-end image; in the programme process, the audience ask matters related to property, but not the sales staff in the field of immediate response; the information collection work although it can achieve certain collection scope, but the accuracy is low, the workload is huge. The 2013 Volvo China open will remain in the "coastal, Lake" held, if Hung company can in detail perfect, event marketing activities will be more successful.*

4.2 Suggestions

1. *Commercial town of recreational facilities and services should be perfected, the commercial town into a "coastal, Lake" recreation center, in order to attract the audience to stay, and this will be more conducive to the information collection work.*
2. *Hung company should strengthen the training of management to the staff, so that employees can operate in strict accordance with the relevant procedures, and it can be helpful to the work of the standardization process.*
3. *Hung company shall ensure that the sales staff immediately, timely appear in front of the audience have the desire to buy.*

4. *A senior hotel, its customers and Hung company's target customers a higher degree of coincidence, Hung company should make full use of this advantage, and more effective to expand new customers.*

REFERENCES

[1] Xiao Linpeng. [M]. Project management of sports events in Beijing: Beijing Sport University press, 2005, 9:136–138.
[2] Hou optical. The sports event marketing study of the nature and the innovation of marketing concept [J]. Journal of Beijing Sport University, 2006 (5):21–24.
[3] Yang Jing. Development of sports marketing marketing channels [J]. Journal of Guangzhou Sports University, 2004, 24 (5):32–35.
[4] Fu Junfang on integrated sports marketing [J]. sports science research, 2005, 94 (4):18–23.
[5] Liu Bao, Li Fawei, Yang Hui. The large-scale sports event management research—event marketing strategy analysis of [J]. Science and technology information, 2007 (32):40–44.
[6] Liu Xindan. The enterprises of our country sports sponsorship [D]. Strategy Research of Beijing: Beijing Sport University, 2010.
[7] Li in full bloom, Pei Rong. Brand communication event marketing marketing based on [J]. journal, 2005 (6):22–26.
[8] He Haiming, Yuan Fang. Sports marketing ten classic case of [M]. Beijing: People's sports press, 2011 (1):54–57.
[9] Yan Jue. Sports scientific research brand communication based on the sponsorship of sporting events[J], 2008 (5):44–48.
[10] Shen Jia. Sports sponsorship objectives and strategy study [J]. Shanghai: Shanghai Institute of Physical Education, 2010 (6:40–43).

Sports Engineering and Computer Science – Luo (Ed.)
© 2015 Taylor & Francis Group, London, ISBN 978-1-138-02650-6

The development status and countermeasures of the ice-snow tourism resources in Jilin province

Lei Wang & Bing Li
College of Foreign Languages, Northeast Dianli University, Jilin, China

ABSTRACT: The ice-snow tourism is one of the new forms of tourism in recent years. In this paper, we take the ice-snow tourism resources in Jilin province as the research object, through the relevant data collection and analysis, we deeply study the development of ice-snow tourism resources in Jilin province. In order to promote the rapid development of the ice-snow tourism industry in Jilin province, at the same time, the study provides a reference for the tourism authorities and tourism corporate decision makers, but also provides a reference for the development of other similar cases with tourism resources in the area.

1 INTRODUCTION

The ice-snow tourism is one of the new forms of tourism in recent years, with broad prospects for development, the ice-snow tourism as a new economic growth point, in our country is in the period of rapid growth. Jilin province has a ski resort more than 20 seats, including five famous ski resorts. In addition to the Changbai Mountain Plateau snow training base, they all are close to the city center, which are rare city ski fields, provide great convenience for tourists. In addition, the Jilin province, convenient traffic and perfect service also provides great guarantee for skiing tourism. The famous "Chinese Changchun Jingyuetan ice and Snow Festival", "Chinese Jilin rime ice and Snow Festival", "Chinese Changbai Mountain ice and Snow Festival" and other festivals for tourists to add the fun. In the face of such popular ice-snow tourism market, how to better combine and use effectively, scientifically and reasonably the potential of ice-snow tourism existing in the Jilin province resources, to achieve economic, social and ecological benefits of coordinated development, is the urgent problem to be solved.

2 ANALYSIS OF JILIN ICE-SNOW TOURISM RESOURCES

2.1 *Ice and snow is rich in resources, regional differences are obvious*

Snow and ice resources in Jilin province is rich in content, which can be played, watched and also we can skate and ski, and there are ice tour, ice show and ice-snowice-snow culture and other art activities which have everything that one expects to find, so that visitors find it fresh and new [1].

Jilin city, Jilin province, is one of the earliest ice-snow sports and tourism city. Songhua River flows southwest, forming a "Pipa bay" in the urban areas flows to the northwest. Even in the coldest days at minus 30 degrees Celsius, Songhua River is not frozen, water is vapored to transpiration fog. The shore along pine willow is frost snow hanging, glittering, which is recognized as one of China's four natural wonders of Jilin rime. Away from the urban 24 km Songhua Lake, there is one of the second batch of national key scenic area, the west side of the lake is China's famous mountains ski. Away from the urban 56 km is the Yongji County Wulihe town, which can be used for domestic and international competition and tourist ski Beidahu ski resort, in 2007 the Asian Winter Games ski competition held here.

Compared with the city of Jilin, capital of Changchun's ice-snow resources with rich cultural atmosphere has held several sessions of ice and Snow Festival, which are relying on the ice-snow tourism products, development of ice and snow culture, art, sports, tourism and trade activities, emphasizing the Changchun City "Automobile City", "Electric City", "forest city", "science and technology in the city" and "the colonial relics City" characteristics, sponsored by the Swedish International Orienteering organizations and the China Changchun Jingyuetan Changchun City government, the Vasaloppet international skiing festival was held annually at Changchun Jingyuetan National Forest Park. Changchun City Lotus Hill ski resort, is a world-class ski competition, domestic competition in most projects, equipment, the most perfect professional athletes ski training base and youth ski training base, is the largest ice and snow entertainment center and sports fitness, leisure and entertainment are the main characteristics of the city landscape resort [2].

Changbai Mountain in winter snow scenery more attractive charm. Because Changbai Mountain is so high, and far away from the industrial towns, the air is fresh, quality of snow white, forming a spectacular silver world. Tianchi around the peaks become a warrior statue, the magnificent waterfall is like a silver belt from the mountain falling into icebergs, clear as crystal, the trees, rocks around both sides of Shirakawa and hot springs, under the function of the rime, are just like blooming pear, like polar animal group of all shapes. The main peak of Changbai Mountain is formed two million years ago after the volcano, Northeast China in the territory's highest volcano, due to its unique geographical structure, every year the snow on the mountain lasts for nine months [3].

2.2 *Characteristics of ice-snow tourism resources, attractive industry*

Originally, winter is the off-season tourism, and the ice-snow landscape make it not a off-season for the tourism of Jilin province, even become a busy season and also rise the reputation of Jilin ice and snow sports, tourism popularity. In 1991 May, the National Tourism Administration launched China's first batch of 14 special tourism lines to meet the 92 friendly—snow scenery tour is one of them; the National Tourism Bureau in 1995 in Jilin province held a national ski tourism seminar, and plan for the deployment of the tourism products of our country skiing planning and development work in the future; rime ice-snow tourism in Jilin City Festival and Changchun ice-snow tourism festival has become a tourism festival of ice and snow in China famous brand; Jilin province held many national winter games, and has successfully fulfilled many of World Cup short track speed skating race to undertake tasks; also successfully held the Asian Winter Games, the fact shows, Jilin province snow sports, tourism is be just developing in a good trend [4].

2.3 *A large number of ski resorts in Jilin province*

As far as now, the province owes a total of nearly 20 more than all types of ski resorts, including snow Game venues Beidahu skiing field, international standard ski—Changchun Lotus Hill ski resort, which is closest to the city, the nation's largest forest Ski Resort—Jilin Lotus Hill ski resort, Changchun city Jingyuetan ski field skiing, alpine skiing—Changbai Mountain ski.

3 ICE-SNOW TOURISM DEVELOPMENT SITUATION AND PROBLEMS IN JILIN PROVINCE

3.1 *Jilin winter tourism development*

Jilin ski tourism started late, to the early 90's, "the beginning of ski tourism". 1992 "Chinese friendly year", the National Tourism Bureau will be the first "ice and snow scenery tour" as one of the 14 special China tourism products, thus officially in the nationwide opened the prelude to the ice-snow tourism development. Jilin province in 2007 becomes the main venue for the Sixth Asian Winter Games, on the other hand, hold various events which led to the ice-snow tourism infrastructure construction. In recent years, the province's ski resort construction scale is more and more big, the standard is getting higher and higher, more and more diverse forms of investment and cooperation. As of now, the province has a total of more than 20 ski resorts in Jilin, Changchun Jingyuetan Beidahu ski tourism, Changchun Lotus Hill ski.

Ice-snow tourism in Jilin province after 20-year development, has been cultivate the Changbai Mountain winter landscape, Jilin rime, Chagan Lake catching in winter and hot spring resort, 5 characteristic brand. Among them, the well-known, Changbai Mountain is the ice-snow tourism brand of the province, natural wonders of Jilin rime ice-snow tourism is my province traditional brands, "the winter ski paradise" in our province is the ice-snow tourism brand advantage, Changbai Mountain winter in the open-air hot spring bath, is the most influential ice-snow tourism projects, Chagan Lake winter fishing in our province is the ice-snow tourism brand characteristics. In recent years, Jilin province has organized the "2009 China Changchun Jingyuetan ice-snow tourism festival and Vassar International Ski Festival", "the first Jilin forest ice and Snow Festival", "2006 China international rime ice and Snow Festival", "2006 Chinese • Jilin ice and snow in Changbai Mountain Tourism Festival", "Jilin China chaganhu Winter Fishing Festival" and more ice-snow tourism festival. These festivals held in Jilin province for the tourism development, which has played an important impetus.

3.2 *Jilin winter tourism problems*

3.2.1 *Lack of an overall strategic planning, iterative project development*

At present, in Jilin ice and snow sports tourism resources the development process, mainly on the development of skiing, ice, snow, winter swimming sports tourism resources, for some ecological tourism and the ice-snow tourism combined resources development is not perfect, for example, Jilin North Lake, Changchun Lotus Hill ski Festival, splashing Snow Festival, lamp exposition will, it is still lacked of these features in the river of ice-snow sports tourism resources development landscape avenue. The majority of Jilin province ski operators lack of advanced management concepts, often only pay attention to the hardware facilities, but not in the governance, environment beautification efforts; they only pay attention to the weight of the input, but are not willing to spend money in operating characteristics, quality of service, which make most ski design mediocre, lack of features, entertainment, and less ski atmosphere is not enough, which also reduces the ski resort in Jilin province attraction.

3.2.2 Backward infrastructure, tourism products with lower grades

A mature ice-snow tourism industry should be the rational allocation of various elements, to create a comfortable and convenient environment for visitors to travel. The ski resort is famous in the world, there are traffic network rail, aviation, highway and convenient support, as far as possible to shorten the visitors in the transportation time. In recent years, although the increase in infrastructure investment, leading to the scenic roads has improved, but there are still many ski field has poor accessibility problems. In addition, some small snow field guide and service function is low, ski, ski panorama introduced the card and sign is not perfect, the parking lot, the visitor center and other hardware facilities outdated, safety protection facilities lack. The short residence time of tourists, tourism economic benefits, tourism revenue in addition to the main attractions tickets, tourism income is related to various consumer accommodation and accommodation to bring. For example, Changbai Mountain ice-snow tourism, is usually a day trip, Changbai Mountain accommodation facilities construction is poor, so only rely on ticket revenue, the tourism economic benefits will be lower if only rely on ticket revenue.

3.2.3 Lack of professional travel services

It is really rapid development of Jilin province travel agency industry, but which has not formed the scale, especially the large-scale international travel, tourism service personnel is relatively poor, in addition, opening various colleges and universities in Jilin province tourism professional training of high level talent is not much, so that all levels of tourism talents have a certain degree of scarcity, which led directly to the development of ice-snow tourism in Jilin province.

3.2.4 Ice-snow tourism value chain effect

The tourism value chain is around tourism wholesalers or tourist attractions, control through the service flow, material flow, capital flow, information flow, tourist flow, business flow and the flow through culture, ask, line, food, health, housing, travel, entertainment, shopping and so on, a series of activities, the tourism suppliers tourism wholesalers, retailers, tourism, tourism as a whole network chain structure model. Jilin province, ice-snow tourism in food, health, housing, travel, entertainment, shopping and so on a series of process, can not effectively links and series, so there is no its tourism value chain effect.

4 THE DEVELOPMENT OF ICE-SNOW TOURISM RESOURCES IN JILIN PROVINCE COUNTERMEASURE

4.1 Increase publicity and promotional efforts, to develop tourism market and domestic

Increasing publicity and promotion efforts are one of the main measures for more domestic and international ice and snow tourist. To make full use of various means carrying out publicity work. We should pay attention to improving the quality of tourism services, do reception service work and tourists, also can invite domestic and foreign travel agency and influential news reporter on-the-spot investigation and experience. So everybody back like a "station", to do publicity for us, can play a multiplier effect. In addition, proactive, multi channel "the sea" is also an important means. On this issue we should be willing to invest, as often participate in various domestic and international tourism fair, fair, send regular outreach promotions group and so on. Can also make full use of domestic and foreign newspapers, magazines, television, radio and other news media and our institutions like propaganda, in order to achieve continuous stimulation, we should repeat promotional purposes. In the promotion of time, we can advance publicity, but also anti season promotion and sales; in the publicity channels, TV and network promotion.

4.2 Follow the principle of market orientation, to play, to improve and enhance the attractiveness of tourism resources

The development of tourism resources is a kind of economic behavior, under the market economic system, tourism resources development must be guided by the market, can not have develop resources randomly, and should be the first to do market research, market development needs to sell tourism products, and handle well the relationship between market and resources. Only in this way can the tourism resources become the real attraction, for tourism services.

Jilin province has the such as the Changbai Mountain and the snow and ice resources richly endowed by nature, therefore, we need to target market, the snow and ice resources and Jilin province and other tourism resources, develop more attractive tourism project.

4.3 For grade market, development is to meet the regional tourism projects of local tourists

It generally believed that the international community, an ideal ski vacation days for 3–7 [5]. December to February is new year's day and Spring Festival two festival, along with the improvement of people's living standard, city residents changed the way that holidays travel has become a new choice of many families, however, long distance travel by traffic, accommodation, time limit and other factors, the short ice-snow tourism of urban residents is potentially more attractive. Population of Changchun, Jilin and Yanji City, has a huge demand for ice-snow tourism. At this time, but also various types of school during the summer vacation, choose suitable for winter sports skiing ground, but also a basic requirement for students rich and colorful life.

In the development of ice and snow two categories of resources, it gives the development of snow resources with the open space, large environmental capacity, participation and strong features, thrilling stimulation, which can cause the majority of ice and snow tourist interests, so we should focus on the snow, may be appropriate to consider increasing the number of strong participation, price low tourism products, to meet the Jilin province on the ice-snow tourism increased demand [6].

4.4 Strengthen the standardization construction

In the development of ice-snow tourism resources in the process, we should pay attention to the standardization of construction, to regulate the market for development projects, establish a good word-of-mouth publicity, encourage tourists to travel and to its surrounding people to travel here again. In addition to the development of a variety of winter tourism products, tourism reception quality is another important factor. Accustomed to high quality service to tourists or reception facilities, service quality is very sensitive. For example the cutlery cleanliness garnish with greasy degree, the hotel shower facilities, public toilet hygiene, are very sensitive. So when the development of ice-snow tourism market should pay particular attention to the quality of the reception facilities [7].

4.5 Defining government functions, strengthen the government's macro-control and industry management

The government of Jilin province plays a very important role in the resource development process, with the development of market economy, the government should transform its functions, to further strengthen macro-control and industry management system, including: government departments or continuous data collection, which is to establish the system, and regularly updated data, so as to better understand the strategic parameters and the plan required; to do tourism complaints and tourism information management, to protect the legitimate rights and interests of consumers, but also to protect the legitimate rights and interests of travel agency. The tourism information management including travel agencies, hotels, restaurants and tourist information. For feedback, we should do a good job in the future development planning. The role of government from direct economic department should be changed the macroeconomic regulation and controlled, and we are supposed to do a good job market forecast, product propaganda work.

5 CONCLUSIONS AND RECOMMENDATIONS

1. *Jilin ice-snow tourism resources development is a long-term and systematic project, not only needs actively promoting and cooperating with tourism enterprises, relevant government departments should act with united strength, also needs support related systems.*
2. *Ice and snow tourist industry is the primary strategic targets to improve the comprehensive competitiveness of ice-snow tourism in Jilin province. Ice and snow Industry is a comprehensive industry chain, supporting the development of ice-snow tourism industry, at the same time the development of ice-snow tourism industry and the related industries play a tremendous role in promoting.*
3. *Jilin winter tourism resources development should increase publicity and promotional efforts, to develop tourism market and domestic; the establishment of different ice-snow tourism products according to different area, should meet the needs of different groups of people.*
4. *Follow the principle of market orientation, for playing, improving and enhancing the attractiveness of tourism resources as a starting point. Build the brand in Jilin ice-snow tourism, a number of domestic first-class, well-known foreign ice and snow tourist resort construction.*

5. *For grade market, development is to meet the need of regional tourism project local tourists tourism, strengthen standardization construction, clear the functions of the government, and strengthen the government's macro-control and industry management.*

REFERENCES

[1] Li Yin, Sun Jianhua, Guo Yujie. Jilin ice and snow sports regional characteristics and market analysis of ice and snow sports [J]. 2004 (7): 62–65.

[2] Compilation group of Jilin Province Tourism Bureau [2] tour guide qualification examination materials. Jilin tour guide [M]. Changchun: Jilin people's publishing house. 2002.56.

[3] Li Guang, Zhang Shouzhi. The west slope of Changbai Mountain sustainable tourism and tourism resources evaluation [J]. Journal of Yanbian University, 2001 (27): 57–60.

[4] Che Xiuzhen. The west slope of Changbai Mountain tourism resources evaluation and sustainable environment and development of [J]. Development, 2001 (16): 1–3.

[5] Lou Jia Jun. Introduction to [M]. Entertainment tourism in Fuzhou: Fujian people's publishing house, 2000.68.

[6] Wang Liang. Development of ice and snow sports tourism thinking [J]. Ice and snow sports, 2004 (5): 96–99.

[7] Li Fuquan, Li Xiaoyan. Research of ice-snow sports tourism network marketing [J]. Ice and snow sports, 2004 (7): 52–54.

Sports Engineering and Computer Science – Luo (Ed.)
© *2015 Taylor & Francis Group, London, ISBN 978-1-138-02650-6*

Volvo China open (Tianjin) analysis of event marketing strategy

Lei Wang & Bing Li
College of Foreign Languages, Northeast Dianli University, Jilin, China

ABSTRACT: With the change of China's economic modes, similar change happens in how to run and arrange sports. It was usually the government that arranges all the sports events before, but now commercial mode is taking the place of the government. Under this condition, how to run and arrange the sports event comes to front stage. To take the most of the profit of a sports event, we need a comprehensive arranging and marketing strategy. In this paper, the Volvo China Open, 2012 is taken as an example to show readers about the marketing strategies and corresponding analysis in sports events. On the ground of the analysis, some advice is provided as references for other sports events.

1 INTRODUCTION

With a history of over 500 hundred years, golf originated in Britain and was first taken to China in 1916. In 80s of last century, it developed well and fast in China. With the increase of the players, other related industries also got progress. And finally as a much more meaningful event, the Volvo Open came to China in 2012, whose final success would attribute greatly to the development of golf in China.

2 SOME CONCEPTS OF MARKETING STRATEGIES IN SPORTS EVENTS

The concept for the sports event marketing general from two angles: one is the definition of sports as a special product or service, by the operator or organizer of marketing directly to consumers. Two is the sports event as a carrier, marketing by events on the sponsoring enterprise service or product. In this paper, from the perspective of Volvo China Open event marketing strategy analysis.

3 MARKETING STRATEGY ANALYSIS OF VOLVO OPEN

3.1 Strategies of events

3.1.1 Choosing the event city and field

As one of the European Tour and the longest international professional Volvo match in Chinese history, Volvo China Open aims to promote the development of golf sports in China. The 18th Volvo China Open will be held in Tianjin and this is the first time for Tianjin.

From the cities over the years, we can see that Volvo Shengshi Management Company uses developed economy, high public income level and internationalized city's image as important standards when choosing the cities. Thus some China's top cities have become the preferred place to hold Volvo China Open.

With the development of Chinese economy and the increasing domestic and international cooperation, in some secondary cities, the economy is developing rapidly, people's income level is becoming higher, nation's cultural life is becoming richer, and international communication is more frequent. Not only the physical infrastructure such as landscape architecture and road building but also soft power such as reserve of talent and people's image have reached the expectant top city level. Thus, more and more investors of investment program abroad, international cultural ceremony and the sponsors of sports events turn their eyes to these cities.

Volvo China Open turned its eyes from top cities for the first time and cheese Suzhou and Chengdu successively. And for the first time it chooses Tianjin in 2012. The reasons they choose Tianjin are as follows. First, the city competitiveness of Tianjin is becoming stronger. It is becoming an international metropolis. Tianjin is in the center of enclosed Bohai Sea. It is also the economic center, logistics center shipping center and the largest open city along the coastline in northern China. As the economy developing rapidly, people's income level is becoming higher. So golf audience is wider. Second, Tianjin has a long history, a developed economy and rich culture. Volvo China Open chooses Tianjin aiming to spread golf around China.

3.1.2 Timing and schedule management

From the characteristic of golf, we can see that it is a kind of outdoors sport. Playing time is long and players are usually requested to play in distinctive time. Thus, the weather has significant influence on the result of the match. The match is often being held in spring or autumn. First, the temperature of some eastern coastal cities such as Beijing, Shanghai and Tianjin is comfortable with little rains in these seasons. To be held in those time can lower the rate of the weather affecting results. Second, people are willing to watch the match in comfortable weather. Meanwhile the audience can watch the high level match in a comfortable atmosphere so that to improve people's visual pleasure.

Volvo China Open will be lasted for one week. The reason is that they want to share two days for players to practice before the match officially started so that players can adapt themselves to the match and try their best. Second, the official match will be lasted for four days. Players can assign their strength rationally. Audience can choose the match days on their own. And this is the character and charm of golf itself. Finally, a series of public relation activities will be held to promote the events and improve the popularity and influence.

3.1.3 *Levels and bonus of the match*

Volvo China Open is held by China Golf Association, which is approved by European Tour and Asia Tour. The bonus of 18th Volvo China Open is up to 20 million RMB (or 3 million dollars). The bonus of the champion is about 3.3 million (or 500 thousand dollars). Apart from the champion's rich bonus, other places' bonus is not cheap. Players of top 70 can get the money. Setting rich bonus is no doubt one of the most efficient marketing approach. Rich bonus can reflect the vastness and high level of the match from a side face. Meanwhile, it can attract plenty of golf stars and they will compete for champion so as to improve the influence and make promotions. It also seizes people's "watch stars" minds and improves the rate of audience coming to watch games.

3.2 *Pricing strategy*

The ticket price is 680 yuan which entitles the holder to watch all the games of Volvo China Open. Roundly pricing strategy is popular with some sports events or cultural performances. In 2012, Volvo China Open uses this strategy and makes the price 680 yuan. Golf is a kind of elegant sport and it's audience are those who have high social status or income level. Round price can satisfy their vanity partly and reflect the committee's attitude of showing a high-level match and supplying high-quality services.

3.3 *Promote strategy*

1. Print media: 《GOLF》 and 《Esquire》
   As the event specified golf magazine, 《GOLF》 will introduce Volvo China Open roundly and deeply from professionally sense of golf. 《Esquire》 is a kind of high-end fashion magazine and shares the similar audience of golf. It will be more efficient to promote this match through this magazine.
2. Internet:
   SINA golf channel is one of the most authoritative professional Chinese golf website.
3. TV media:
   CCTV is one of the largest official TV media in China, which has the biggest television viewership. Its programs are broadcasted in Mainland China and the whole world through satellite and Internet.
4. Outdoors advertising is a kind of early ads form and has its own advantages. It has the character of forced appeal which means no matter you accept it or not, it would be showed for you. Even the passers-by who is hurry on with his journey can be impressed by the ads.

4 CONCLUSION AND SUGGESTION

4.1 *Conclusion*

As the earliest international professional event, Volvo China Open has past 18 years and becomes one of the high-level international golf events in China even worldwide. It makes indelible contributions to the promotion of golf in China and the improvement of golf level. Through the marketing strategies of 2012 Volvo China Open, we know that the strategies are systematic and comprehensive. The advanced international event management system is the key to the success of this match.

4.2 *Suggestion*

4.2.1 *Customer oriented and improving the quality of experiencing*

Volvo China Open makes many kinds of marketing strategies, which offer various experiences. The committee got a great affirmation. But they should pay more attention to details and customers, improve every tiny segment and the quality of experiencing so as to make the match level match the quality of service. For example, through adding souvenir give-out windows and scattered give-out places, they can shorten waiting time.

4.2.2 *Popularizing knowledge of golf and promoting the development of golf*

At present, golf belongs to narrow sports. Common people can't reach the area and they know little about golf. Volvo China Open dedicates to the deepen the promotion of golf in China. Thus they should not only attract more people to know golf but also popularize golf and match rules, introduce watching etiquette so as to let audience know about golf correctly.

4.2.3 *Strengthening propaganda and make event brand*

One of the advantages of the country where golf develops well is getting rid of conventional limits. There is no self-imposed mind. There is a lot of room to develop golf. Thus, the committee can create aggressively during promoting, try to cooperate with different partners to promote together, expand the audience. Meanwhile they must strengthen the propaganda, hold public relation activities in multi-areas, attract the media, promote the event popularity and influence and let Volvo China Open to become a grand sporting event among audience.

REFERENCES

[1] Wang Shouheng, Ye Qinghui, the definition of sports events and classification of [J], Journal of Capital Institute of Physical Education, 2005. (6:54–58).

[2] Liu Qingzao, sports event operation and management of [M], People's sports press, 2006. 123–126.

[3] Chen Yun, character and basic strategy of modern sports marketing [J], Journal of Shanghai Institute of Physical Education, 2003. (4):41–43.

[4] Zhu Renkang, Wang Zhixue, Guo Aimin, [J], operation and development of sports events, Journal of Nanjing Sport Institute, 2007. (09):25–27.

[5] Lu Feng, development and problems of China's sports market analysis [J]. Journal of Wuhan Sports Institute, 2001, (7):36–39.

[6] Fu Junfang [J], on integrated sports marketing, sports science research, 2005. (05):18–21.

[7] Huang Lu, Fu Xiaochun, the connotation of [J], on sports events. Journal of Capital Institute of Physical Education, 2005 (6):51–54.

[8] Zhong Tianlang, and the practice of [M], a theory of sports management Fudan University press, 2004:102–105.

[9] Yang Jing, sports business marketing channel development [J], Journal of Guangzhou Sports University, 2004. (3):22–26.

Sports Engineering and Computer Science – Luo (Ed.)
© *2015 Taylor & Francis Group, London, ISBN 978-1-138-02650-6*

Volvo China Open (Tianjin) internal and external research

Lei Wang & Bing Li
College of Foreign Languages, Northeast Dianli University, Jilin, China

ABSTRACT: This paper chooses the 2012–2013 year Volvo China Open as the research object, we use external knowledge to make 2012–2013 years Volvo China Open operation subject, participants and with the study of non participating subject clear, and form a scope of events positive externalities of production subject and reception subject; then based further on all positive impact and in the analysis of events external and market compensation of the existing mechanisms, we describe and analyze the positive external events on coastal lake project villa owners, entertainment, tourism and other industries and the whole city of Tianjin built; finally, respectively, from the perspective of the government and the market perspective, we carry out the ways and methods of events positive externalities internalization.

Keywords: Volvo China Open; internalize positive externalities; government; market

1 INTRODUCTION

From the 1995 event was born until 2011, Volvo China Open has been hosted 17 games in Beijing, Shanghai, Shenzhen, Suzhou and Chengdu four city. 2012 and 2013 Volvo China Open hosted by the Tianjin Binhai Lake Golf Club. At present, it is the European tour and Asian tour with a jointly approved event, and Chinese has the longest history, which is the international influence of the largest top occupation golf. As a commercial operation is getting more mature world class golf occupation competition, the market mechanism plays a dominant role in the allocation of resources China race Volvo Open.

2 OVERVIEW

At present, domestic research on sports externality are as follows: Wang Lang's "western sports tourism externality study", Deng Chunlin "China sports externality and internalization", Zheng Zhaoyun "occupation in sports market externality problem analysis and research", Huang Haiyan "the large-scale sports event the positive externality and internality approaches", Zhang Xinping's "athlete human capital externalities internalization path and "Chen Yuanxin", such as large-scale comprehensive sports facilities in the externality", Chen Yuanxin's "body model—large sports facilities external correction of ideal mode".

Generally speaking, from the domestic perspective, the value of sports is causing more and more attention, in the adjustment of economic structure

and the background of globalization, the external spillover effect of sports industry on the development of regional economy has become a new focus of research of the domestic scholars. Research results of the above aspects, developed for a specific sports, athletes, venues and facilities, sports teams, sports resources and use of the process of externality is obviously insufficient. This paper discusses the research and analysis of the externality problems in the specific sport events.

3 THE CONCEPT OF EXTERNALITIES AND THE THEORY

3.1 *The definition and characteristics of externalities*

An externality is a basic concept of economics, it is a private and social benefit, private and social costs inconsistent phenomenon, is an economic subject to imposed another economic main body, can not fully reflect the effect of the external price and market transactions. Divided into subject to accept the production subject and externality. Externalities: first, non participatory decision-making external subject. Second, external reception subject lack "efficient feedback mechanism".

3.2 *Inner externality*

No efficiency externalities cause markets to allocate resources, need to be corrected, namely "internalization". Externalities can be internalized by the two angles of the government and the market, the relevant economics theories respectively Pigou tax and Kos theorem.

4 MAIN RESEARCH CONTENTS

4.1 *Volvo China Open (Tianjin) related subject*

4.1.1 *China Open (Tianjin) the main operation*
In 2012–2013 year Volvo China Open positive externality generating body as shown in Table 1.

4.1.2 *China Open (Tianjin) participants*
Participants include golfer, referee, player entourage, volunteers, audience, media, sponsors and key figures, 2012 and 2013 Volvo China Open and the positive externality of participants include sponsors and media.

4.1.3 *China Open (Tianjin) non participation of beneficiaries*
In 2012–2013 year Volvo China Open Seyfert participation of beneficiaries has a lot, affected by non participants mainly include the following aspects: (Table 2).

4.2 *Volvo China Open (Tianjin) positive impact on the external and the existing market compensation mechanism*

Established in 2012–2013 years Volvo positive externalities Chinese open, first of all should be clear of all the positive impact of the event of

Table 1. Volvo China Open operation organization.

The name game operation	Name
Sponsoring organizations	Chinese Golf Association
Events management and Promotion Agency	Volvo event management company
Host Stadium	Tianjin Binhai Lake Golf Club
Co organization	Tianjin Municipal Sports Bureau

Table 2. Volvo China Open Seyfert participates in the body.

Non participating subject	Name
Personal	Villa owners coastal lake project
Organization	Traffic transportation enterprises
	Coastal lake surrounding tourism enterprises
Sociology	Tianjin city

"external"; second, the competition of business model for the "external" compensation market feedback mechanism; finally, we should have been compensated external positive effect, and get all the external Volvo China Open final.

4.2.1 *China Open (Tianjin) positive effect on externally generated*
In 2012–2013 year Volvo China Open on the "outside", namely all the participants and the positive effect of non participation of beneficiaries generated can be summarized as the following four aspects:

First, the economic growth effect. Showing increased coastal lake construction surrounding public infrastructure investment, stimulating the city of Tianjin Binhai Lake and the surrounding accommodation, transportation, entertainment and so on.

Second, the industrial structure effect. Performance to promote the development of golf industry Tianjin city, and other related industries, such as the development of the news media industry and tourism industry.

Third, city brand effect. To enhance the brand image Tianjin city, to expand the visibility and influence.

Fourth, personal benefit. Performance optimization of coastal lake villa owners living environment, improving the real estate value and so on.

4.2.2 *China Open (Tianjin) compensation mechanism is outside the existing market*
In 2012–2013 year Volvo China Open continues to use some traditional sports business operation mode, this eliminates some positive external events resulting in a certain extent.

Specifically, the Volvo China Open uses the same business model of sponsorship, sponsorship of enterprises in the enjoyment of sports events brought the "right" at the same time, also must pay and pay back the corresponding sponsorship fees. This is accepted by the sponsoring enterprise external positive effect will no longer belong to positive externalities. Similarly, the Volvo China Open also ask the broadcasting and media coverage of the games for a fee, the compensation mechanism is internal and external events to the media of the original.

Through the above sponsorship, TV relay marketing events such as business process, Volvo China Open generated external positive effect through market self internalization.

4.3 *Volvo China Open (Tianjin) positive externality*

In the case of that Volvo Chinese Open commercial operates smoothly and effectively, the external part of the positive effects can be internal, external events of the external positive effect is surplus, the positive externality is a match of the existing mode and operation that cannot be internalized.

4.3.1 *To the villa owners will bring a positive impact on coastal lake project*

Volvo China Open for two consecutive years in the organization will increase the investment value of the villa owners, and bring direct economic benefits; and the games will also promote the transformation of the surrounding infrastructure upgrade, convenient coastal lake villa owners can get around, but also gives them a better living environment, these all belong to the Volvo China Open the positive externality that brought to coastal lake villa owners.

4.3.2 *Led to the development of related industries*

The Binhai Lake Golf villa project relies on the yellow port base and two base two reservoirs, lake area of 30000 acres, birds and insects fish is very rich in resources. A large area of wetlands surrounding the project has been listed as Tianjin City Hall ecological wetland reserve, which constitutes the water ecological system structure and characteristic. Volvo China Open give the world an opportunity to understand Tianjin Binhai Lake for the development of tourism areas around the coastal region has opened up a vast space. Moreover, the Volvo China Open, the domestic top golf events not only attract Tianjin local audiences to watch the scene, around Beijing, Hebei and other regions of the audience in a continuous line, which will drive the development of transportation industry.

4.3.3 *Volvo China Open (Tianjin) is helpful for the establishment of Tianjin city brand*

In Tianjin city, previously, Volvo China held in city like Beijing, Shanghai, Shenzhen, Suzhou, Chengdu, they are the domestic economy developed, the international high visibility of the city. Tianjin city gets the right for 2012 and 2013 to host, in a degree, it is a show of strength.

Effect of media report on Volvo China Open and the participating players will expand visibility, which plays a key role for Tianjin city. Volvo China Open is the European tour and the same Asian Federation approved games, it naturally also caused world concern, domestic and foreign media have come to Tianjin, four days live recording and reporting on the event, the world 600000000 target audience. Occupation golfers from different countries and regions gathered at the Tianjin, through word of mouth effect, can expand Tianjin's visibility and influence, to promote the shaping of promotion and brand image of Tianjin city.

4.4 *Volvo China Open (Tianjin) positive externality internalization*

Positive external effects above will produce great benefits to the whole society, but for Volvo event operation subject, deviated from the private net profit and social profit, which will damage the enthusiasm of the main supply of game operation, it is not conducive to the development of sustainable health events. Therefore, on the existing basis, it is necessary to further Volvo positive externalities internalization.

4.4.1 *China Open (Tianjin) market approach is to internalize the externality*

4.4.1.1 Expand visibility to Tianjin Binhai Lake Golf Club bring more revenue

Undertaking the Volvo China Open to Tianjin Binhai Lake Golf Club awareness has been an unprecedented increase, which provided the possibility of more population inflow, more population flows to more consumption, and coastal lake ball will bring the private benefit more.

4.4.1.2 Expansion in similar projects in Tianjin Binhai Golf Club's influence to bring more revenue

Tianjin Binhai Lake Golf Club as a host for two years the highest level Tournament Golf Club, the coastal lake Club advantage before the world, especially attracts the Tianjin Municipal Golf amateur eye, and expands the coastal lake club in the same type of golf club's influence, and further improves the ball golf membership sales. At the same time, in the Binhai Lake Golf Club, and the supporting construction villa, expands the Tianjin Binhai Lake Golf villa project influence, which will also improve the villa sales.

4.4.1.3 Through the innovation of the present event operation mode for Volvo event management, it brings more revenue

Volvo China Open commercial operation is closely connected with the events of the enterprise or organization, and it is divided into the first class. And levy and right size consistent with the expense, these enterprises or organizations are just sponsor and media; and we also prohibit the second categories that did not pay the enterprises and organizations to exercise the corresponding rights, namely "exclusive". The first kind of enterprises and organizations more, bring the Volvo event management, economic income is large, private and social benefit gap will be reduced accordingly. Therefore, Volvo event management, through the innovation of the present event operation mode, the establishment of a more effective and reasonable market compensation mechanism, the positive externalities are more internalization.

4.4.2 China Open (Tianjin) government approach is to internalize the externality

Golf tournament for the high requirements of venues, facilities for, and high competition bonuses and other reasons, the golf tournament is costly, in order to keep Volvo China Open evergreen and gradually expand its influence, relying on market forces is not enough. This is relevant to the needs of all levels of government and the sports administration department DO SOMETHING PREVIOUSLY UNRELEASED. The main methods are:

4.4.2.1 Government to give some subsidies to match operation economy

The government and the sports administration department can match operation subject must give subsidies to support Volvo China Open business, this approach can expand the sources of income, by increasing the private benefits of the main part of the game operation main directly, and then improve the private net profit, net profit from to eliminate the private and social net profit.

4.4.2.2 Government related businesses to give some tax relief policy

The government can also be given to related business tax relief policy, such as hosting stadium, sponsored enterprises, this method reduces the private costs of Volvo China Open for consumption, also to increase private net profit, eliminate private net profit and social profit from the results, the positive externality internalize Volvo China Open in a certain extent.

5 SUGGESTIONS

1. *Positive externality is beneficial to the whole society, positive but it hit the private supply of events, resulting in sports market failure. In order to ensure the healthy development of sports industry, the main part of sports event operation and the government should pay enough attention to the positive externalities by the way of internalizing positive externalities.*

2. *The government in support of game development should make clear of its functions and powers. The government approach itself cost, at the same time, government intervention may also lead to rent-seeking activities, leading to distortions in the allocation of resources, and the formation of government failure situation.*

3. *Events business determine the market compensation feedback mechanism, to spend a lot of money, but excessive commercialization, will dampen the enthusiasm of others and the society to participate, which cause damage to the social welfare. Therefore, the internalization of externalities, reasonably match the main part of operation, we should pay attention to market the establishment of compensation mechanism.*

REFERENCES

Chen Yuanxin, body model—[J] ideal mode of large sports venues and facilities external correction, universal sports market, 2008 (5).

Chen Yuanxin, Huang Aifeng, Wang Jian, large-scale comprehensive sports facilities in the external [J], Journal of Shanghai Institute of Physical Education, 2007 (1).

Deng Chunlin, Zhang Xinping, Wang Lijun, Chinese sports external and internal [J], Journal of Wuhan Sports Institute, 2010 (9).

Huang Haiyan, Zhang Lin, Li Nanzhu, positive externality and internality approaches the large-scale sports event [J], Journal of Shanghai Institute of Physical Education, 2007 (1).

Wang Lang, externality of the sports tourism in western minority internalization of [J], Journal of Xianyang Normal University, 2009 (11).

Yu Shouwen, sports industry impact on the city of [D], Fudan University, 2007.

Zhang Xinping, Liu Di, Deng Chunlin, externality and internalization of athlete human capital property rights and transaction costs—Based on the theory of [J], the Military Institute of physical education, 2011 (1).

Zheng Zhaoyun, Zhao Liguang, Yang Jichun, Wang Jian, Han Kaicheng, analysis of external problems of occupation in sports market and explore [J], Journal of Tianjin University of Sport, 2004 (12).

Sports Engineering and Computer Science – Luo (Ed.)
© 2015 Taylor & Francis Group, London, ISBN 978-1-138-02650-6

Social sports instructors should have the qualities of beauty

Jian Liu & LiJun Xu
School of Physical Education, Northeast Dianli University, Jilin City, China

ABSTRACT: The social sports instructor is the transfer of the angel of beauty, they should have a good comprehensive quality. Including the good moral character, physical beauty, technology, skills of beauty; beauty rituals and comprehensive quality. Being taught by precept and example, very intuitively, concretely delivery to the pursuit of beauty through the guidance activities not only enable students' bodies tend to beautiful, but also gradually achieve the real healthy of physical, psychological, social interaction in the beautiful atmosphere.

Keywords: social sports instructor; quality of health beauty

1 INTRODUCTION

This beauty is not aesthetics but beautiful. The responsibility of social sports instructors is the guidance, so, as an instructor, only by having better quality they can guide the social sports participants who have healthy body, soul and beautiful body. The nature of work: on the one hand, the social sports instructors are active, they should actively take the initiative to do science, and correct guiding; on the other hand, they are passive, they accept the selection of participants passively. This choice sometimes is more ruthless. We must change from passive to active, it is necessary to strengthen their professional training, and improve the comprehensive quality. The person who has a good literacy is charm.

2 BEAUTIFUL LOVE

2.1 Loving the business and being dedicated. We must have the selfless dedication, having a high sense of responsibility for the industry and every student. Know about the nature of each student, their physical health and the purpose of what students learn to fitness for. According to the different occupation, different levels, different training levels, different physical quality, age, gender, hobbies, differ from man to man, step by step guide, efforts to help them achieve their goals, to achieve their ideal.

2.2 Love each person who accepts our instruction. Sincere love will penetrate to their own behavior and language. A look, a movement or a word will convey warmth, encourage, moved to generate power and attract. Improve health and improve body shape are not accomplish at one stroke thing, they need love and encouragement, their perseverance and instructor sincere persevere and unremittingly efforts to achieve. The instructor love in heart and respect, care, sincerity love each student. And they need to make the student feel love, attention and to have the infinite power. Then the guiding process will be more attractive. And between the instructor and students, students and students pass love to each other and mutual infection in this full of fitness loving atmosphere. So they are happy, healthy, and beautiful.

3 PHYSICAL BEAUTY

One of the characteristics of social sports instructor occupation is the great demonstration, so we should have the beauty of body and the power of example. Both collective aerobics lead operator or muscle instructors need to set a direct and specific example in the guidance process for students. The body-building of instructors will make the students to appreciate, admire, yearn, and strive to achieve. Then establishing credibility of instructors, increasing the charm for them, improve trust then the fitness guide process will be harmonious and smooth. Conversely, a fitness instructor who does not have a beautiful body is not going to make students desire and trust. Therefore, to become a qualified social sports instructors, first, they should train hard, plastic body scientifically, constantly improve their own body good image, to make the target of the participants more clearly, so as to participate in the healthy physical and mental activities more actively and be willing to accept the guidance.

4 KNOWLEDGE BEAUTY

4.1 The social sports instructors are service personnel, to make scientific guidance for public health, but also bear the work of teachers. To teach students to enhance physical fitness, knowledge and professional knowledge of physical and mental health, to enable students to master scientific knowledge and skills, benefit for life. Science, correctly guide to theoretical knowledge of social sports specialty of sports physiology, sports anatomy, sports psychology, sports nutrition and so on the basis of theoretical disciplines and related disciplines and teaching requires instructors, as a scientific theory and fitness guide, rigorous scientific instruction. The specialized theory knowledge fitness instructor is not rich, the guidance is not rigorous, as teachers, teaching is not a good student.

4.2 To participate in fitness activities groups from different level, with different quality. Guidance of such groups, not only to the rigorous scientific guidance, scientific fitness. But also continue to broaden the knowledge, has the multi discipline rich knowledge, so that it can help with the fitness communication, interaction, establish harmonious relationship. Otherwise easily be despised as a "simple minded, well-developed limbs". Despise the basis is not attractive, but what about the "attract" "keep" the consumer group.

5 SKILLS

Whether collective aerobic training or muscle training instructors must have a high level of professional skills. φ demonstration: technology moves norms, beautiful generous, play a model role. Can accurately grasp the position for demonstration, the characteristics and the degree of difficulty can skillfully use the back and side mirror model demonstration, demonstration, and at the right time and flexible conversion between various demonstration surface finish; κ explanation: standard pronunciation, language easy to understand, is rich in content. The correct explanation of the action, action, easy difficulty, more should be made clear is apt to make mistakes, so that all kinds of damage to the body of wrong action is easy to produce; methods: flexibility, diversity. According to the different health status, training base, targeted, flexible use of teaching methods, and a variety of teaching methods, to enable students to master the correct action as soon as possible, improve fitness, fitness level; μ scientific guidance: the use of professional theory knowledge, scientific guidance. Guide the students to understand the scientific principles of action training in various forms, different action, different amount of training and different training

intensity to promote body health, physical fitness function. Guide the students "safety" training, master the knowledge self supervision, self supervision system. Aerobic training instructors have set times for different age, different physical crowd aerobics skills. Language and non language tips and motivational skills level, but also the instructor guidance skill.

6 ETIQUETTE BEAUTY

Etiquette is refers to people in various societies used to beautify themselves, respect other people's behavior standard and criterion. Specific performance for the courtesy, etiquette, ceremonies, instrument.

6.1 Of our nation is the etiquette of helping, etiquette in respect of information transfer. Respect for others is the bottom line of our fitness instructors should possess, most human accomplishment. Do not wealth and rank can with due respect, trust each other, make each other feel respected, warm. At the same time through the etiquette also shows a fitness instructor temperament, accomplishment, and win the respect. Therefore, instructors should master the basic knowledge of etiquette, etiquette training, standardize our every word and action, every act and every move, speech and deportment infiltrating our higher perfect self-cultivation. By expressing our polite language, behavior, convey our initiative, enthusiasm, kindness, love and respect to each other, as the spiritual foundation of harmonious guidance, service and lay a good.

6.2 Fitness guide behavior mostly through fitness instructor and tutee mutually close guidance form completion, the exchange and communication for technology, psychological and emotional interaction. So do to us guidance, service of fitness instructors professional etiquette requires higher. Professional standing force, professional model, professional help and protection and professional technical guidance and so are we in this industry etiquette forms.

7 SUGGESTIONS

7.1 Training the social sports instructors as far as possible in the colleges and universities, especially in sports professional students. Because they have learned the professional theory at courses, such as: exercise physiology, anatomy, nutrition, sports injury and repair and so on. At the same time, they have the personal experience and practice experience on the aerobics, aerobics, ball games, track and field sports training. In addition, edify sentiment in education and university education is a good character, only strengthening the cultivation of comprehensive quality comprehensively in the teaching. The university also can be benefit by constantly exploring and practicing, training mode form practical college, develop a high level of fitness instructor.

7.2 The government and the society of social sports instructor training course content added to class etiquette. To overcome the shortage of only pay attention to short-term training techniques and skills. Not only needs to have the professional etiquette training, but also strengthen our fitness guide industry etiquette. Such as: start from the most basic stand. Not to stand in the way. Show out the self-confidence, station-respect; help and protect people; on the site, be with equipment-respect.

7.3 Quantitative fitness instructor physique evaluation index. And make it as a fitness instructor qualification will be one of the factors, and the number of annual inspection. This is conducive to aspiring to become a fitness instructor or have been engaged in the work of fitness instructor pay more attention to their own image shaping. To keep healthy, beautiful form is not only plays the role of an example, set up the visual image, but also can prolong the working life of the fitness instructor.

Fitness instructor is the beautiful angel. Transfer the information through the guidance activities, create beautiful atmosphere, building good results.

It is a practitioner of beauty, guiding the healthy beauty, form beauty. Transfer of industry to establish a beauty, the beauty of nature. Beautify human, beautify the society and nation, the sacred and great.

REFERENCES

[1] Wu Xiaoling et al. Study on the current situation and Counter measures of social sports instructors [J]. 2004–09–20 Journal of Beijing Sport University city.

[2] Li Xiaoyan; Peng Yunzhao et al. Give full play to the advantages for the cultivation of social sports instructors service [J]. 2003–02–20 Journal of Wuhan Sports Institut.

Sports Engineering and Computer Science – Luo (Ed.)
© 2015 Taylor & Francis Group, London, ISBN 978-1-138-02650-6

Explanation and demonstration of the football course

Jian Liu & LiJun Xu
School of Physical Education, Northeast Dianli University, Jilin City, Jilin Province, China

ABSTRACT: Soccer special lesson explanation and demonstration is different from public class. Their requirements are for a higher standard. This paper puts forward the explanation to "finish, transparent, quasi" and demonstration to "quasi, fine," in order to improve the quality of explanation and demonstration to ensure the effectiveness of teaching.

Keywords: soccer special lesson teaching; demonstration

1 INTRODUCTION

Football specialized course is an elective course for college students, and also the largest one loved by students, and the teaching effect, teaching quality requirements will be higher. According to the actual situation of students, innovation is an important topic in football teaching. Based on years of teaching experience, sum up on the soccer special lesson teaching and demonstration of the "six words", and these are quite big to be used in football teaching gains. [1]

2 THE EXPLANATION OF "FINISH, TRANSPARENT, QUASI"

The explanation is the teacher in the football course is in the use of oral language teaching task, to the student the action name, technical essentials and operation requirements. It is a means of teaching practice to guide the students to master the basic techniques and skills, action, so that students can accurately understand the whole process of the football action and ensure good teaching effect.

2.1 *The explanation of "refining"*

In the explanation of the teachers according to the different items and requirements, grasp the teaching key, outstanding key and difficulty for the key links technology action part of concise and comprehensive. In the idea, make clear that students should firmly grasp the key technology. For example, the positive dribbling or outside instep dribbling, the focus is to require students to master the support leg (leg) and kicking foot (leg) effect. I'm in a lecture requires only beginners to stand up, instep positive or lateral gently push the ball.

And pay attention to speed, running speed to be consistent on the line. The slow to learn, dribbling dribble teaching objective. Let the students imperceptibly master support leg (leg) and kicking foot (leg), and then teaching any kind of football footwork becomes easier.

2.2 *The explanation of "thoroughly"*

The key point for teaching basic technology to speak clearly, should step further explanation. The breakthrough refers to the chart. Demonstration, metaphor and humorous language are necessary. It helps students to deepen understanding of key parts, and infer other things from one fact. For example, several football footwork just kick foot ball position (each position refers to the ball and foot) different. The general play basic technology contains the run-up and run point of view, the last step of leg supporting, kicking leg swing, foot ball and ball feet of their respective positions. They are playing the parts of common technology. Make the students clear a playing method, then teach way to play another, through the common point to analyze the different points. It is easy to master all kinds of footwork.

2.3 *The explanation of "accurately"*

Explaining to accurately refer to the explanation of the contents, and it must be correct, scientific, and consistent with the requirements of the syllabus. Use the most concise language (jargon), the shortest time, the concept, action essentials, visually clear, level and explain. Enable the students to listen to understand, easy to understand and remember it. For example, in the study of the "two a" tactics teaching, I was the first to make clear to students what is the "two a" tactical purpose and intention, whether to take "inclined inserted straight biography" "cross" protective "wall kicking" in line "s" kind of tactics, is to get rid of each other's defense, never to foul. Especially in the last line of defense, you must accomplish quick passing, and there can be no slow and hesitant. Otherwise, it may cause the offside. I often combine the teaching and training examples to give students the offside, analyze, and explain. Let the students in the "two a" tactical exercises at the same time, strengthen the flexibility and agility exercises and improve their level of skill and tactics.

3 DEMONSTRATION OF "QUASI, FINE, AS"

In football teaching process, the teacher puts the whole skill demonstration as the complete specification to the students to demonstrate the use of body language, so that students master the theoretical knowledge to explain the image of the digestive process, understand the action of the structure, sequence, and image as well as the essentials and methods, so as to better imitate and form the correct movement imagery. The teacher's demonstration is the most vivid; it is most suitable for football teaching in a visual means of teaching.

3.1 Demonstration of "accurately"

The purpose of the demonstration is to enable students to understand some technical movement image. It tells the student what this skill is. Therefore, in the process of teaching, teachers should carry on the demonstration according to the actual situation of the teaching tasks and the students, such as learning materials. In order to enable students to establish a complete concept of action, teacher should do first complete technical action demonstration, repeat the action key technology demonstration before the teacher must tell students to watch demonstrations direction, location and time, and enable the students to understand how to observe the teacher as soon as possible to complete the action demonstration according to their situation, and seize the key. The teacher's demonstration action priorities for specification, coherence, complete, beautiful. After years of practice I felt that accurate, skilled teachers, a beautiful demonstration can eliminate the psychological barrier of students, to stimulate students interest, and mobilize the enthusiasm of the students, which is to promote students to establish the correct action essentials. It is very important. For example: teach instep kick, I took this action essentials specification, coherence, complete, beautiful show to students: oblique run-up stride, a foot on the ball

side, kicking leg knee to turn, toes pointed out the ball, instep shot. Help students to stimulate the students' interest, intuitive, and make it easy to deepen the understanding of the correct action.

3.2 Demonstration of "refinement"

Demonstration of "refinement" is a skill or technology demonstration, according to the actual situation of the students, teachers should take the decomposition model and normal speed and slow down the speed of the demonstration, to promote students' error correction and deepen understanding. As for the new curriculum, teachers should use the normal velocity model as a complete technical movements, and it can complete technical structure and enable the students to understand the textbook, and then according to the contents of the lectures in slow decomposition model, enable students to understand the action essentials, requirements, and the establishment of a complete action representation. [2] In addition, teaching video available football normal playing and slowly playing demonstration make up for the deficiencies and increase the effectiveness of the model to explain. And in the course of practice, teachers should master the technique movements good students practice according to the specific situation of students problems, then the teacher should analyze when is necessary. Teachers can imitate the student of wrong action to be compared. In this way, the correct technical movement will have a more profound impression on the students' mind. Thus it can improve the teaching effect.

3.3 Demonstration of "visual"

Demonstration teaching is to give students example. This must let all students see. Therefore, the teacher's demonstration is not only to regulate, but also pay special attention to the position and the aspect of demonstration. Visibility has a full range of it. Demonstration of the position to choose according to their formation, action properties and safety requirements to the best position for demonstration and lead the students to practice. As in basic action teaching professor, the line formation, teachers should stand in the equilateral triangle vertex position demonstration line. And as the "ball shooting practice" teaching, let students stand on either side of teachers' demonstration, which enable students to line of sight is always with the teacher's demonstration action mobile. In actual teaching, demonstration method the position of the most can let students to observe the teacher's demonstration action. In the "visual" must pay attention to the direction of the demonstration, should according to the structure and requirements of the student movement, observe the movement's location. Teachers should try to demonstrate the direction of motion, route and students and do the direction, route consistent, can adopt the positive demonstration, side demonstration and back side mirror model and so on. If the location and direction of teacher demonstration improper selection, will affect the part of the students due to not see complete, correct motion coherence and illusions, forming technology concept wrong, lose the exemplary role, directly affects the teaching effect.

4 THE EXPLANATION AND ORGANIC COMBINATION OF DEMONSTRATION

In the football class, teacher's demonstration is the foundation of students' perception action external image, and the explanation is an important tool to enable students to understand the inherent law of action, both used to shorten the process students understanding of technical action, improve the effect of classroom teaching.

4.1 *Need to explain the demonstration match*

Football is a theory and practice, knowledge and technology combined with the course, need the teacher's spoken language and body language to complete, class, teachers should on the edge edge model, students can understand, see, and learn.

4.2 *The model needed to explain the follow up*

The demonstration aims to understand and master the knowledge and skills. Especially in decomposition model, to do an action essentials must also explain the requirements; in the correct action requires, demonstration shouldn't be out of sync with the explanation Practice has proved that in the teaching process, only the explanation and demonstration are in combination with the concept, will students establish a complete, correct technical movement and form the correct representation to improve the practice effect.

In short, in the soccer special lesson, teachers adhere to "ensure plenty of practice" principle, give full play to the role of oral language and body language in teaching, the explanation and demonstration closely, promote students' intuitive feeling and thinking activities organically, to good effect, to improve the quality of football teaching.

REFERENCES

[1] "In modern football". The people's sports press [M]. 2000.6.
[2] Xu man. In PE teaching explanation and demonstration. Education in the [J]. 2013.08 Era.

Sports Engineering and Computer Science – Luo (Ed.)
© 2015 Taylor & Francis Group, London, ISBN 978-1-138-02650-6

Feasibility analysis of deepening the development of winter tourism in Jilin province

Jian Liu & LiJun Xu
School of Physical Education, Northeast Dianli University, Jilin City, Jilin Province, China

ABSTRACT: Jilin province is located in the center of Northeast China, where winter tourism resources are very rich. But the winter travel in Jilin province is not prospect and has not been developed completely, compared with the Heilongjiang province and Liaoning province. Through analysis, we have come to the conclusion that there are advantages and disadvantages of deepening the Jilin province's winter tourism. Thus, we shall seize the opportunities, and face the challenge, then promote the development.

Keywords: Jilin province; winter tourism; feasibility

1 INTRODUCTION

Jilin province, which the geographical position is superior, has the better natural winter tourism resources than the other two provinces, and has abundant resources of snow and ice. While the meantime, the winter tourism has lagged far behind the neighbor, Heilongjiang province and Liaoning province. During the Spring Festival of 2009, the "Golden Week" tourists in Jilin province reached a total number of 2,541,500, and the tourism revenue reached 1,398,000,000 Yuan; while the same period in Heilongjiang province reached 6,000,000 in tourists and 4,050,000,000 Yuan in tourism revenue; and Liaoning, has welcomed more than 5,559,000 tourists and received 2,730,000,000 Yuan. It's good in the first stage of the development, but the resources development, utilization and development are far from enough, not strong enough. [1]

2 FEASIBILITY ANALYSIS

2.1 *The advantage analysis*

2.1.1 *Unique advantages in natural resources*
There are Changbai Mountains, Songhua River, Songhua Lake and rime ice and snow spots, such as: the magnificent Linhaixueyuan, beautiful pure white snow, Matsuyama snow and ice in the open-air hot spring. Changbai Mountains is one of the world's longest snow period resort. The snow period is 150 days or so. During November, people can ski. The only no frozen river, Songhua River, is located at the northeast Jilin. The "Jilin rime" on the riverbank is one of the four natural wonders in China. Mist filled, Song flowers blossomed, like a "fairyland." The Songhua Lake is vast in area, numerous in lake arms, like a dragon lying in the mountains. In winter, the Songhua Lake is surrounded by mountains covered with snow. People can go hunting or ice fishing in this season. The Chagan Lake and other lakes are the vital resources of developing winter tourism in Jilin province.

2.1.2 *Historical and cultural resources*
Having lived with ice and snow for a long-term, Jilin people gave birth to the traditional national snow and ice culture and has a strong national characteristics and regional customs—Kanto cunstoms—in clothing, food and shelter. In addition to skiing, skating, ice, ice skating, luge, bobsleigh, car, pumping ice monkey, the ancient and unique customs of ice fishing and hunting are not inherited. "Ice lake Teng fish"—a kind of winter fishing in Songyuan Chagan Lake which started from Liao Dynas—has now been listed as a national intangible cultural heritage; Yanbian Korean Autonomous Prefecture is China's largest Korean nationality, folk culture and folk unusually are rich. Siping Yehiel Manchu folk culture, Jilin city of Jurchen folk culture and so on have great potential development.

2.1.3 *Traffic resources*
On the one hand, Jilin province is located in the central part of Northeast China. It's an important channel towards north and south. Aviation, railways and highways form a three-dimensional traffic network, Changchun is the center. On the other hand, with the construction in air route, expressway, trunk road, traveling road and other major transport projects, Jilin gradually

forms a three-dimensional transport network. In September 28th, 2008, expressway go through Jilin to Yanji is opened to traffic, it is 420 km long and it only takes 3 hours. The distance between Jilin province and the centre Changbai Mountains region is shorter. From August 2008 Chinese first forest tourism airport—Changbaishan Airport, visitors are easier to get to Changbai Mountainss. In addition, high speed train was open in January 11, 2011. Jilin—Beijing, Shenyang north, Harbin, Changchun have opened also. The intercity railway between Changchun and Jilin has a 20 round-trip daily, train departure time interval is only 50 minutes. It also goes through Changchun Longjia airport. It's good for winter tourism.

2.2 Analysis of problems of restricting the development

2.2.1 Lack of management system

The enterprise owns strong randomness. The development and construction are repeating a similar theme. Lack of management results in a waste of resources and destruction of the natural resources, and then leads into the situation of scattering, poor and weak. Thus, the development of winter tourism cannot form scale. It is difficult to form the tourism industry system, not to mention a bigger and stronger tourism system. This situation has seriously affected the overall competitive strength of winter tourism industry in Jilin province, and restricted the rapid, healthy development of Jilin winter tourism.

2.2.2 The lack of a strong image of ice and snow tourism brand

When it comes to the Heilongjiang ice-snow tourism, you will think about Harbin ice landscape. The ornate, fascinating, snow world looks attractive December to January in Harbin city.

Our Jilin winter tourism is still not an impressive tourism brand. The development of Changbai Mountains is not thorough enough. Jilin Beida Lake skiing field has the better software and hardware facilities than Yabuli skiing field. However, after 2007 the Asian Winter Games, the owner of the field has changed several times. Moreover, this bad business situation weakens the competitive of the filed. Changchun Jingyuetan ski field, whose has only one snow road, restricts by the lack of snow road. Chaganhu winter fishing, which has been broadcast by CCTV this year, is known to every family now, but the project is single, extension isn't enough. "Jilin rime" is a spotlight, but the uncertainty of rime makes it less attractive to tourists from far away. [2]

2.2.3 Lack of professional talents cultivation of winter tourism depth

To build Jilin's winter tourism needs professional support. On the one hand, because of the strong seasonal characteristic of winter tourism, the practitioners themselves have no push to promote their specialties in business and improve their professional level; on the other hand, staff's service quality and the understanding of winter tourism industry are not enough. Then the practitioners of great professional acknowledgement are in need. Some of the employed are often know only some aspect of winter tourism. Most of the winter sports professionals' technical guidance abilities are not enough, and they lack professional theory knowledge.

2.2.4 Lack of humanistic foundation

The popularity of the winter sports is the basic humanistic development of ice-snow tourism. The current number of participants of winter sports in Jilin province is small. Professional winter sports team has been in decline in recent years, so have the athletes. At the same time, there are not enough sports participants. Even in severe schools, the winter sports facilities are difficult to see. There are few schools poured ice in winter. Missing of the representative ice and snow sports in school causes the traditional winter sports such as skating, ice monkey car pumping activities can't inherit and develop. The winter sports teaching is like a sowing machine, it had the opportunity to promote itself to the students from all over the country. Unfortunately, there is not enough ice field. [3]

2.3 Conducive to the development of the opportunity analysis

2.3.1 The strategy of rejuvenating the old industrial base in Northeast China implementation provides an opportunity for the deep development of ice snow tourism in Jilin province

Since 2003, a series of measures and planning to revitalize northeast old industrial base are pointed out: "fully employ the rich tourism resources and the unique advantages in the northeastern region, vigorously develop the tourism industry," the development of winter tourism industry in Jilin province provide a hitherto unknown opportunity. Government at all levels in Jilin province has given great attention to the tourism industry. Government's support is increasing, which in turn will further the development of winter tourism. [4]

2.3.2 The comprehensive development of national winter tourism is the challenge and chance

Jilin province is located in the northeast of intermediate position; the topography determines the winter natural resources in Jilin province are better than those of Heilongjiang province, Liaoning province. Liaoning province hits the brand of "snow tourism in the first station." Heilongjiang ice-snow tourism leading to show. These are strong challenges to the Jilin province winter travel. In the "Golden Week" during Spring Festival of 2009, the total tourism income of Jilin province is 1/3 of Heilongjiang province, 1/2 to Liaoning province, which has brought Jilin winter tourism tremendous pressure. Challenges and opportunities coexist. In 2009 the Jilin Provincial Committee, the provincial government hired a Beijing Davos Summit Tourism planning and Design Institute of planning to design the ice and snow tourism industry in Jilin province, promoting the in-depth development of the winter tourism.

2.3.3 *Rapid development of national economy is a good opportunity for winter tourism development*

With the rapid development of the national economy and national strength, people have improved their living standard. People's concept of life, way of life is changing. Tourism, vacation, leisure and entertainment have become an indispensable part of people's life. Pay attention to the tourism information, understand the tourism hot spot, and change direction of tourism. With all these methods, the northern winter travel has grown strong. In winter, you can go southward to Hainan, and northward to snow and ice. Winter culture has become hot, and provided the opportunity for Jilin province to develop winter tourism in depth.

REFERENCES

[1] Gan Jing. Study on the ice and snow tourism development in Jilin province [D]. Northeast Normal University, 2009.
[2] Zhang Xin, Liu Yifei. Jilin formed Changbai Mountainss three-dimensional transportation network [N]. Chinese traffic report, 2009-8-5.
[3] Zhang Hongjiang. Ice and snow culture in modern sense [N]. Jilin daily, 2010-1-21.
[4] Sun Daowei, Chen Tian, Jiang Ye. Changbai Mountainss nature reserve of tourism resources exploitation and ecological environment protection measures of [J] Journal of Northeast Forestry University, 2005.5:97–99.

Sports Engineering and Computer Science – Luo (Ed.)
© 2015 Taylor & Francis Group, London, ISBN 978-1-138-02650-6

On the development of Ethnic Traditional Sports from a sociological perspective

Na Cheng
Jilin Institute of Physical Education, Jilin, China

ABSTRACT: In the new century, the social transition of the mode of production and the increasing demand of the health require that ethnic traditional sports constantly makes reformation, innovation and active adaption. At the same time, the development of ethnic traditional sports also indirectly promotes the development of many aspects about politics, economy, culture, etc. The essay analyzed the development orientation of the ethnic traditional sports in China in the future from the perspective of sociology view. Finally, it also had an important realistic meaning for protection, promotion and development of ethnic traditional sports.

Keywords: sociology; Ethnic Traditional Sports; development

1 INTRODUCTION

With the tide of globalization in 21st century, there is a cultural trend of different countries and regions towards unification of ethnicity and world. Ethnic Traditional Sports, as a part of this integrated whole, is no exception. Ethnic Traditional Sports are the window and platform of communication of sports culture between our country and the world. We will lose this platform of communication if we don't have Ethnic Traditional Sports.

On the contrary, if we don't develop Ethnic Traditional Sports in the category of the world's sports and enrich and develop the sports career of mankind, Ethnic Traditional Sports will lose the power and direction of its development.[1] Ethnic Traditional Sports play an important role in the culture and life of modern people. It is necessary to study Ethnic Traditional Sports from the perspective of sociology. Thus it can promote the development of Ethnic Traditional Sports in a healthy and scientific way.

2 THE DEVELOPMENT OF SOCIETY PROMOTES THE DEVELOPMENT OF ETHNIC TRADITIONAL SPORTS

2.1 *The change of the social mode of production is the impetus of the development of Ethnic Traditional Sports*

Ethnic Traditional Sports need to adapt the demand of the modern society because of the improvement of lifestyle and the quality of the life. The development of market economy in Well-off Society will promote the development of productivity and the improvement of material living conditions. Meanwhile, it also enhances social competitive. While people enjoy the rich material life, they have to handle larger stress and more stressful work. People can't keep healthy without sports. So local Ethnic Traditional Sports have full of vigor and is deeply loved by the majority. Ethnic Traditional Sports differ with Competitive Sports, because the participation constraint condition can be easily satisfied and on the aspect of restrict rules is not perfect. It also promotes the evolution and innovation of the some parts of Ethnic Traditional Sports items.

In the progress of building a well-off society in an all-round way, the combination of Ethnic Traditional Sports and Entertainment reflects people's pursuit of the happiness life after the economy development reached a level. People are eager for the Leisure Sports when well-off society completely comes. It shows that Ethnic Traditional Sports are concerned about humane and unrolled the real charm of Ethnic Traditional Sports as Chinese culture. It is perceived that the change of present society and lifestyle shows the higher price of exercise in Ethnic Traditional Sports.

2.2 *The development of society promotes the modernization of Ethnic Traditional Sports*

Chinese Ethnic Traditional Sports are a product of Chinese traditional agricultural society in Han Nationality and nomadic society and hunting society in ethnic minorities. It is a cultural form adapts to natural, economic environments and the cultures of multinational folk customs. Many of Ethnic Traditional Sports are combined with local production and life, like "Beating shoulder poles" of the Zhuang Ethnic Minority, "Jumping bamboo poles" of the Li Ethnic Minority and so on. Some are connected with festival manners and religious rituals, such as "Flying kites" in South of the Five Ridges, "Beating Lotus" in some areas of Hakka. It has own tradition.

With the constantly deep process of industrialization and the coming of globalization, the environment that Ethnic Traditional Sports depend on is changing gradually even disappear. If Ethnic Traditional Sports want to keep on existing and develop in modern society, the most important thing is the modernization of Ethnic Traditional Sports.

It is except for the protection of it from the perspective of culture. On one hand, we should make Ethnic Traditional Sports separated from religion and folkways. Besides, we should make it weaken its functions of Religious rituals and highlighted its functions of athletics, exercise, entertainment and education. On the other hand, with the theories and methods of modern sports science, we should draw on the experience of the excellent consequences and reform Ethnic Traditional Sports. In addition, we should also make the way and rule of sports and the methods of judgment adapt to the operation of law about modern sports and the Olympic spirit. Let it won't lose its national tradition completely. Finally, let it make it be equipment with the modernization of modern sports, adapting to the change of new environment.

3 THE DEVELOPMENT OF ETHNIC TRADITIONAL SPORTS MAKE THE DEVELOPMENT OF SOCIETY MORE PERFECT

3.1 *Ethnic Traditional Sports have been an indispensable part of social lifestyle*

Ethnic Traditional Sports items were created in the long common social life and productive labor of all nationalities. It is a form of exercise from generation to generation. A form was created by own nationality, because it has most national characteristics, and was connected with their ways of life and production. Its roles and contents of the activities are spread in the range of own nationality. It is strong self entertaining and seasonal, so many activities are held in the most important festivals. It is a conservative and complacent mode amusing itself, so we have to take action to reform and innovate and follow the mode of the whole life and the

localization. With the development of economy in the present society, the improvement of people's living standard the more requirements of health, people show an active attitude to the sports. But because of the restrictions of many reasons like economy and culture, there are many problems about the development of mass sports in China. The development of Ethnic Traditional Sports is the urgent requirement of public.

How to make the native of Ethnic Traditional Sports combines with the popularity of Mass Sports organically. It is an important realistic meaning to the development and popularization of Mass Sports in national areas.

3.2 *Ethnic Traditional Sports have been an effective gateway to promote the development of local economy*

The cultural anthropologists found that 20th century is a century that was inherited found and created consciously. But 21st century is a consciously creative time and culture passes for a kind of resources to be used to develop local economy. Nowadays, "Culture sets up the stage, economy put on a show" is a most popular mode of development of local economy. This economic development model promoted the development of local economy subjectively, but it retained, inherited and carried forward local traditional cultures.[4]

Ethnic Traditional Sports are as a kind of existing form of Ethnic Traditional Cultures. With the help of this kind of mode of development of local economy, Ethnic Traditional Sports can also combine with economy and market and look for the own space for development. We can do as follows: a) select the traditional sports items of distinctive characteristics and effective exercise to hold technical trainings. The traditional sports items are training classes of fitness martial arts and Chinese shadow boxing; b) select some items of good ornaments and participation to combine with the resources of national tourism, curriculum tourism of Ethnic Traditional Sports. The sports are shows of Traditional Sports culture in areas of tourism, rowing the dragon boat, Bamboo Dance of the Yao Ethnic Minority and so on; c) select some items with good skills and competitiveness to explore the market of competitive performance, such as "Darwaz" of Uygur, "Climbing a mountain of swords" of Hakka, etc; d) perform the culture of Ethnic Traditional Sports through the different ways of the market development. For example, Erdos international Nadam Fair was held by the Inner Mongolia; e) explore the markets of invite business and tourism to promote the communication of cultures in different areas.

3.3 Ethnic Traditional Sports have been an important method of promotion to social stability and unity

One of the main foundations of national culture is producing and strengthens the social cohesion power of a nation. The culture of Ethnic Traditional Sports contains the large concentric power and national cohesion power. It is a bond to link the emotion of nations. For example, there are large and grand traditional festivals in non-governmental circles of some national areas. The festivals are called "Guizhou Plateau" "nongx mol" of the Miao nationality, "Ox King Festival" of the Bouyei nationality, "Da's day" of the Yao nationality, "Duan Festival" of the Shui nationality, etc. These Chinese traditional festivals are a part. There are Chinese traditional competitive sports, national technology and national customs.

In fact, with the show of national songs and dance, they make brothers of different nations have a chance to come together and communicate with each other. Then know about each one deeply. The communication of life style and aesthetic of every nation sets national festivals as a sign, and put our national culture into the national festivals, from generation to generation. At the same time, it can illuminate the ambiance to achieve the aim of enhancing national unity and cohesion power.

4 TREND OF DEVELOPMENT OF SOCIAL ETHNIC TRADITIONAL SPORTS IN THE FUTURE

4.1 International trend

The functions of Ethnic Traditional Sports become stronger and stronger, and it make Chinese government pay most attention to. It unearthed and sorted the excellent culture of Ethnic Traditional Sports, and took all kinds of measures to strengthen international communication, for example, International Nadam Fair was held in Mongolia Erdos in 2012 was an typical case of international trend. With the development of new technical revolution in the world, longer leisure time and vigorous development of Mass Sports, the development of international communication and academic research about Ethnic Traditional Sports will develop a lot.

4.2 Scientific trend

In 2005, "CPC Central Committee and the State Council on promoting the construction of a new socialist countryside views" puts forward to lead farmers to do the next things: a) advocate science; b) boycott superstition; c) transform social traditions; d) break bad habits; e) and set up advanced ideas and good moralities, advocate scientific healthy and civilized lifestyle.

As a result, scientific exercise is the development orientation of Mass sports in the new times. In order to adapt to the wide requirement of exercise from people in China, Ethnic Traditional Sports make a positive difference. At the same time, Ethnic Traditional Sports use the theories and methods of modern technologies to raise the technical level, appreciation and rationality of Ethnic Traditional Sports. Meanwhile, the theories and methods include policy decision, management, and scientific research of Ethnic Traditional Sports and so on. As it stands of every delegation that participating many kinds of Ethnic Traditional Sports, some new philosophies, new training methods and new technologies from competitive sports will be applied on in time.[6] The functions, position and effects of Ethnic Traditional Sports become remarkable day by day. The development of Ethnic Traditional Sports will go to the scientific direction soon in the future.

4.3 Living trend

When people enter well-off society, the position of health will be raise in people's mind. The attitude that passive exercise was changed to be active. Ethnic Traditional Sports will be an indispensable part in people's lives and promote living deeply. In 2005, "CPC Central Committee and the State Council on promoting the construction of a new socialist countryside views" puts forward to promote the project of exercise for farmers and hold various forms of popular culture and physical activities in teaching through entertaining.

Community is an epitome of society, also is a station of life. The developments of society and economy realize the functions of community. People depend on community highly, so social sports are not the sports of retired people in the traditional conception. It belongs to all the people in community. Meanwhile, we can see the increase of the leisure time and high attachment to family, the rise of mental position and obvious strength of participate actions. Thus, we have to pay attention to social sports adequately to realize living sports. Social sports can promote living sports become deep further.

4.4 Industrialization trend

With the good foundation of the public, Ethnic Traditional Sports propagandize widely step into industrialization gradually and promote the development of local economy.

Erdos international Nadam Fair is a typical example. In June 28, 2010, Erdos held an introductory conference of Erdos international Nadam Fair. The aim of it is to dig out the market potential of thousands of Erdos international Nadam Fair fully. Besides, the aim is to strengthen the reputation and influence as well as activate the culture, economy and resources of Erdos international Nadam Fair. Furthermore, the aim is also to realize the communication and interaction between government and corporation, and propel the market operation of local culture brand. In the introductory conference, Erdos international Nadam Fair's executive committees of Merchants Department represent the People's Government signed a cooperation agreement with Mengniu Group, Wulan Group, China Mobile, Erdos Wines and other tens of corporations. This introductory conference had important reference meaning to Erdos, autonomous regions and even whole country. Especially for Erdos, it was a large international magnificent sports meeting. It also was a window of Erdos advances towards the world, and let the world knows about Erdos more.

Erdos international Nadam Fair was using "International Erdos, Grassland Carnival" as a medium. The purpose is: a) to build a national characteristic brand of international culture sports; b) promote international broad; c) profound grassland culture; d) develop the colorful career of Ethnic Traditional Sports; e) try to enlarge and strengthen the soft power of harmonious development about economic society of Erdos.

Erdos international Nadam Fair is a grand Nadam meeting with the highest specification, largest scale and most colorful activities. It is a grand meeting of culture sports, and a grand meeting of a national union. In addition, it is a grand meeting to realize the largest economic returns and social returns. The government of Erdos has registered and set up Global Nadam Sports Culture Co. Ltd to take great pains to operate the activities of Erdos international Nadam Fair. The company is also to promote it as an international broad of culture sports. Global Nadam Sports Culture Co. Ltd

is a company that prepared to be a listed company in Growth Enterprises Market.

At that time, many activities of Erdos will be in motion by the Global Nadam Sports Culture Co. Ltd. Finally, the activities can make positive contributions to Chinese Ethnic Traditional Sports advance towards the World.

5 CONCLUSION

From the perspective of sociology view, Chinese ethnic traditional sports displays its own price of function, and tries to integrate with the world sports. At the same time, the social development also needs to promote the excavation, arrangement and popularization of ethnic traditional sports. Ethnic traditional sports have to take a chance from the times. In the end, the sports can adapt to the current social situation positively and achieve the own value of politics, economy and culture widely. At last, ethnic traditional sports make positive contributions to democratic sports.

REFERENCES

[1] Xiao Rong Xu. The Sociological Thinking of The Tendency of 21-Centry Chinese Traditional Sports[J]. Jia Ying College Academic Journal Natural Science), 2005, 2(3): 124–126.
[2] Pi Xiang Qiu. Introduction of Traditional Nation Sports[M]. Beijing: Higher Education Press, 2008.
[3] Xiao Ming Hu. Nation Sports[M]. Guilin: Guangxi Normal University Press, 2005.
[4] Yuan Rao, Bin Chan. Sports Anthropology[M]. Kunming Yunnan university press, 2005.
[5] Yi Ke Ni. On Chinese Traditional Sports[J]. Sports Science, 2004, 24 (11): 54–61.
[6] Zong Hu Qu. Traditonal Nation Sports in School[M]. Beijing People's sports press, 2002.
[7] Wa Gao. The Research of The Modernization of Traditional Mongolian Sports[J]. Shandong Physical Education Institute Academic Journal, 2009, 25(11): 21–25.

Sports Engineering and Computer Science – Luo (Ed.)
© *2015 Taylor & Francis Group, London, ISBN 978-1-138-02650-6*

Regional sports resources integration research on influence factors and mechanism

BaoBin Duan

Physical Education College, Jiujiang University, Jiujiang, Jiangxi, China

ABSTRACT: By using the methods of literature, logical reasoning, this paper analyses the influence of the system integration of sports resources in the area. According to the result, there are five main factors affecting the region integration of sports resources. These are the natural and geographical conditions, the level of economic development, land factors, government policy and the demands of social and cultural, sports, leisure sports. And two aspects of power driven mechanism of integration of sports resources have been put forward in the area, namely the dynamic factors and constraints.

1 INTRODUCTION

According to the fifth national sports field survey, compared with those in the early days, the number of our country sports venues has increased by 170 times, the area of each capita sports venues has grown 103 times. But the relative lack of fitness and people's increasing demand of sports are still the principal factors restricting the development of Chinese sports. With the speeding up of urbanization process, many changes have taken place in the city. Spatial structure, the size and distribution of population, the structure and quality the sports venues area and space layout can meet the needs of the construction and development of city. To meet the people's increasing needs of fitness becomes a real problem for the future construction. Therefore, in order to construct the public service system of national fitness, it is of great significance to explore sports resources integration of the region's influence factor and mechanism, to construct the regional sports mode of resource integration and promote the coordinated development of sports and the area.

2 THE CONNOTATION OF SPORTS RESOURCES INTEGRATION WITHIN THE REGION

About the sports resources in the area integration problems, its focus is actually the problem of resource allocation and planning. The distribution of regional sports resources is a chaotic system. Therefore, according to the basic principle of system theory, the so-called "integration of sports resources in the area" refers to a certain region with the existing sports venues resource allocation on the basis of the regional natural resources and social and economic development level and people's physical needs. Its aim is to make the area sports resources to achieve the best running status in the integration of the system and the measures taken. The essence and meaning of physical education resources integration within the region should be the regional allocation of sports resources rationally and highly.

2.1 *The rational allocation of sports resources*

The so-called regional allocation of sports resources rationalization, an allocation of resources, is an organic link between elements of polymer with good quality. The two refer to the allocation of resources to the nearest. It meets the physical needs of the people. The comprehensive benefit can produce good sports facilities (including social benefit and economic benefit). From the angle of the system theory, due to the organic connections within the system and orderly combination, various elements in the system constitute the overall structure of the system. And the system is more than a simple arithmetic. Instead, in this system, the whole is greater than the principle of rotation parts. Therefore, the combination type elements of sports resource allocation structure are different, and the system structure is also different. Consequently, according to the regional natural resources, population, physical needs and preferences, as well as regional social and economic development level, city development goal, sports resources allocation and the essence of sports resources integration have the following characteristics. Firstly, the sports resources structure elements allocation to regional entity structure, Secondly, the allocation of sports resources in the area to form different combination types. Thirdly, the physical and social benefit and sports economic benefit. In general, the social benefit and economic benefit of the sport resource allocation can't be ignored. The higher the aggregate quality resource allocation is, the higher the level of the rationalization and optimization of utility is.

2.2 *Allocation of sports resources height*

The allocation of sports resources in the area of process structure refers to the development of sports resources from the low level to the high level. The spatial structure of the level of complexity, diversity, structure and function structure are closely linked. The ordered structure state is spatial structure from lower form development basic symbol. And as for the need to structure level, most people are progressive, dynamic and continuous regenerative. Obviously, along with the continuous improvement of the socio-economic level, people physical needs also gradually change, from simple to complex, from single to multiple, from elementary to advanced. Changes in the law of development and resource structure of human physical demand determine the structure of sport resource and characteristics of resources in the development.

The importance of the integration of sports resources in the region lies on rational allocation of sports resources in the region. Rational allocation of sports resources is the element task. In fact, the high level of rational allocation of sports resources is interaction. In fact, the growth process of this interaction is the process of integration of resources and development.

3 FACTORS AFFECTING THE INTEGRATION OF SPORTS RESOURCES IN THE AREA

Many factors affect the sports resources integration in the region, such as geographical factors, population factors, land factors, location and traffic conditions, economic development level, historical reasons. The natural and geographical conditions, the level of economic development, land factors, government policy, social and cultural, sports, and leisure sports demand are the main factors affecting the region integration of sports resources.

3.1 *The natural geographical environment*

For any country or region, the growth and development of sports resources structure mostly result from the most powerful topography and climate conditions. For example, because of the climate condition of the thick ice and snow all day and night in Harbin, ski field is the main sports venues. While in China's coastal city, Dalian, backed by the Northeast river resources, facing the blue sky blue sea, has 1900 km of coastline and numerous islands of natural resource advantage. Along the coast, water sports venues cluster. Apparently, the natural geographical environment plays an important role in territorial sports resource distribution. Therefore, it influences the scale of the sports activities. Of course, natural condition is the material basis for survival and development, which has certain influence on the emergence and development of sports. But it is not the decisive factor. Society contains the various social relations. Sports play an important role in its development. The natural and geographical factors of sports resources integration are only the rigid constraint factors. Based on social development conditions and their own needs, human beings build new sports environment by their strong activity and technology.

3.2 *Economic factors*

Sports resources are the carrier of the development of sports. They influence the economic environment. The importance of integration of sports resources is self-evident. It can be shown in national economic development. National economic development mode is different from that of the political system. It is the inevitable result of different economic systems functioning together. Sports system is the product of the political system and economic system. Therefore the different economic system operation mode determines the different operating modes of different sports system. Accordingly, sports system operation mode determines the allocation of sports resources in different ways. Therefore, the regional economic strength is abundant. The higher the demand is for sports and sports resources in the region, the higher the density is. And the sports will be more advanced. With more complete and more reasonable structure, development is relatively perfect. So more support should be given to the development of the regional economy for the construction of sports resources integration.

3.3 *The factors of land*

Land is a non-renewable resource. The achievement of China's rapid economic growth and the fast urbanization is at the cost of land resources. For all the industries, the demand on land is growing. The sharp contradiction between people and land become the bottleneck of development of all industries. As we all know, the optimization of sports venues and the layout of facilities are based on a range of land. Therefore, how to obtain the land is a key problem for regional sports venues and the optimization of the facilities layout.

3.4 Human factors

3.4.1 Governmental sport policy factors
The construction of sports resources belong to public investment field. So the government expenditure, public investment and national macro policy all have great effects. Secondly, various statutory planning of city function zoning is the main guide for city sports resources. The location of sports facilities and the pattern of resource distribution should meet the requirements of the development strategy of the city. Especially, the construction of some large sports facilities needs to be consistent with the overall planning and construction throughout the city and its region. In short, the spatial distribution and optimization of the regional sports field is a kind of artificial intervention for the purpose of activity. Physical space layout area is not only good for its own future development. The fusion control city planners should have sports consciousness.

3.4.2 Sports factors
A city develops because of the inner and the external strength. Therefore, the large-scale sports event has the characteristic of the relative concentration in time and space. It is to break the continuity of the rhythm of city development. It also stimulates the city stadium, transportation, services, and infrastructure investment. With the impact on city development and physical changes in resource allocation, it is several times higher than the normal development of city. It not only makes the name of the sports reshape modern city but also promote the city sports space development and the space pattern planning models. Practical experience shows that it is an important role for the provinces and cities to host major sports events in China.

3.4.3 Factors of sports leisure demand
With the continuous development of productive forces, the workers are gradually liberated from the labor practice and need more leisure time. Obviously, great changes have taken place in culture, science, education and the way of people' life. Leisure demand has increasingly become higher and higher. As an important way of leisure, Leisure sports have become fashionable. They mean standard modern and civilized way of life and have played a more and more important role in human's social and cultural life. Sports resources are the material carrier of people leisure sports. Sports and leisure life depend on the allocation of sports resources. Sports resources of different types, grade and functions have become an important problem the town or city builders have to face. Sports places have become an indispensable part for a town or city. In our country, lot of cities have unique climate, environment, culture and other resources. So there is a lot of space for people's sports leisure vacation.

3.4.4 Sports market factor
The shift of supply and demand is the eternal theme of market changes. Sports supply market is no exception. So, when we consider the sports resources integration, besides the demand of the sports market. It is necessary to consider the natural resources such as land resources and supply factors. Because of the improvement of people's life, commercialization and popularization of sports, and the consumption in the process of urbanization, the office of the State Council in 2011 promulgated the "guidance on speeding up" of the sports industry". Sports fitness entertainment industry grows rapidly in China. Public health services have been just unfolding. A superb collection of beautiful things will satisfy the people's demand of sports. The development of sports industry, sports market, sports management and project facilities, economic effects of stadiums, sports venues, sports leisure place is becoming more and more important for local governments. Therefore, the sports market demand for construction of sports venues have gradually become the factors affecting physical space layout optimization.

4 MECHANISM OF SPORTS RESOURCES INTEGRATION WITHIN THE REGION

For different interest, they move and pursue the interests. To satisfy the need in the process is a dynamic process. In the course of the regular pattern, the differentiation and conflict of the interests, also a variety of dynamic interaction between the main and interaction, is the dynamic mechanism of so-called development.

Generally speaking, dynamic mechanisms drive regional sports resources integration from two aspects: dynamic factors and constraints. Rigid binding constraint factors mainly come from the regional natural geographical environment, land factors and the level of economic development. The power factor is mainly from government policy, sports events and sports leisure demand factors. The factors of sports market factors are flexible guiding force. The two forces interact in different development backgrounds. The influential factors of the integration of sports resources in the area will change with time and other things.

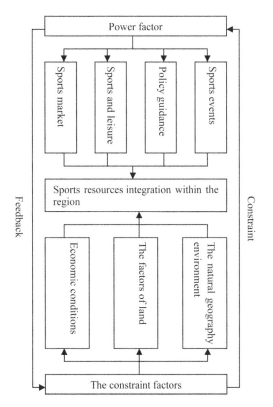

Figure 1. Regional sports resources integration mechanism map.

But in essence, the final status of integration of sports resources in the area is not completed in a driving force, but in the balance of each factors. (Fig. 1).

5 SUMMARY

The importance of rational allocation of sports resources in the region lies on the integration of sports resources in the region. Rational allocation

of sports resources is the element task. In fact, the rational allocation of sports resources and upgrading interact with each other. To realize the allocation of resources, we must make it reasonable. And to realize the rational allocation of sports resources, there must be a dynamic process. But the process of sports resources integration has the characteristics of relativity, dynamics, arduous, correlation, uncertainty, integrity. It is not only affected by the regional natural geographical, economic development, land resources and other factors. Regional sports humanistic social factors and the sports market factors are also the important factors. Therefore, the driving mechanism drives the integration of sports resources in the area from two aspects: dynamic factors and constraints. Rigid binding constraint factors mainly come from the regional natural geographical environment, land factors, such as the level of economic development. The power factor is mainly from government policy, sports events, sports leisure demand factors, factors of sports market. The two leading forces interact with each other in different development backgrounds. The influential factors of the integration of sports resources in the area will change with time and other factors.

REFERENCES

[1] Nie Hualin, Zhao Chao [M]. Introduction to regional spatial structure of Beijing: Chinese Social Science Publishing House, 2008.
[2] Xiong Maoxiang. Sports environment [M]. Beijing: Beijing Sport University press, 2003.
[3] Xiong Maoxiang. Sports environment [M]. Beijing: Beijing Sport University press, 2003.
[4] Huang Yaping. The theory and space city spatial analysis [M]. Nanjing: Southeast University press, 2002.
[5] Wang. China's regional sports development research [J]. Journal of physical education, 2010, (9).
[6] Guo Hongyu, Cai Yunnan. In order to reshape the city sports name [J]. Planner, 2009, 25 (11).
[7] Xu Yueyun. On the sports in the modern city in the development of the value and function of [J]. Journal of Shandong Sports Institute, 2011, 27 (8).

Sports Engineering and Computer Science – Luo (Ed.)
© *2015 Taylor & Francis Group, London, ISBN 978-1-138-02650-6*

Discussion on public healthcare under the perspective of lifelong sports

Fei Xu
Department of Physical Education, Wuhan Textile University, P.R. China

ABSTRACT: This paper describes the spirit-body combination and body-mind combination of healthcare, with emphasis on the interrelated relationship between healthcare and lifelong sports, and allows the public to implement lifelong sports through healthcare.

Keywords: healthcare; lifelong sports; tradition; public

1 INTRODUCTION

The development process of Chinese traditional culture of healthcare reflects the art of ancient Chinese working people continuously exceeding the realistic in life. Daoyin Health Preservation Exercises was appeared as early as the Western Han Dynasty which was given priority to medical and which the representative was Wuqinxi. The word of "Healthcare" was first put forward in the "Chuang-Tzu Health Master" which was appeared in Warring States Period. The definition of healthcare in "The encyclopedia of China" is as the principle of adjusting Yin-Yang, Qi and blood, protecting spirit, using Tiaoshen, Daoyin, expiration and inspiration, health maintenance in four seasons, dietary regimen, health preserving with drugs, abstinence, Bigu and other means to achieve the purpose of longevity and health. The definition of healthcare is divided into two categories, one is static healthcare without sports, and the other is dynamic health methods based on physical exercise.

2 HEALTHCARE NEEDS BOTH SPIRIT AND FORM

2.1 *The romantic charm of healthcare*

The theory of Yin and Yang is the foundation of healthcare in Traditional Chinese Medicine. "Qi and Yuan theory" mentioned in ancient Chinese philosophy played an important role for the formation and development of traditional healthcare. "Inner Canon of Yellow Emperor" pointed out "Man's life if the accumulation of Qi" which means Qi is the original composition of life. "Taiping Jing" has said "people will be dead without Qi". Therefore, the ups or downs of human body are determined by the surplus or empty of Qi and the balance of Yin and Yang in human tissues. If Yin and Yang is harmonic, human will is longevity, on the contrary, human will be short-lived. So keeping a harmonic internal environment of Yin and Yang has become an important basis for the balance of physical health. According to this principle, people should not only keep a certain movement, but also prevent the excessive and extreme sports. The healthcare thinking of Traditional Chinese Medicine emphasizes keeping a relatively stable movement of Yin and Yang to reduce the loss of "Yuan and Qi", so as to achieve the purpose of health and longevity.

Regulating viscera, dredging meridians, complying with timing, regulating Qi are the connotation of healthcare. Human body as viscera—centered runs through the meridians to make the Qi and blood connect with body. Throughout the human body, viscera, meridians, Qi and blood, they are not only connect with each other, but also restrict each other. Healthcare actions and practice methods according to Yin and Yang, viscera, meridians, and timing highlight the communication between action and human body meridian point. Yin and Yang theory in Traditional Chinese Medicine thinks that pubic region is central, it affects viscera and meridians. Chinese ancients said, pubic region is Qihai point which can eliminate the disease. So PEIYUAN, spermatogenesis and repose for external Qi always though the acupuncture point. Qi turns and flows to all parts of human body to make the body fluid. Regulating body needs to relaxation, standup and strength. Relaxation demands body relaxes from inside and outside and to be static and nature; standup means can support all directions; strength needs the coordination of the whole body and integration from internal and external. The objective of complying with timing and regulating Qi is to make human body adapts the changing external natural environment by internal adjustment, so as to maintain normal physiological functions. The so-called germination in spring, growing up in summer, harvest in autumn, storage in winter is the best summed up for this assertion. If people follow the rule, healthcare can make strengthen heart and liver, supplementing spleen and lung, reinforce kidney and brain, improve body and mind.

Essence, Qi and Spirit are the essence of healthcare. Essence, Qi and Spirit were concepts in ancient Chinese philosophy, they mean the original material and elements that formed the universe. Traditional Chinese Medicine believes that Essence, Qi and Spirit are the fundamental of human life activities. People who paid attention to healthcare in ancient times called Essence, Qi and Spirit the three treasures in human body, as people often said: "sky has three treasures which are sun, month and stars; earth has three treasures which are water, fire and wind; human has three treasures which are Essence, Qi and Spirit." So maintaining Essence, Qi and Spirit is the main principle of fitness and anti-aging, especially when Essence, Qi and Spirit are declining gradually and people getting old, they should more cherish the three treasures, ancients attached great importance to this. Xunzi has said: "Carrying out production activities with adequate material according to weather changes, then nothing can make people sick; lacking adequate material and being lazy to carry out production activities, you cannot be healthy." It has two meanings: the first, we must pay attention to the supplement of Essence, Qi and Spirit, the second, we should not abuse the three treasures.

2.2 The body releasing of healthcare

After thousands of years, traditional Chinese healthcare methods such as Tai Chi, Tai Chi Sword, Wuqinxi, Baduanjin, martial arts, QiGong massage and so on were perfected in constant change and development, so as the our traditional healthcare can be inherited. Some domestic literature points out that Tai Chi, Tai Chi Sword, Wuqinxi, Baduanjin and other healthcare methods can increase the content of serum interferon in blood test for practitioners and drive the cerebral cortex into a rest and relaxation mode, thus reducing the oxygen consumption, increasing the oxygen storage and making the brain get adequate oxygen supply. [7] The most prominent performance after the brain getting adequate oxygen and the brain nervous system getting decompression is the enhanced—memory, eliminating tension and fatigue. Practitioners must keep meditation, focus on maintaining interior, and pay attention to the coordinated development of body and mind in the process of practice, and through such thoughts to

reach an equilibrium state between body and mind. In such a situation, it has the function of eliminating the heart-fire, dredging meridians, promoting the metabolism and blood circulation, and has a better practice effect for each organ system, thus it is conducive to health and longevity.

Chinese traditional healthcare based on heart, at the same time Chinese culture puts introspection, insight and internal observation as the core culture. Based on heart and standard by body and mind, achieving the body-mind combination under the unifying minds and though overall body exercise is a major feature of Chinese traditional healthcare. Chinese traditional culture pays more attention to the culture of inherent temperament, character and spirit, regards the human body as the spiritual home. Therefore, Chinese traditional healthcare advocates outside-to-inside through physical exercise, promotes the invisible spiritual sublimation from tangible physical activities, so as to figure the ideal personality. The emphasis of many traditional sports which is health-oriented is fugue.

3 PUBLIC HEALTHCARE UNDER THE PERSPECTIVE OF LIFELONG SPORTS

3.1 Lifelong sports accompanying a lifetime, along with lifetime healthcare

A lifelong sport is a new concept which was put forward in the reform and development of sports since 1990s. It means that a person accept lifelong physical exercise and physical education. The meaning of lifelong sports, there are two ways: the first refers a person learning and taking part in physical exercise, having a clear purpose for whole life, so that making sports become an important and integral part of life; the second, under the guidance of lifelong sports thought, providing practical opportunities to participate in sports activities at different times and in different areas of life which aiming at systematic and integrated sports.

There are rules to follow for the development of human body. Human life can be divided into three stages which are Growth Period, Mature Period and Decline Period. Physical exercise can improve health and enhance physical fitness, so it has a positive impact for different periods of the human body. Therefore, we must put forward the corresponding requirements for physical exercise according to the characteristics of different periods of human development. In Growth Period, the emphasis is promoting the body's normal growth and development; in Mature Period, maintaining vigorous and energetic strength must be put in the first place; in Decline Period, it is important to delay the recession, extend working life and longevity. There are different requirements, exercise content and method for different stages. In other words, human life should be accompanied by physical exercise and has different goals, requirements, content and method at different times.

Lifelong sports are the goal and healthcare is the method. The purpose of pubic establishing lifelong sports is healthy; healthcare provides not only sports and fitness methods, but also the healthy lifestyle. Therefore, healthcare will accompany pubic a lifetime inevitably from theory to practice.

3.2 *Lifelong sports and healthcare can improve each other*

Lifelong sports and healthcare can promote and improve each other. The proposing of lifelong sports is most of theoretical aspects as the guiding ideology; it is a very large range that we should do meticulous work to implement it. We need to develop public the consciousness of lifelong sports, from this concept, the proposing of paying attention to diet and adjusting schedule will be closer to the public. We should guide them to keep a watchful eye on this, on this foundation, imparting health knowledge and traditional method let them blend in healthcare and toward lifelong sports step by step. Healthcare is the method, and lifelong sport is the goal. The proposing and learning of healthcare make the lifelong sports more practical. Similarly, the persistence of the lifelong sports also point out the direction for healthcare.

Lifelong sports were originated in the West; healthcare was a native product of East. Watching healthcare from the perspective of lifelong sports, it is a kind of culture communication of Chinese and western in fact. Culture communication was the product of human interaction and it is an important approach of cultural development. Along with the development of the breadth and depth of human interaction, the scale of culture communication is bigger and bigger, the speed is faster and faster, and the level is deeper and deeper, which is the law of cultural development. Culture communication is the inherent requirement of culture development; it is determined by the universality and particularity contradiction of culture. There are both particularity and universality for different ethnic culture; it is the unification of individuality and commonness. The personality and particularity of culture determines the necessity of cultural communication. The commonness and universality provide the possibility of cultural communication. It is the contradiction between cultural particularity and universality, which makes the culture colliding and communication constantly among different countries and nationalities. The culture of any

country and nation is the product of a certain social practice, it has strengths and limitations. Only absorbing nutrients from other cultures, it can be eternal youth and vigor. In comparison, the goal of lifelong sports is stronger; the effectiveness of physical exercise shows on the muscles shape and body curve. Healthcare can combine physical exercise and life, and has a close relationship with the growth and development of public. Healthcare and lifelong sports are complementary and can promote each other.

3.3 *The inheritance of eastern philosophy*

Healthcare is a kind of method and habit; it comes from public life and then service the public. Healthcare left by our predecessors is the crystallization of wisdom and the precipitation of history; it has stood the test of time and the confirmation of millions of people, the public need to pass it on continually. Getting recognized by modern people, ancient things must be linked with today's things, we can use the opportunity of developing the national fitness and the timing of advocating lifelong sports to let the height of lifelong sports bear the depth of healthcare.

The development of modern society requires people in different social strata, age and occupation take part in learning, and the learning society requires lifelong sports for public. It is a an inevitable trend for lifelong sports, along with the increasing leisure time and the improving living standards, the public will target on happiness and health for one life instead of materialism as the first choice. Under this condition, healthcare will be socialization increasingly and become an important part of public life.

It's worth mentioning that contemporary college students as a reserve force should not be underestimated. First, contemporary college students have a better understanding for traditional culture under China's education system, so that if they contact related and more systematic healthcare, they will be easy to understand from knowledge level, combining with their own education degree, they can understand by analogy. Healthcare is a treasure that ancestors left to us; there is not only the exercise method of physical fitness and prolong life in it, but also the philosophical thought which can strong heart, self-improvement and self-enriching. Contemporary college students learn Chinese and foreign, ancient and modern, we hope that they can discover new flashpoints in the eyes of modern people and promote healthcare combining the big premise of lifelong sports, so that the traditional culture can be inherited and developed, more and more people will be close to lifelong sports and healthy life by oriental philosophy. Second, contemporary college students are the masters of future. How can the future masters do not understand the traditional healthcare at all? We hope that we can retain some ancients' cultures at the society progress and have a better comprehension and summary at the basis of predecessors.

4 CONCLUSIONS

Chinese healthcare has a long history. Many health scientists in history mastered medicine, martial arts, literature and history, and philosophy. They emphasized not only the robust body, but also health psychology. The function of healthcare is not only improving immunity and preventing disease, but also advocating a kind of life which person and nature are harmonious and unified, so that to achieve spirit-body combination and body-mind combination. We hope that the new generation of Chinese will learn and accept healthcare, and inherit and carry forward the treasure of Chinese splendid culture.

Many view of lifelong sports coincides with healthcare, they have something in common and their own characteristics. The proposing of lifelong sports enhanced the consciousness of sports; the advocating of healthcare laid the foundation of sports. They are both the communication of Chinese and western cultural, and the exchange of experience about the physical exercise. We can carry out healthcare by the headline of lifelong sports, and use healthcare to promote lifelong sports; they can promote each other and develop harmoniously.

ABOUT THE AUTHOR

Born in 1984, Xu Fei, female, a lecturer coming from Wuhan, Hubei province. Her research direction is theory of sports pedagogy and training.

REFERENCES

[1] Encyclopedia of China. Encyclopedia of China Publishing House, 2009 (in Chinese).
[2] Inner Canon of Yellow Emperor. Publishing House of China, 2009, 7 (in Chinese).
[3] Yang Jilin. Taiping Jing. Publishing House of China, 2013, 4 (in Chinese).
[4] Chen Sheng. Research on the Combination of Healthcare and Lifelong sports. China Adult Education, 2008, 1 (in Chinese).
[5] Nan Huaijin, Xu Qinting. Note and Translation on Zhouyi. Chongqing Press, 2011, 1(in Chinese).
[6] Liu Weiguang. Lifelong Education Curriculum about Sports and Health. Beijing Jiaotong university Press, 2010, 7 (in Chinese).
[7] Xu Jingwen. Fitness Exercise and Traditional Healthcare. Journal of Jindongnan Teachers College, 2001, 12 (in Chinese).
[8] Shao Jun, Lin Xiufeng. The Development of Traditional Healthcare in Physical Education. Journal of Shanghai Institute of Technology Natural Science, 2001, 11 (in Chinese).
[9] Xiong Xiaozheng. The Meaning of Chinese Traditional Healthcare Philosophy for the future development of sports. Sports history, 1991 (in Chinese).

Sports Engineering and Computer Science – Luo (Ed.)
© *2015 Taylor & Francis Group, London, ISBN 978-1-138-02650-6*

A study on festive sports tourism development in Lijiang city of Yunnan province

Zhao-Long Zhang & Ming-Ya Zhang
School of Tourism Culture, Yunnan University, Lijiang, Yunnan, China

ABSTRACT: Owing to the various festival activities of ethnic minority in Lijiang, it can not only enrich the travel contents but also break the bottleneck of development currently to develop the festive sports tourism. Considering the current development situation of festive sport tourism in Lijiang, this paper proposes the mode of market operation to develop and shape the festive sports tourism in Lijiang based on the local cultural resources in festive sports events, expecting to open a new chapter of Lijiang tourism industry.

1 FACTUAL BASIS FOR FESTIVE SPORTS TOURISM DEVELOPMENT IN LIJIANG

Lijiang is home to many ethnic groups such as Naxi, Bai, Lisu, Yi, Mosuo, Tibetan, Nu, Pumi and Miao nationalities and so on; all of them have their own various traditional festivals, for instance, the Xun Tianba Festival (Terraced Field Parading Festival, held on the 13th day of the first lunar month to urge the villagers to conduct farming activities) of Dai nationality and Spring Planting Festival of Jingpo nationality. The various sports activities representing their own cultures held in these festivals are collectively called festive sports events. On the occasion of Lijiang tourism industry coming into maturity, it has been a critical problem to the development of Lijiang tourism industry that how to prolong the life cycle of the industry and maintain its strong vitality. According to researches by tourism scholars both at home and abroad, the brand extension management can be used to render the old tourism brands new brand features and thus revitalize the tourism brands, which is an inevitable choice of sustainable tourism development[1]. Based on the development theories of Festival & Special Event Tourism (FSE Tourism) and the cultural traits of Lijiang tourism industry, it puts forward the development concept of festive sports tourism in Lijiang on the principles of creating, exerting and magnifying the long-term impacts of festivals.

1.1 *The development of festive sports tourism can enrich the tourism brands in Lijiang*

The ethnic festival activities with traditional sports activities of ethnic nationalities as the theme are

of strong initiative owing to its diversified forms and traditional sports contents, so it can arouse the enthusiasm of involved tourists and make their journey more authentic, which can satisfy their demands for experiencing the folk cultures of ethnic nationalities like the sacrificial festivals such as Zhuanshan (Round-Mountain) Festival of Mosuo people that put tourists in awe of what they see. Second, the development of folk festive sports tourism activities also presents the local cultures dynamically and brings more vitalities and energies to the local tourism development and external publicity of ethnic cultures, for example, the social festival of the Double Third Festival (held on the third day of the third lunar month) of Bai people has become a platform for financing negotiation. Third, considering the requirements of developing hallmark tourism brands, the thick national features of festive sports activities of ethnic groups can further intensify the symbolic themes of folk festivals, which can facilitate the construction of local tourism brands, such as the Sanduo Festival (on the eighth day of the second month in Chinese lunar calendar), Torch Festival (on the 24th day of the sixth lunar month) and Ox-slaughtering Ceremony (held in the autumn to celebrate harvest by sacrificing cows to the gods) which are full of national characters.

1.2 *The development of festive sports tourism can drive the local economic growth*

To enhance the influences of hallmark or thematic festival activities and prolong its time of impacts, the basic strategy is to carry out relevant FSE's serial operation. In reference to Dai Guangquan's FSE serial researches designed from the perspective of Destination Life Cycle (DLC), it finds that the

---

Foundation Project: 2014 Scientific Research Foundation of Yunnan Provincial Department of Education (2014C198Y).

development of serial FSEs based on the regional hallmark festivals can realize the sustainable development of its local economy. According to the classification of festivals and events by International Festival and Event Association (IFEA), the FSEs include arts festivals, sporting events, fairs, meetings and conventions as well as expositions, in which meetings and conventions, sports tourism and sports events are regarded as the top three fast-developing industries in the world[2]. What's more, owing to the diversified sporting events full of ethnic features in the ethnic minority areas, the sports industry has a bright prospect. Therefore, it is rather feasible to develop festive sports tourism in ethnic minority areas where the economy shall be developed urgently.

2 ADVANTAGES IN DEVELOPING FESTIVE SPORTS TOURISM IN LIJIANG

2.1 Rich experimental resources of festive sports tourism

Through the development of FSEs with traditional sports activities of ethnic nationalities, the ethnic sports activities full of competitive and recreational characteristics as well as thick colors of religious ceremony provide the tourists opportunities to feel and experience the vitalities of ethnic nationalities by participating in the sports activities such as horse racing in the Sanduo Festival of Naxi people, and climbing the knife pole in the Knife-pole Festival of Lisu people. As a result, the tourists can take part in the competitions and feel pleasant, and also will be attracted by the splendid natural scenes and local cultures full of regional characteristics, thus they may get satisfying reactions easily and then appreciate with their inner heart and refine their mind, by which it can inspire the in-depth emotional experiences of tourism.

2.2 Bright market prospect of festive sports tourism

The sports events rely on the unique natural resources and colorful national features, so the sports tourism has a broader prospect in the market. Sports industry has become a sunrise industry for the economic development of our country, and the various sports competitions and events held in the ethnic sports events by integrating the economic activities with cultural connotation of ethnic sports can show the distinguishing sports cultures of the nation[3]. However, under the influences of traditional tourism habits, most tourists still stay at the phases of sightseeing tours and shopping tour but take a wait-and-see attitude to the new

emerging tourism approach. With the promotion and popularization of thoughts of life-long sports, lots of people have transformed their philosophies on travel and health, and more and more people join in the team of sports tourism. Considering the enormous tourist source market and sports tourism resources in Lijiang, it will be a great chance for Lijiang to develop sports tourism; besides, it can also utilizes its exclusive cultural resources to develop both the economy and culture in Lijiang.

3 MEASURES FOR DEVELOPING SPORTS TOURISM IN LIJIANG

3.1 The development of sports tourism shall rely on the native cultural resources

From the current situation of festive sports tourism activities held by various ethnic groups all over the country, it can be seen that the sports events impacts in most areas and ethnic groups are not so desirable except several sports events of higher reputation such as horse racing festival of Tibetan group in Tibet and the Nadam Congress of Mongol nationality[4]. The reasons are that the integration of FSE activities with local cultural is not enough and lacks of deeper elements to attract tourists, and tourists are more likely to experience the profound cultural charms of different ethnic groups apart from appreciating the competitive minority national sports. As a result, the development of festive sports tourism in Lijiang must dig the connotation of folk cultures, since only products full of unique features in the tourism market can form long-term appeals to tourists outside the area. Besides, the activities contents of festive sports events shall retain its regular programs and also add with some innovative ones full of characteristics of different time, so that the tourists can receive different experiences and feelings every time. For instance, new items are added in the sports events of Double Three Festival of Zhuang nationality in Guangxi province and women's events are also developed for the horse racing festive in Shangri-La of Yunnan province, all of which are designed to better develop the sports tourism and thus better integrate the cultural connotation and sports events creation[5].

3.2 Introduce market operation mechanism to develop FSE tourism

It is the common modes in the country that the planning and operation of FSES are organized by governments. The advantages of operation by noncommercial institutions lie in the powerful appealingness, high accountability and beneficial to the national unity. On the contrary, from the perspective of the requirements of market economy system, these institutions often ignore the economic effectiveness, but too much attention to the political effectiveness does more harm than good. To be specific, although the central government advocates market operation, there are still situations that the institutions of upper levels conduct event planning via executive orders, making the FSE planning activities unable to develop according to the market law in the end. In Lijiang, the ethnic festival and events with larger impacts and higher reputation such as Sanduo Festival and Dongba Festival are all undertaken by the government. The governments spend great efforts and energy in the events but can not get considerable economic effectives, so if these events can be undertaken by enterprises or companies, it can not only reduce the financial risks of the governments but also increase the enterprises' enthusiasms and initiatives significantly, by which it can form a complete mechanism of invest and return to protect the economic returns of developing festive sports tourism, and also avoid excessive government intervention and the occurrence of undesirable festival effects.

3.3 Publicize and create the brands of festive sports tourism in Lijiang

To develop festive sports tourism, the primary thing is to make tourists understand, learn and know that we are organizing sports tourism, so it is critical to expand the publicity of festive sports tourism. However, the brand impacts of FSE tourism in Lijiang are clearly adequate at present, for example, there are few reports about horse racing and material arts competition during the Sanduo Festival of Naxi nationality, but most people taking part in the events are local residents and only a few are individual travelers, showing that current events undertakers have serious weak awareness of developing festive sports tourism. For Lijiang, it shall fully utilizes the advantages in existing FSE tourism resources, uphold the strategy of creating diversified tourism brands in Lijiang, and take all-round publicity and promotion for Lijiang's serial festive sports events, so as to foster a new Lijiang tourism market and create a new label for Lijiang tourism.

REFERENCES

[1] Wang Zhe. A Primary Investigation on Tourism Brand Management in Shanghai and Suzhou River from the Perspective of Brand Equity [J]. Management Observer, 2012 (36):191.

[2] Zhang Xiaolin. Study on Development of Sports Tourism Industry in Western Minority Nationality Region [J]. Journal of Xi'an Institute of Physical Education, 2004(3):32.

[3] Wei Xiaokang. Ethnic Minority Traditional Sports and Culture Inheritance [M]. Beijing: The Press of the Central University of Nationalities, 2009:13.

[4] Qin Weifu, Zhang Zhaolong. Ethnic Minority Traditional Sports Events on the Regional Impact of the Economic and Social Development [J]. Journal of Qinzhou University, 2013, 28(5):51.

[5] Chen Wenhua, Zhang Zhaolong. Marketization of Ritualistic Festivals Sports Based on Ritual Theory in Tourism [J]. Journal of Shenyang University, 2014, 16(3):309.

Sports Engineering and Computer Science – Luo (Ed.)
© 2015 Taylor & Francis Group, London, ISBN 978-1-138-02650-6

The research on interactive development of the horse racing festival exploitation and tourism in Shangri-La

Zhao-Long Zhang & Ming-Ya Zhang
School of Tourism Culture, Yunnan University, Lijiang, Yunnan, China

ABSTRACT: This thesis aims the research of Shangri-La's horse racing festival sports activities and the discussion on the sporting racing events impact on the development of tourism in ethnic minority areas. The research results show that the sports events exploration has some similarities with the tourism development. Holding the national sports racing events can promote the construction of local infrastructure, increase the income of local tourism economy and improve the quality of tourism services. It is very necessary to break the barriers in the management of sports, culture and tourism and to promote the mutual development of the sports racing events and tourism in a merging thinking way.

1 THE CONNOTATION OF THE HORSE RACING FESTIVAL EXPLOITATION AND THE TOURISM ECONOMIC DEVELOPMENT

In 2009, the state council of China promulgated the "State Council's Opinion about accelerating the development of tourism". The document points that we should vigorously promote the fusion development of the tourism and sports and cultivate the new tourism consumption hotspot with the sporting events as a platform[1]. Throughout the domestic and foreign research on sports economy, a considerable amount of research can be found in large sports events such as the international events: Olmpics and World Cup, and the research on folk traditional sports activities is relatively rare. Shangri-La is the holy land of the world tourism culture with a lot of folk traditional sports activities, which provides a good condition of the combination of sports and tourism.

On December 17, 2001, approved by the state council, the Zhongdian County in northwest Yunnan Diqing zhou changed its name to the Shangri-La, the world-seeking fairyland finally appeared in front of people. Shangri-La locates in the communications hub on the Tea-Horse Ancient Road which connects the Mainland and the Tibetan area. The national culture presents a multi-ethnic and multi-religious, diversified ethnic cultural symbiosis and the communion of cultural community due to the special geographical position. There are 14 indigenous minorities

and more than 100 ethnic traditional festivals. For example, the larger ones such as the Sanduo Festival of Naxi Nationality, the Torch Festival of Yi Nationality and the Tibetan Horse Racing Festival, among them there is the people's participating in the largest scale—Horse Racing Festival. Shangri-La horse racing festival competition has developed into the Men's 1000 m, 3000 m and 5000 m, the women's 1000 m, 2000 m and 3000 m horse speed race, the men's 1600 m walking horse, running horse and picking hada, running horse and shooting and other man horse skill; 800-meter yak race; mountain climbing, archery, yenu, cattle Rafah and many other big events of national traditional sports. Now in order to enrich the traveler's racing events experience, the exhibitions races are added to the horse racing activities such as the Guozhang Dance, Jump string and Reba Dance and so on[2]. The sports and tourism appear as two less relevant industry, but from the tourism situation during the Shangri-La horse racing festival, sports have a lot of coordination with tourism. From the nature and feature of both, sports and tourism have some similarities. They are the social activities which have the aim to meet people's spiritual and cultural needs. They also have many similar characteristics such as the untransferability, nonproductivity and the simultaneousness between the product production and consumption. Moreover, the national sports racing events with a sharp theme characteristic can be converted into the easily recognizable tourism logo, and a high-level tourism service quality can provide support for the racing event holding. From this point of view, the mixed development of both will produce a lot of comprehensive benefits.

Foundation Project: 2014 Scientific Research Foundation of Yunnan Provincial Department of Education (2014C198Y).

2 HORSE RACING FESTIVAL TOURISM EXPLOITATION AND SHANGRI-LA ECONOMIC DEVELOPMENT

2.1 *The exploitation of racing events directly increases Shangri-La's tourism revenue*

With the continuous expansion of Shangri-La's sports racing events scale, the number of race watchers is increasing year by year. Since 2001, there has been nearly 100000 person/time of Shangri-La county tourism reception during the horse racing festival. During this period, the third industry, the hotel industry income is very considerable, and during the small holiday of Dragon-Boat festival, the income of the horse racing festival reached to 19.7738 million yuan. With the rising of the Shangri-La's racing brand, the investment promotion is very great, for example, during the period of horse racing festival holding in 2012, the investment capital reached to 460 million yuan and the yearly output value reached to 5.812 billion yuan, 31 times that of ten years ago. The admiringly coming businessmen of newly building hotels and restaurants are more and more. Only in the first half of 2013, nine hotels were added with a vestment of 306 million yuan. It made a very great contribution to the development of the third industry in Shangri-La[3]. It is obvious to see that the horse racing has become the important tourism brand in Shangri-La from the horse racing festival's continuous success and has stimulated the local economy with the unique charm. For this point, the horse race stretches one day into three days, and the women's race has been added in recent two years. The racing events highlighted the economic benefits, for example, only during the Shangri-La horse racing festival in 2014, it has a state of total 130250 tourists and the tourism revenue of 51.0516 million yuan. Therefore, the development of sports racing events especially the ones with the sharp speciality and the entertainment will produce the great tourism combined effect. The horse racing festival has become a recommendation business card of Shangri-La's new tourism watch spot.

2.2 *The racing event exploitation promotes the Shangri-La tourism infrastructure construction*

Holding sports racing events can promote the perfection of the local tourism support system, it has become the consensus of the development of city. Especially major international sporting events holding is obvious such as the Olympic Games, the World Cup. Take the 2008 Beijing Olympic Games for example, from 2002 to 2007, Beijing directly invested 250.7 billion RMB yuan into the city infrastructure construction such as the urban transportation, municipal administration, information, and environment and so on. Also, in order to develop the horse racing festival, Shangri-La increased the construction of tourism infrastructure, for example, for the Dragon-Boat Festival visitors surge during the races, the state government organized and built the Longtan Park, which can not only meet the needs of race-watching tourists but also improve the local living environment and atmosphere and provide the better entertainment facilities and sites for the local residents. The five-phoenix mountain in Shangri-La is the traditional racing venues since the ancient times. In view of the needs of the development of sports events and state sports construction, the new five-phoenix horse racing center was built on the original location in the Shangri-La county in 2007. The racecourse design highlights the multi-function, in the middle is a standard football field, and then the track and field runway, go towards outside is the horse racing track, the watch platform designs for sun shading which can satisfy the needs of all kinds of sports, concerts and the large gathering[4]. In addition, in order to build the national traditional sports tourism brand, the county urban construction has been carried out on the comprehensive reform, the city's roads, sanitation, landscaping and other facilities have been increasingly improved. Overall, the transportation in every big scenic sites has been accelerated. Now Shangri-La has applied to become the county health city with the increasing development of the horse racing festival.

2.3 *The exploitation of racing events promotes Shangri-La tourism power*

Holding sports racing events can improve the hard power further in the host area. For example, when the horse racing festivals were held first, the surge tourists'accommodation became the first question, the county government had to announce on TV to mobilize the residents to let the travelers to live. With the rising of the horse racing festival tourism brand, the more and more foreign merchants came Shangri-La to build the hotels and restaurants. In 2013 the whole state has 508 hotels with the reception ability, 38500 beds. The tourists daily capacity is 40000 people (time)[5]. There is a total of more than 1240 of the tourists all the year round, year-on-year growth of 22.61%. The income of tourism is more than 12 billion yuan, up 25.76% from a year earlier. Holding sports racing events can promote the tourism industry upgrading, for example, at the end of the event, the related tourism products such as the facilities, fields and environments can be used continually. Another example, some big sports are to attract tourists such as wrestling, horse racing, archery and a series of challenging performances. Horse racing festival competition also leads to the development of the sports tourism economy, such as the Shangri-La rock climbing, mountain skiing, walking through the woods, bungee jumping and so on. These unique tourism resources attracted deeply the masses of tourists with a stimulus and a thorough release of heart[6]. Holding sports racing events can improve the connection between tourism and the outside world, which can improve the quality of tourism services to promote tourism service level to the mature development. For example, the Shangri-La state government carried out the standardization of the Tibetan family visiting to make the tourism service product safety, environment and services better and better constantly.

3 THE INTERACTIVE DEVELOPMENT MODEL BETWEEN THE RACING EVENTS EXPLOITATION AND SHANGRI-LA TOURISM

In this study, horse racing festival holding and Shangri-la tourism have connected with each other, which determines the inevitable existing complex interaction between them. The composition of both is nothing more than three factors: the first one is contestants and tourists, they are the main body of the racing events and the tourism activities. The second one is the object of tourism activities such as the events, tourism resources and so on. The third one is sports tourism, it is the medium body of event tourism activities. So only two fusion development can realize each other's interests[7]. First of all, the horse racing festival events must be put into the local tourism development overall planning. According to the time and the content of the racing events, the superior department provides the required tourism infrastructure and organizes the related tourism enterprises to provide sports competitions holding relatively perfect supporting services and provide security for the interactive development between the tourism and the horse racing festival.

Second, in the horse racing festival marketing stage of development, the related tourism management departments and enterprises should invest a lot of money and other resources for a particular sports tourism market and realize the active marketing for the target tourist market and make an effort to attract as many tourists as possible for the large horse racing festival and tourism. Finally, in the period of the horse racing festival, the city tourism whole services level will be improved through providing the contestants and watchers the better supporting services. The good momentum of development will in turn be advantageous to the events to attract more audience. In addition, by exploring the related tourism products and derivatives, the tourism enterprises can also obtain a more considerable income and more stable customers and contribute to

the win-win of the interactive development between the racing events exploitation and tourism.

4 CONCLUSION

The comprehensive planning of the sports events holding is the first step in the whole horse racing festival marketing. It will be related to the success of the event. Therefore, the horse racing festival organizers must cooperate with Shangri-La tourism management department, and the related tourism enterprises can use the culture content of the event to publicize the Shangri-La new tourism image. Next, the horse racing festival can be as a tourism resource to develop and as a tourism product to run. We should try to explore the activities around the festival and create the rich and related tourism products. The event marketing operation institution can communicate and cooperate with the travel agency to make an integration of tourism resources and to optimize the configuration and to launch a new tourism project to attract more tourists. Finally, in order to more effectively promote the interaction and fusion development between the horse racing festival events and the Shangri-La tourism. It is necessary to formulate relevant regulations or mechanism to implement effective system arrangement for industry development and planning guidance.

REFERENCES

[1] Pang Xuwei, Gao Wenqian, Zheng Yue. The Interactive Development Model between the Large Sports Racing Events and the City Tourism in Shanghai [J]. P.E. Research, 2011, 32(6):14.
[2] 2014 the Successful Closing of the Shangri-la's May Horse Racing Festival [EB/OL]. http://ynjjrb. yunnan.cn/html/2014–06/05/content_3236118.htm, 2014–06–05.
[3] The National Bureau of Statistics. Diqing Zhou Statistical bulletin in 2013 the National Economic and Social Development. [EB/OL]. http://my12340.cn/ article.aspx?ID = 3162, 2014–03–25.
[4] Qin Weifu, Zhang Zhaolong. The Traditional Minority Sports Events' Impact on Regional Economic and Social Development [J]. Journal of Qinzhou University, 2013, 28(5):51.
[5] DiQingZhou tourists break through 12.4 million person-time in 2013 [EB/OL]. http://yn.people.com.cn/ news/yunnan/n/2014/0108/c228496–20331079.html, 2014–01–08.
[6] Chen Wenhu, Zhang Zhaolong. The Thinking about the Ritual Festivals Sports Marketization-Based on the Theory of Tourist Ceremony [J] Journal of Shenyang Universtiy, 2014, 16(3):309.
[7] Li Xiaochun, Zhang Zhaolong. Guangxi Ethnic Traditional Festival Sports Business Development Research [J]. Journal of Xi'an Institute of Sports, 2010, 27(4):42.

Sports Engineering and Computer Science – Luo (Ed.)
© 2015 Taylor & Francis Group, London, ISBN 978-1-138-02650-6

Cases and reflection on the cost-sharing arrangement of intangible assets—from cases of VERITAS in the USA[1]

Zhanxia Wu
Professor, Shanghai University of International Business and Economics, Shanghai, China

Xueli Chen & Manjiao Liu
Students, Shanghai University of International Business and Economics, Shanghai, China

ABSTRACT: In 2008, US Internal Revenue Service made the judgment on transfer pricing of intangible assets between VERITAS US and its subsidiary US VERITAS Netherlands: VERITAS US should pay taxes in arrears of $1.244 billion. This essay compares laws and regulations of intangible assets and cost sharing arrangement between the USA and China based on cost sharing arrangement in the essay and make suggestions on China's laws of cost sharing arrangement.

1   INTRODUCTION[2]

1.1   *Case background*

VERITAS is a company registered in Delaware, developing and selling software products. It expanded its business overseas and become a global parent company[3] in the late 1990s. Before the case occurred, its organizational structure is showed in Figure 1.

VERITAS US and its Subsidiary VERITAS Ireland entered into the Cost-Sharing Arrangement (referred to as CSA) including the Agreement for Sharing Research and Development Costs and the Technology License Agreement on November 3rd, 1999 when VERITAS US became the global parent company. On the same day, the US company transferred the existing intangible assets to VERITAS Ireland and VERITAS Ireland paid buy-in payment to VERITAS US.

The Irish subsidiary paid 63 million dollars to the US company on the above-mentioned existing intangible assets in 1999. In 2000, the Irish subsidiary paid the one-off buy-in payment 166 million dollars and in 2002, the payment adjusted to 118 million dollars.

US Tax Department questioned fairness of the above-mentioned payment based on Section 482

afterwards. After examining the earnings in 2000 and 2001, U.S. Tax Department drew the conclusions: the Cost-Sharing Arrangement did not reflect VERITAS US's income clearly; the transfer value of intangible assets depends on the estimated value and tax authority regarded this transfer value as 1.65 billion dollars through measurement; finally, in 2008, the court adjusted the buy-in payment of above-mentioned intangible assets through profit split method and imposed VERITAS US with a hefty fine.

1.2   *Key content of the case*

1.2.1   *Content of the cost-sharing arrangement*
The Cost-Sharing Arrangement includes the Agreement for Sharing Research and Development

---
[1]This study is the phased research result of 2012 Shanghai educational committee's major projects of scientific research innovation "(12ZS165)".
[2]https://www.ustaxcourt.gov/.
[3]On July 2nd, 2005, VERITAS US was purchased by Symantec Corp and became one of Symantec's wholly owned subsidiaries. (Symantec's).

Figure 1.   Structure of VERITAS.

Costs and the Technology License Agreement. In VERITAS case, the signatories agreed to pool their respective resources and R&D efforts related to software products as well as software manufacturing processes and share the costs and risks of such R&D on a going-forward basis based on the signed Agreement for Sharing Research and Development Costs (referred to as RDA).

VERITAS US, the parent company provided VERITAS Ireland, the subsidiary company the exclusive and perpetual right to manufacture products utilizing, embodying or incorporating the Covered Intangibles; VERITAS US can provide the nonexclusive and perpetual right to otherwise utilize the Covered Intangibles worldwide, including in the marketing, sale, and licensing of products utilizing, embodying or incorporating the Covered Intangibles, and in further research into similar technology.

1.2.2 *The agreed intangible assets by signatories*

The RDA defined "Covered Intangibles" as: any and all inventions, patents, copyrights, computer programs (in source code and object code form), flow charts, formulae, enhancements, updates, translations, adaptations, information, specifications, designs, process technology, manufacturing requirements, quality control standards, and other intangible property rights arising from or developed as a result of the Research Program.

Pursuant to the TLA, VERITAS US granted VERITAS Ireland the right to use certain "Covered Intangibles", as well as the right to use VERITAS US's trademarks, trade names, and service marks in EMEA and APJ. The TLA defined "Covered Intangibles" as: any and all inventions, patents, copyrights, computer programs (in source code and object code form), flow charts, formulae, enhancements, updates, translations, adaptations, information, specifications, designs, process technology, manufacturing requirements, quality control standards, and other intangible property rights arising in existence as of the Effective Date of this Agreement, relating to the design, development, manufacture, production, operation, maintenance and/or repair of any or all of the products.

2 THE CRITICAL ANALYSIS OF CSA IN VERITAS CASE BY US TAX AUTHORITIES

2.1 *The comparability between controlled and uncontrolled transactions[4]*

The use of CUT method requires the controlled and uncontrolled transactions involve the same or comparable intangibles[5]. Section1.482–6, Income Tax Regs., provides that the following factors shall be considered in determining comparability between controlled and uncontrolled transactions: functions, contractual terms, risks, economic conditions, and property or services. An analysis employing these factors confirms that VERITAS US' unbundled OEM agreements are sufficiently comparable to the controlled transaction.

The first factor, functional analysis, compares the economically significant activities undertaken, or to be undertaken, in the controlled transactions with the economically significant activities undertaken, or to be undertaken, in the uncontrolled transactions. VERITAS Ireland and the OEMs undertook similar activities (e.g., manufacturing and production, marketing and distribution, transportation and warehousing, and so on.) and employed similar resources in conjunction with such activities. Respondent contends, however, that the OEM agreements and the controlled transactions are not functionally comparable because the R&D function is important. Thus, the focus of the buy-in payment analysis should be on transactions involving preexisting intangibles. For the products in existence on November 3, 1999, there are no significant differences in functionality.

The second factor is the comparability of contractual terms. Respondent contends that the contractual terms of the OEM agreements are not comparable with the controlled transaction for two reasons. First, respondent contends that the OEMs often provided VERITAS US with APIs, source code, or information about their hardware so VERITAS US could adapt VERITAS US products to the OEMs' hardware and operating systems, whereas VERITAS Ireland did not have an operating system, APIs, or source code. Second, respondent contends that the OEMs provided engineering assistance to VERITAS US in connection with the development of VERITAS US bundled products, whereas there is no evidence that VERITAS Ireland was in a position to provide engineering assistance to VERITAS US. Thus, there are no significant differences in contractual terms.

The third factor compares the significant risks borne by the parties that could affect the prices charged or the profit earned in the controlled and uncontrolled transactions. The parties to the controlled and uncontrolled transactions bore similar market risks, similar risks associated with R&D activities, similar risks associated with fluctuations

---

[4]INCOME TAXS—26 CFR 1.482–6 Profit split method.

[5]As the reference of Section 1.482, the comparability of the controlled transaction and the uncontrolled transaction demands the intangibles involved are in the same field or market, and own similar potential profit.

in foreign currency exchange rates and interest rates, similar credit and collection risks, and similar product liability risks. Respondent contends, however, that the risks borne by VERITAS Ireland and the risks borne by the OEMs are not comparable because the OEMs were subject to the risk that the version of technology they licensed would not do well in the market.

The fourth factor compares the significant economic conditions that could affect prices or profit in the controlled transaction to the significant economic conditions that could affect prices or profit in the uncontrolled transactions. The court note, however, that both the OEMs and VERITAS Ireland competed in similar geographic markets, incurred similar distribution costs, marketed products that faced similar competition, and were subject to similar economic conditions. While certain economic conditions (e.g., interest rate fluctuations, general vicissitudes of the market, and so on.) affect prices and profits for both startups and established businesses, the impact on a particular business may certainly depend on the business' economic stability and market position. The analysis of this factor narrowly weighs against a finding of comparability.

The fifth factor compares the property or services provided in the controlled transaction to that provided in the uncontrolled transactions. Respondent contends that under the OEM agreements, VERITAS US generally contracted to provide only the development work necessary to ensure its products would work with the OEMs' products, whereas under the CSA, VERITAS US provided make-sell rights and preexisting intangibles for research to produce future generations of technology. With respect to the controlled transaction involving the transfer of preexisting intangibles and the uncontrolled transactions involving VERITAS US' unbundled OEM agreements, there are no significant differences in property or services provided.

2.2 The valuation of intangible

In this case, the court applied two kinds of valuation methods, but neither of them was accepted.

3 THE COMPARABLE ANALYSIS OF CHINA'S INTANGIBLE CSA

3.1 The definition of intangible in China's income tax[6]

As in People's Republic of China on Enterprise Income Tax Provisional Regulations, the intangible properties are no physical form and non-monetary long-lived assets that enterprises held for production, providing services, rental or administrative purposes, including patents, trademarks, copyrights, land use rights, non-patent technology, goodwill, and so on.

3.2 Special tax adjustments in the income tax act provisions in China[7]

Business transactions between the enterprise and its related parties, the transaction does not comply with the principles of independence and reduce enterprise or its affiliates taxable income or income, the tax authorities are entitled to a reasonable adjustment method. The cost occurred in enterprise and its related parties jointly developing, transferring intangible assets, or jointly providing, accepting labor, shall be apportioned in accordance with the arm's length principle when calculate the taxable income.

An enterprise may propose pricing principles and calculation methods associated with business dealings with its related parties, after negotiating, confirming with the tax authorities, the APAs comes. (Advance Pricing Arrangement, is an agreement reached between tax authorities and enterprises, on enterprises' future pricing principles and calculation methods in controlled transactions in accordance with the arm's length principle).

Companies can reach CSA in accordance with the principles of independence and the cost shared with its related parties. When cost-sharing shall in accordance with the principle of matching costs with expected revenues, and submit relevant information within the specified period, including: 1. Contemporaneous associated with controlled transactions prices, standard costs, calculation methods and notes; 2. Contemporaneous associated with controlled transactions resale(transfer) price or the final sales price (negotiable) prices of property, the use of property rights, and labor; 3. Information other enterprise should provide about product prices, pricing and profit levels comparable with the surveyed enterprises related to controlled transaction investigation; 4. Additional information relating to controlled transactions.

3.3 Intangible assets intangible assets GAAP measurement requirements in China[8]

Accordance with Accounting Standards No. 6—Intangible Assets (2006) Cai Kuai [2006]

[6]http://www.chinaacc.com/new/287/288/304/2006/1/ad522118473517160026672.htm.

[7] http://news.qq.com/a/20070319/001580_1.htm.
[8]http://www.chinaacc.com/new/63/64/77/2006/2/ma277421946182260021836–0.htm.

Table 1. The definition comparison of intangible between China and America.

	America	China
Held purposes	To design, develop, manufacture, product, operate, maintain and repair any or all products	Production, providing services, rental or administrative purposes
Form	No physical form	No physical form and non-monetary long-lived assets
Types	Any and all inventions, patents, copyrights, computer programs (in source code and object code form), flow charts, formulae, enhancements, updates, translations, adaptations, information, specifications, designs, process technology, manufacturing requirements, quality control standards	Patents, trademarks, copyrights, land use rights, non-patent technology, goodwill, and so on

No. 3, valuation of intangible assets is as follows: (A) Intangible assets shall be initially measured at cost: The cost of purchased intangible assets, including the purchase price, relevant taxes and other directly expenses attributable to bringing the asset to its intended use., The cost of the purchase price of intangible assets deferred payments beyond normal credit terms, has financing nature, and the purchase price should be determined on the basis of the present value of the intangible assets. (B) The cost of self-developed intangible assets, including the total expenditures during meet the criteria of the Article IV[9] and Article IX[10] and reach their use, but prior periods expenditures do not adjust. (C) Cost invested into intangible assets should be determined by value in the investment contract or agreement, but excluding the value of the unfair contract or agreement. (D) The cost of intangible gained by non-monetary assets exchange, debt restructuring cost of intangible assets, government grants and merger shall be determined respectively by, "Enterprise Accounting Standards No. 7 Non-monetary assets exchange", "Enterprise Accounting Standards No. 12 Debt restructuring","Enterprise Accounting Standards No. 16 Government grants "and" accounting Standards for Enterprises No. 20 business Combinations".

---

[9]The intangible can be recognized if the following conditions have been met: (A) the economic benefits related to intangible assets are likely to flow to the enterprise; (B) the cost of the intangible can be measured reliably.

[10]Expenditure on internal research and development projects can be recognized as intangible assets, if all the following conditions have been met: (A) the intangible asset has technical feasibility in use or sell. (B) the intangible is held in order to use or sell. (C) the way the intangible conduct economic profit can show its feasibility. (D) there are adequate technical, financial resources and other resources to support the development of the intangible assets, and there are adequate ability to use or sell the intangible asset. (E) the cost of the intangible can be measured reliably.

Provisions of China's intangible accounting standards, is far short of the need to assess the value of intangible assets.

3.4 Management of CSA in China

China's "Implementation Measures for Special Tax Adjustments (Trial)" National Tax [2009] No. 2, on a cost sharing agreement made with the following requirements:

Enterprise and its related parties signed the cost-sharing agreement to jointly develop, transfer of intangible assets, or jointly provided services received, shall comply with the provisions of this chapter.

The participating parties of CSA own the right to share the benefit of the intangible developed and transferred or the labor activities involved, and bear the cost of the corresponding activities. Costs borne should be associates between related parties and unrelated parties under comparable conditions in order to obtain the above-mentioned benefit. Participants using the CSA's developed or transferred intangible assets do not need other royalties. Beneficiary of intangible assets or services in CSA should be reasonable, predictable revenue metering, and on basis of commercially reasonable assumptions and business routine.

CSA should include: (A) name for the party, the country (region), relationship, rights and obligations in the protocol; (B) content and scope of the intangible assets or services in CSA, and the specific participants commitment to R & D activities or services and their responsibilities, tasks; (C) duration of Agreement; (D) the calculation methods and assumptions participants use to expected gain; (E) The determination method of the initial investment and subsequent pay's amount, form, method and description in line with the principles of independence.(F) description of participants use and change in accounting method; (G) program and handling requirements for participants to join or withdraw; (H) conditions and handling requirements

Table 2. Comparing of CSA between China and America.

	America	China
Provisions of the agreement	Include two or more participants;	Name for the party, the country (region), relationship, rights and obligations in the protocol
	Provide a method to calculate each controlled participant's share of intangible development costs, based on factors that can reasonably be expected to reflect that participant's share of anticipated benefits	The determination method of the initial investment and subsequent pay's amount, form, method and description in line with the principles of independence.(F) Description of participants use and change in accounting method
	Provide for adjustment to the controlled participants' shares of intangible development costs to account for changes in economic conditions, the business operations and practices of the participants, and the ongoing development of intangibles under the arrangement	
	A description of the scope of the research and development to be undertaken, including the intangible or class of intangibles intended to be developed	Content and scope of the intangible assets or services in CSA, and the specific participants commitment to R & D activities or services and their responsibilities, tasks
	A description of each participant's interest in any covered intangibles	The calculation methods and assumptions participants use to expected gain
	The duration of the arrangement	Duration of agreement
	The conditions under which the arrangement may be modified or terminated and the consequences of such modification or termination.	Program and handling requirements for participants to join or withdraw
		Conditions and treatment requirements of changing or terminating the agreement
		Conditions and handling requirements of compensation payment between the parties
		Provisions of non-participants using the protocol outcomes
		Description of participants use and change in accounting method
Requirements for participants	Uncontrolled taxpayer	Should have a legitimate business purpose and economic substance
	Controlled taxpayer who can reasonably anticipates benefits from the use of covered intangibles	Should comply with the principle of an independent transaction
		Should follow the principle of matching costs and benefits
		Should record or prepare, preserve and provide the contemporaneous documentation of CSA
		The time between the date signing the CSA and the operating period should be more than 20 years

of compensation payment between the parties; (I) conditions and treatment requirements of changing or terminating the agreement; (J) provisions of non-participants using the protocol outcomes.

Enterprise and its related parties signed a CSA with one of the following circumstances, its share of the costs could not be deductible: (A) does not have a legitimate business purpose and economic substance; (B) does not comply with the principle of an independent transaction; (C) did not follow the principle of matching costs and benefits; (D) fail to record or prepare, preserve and provide the contemporaneous documentation of CSA; (E) the time between the date signing the CSA and the operating period is less than 20 years.

4 SUGGESTIONS TO PERFECT COST SHARING ARRANGEMENT[11]

According to the above table "comparison of the cost sharing arrangement between China and US", Chinese cost sharing arrangement is more detailed than American's. However, except the content of cost sharing arrangement, other clauses still can be improved.

1. Add "participants' economic conditions, the business operations" to the main content of the cost sharing arrangement. This element can predict the participants' capability of on-going investment, that is the durability of the CSA.
2. If there are comparable transactions, estimate whether the newly signed cost sharing arrangement can refer to the comparable transaction needs specific regulations.

   In China, the income tax law and enforcement regulations do not clearly present the related reference of comparable cost sharing arrangement. But according to US section 1.482–1(d)(1), comparability can be considered through the factors below: functionality, contract terms, risks, economic conditions, property or services.

   US income tax law regulations are more intact. Our country can consult the US laws to formulate factors of cost sharing arrangement.
3. If there is no previous comparable transaction when signing the new cost sharing arrangement, clauses of guidance can be made to discriminate if the calculation of sharing amount is within arm's length.

Chinese tax laws only mention that: when company sharing cost with related parties, it should share the cost based on the principle of matching costs with expected revenues and submit relevant information within the ruled period. If companies violate the regulations one and two when sharing the costs, the shared costs cannot be deducted when calculating taxable income.

US tax laws mention that: provide a method to calculate development costs of intangible assets based on the factor of rational expectation reflecting participants' share of the expected revenues.

In a word, tax laws of both countries don't mention clearly whether the sharing amount is fair, that is, the specific evidence of estimating the expected revenues are fair or not.

As a result, it is suggested that clauses of guidance can be made to regulate the fairness of sharing amount. For example, regulate the company should include the following prediction proofs when estimating the future revenue: (1) company's evidence of estimating the future operation conditions; (2) company's evidence of revenues of transferred intangible assets; (3) company's evidence of estimating the prospect of transferred intangible assets; (4) company's evidence whether to continue to develop transferred intangible assets in the future.

---

[11]http://www.chinaacc.com/new/63_67_/2009_1_22_wa080422725122190029988.shtml.

Section 3: Computer science and applications in sport

Sports Engineering and Computer Science – Luo (Ed.)
© 2015 Taylor & Francis Group, London, ISBN 978-1-138-02650-6

Applied research of 3D technology in the women shape-building dance

Yan Zhang
Sports Department of Guangxi University of Chinese Medicine, Nanning, Guangxi, China

ABSTRACT: *Purpose*: In view of a series of problems that exist in the women shape-building dance teaching fitness club which opened by the woman, we have probed into the application value and effect of 3D technology in shape-building dance teaching with shape-building dance coach and experts.
  Methodology: With two experimental groups (22 people) and control group (28 people) of woman shape-building dance in Beijing and Shanghai fitness club as experimental subjects, some methods are used, such as literature material method, interview method, measuring method, comparative method and mathematical statistics, to study their relevant data (age, body mass index, seating forward bend, step test, sit-ups, motor skills evaluation results and routine skills evaluation results), etc. Both the 3D technology and the conventional teaching can improve students' motor skills and routine skills evaluation results significantly, enhance and maintain their weight index value, seating forward bend, step test and sit-ups; Compare to conventional teaching, the 3D technology can significantly improve students' motor skills results, seating forward bend, step test, sit-ups and routine skill evaluation results, at the same time, increase weight index value to some extent; It can not only urge students to take the initiative to learn the shape-building dance movements, but also strengthen the comprehensive physical quality education, and promote their physical and mental health, with an obvious effect. So the 3D technology is worth popularizing because it can effectively supplement and improve the conventional teaching.

Keywords: 3D technology; woman; shape-building dance; applied research

1 INTRODUCTION

The shape-building dance teaching is a newcomer in the fitness club in China. In the early mode of shape-building dance teaching, more attention is paid to the phase of imitation and strengthening practice, but the actual effect is not good. With the rapid development of global 3D technology, China has been gradually entering into the 3D digital age. At present, 3D technology has penetrated into every corner of our life, so the industry related to 3D has developed rapidly in the shape-building dance. The 3D technology (short form of "Three Dimensions") refers to three dimensions or three coordinates, namely, length, width, height or depth. In other words, it is three dimensional, relative to the plane (2D) with length and width. Why the 3D can be realized? Because people have two eyes, a left eye and a right eye, and there is a subtle difference when seeing the same object, that called "parallax". Thus a stereo image is formed in the brain. The 3D display technology reconstructs the parallax in an artificial way. Generally, through the display technology, the left eye only deals with the images which received by the left eye, and the right eye only handles the images received by itself. Then the brain composite two images into a complete stereo image, making people produce an immersive feeling. In this paper, to ensure the research is realistic and believable, we test the angles of age, Body Mass Index (BMI) = weight (kg)/height (m)^2, seating forward bend, step test and sit-ups, motor skills (scores) and routine skills (scores) and evaluated them after the experiment, to analyze the teaching effect of the 3D technology in the women shape-building dance of the fitness club for reference.

2 RESEARCH OBJECT AND PRINCIPAL METHODS

2.1 Time, site and object

Experimental period: from January 1st 2012 to December 31st 2012. Site: indoor gyms of shape-building dance of Beijing and Shanghai fitness clubs.

Objects: 4 groups of shape-building dance students in Beijing and Shanghai fitness clubs, 50 people in total, are divided into two experimental groups (22 students) and two control groups (28 students) reasonably. The coach of experimental groups uses the general teaching methods and the 3D technology, and the coach of control groups uses the conventional teaching methods at the same level.

2.2 Literature material method

In the library of Beijing Sport University and Guangxi University of Chinese Medicine, by using China Journal Net, Wanfang Database and the rest, we have collected related documents, literature and academic papers of shape-building dance and 3D technology, for the early theoretical preparation to study the application of the 3D technology in the shape-building dance.

2.3 Interview method

By visiting or interviewing experts from sports schools of China at all levels of shape-building dance and 3D technical, we understand their main views about the shape-building dance training and teaching.

2.4 Measuring method

The age, Body Mass Index (BMI) = weight (kg)/height (m)², seating forward bend, step test and sit-ups, motor skills evaluation results (scores) and routine skills evaluation results (scores) are all measured before and after the experiment.

2.5 Contrast method

The contrastive analysis is conducted on experimental groups and control groups before and after the experiment.

2.6 Mathematical statistics

Statistical data are handled three levels: SPSS16.0. $P > 0.05$ means there is no difference between two groups of data, $P < 0.05$ means there is a significant difference between two groups of data, and $P < 0.01$ means there is a high significant difference between two groups of data.

2.7 Experimental Design Ideas for 3D Technology

1. *Control of experimental objects: 50 female students who are in the same level in four classes are selected from the fitness club in Guangxi randomly. They are experimental groups that use the 3D technology and the conventional teaching methods, and others are control group use the conventional teaching methods.*
2. *Control of experimental conditions: The experimental period is one year. The teaching progress, site equipment, teaching period and times, teaching contents and sports intensity, teaching sports load of experimental groups are basically consistent with control groups.*

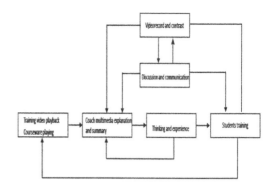

Figure 1. Coach experimental design ideas of 3D technology.

3. *Control of experimenters: The shape-building dance coaches who are at the same level assume the teaching responsibility for women shape-building dance of experimental groups and control groups, respectively.*
4. *Experimental design for 3D technology: The coaches or experts design a scheme combining the 3D technology with the practical technology learning. Guide students to participate in the experiment and study actively, fully arouse their learning enthusiasm, initiative and creativity, and improve the effect of in-depth and full understanding of the learning content. The students develop and enhance their skills repeatedly and improve from theory to practice and then to theory continually. The specific learning, practice, practical operation and design ideas are as follows (Fig. 1).*

3 ANALYSIS OF EXPERIMENTAL RESULTS

3.1 The evaluation results of female students' age, body mass index, seating forward bend, step test and sit-ups, motor skills and routine skills before the experiment are analyzed for comparative study.

In order to achieve the realistic research effect, female students' age, Body Mass Index (BMI) = weight (kg)/height (m)², seating forward bend, step test and sit-ups, motor skills and routine skills evaluation results before the experiment are effectively selected, measured and handled statistically by the SPSS software. The results of Tables 1 and 2 show us: age (D-value = 0.5, F = 0.16, P = 0.69 > 0.05); BMI (D-value = 0.31, F = 0.26, P = 0.61 > 0.05); seating forward bend (D-value = 0.22, F = 0.22, P = 0.74 > 0.05); step test (D-value = 0.82, F = 0,12, P = 0.73 > 0.05); sit-ups (D-value = 0.82, F = 0.82, P = 0.66 > 0.05); the

Table 1. Contrast analysis before the experiment of data of age, body mass index, seating forward bend, step test and sit-ups, motor skills and routine skills results of experimental groups and control groups.

Indexes	Age (years old) X±S	BMI X±S	Seating forward bend X±S	Step test X±S	Sit-ups (piece) X±S
Experimental group	41.50±4.24	22.98±1.92	9.40±2.31	52.14±8.08	30.36±3.93
Control group	41.00±4.57	22.67±2.16	9.40±2.31	52.96±8.65	30.88±3.83
D-value	0.5	0.31	−0.22	−0.82	−0.52
Testing F value	0.16	0.26	0.11	0.12	0.20
P value	0.69	0.61	0.74	0.73	0.66

Table 2. Comparison table for motor skills scores and routine skills scores of experimental groups and control groups before the experiment.

Indexes	Motor skills evaluation results (score) X±S	Routine skills evaluation results (score) X±S
Experimental group	64.46±8.02	56.27±5.87
Control group	64.93±7.32	57.07±5.00
D-value	−0.47	−0.8
Testing F value	0.05	0.27
P value	0.83	0.61

evaluation results of motor skills (D-value = 0.47, F = 0.47, P = 0.83 > 0.05) and routine skills (D-value = 0.8, F = 0.8, P = 0.61 > 0.05), both of which the D-value difference is small and P value are more than 0.05.

It indicates that in statistical significance, there is no significant difference between the evaluation results of two groups of students in terms of age, Body Mass Index (BMI) = weight (kg)/height (m)^2, seating forward bend, step test and sit-ups, motor skills and routine skills evaluation results, and two groups of students basically have the same conditions of physical qualities and sports skills, it provides guarantees for the research on the experimental results.

3.2 Comparative research on the analysis of body mass index, seating forward bend, step test and sit-ups, motor skills and routine skills evaluation results of experimental groups and control groups before and after the experiment

After handled statistically by the SPSS software, the results in Tables 3 and 4 show us: The BMI results of experimental groups before and after the experiment are reduced by 1.11 on average and the statistics P = 0.038<0.05, indicating there is a significant difference in the BMI data before and after the experiment. Thus it can be seen that

Table 3. Contrast analysis of data of body mass index, seating forward bend and step test of experimental groups or control groups before and after the experiment.

Index	Experimental class	Control class
BMI X±S		
Before the experiment	22.98±1.92	22.67±2.16
After the experiment	21.87±1.49	22.60±2.12
P value	0.038	0.901
Seating forward bend (cm) X±S		
Before the experiment	9.40±2.31	9.46±2.29
After the experiment	11.16±2.07	9.75±2.36
P value	0.011	0.851
Step test (score) X±S		
Before the experiment	52.14±8.08	52.96±8.65
After the experiment	59.73±8.93	54.04±8.97
P value	0.005	0.651

Table 4. Contrast analysis of data of sit-ups, motor skills and routine skills of experimental groups or control groups before and after the experiment.

Index	Experimental class	Control class
Sit-ups (piece) X±S		
Before the experiment	30.36±3.93	30.88±3.83
After the experiment	33.23±3.85	30.93±3.75
P value	0.019	0.944
Motor skills evaluation results (score) X±S		
Before the experiment	64.46±8.02	64.93±7.32
After the experiment	84.05±4.80	78.93±6.65
P value	0.000	0.000
Routine skills evaluation results (score) X±S		
Before the experiment	56.27±5.87	57.07±5.00
After the experiment	79.77±2.64	75.79±7.63
P value	0.000	0.000

the application of 3D technology can significantly arouse the enthusiasm and initiative of student learning, boost students to do exercises voluntarily, and lose weight greatly.

BMI results of control groups before and after the experiment are reduced by 0.07 on average and the statistics P = 0.901 > 0.05, indicating there is no significant difference in the BMI data before and after the experiment. Thus it can be seen that the conventional teaching method can improve and maintain BMI data of female students, but the actual effect is not ideal.

Seating forward bend results of experimental groups before and after the experiment are increased by 1.76 cm on average and the statistics P = 0.011 < 0.05, indicating there is a significant difference in the seating forward bend data before and after the experiment. Thus it can be seen that the application of 3D technology can make students fully display their own body and figure, graceful posture and good flexibility.

Seating forward bend results of control groups before and after the experiment are increased by 0.35 seconds on average and the statistics P = 0.851 > 0.05, indicating there is no significant difference in the seating forward bend data before and after the experiment. Thus it can be seen that the conventional teaching method can improve and maintain the seating forward bend results of female students, but the actual effect is not ideal.

Step test results of experimental groups before and after the experiment are increased by 7.56 on average and the statistics P = 0.005<0.01, indicating there is a significant difference in the step test data before and after the experiment. Thus it can be seen that the application of 3D technology can mobilize students' sports load and exercise density while learning, and continuously improve students' heart pump function.

Step test results of control groups before and after the experiment are increased by 1.08 on average and the statistics P = 0.651 > 0.05, indicating there is no significant difference in the step test data before and after the experiment. Thus it can be seen that the conventional teaching method can improve and maintain the step test data of female students, but the real effect is not ideal.

Sit-ups results of experimental groups before and after the experiment are increased by 2.87 on average and the statistics P = 0.019<0.05, indicating there is a significant difference in the sit-ups data before and after the experiment. Thus it can be seen that the application of 3D technology can dramatically improve students' graceful posture while learning, keep physical coordination, and practice the strength in the waist.

Sit-ups results of control groups before and after the experiment are increased by 0.05 on average and the statistics P = 0.944 > 0.05, indicating there is no significant difference in the sit-ups data before and after the experiment. Thus it can be seen that the conventional teaching method can improve and maintain the sit-ups data, but the real effect is not ideal.

Motor skills results of experimental groups before and after the experiment are increased by 19.59 on average and the statistics P = 0.000<0.01, indicating there is a significant difference in the motor skills data before and after the experiment. Thus it can be seen that the application of 3D technology can significantly enhance the learning foundation and difficulty of students' motor skills, greatly improve the students' motor skills scores.

Motor skills results of control groups before and after the experiment are increased by 4 on average and the statistics P = 0.000<0.01, indicating there is a significant difference in the motor skills data before and after the experiment. Thus it can be seen that the conventional teaching method can significantly improve the data of motor skills, achieving a desired result.

Routine skills results of experimental groups before and after the experiment are increased by 23.5 on average and the statistics P = 0.000<0.01, indicating there is a significant difference in the routine skills data before and after the experiment. Thus it can be seen that the application of 3D technology can mobilize students to learn the routine skills art of shape-building dance, greatly improve students' routines skill performance of shape-building dance, achieving an ideal effect.

Routine skills results of control groups before and after the experiment are increased by 18.72 on average and the statistics P = 0.000<0.01, indicating there is a significant difference in the routine skills data before and after the experiment. Thus it can be seen that the conventional teaching method can significantly improve the evaluation results of routine skills, achieving an estimated result.

3.3 *Comparative research on the analysis of body mass index, seating forward bend, step test, sit-ups, motor skills and routine skills evaluation results of experimental groups and control groups after the experiment*

After handled statistically by the SPSS software, the results in Tables 3 and 4 show us: Comparative to control groups, the BMI results of experimental groups after the experiment are increased by 0.73 on average, and the statistics F = 1.91, P = 0.174 > 0.05, indicating there is no significant difference but a certain rise in the BMI data of two groups after the experiment, and the determinants of MBI value are very complicated. Thus it can be seen that compared with the conventional teaching, the 3D technology can increase students' MBI value to some extent, and female students can lose their weight and get healthier.

Table 5. Contrast analysis of body mass index, seating forward bend, step test and sit-ups results of experimental groups and control groups after the experiment.

Index	BMI X±S	Seating forward bend (cm) X±S	Step test (score) X±S	Sit-ups (piece) X±S
Experimental group	21.87±1.49	11.16±2.07	59.73±8.93	33.23±3.85
Control group	22.60±2.12	9.75±2.36	54.04±8.97	30.93±3.75
D-value	−0.73	1.41	5.69	2.30
Testing F value	1.91	4.96	4.98	4.52
P value	0.174	0.031	0.03	0.039

Table 6. Contrast analysis of motor skills and routine skills evaluation results of experimental groups and control groups after the experiment.

Index	Motor skills evaluation results (score) X±S	Routine skills evaluation results (score) X±S
Experimental group	84.05±4.80	79.77±2.64
Control group	78.93±6.65	75.79±7.63
D-value	5.12	4.98
Testing F value	9.25	5.48
P value	0.004	0.023

Comparative to control groups, the seating forward bend results of experimental groups after the experiment are increased by 1.41 cm on average, and the statistics F = 4.96, P = 0.031<0.05, indicating there is a significant difference in the seating forward bend data of two groups after the experiment. Thus it can be seen that compared with the conventional teaching, the 3D technology can greatly increase students' seating forward bend value, and make their bodies and figures more graceful.

Comparative to control groups, the step test results of experimental groups after the experiment are increased by 4.98 cm on average, and the statistics F = 4.98, P = 0.03<0.05, indicating there is a significant difference in the step test data of two groups after the experiment. Thus it can be seen that compared with the conventional teaching, the 3D technology can greatly increase the step test data and enhance their heart function.

Comparative to control groups, the sit-ups results of experimental groups after the experiment are increased by 4.52 on average, and the statistics F = 4.52, P = 0.039<0.0, indicating there is a significant difference in the sit-ups data of two groups after the experiment. Thus it can be seen that compared with the conventional teaching, the 3D technology can greatly increase the sit-ups data, enhance their strength in the waist, make the shape-building dance movements of their upper and lower parts of the body more elegant and coordinated. Comparative to control groups, the motor skills results of experimental groups after the experiment are increased by 5.12 on average, and the statistics F = 9.25, P = 0.004<0.01, indicating there is a significant difference in the motor skills data of two groups after the experiment. Thus it can be seen that compared with the conventional teaching, the 3D technology can greatly increase the motor skills data, enhance their basic movement, make their movements complete and unified.

Comparative to control groups, the routine skills results of experimental groups after the experiment are increased by 4.98 on average, and the statistics F = 5.48, P = 0.023<0.05, indicating there is a significant difference in the routine skills data of two groups after the experiment. Thus it can be seen that compared with the conventional teaching, the 3D technology can greatly increase the motor skills data, and make their routine movements more elegant and complete.

4 CONCLUSIONS

1. *Both the 3D technology and the conventional teaching can significantly improve students' motor skills and routine skills evaluation results, enhance and maintain their weight index value, seating forward bend, step test and sit-ups.*

2. *Comparative to the conventional teaching, the 3D technology can significantly improve students' motor skills results, seating forward bend, step test, sit-ups and routine skill evaluation results, at the same time, increase weight index value to some extent.*

3. *The 3D technology can effectively supplement and improve the conventional teaching. It can not only urge students to take the initiative to learn the shape-building dance movements, but also strengthen the comprehensive physical quality education, and promote their physical and mental health, with an obvious effect. So the 3D technology is worth popularizing.*

REFERENCES

[1] Zhong Li. The Experimental Research on the Model of Inquiry Teaching of Body Building Exercise [J]. Journal of Chengdu Physical Education Institute, 2004, (2).

[2] Dong Yeping. Application of "Self-description Starting" Method in the Eurhythmics Teaching in P.E. Specially in Normal Colleges and Universities [J]. Journal of Beijing Sport University, 2005, (7): 952–954.

[3] Ma Zhenhai. Quality-oriented Education: Concepts and Practice [M]. Kaifeng: Henan University Press, 2003.

[4] Shan Yaping. Research on the Aesthetic Value of Dance Sports [J]. Journal of Beijing Sport University, 2004, 27 (1): 144.

[5] Gong Zhengwei. Sports Teaching Theory [M]. Beijing Sports University press, 2004.

[6] Guo Xiuwen, Application of SPSS Statistical Software in the Sports [M]. Beijing: Peoples Sports Publishing House of China, 2007: 185–196.

[7] Li Shiming, Multivariate Analysis of Practical Sports [M]. Beijing: Peoples Sports Publishing House of China, 2007: 251–271.

Sports Engineering and Computer Science – Luo (Ed.)
© *2015 Taylor & Francis Group, London, ISBN 978-1-138-02650-6*

An exploration of the ways to computer aided college English teaching and learning

HaiYan Zhao

College of Foreign Languages, Northeast Dianli University, Jilin, China

ABSTRACT: With the development of computer technology and the widespread access of computers to the general public, Computer Aided Instruction (CAI) is playing an increasingly important role in the field of education. The deepening of college English teaching reform and promotion of quality education happens widely and fast. Thus CAI is greatly preferred by English teachers and English learners of all ages because of its unique advantages. This paper focuses on exploring the ways in which computer technology assists college English teaching and learning. Actually, the ideas to write this paper comes from trying to provide some reference for college English teachers and students.

1 INTRODUCTION

Since 1999, there has been a relatively shortage of college English teachers due to the expansion of college enrollment. In order to ease the contradiction caused by teacher-student ratio imbalance, some schools have taken large-class lectures. As a result, teachers cannot satisfy the specific needs of each student in class.

Therefore, autonomous English learning after class contributes a lot to the language learning efficiency of students. How to improve students' autonomous English learning ability is the key and difficulty of college English teaching. In addition, a considerable amount of knowledge is intended to be imparted in each class period, which is hard-pressed for teachers to accomplish only by relying on verbal teaching and blackboard writing. The above-mentioned can be solved by CAI. This is also in line with the College English Curriculum Requirements formulated by the Ministry of Education.

2 THE ADVANTAGES OF COMPUTER-AIDED COLLEGE ENGLISH TEACHING AND LEARNING

Computer Aided Instruction refers to a teaching method in which part of the teaching mission is performed by computer or by the teacher with the help of computer technology. As a result, knowledge and skills can be imparted to students.

2.1 Computer-aided college English teaching and learning is conducive to increasing knowledge intensity

Since time is limited in each class, how to improve the efficiency of classroom teaching is a fundamental subject for all disciplines. The traditional monotonous teaching form, the verbal presentation and blackboard writing, can hardly guarantee the accomplishment of the teaching content in each period. Whereas, the multimedia courseware, made ahead of the class time, can be used to demonstrate teaching content to students. By this way, it saves the time wasted by writing on the blackboard so as to improve classroom knowledge density. Thus, the time taken by explaining one word in the traditional way is sufficient for explaining two or more words. Besides, students' course books are accompanied by self-learning CDs, with the aid of which students can conduct an in-depth preview before class. Therefore, teachers have more time to explain more and more extra-curricular knowledge. Furthermore, the multimedia courseware can reproduce any part of the lesson, which can be taken by students at any time in comparison with new knowledge.

2.2 Computer-aided college English teaching and learning can help mobilize the students' interest in learning and enhance students' memory of knowledge

Computer-aided college English teaching and learning combines text, images, sound, animation, video, and other media information. Meanwhile, it makes monotonous English teaching vivid and attractive so as to fully mobilize students' vision and auditory sense and stimulate students' desire to express themselves in English. Besides, students' anxiety caused by the traditional teaching way can be eased to some extent their enthusiasm to participate in English teaching activities can be inspired. On the other hand, English language teaching software is becoming humanized. A few Flash or small skits, which sometimes simulate real-life situations, are inserted into the software. In this way, students can fix knowledge in their memory.

2.3 Computer-aided college English teaching and learning can simplify the teachers' task of preparing lectures

Currently, almost every textbook is equipped with a CD-ROM, which can be directly used in teaching. For those without CD-ROM, teachers who teach the same course can coordinate their efforts to make a courseware so that students will learn knowledge from several teachers instead of a single one. First line of text or heading.

2.4 Computer-aided college English teaching and learning can promote students' autonomous English learning

With the rapid development of information technology nowadays, networking has become an important trend in the development of computer-assisted teaching. English learners can communicate with and learn from each other. In the meanwhile, they can download and share learning resources through network and then study English at their own pace and in their own way.

The above-mentioned advantages possessed by computer-aided college English teaching and learning determine that will play an important role in the teaching and learning of English. Therefore, it is necessary to start and strengthen college English teaching with the aid of computer technology. The ways, college English teaching and learning with the aid of computer technology, will be explored in the following part. Layout of text headings to the left of their column and start these headings with an initial capital. Type the caption above the table to the same width as the table (Table caption tag). See for example Table 1.

3 PHOTOGRAPHS AND FIGURES

3.1 In-class teaching aided by computer technology

3.1.1 Teaching process aided by computer technology

Courseware has become one of the most effective auxiliary teaching tools for English teachers.

With vivid images and auditory effect, teaching content can be imparted to students in a more attractive and convenient way.

3.1.2 Facilitating testing and effective assessment implementation

Examination questions stored in the test database software are designed by excellent education experts and teachers based on teaching syllabus. The software is designed by computer software specialist. It takes no more than several minutes for teachers to output the exam questions, which not only saves time, but also ensures the quality of the questions. Upon the completion of the test, computer can analyze the results quickly, calculating the average score, students' ranking in no time.

3.2 Autonomous English learning with the aid of computer technology

Students are different in their levels of English proficiency. So learning paces and learning methods and teachers cannot satisfy the specific needs of all the students. Therefore, autonomous English learning after class is an important way to improve their English proficiency. Autonomous learning, a learner's self-management of his study, is also defined as the ability and willingness to make his own choice with regard to study. Thus it is both a learning attitude, but also an ability to learn on one's own. Advocating autonomous learning in foreign language teaching requires that students, under the guidance of teachers, choose their own materials and methods. They are also required to draw out an appropriate learning plan, monitor and evaluate their own learning process, so as to gradually improve their English learning efficiency. Computer network provides new and advanced conditions for autonomous English learning.

3.2.1 Creating a QQ group to establish a small autonomous English learning environment

Autonomous English learning not only involves individuals, it is social in essence. Interaction, negotiation and cooperation among learners are important factors in promoting learners' autonomy. Support from fellow learners can ensure the autonomous learning Effect. A QQ group can be created through a computer network for English learners to communication online. Firstly, students can discuss the problems they encountered in their study with one another. During the process of discussion, their knowledge will be commonly increased; Secondly, students share learning materials with each other; also, learners can learn from each other.

3.2.2 *Creating English knowledge bases*

The use of multimedia and network resources is one of the differences between traditional self-learning mode and the autonomous learning mode under computer network Environment. Firstly, teachers can provide students with English information on a variety of network resources, such as URLs (www.putclub.com, CNKI, etc.) and online courses (Harvard University online courses, etc.); secondly, students and teachers can work together to take advantage of multimedia technology to create various English knowledge bases to facilitate students' autonomous learning. The bases can be classified into listening material base, spoken language base, base of foreign culture, base of grammatical knowledge and base of linguistic theory etc.

3.2.3 *Learning vocabulary through online corpus*

Vocabulary acquisition is the basis of language learning. The more words a learner masters, the stronger his listening, speaking, reading and writing ability is. Traditional English word learning concentrates on the sound, form and meaning of words, while ignoring their use in specific context. As a result, some learners are good at reading, and translating of English into Chinese, but poor at communicating both in written form and orally. Numerous ungrammatical or unacceptable expressions appear in learners' communicating process.

As a database of storing language material in real life, corpus is equipped with the functions of word retrieval, contextual co-occurrence, synonyms discrimination, syntax inquiry, word frequency retrieval. It is increasingly applied to language teaching and learning. English learners can grasp the grammatical meaning, connotation, stylistic meaning and affective meaning of a word by using online corpus so as to avoid learning its conceptual meaning in isolation from its use in concrete contexts.

4 ISSUES THAT DESERVE NOTING CONCERNING COMPUTER-AIDED COLLEGE ENGLISH TEACHING AND LEARNING

4.1 *The role played by computer in college English teaching should not be overstressed*

In CAI, the leading role of teachers and the subject status of students should not be ignored. CAI should not be converted into electronic traditional teaching and the computer should not become electronic blackboard. Computer technology is applied just for a better teaching effect. Teachers should not type all the teaching content on computer and turn themselves into computer operators. Communication and interaction between teachers and students is the most effective way of language acquisition.

4.2 *Knowledge capacity in one period should be appropriate*

Be sure not to convert CAI into a modernized cramming teaching mode by including an excessive amount of knowledge in one period due to the advantage of high density of information exchange possessed by CAI can achieve large capacity. Instead, teachers should concentrate on helping students deepen their understanding about knowledge, expand their knowledge structure and cultivate their learning skills in combination with the teaching emphases and difficult points in each class.

4.3 *Identifying teachers' role in computer-aided autonomous English learning*

Autonomous learning does not mean either self-study or laissez-faire attitude of teachers to learners. In fact, the effect of learners' autonomous learning depends largely on the role played by teachers. Therefore, it is very important to identify the role of teachers in autonomous English learning.

5 CONCLUSION

Computer technology is an effective auxiliary English teaching means. It is conducive to stimulating students' interest in English learning and provides a new way for college English teaching and learning. It makes great contribution to college English teaching and learning. Applying CAI to college English teaching and learning is not only the trend of modern education reform, but also the inevitable development of the times. English educators and learners should improve their ability to use modern educational technology, making use of computer technology to improve teaching and learning.

REFERENCES

Heloc, H. 1981. Autonomy and Foreign Language Learning. Oxford: Pergamon Press.
Jia, Yan. 2000(12). Autonomous Learning Mode and Foreign Language Teaching. Journal of Zhengzhou Institute of Light Industry: 56–57.
Li, Bingde. 1990. Teaching Theory. Bejin: People's Education Press.
Wang, Yan. 2007. Expectations of Autonomous Learners on the Role Played by Teachers. Foreign Language World: 43–44.
Yuan, Changhuan. 2001. Computer Assisted English Learning. Beijng: People's Education Press.
Zhang, Guoying. 1995. Computer Assisted English Learning and Research. Shanghai: Shanghai Foreign Language Education Press.

Sports Engineering and Computer Science – Luo (Ed.)
© 2015 Taylor & Francis Group, London, ISBN 978-1-138-02650-6

Design of student employment information management system based on ASP

YangYang Ge

The Graduate School, Harbin University of Science and Technology, Harbin, Heilongjiang, China

ABSTRACT: With the continuous development of higher education and the extensive application of computer technology, it is necessary to establish a sound network information management for colleges to improve the working level and the efficiency. Combined with the actual needs and the specific requirements of college employment management, the student employment information management system based on ASP has been developed to meet the demands of colleges students, managers and employers. In this paper, the system requirements, the system functions, the whole structure of the system and the hierarchical design of the system have been analyzed in detail, which plays an important role in realizing the dynamic management of student information and promoting the information construction of employment. At the same time, it is also beneficial to provide the operable platform for the efficient and systematic management.

Keywords: ASP technology; employment information management system; analysis and design

1 INTRODUCTION

Employment work is the last link of higher education, which is an important component of higher education management. With the sharp increase in the number of graduates and the continuous process of school management digitization, it is required to actively promote the information management of employment work, design the graduate employment information management system and make full use of the computer technology to improve the efficiency of employment while strengthening the standard management of employment.

2 SIGNIFICANCE OF THE RESEARCH ON EMPLOYMENT MANAGEMENT SYSTEM

2.1 *Realizing the standard management and the network office*

For the student employment information management system, the available employment information management mechanism and functions can be integrated, the computer technology can be fully used and the graduate employment information management can be systematized. Afterwards, the standard management of information and the scientific statistics can be realized. At the same time, the automatic and networked college employment management can be also achieved so as to improve the working level of graduate management.

2.2 *It is convenient and the efficiency is improved*

The establishment of student employment information management system is beneficial to the employment staffs who can get out of the heavy manual operation. Then, the management workload will be reduced. The employer can screen the student resumes which meet the recruitment requirements of company from the mass resumes so as to better choose the best talents and save time in reading a lot of paper. Meanwhile, the students can also learn more recruiting information from this system and rapidly know their suitable positions. The development of employment information management system is beneficial to the students, the colleges and the enterprises, which greatly improves the efficiency of college employment work.

2.3 *Achieving the dynamic management and the real-time statistics*

The implementation of graduate information management system is conducive to perform the effective dynamic management for students. Then, the college can macroscopically grasp the graduate employment information, provide the targeted guidance and services and real-timely collect the employment data so as to provide the quick and effective decision making for colleges to adjust to the employment work. Meanwhile, it can also provide the efficient and reliable basic data support for provincial college career center and promote information exchange and information sharing among colleges and universities.

2.4 Realizing the information sharing and facilitating the tracking

The establishment of student employment information management system is beneficial to the uploading, the modification, the checking and the sharing of the employment data and the student resources between the departments and the colleges. It is advantageous for the implementation of college graduate employment management system and the department employment management system. At the same time, it is also conducive to establishing the three-dimensional employment service system.

3 DESIGN OF EMPLOYMENT MANAGEMENT SYSTEM

With the continuous update and development of network technology, Web development innovation technology has been appeared. ASP.NET developed by Microsoft is the leader in these new technologies. ASP. NET[1] is a part of.NET Frame-Work and a server-side scripting technology that can promote the Internet server to execute the script embedded in web pages. The efficiency and the safety of ASP.NET are far more than ASP. It can dynamically create the documents on a Web server while requesting documents via HTTP. It provides a safe, convenient, powerful and reliable cash programming environment for the development and the application of web[2].

The employment management system takes the graduate employment management as the center, integrates many years of experience in employment management and reform and benefits the students, the administrators and the employers. At the same time, it focuses on the improvement of working efficiency and the humanization design as well as the realization of the data sharing and the rapid search functions. As a result, it can provide the detailed data for school decision-making and avoid the loopholes in management. The system adopts B/S mode structure, ADO data access technology and network technology and Microsoft SQL Server 2000 (compatible with SQL Server 2005) relational database system in order to ensure the security of data. Meanwhile, C# language is matched with ASP.NET on.NET platform to achieve the design of graduate information management system. Then, the design of system will be more rationalized.

3.1 Analysis of system demands

For the graduate employment information management system, it is required to design the foreground support and the background support. The functions of foreground include: providing the information query functions of student sources, facilitating the employers to find the appropriate student data, publishing the recruitment information, providing the query interface of recruitment information and the employment opportunities for students and offering the link function to facilitate students to understand more employment information as well as affording the employment policy guidance services and the quick registration function to facilitate the landing of students and employers. The functions of background include: realizing the operations such as the input, the update, the editing, the deletion and the presenting of student information, providing the addition and the update of employment information as well as the propaganda, the guidance and the answering work of employment policies[3].

3.2 Analysis of system functions

According to the demand analysis, the student information management system in employment website should provide the following functions for graduates, employers and administrators.

3.2.1 Functions provided for students
The functions include: browsing and searching the latest job recruitment information, the employment policies and regulations, providing the query and the information retrieval services with various conditions, publishing the job resumes of students, browsing the relevant information of employers and realizing the employment counseling function. The students can ask the guidance teachers some questions about the employment and they can also communicate with the employers on line.

3.2.2 Functions provided for employers
The functions include: registering the employer information, publishing the corporate presentations and job information, realizing the information browsing and search function such as the employment query of students, sending the relevant information such as inquiry, interviews or written tests to students, browsing or viewing the basic information of students and the resumes that are delivered to the company, giving the feedback of the employment counseling of students and browsing the students sources, the majors and the teachers of college.

3.2.3 Functions provided for administrators
The functions include: providing relevant consulting and the information publishing such as the career guidance information, the recruitment information, the job information, the position information and the recommendation information,

checking and certifying the relevant information of graduate students as well as the accuracy of employer information, realizing the student data management such as the number of employed students and the unemployed students, answering the questions provided by students on line, analyzing the employment status such as the employment rates, the employment destinations of students, the popular majors and the demands for students' abilities. According to these information, the colleges can appropriately adjust the majors to cultivate the talents who can meet the needs of society[4].

3.3 Design of the whole structure of system

According to the results of function analysis, the whole structure of the student employment information management system can be designed. This system is divided into three login user permissions, namely the administrators, the students and the employers. The application requirements of system are different accordingly. The whole structure of this system is shown in Figure 1.

3.4 Hierarchical design of system

3.4.1 Design of student module
The main functions of student module include the management of the basic information of students, the searching of employer information, the browsing of public information, the online consultation and the job answering, etc. The students can search and maintain their own basic information

through the personal information management, including the input of resumes and natural conditions of students. The students can ask the guidance teachers or the employers some questions about employment. The employment management includes the delivery of electronic resumes, the recovery of job information and the collection of job information, which is shown in Figure 2.

3.4.2 Design of employer module
The employer is one of the three users in the student employment management system. The employment and the employment management will be impossible if the employers are not involved. In the design of the system functions, when the employer logins the system, they can publish the recruitment information, search talents, manage their basic information, communicate with graduates, release the interview notice and answer the questions, etc. For example, when the employer or the company finds the appropriate graduates who are suitable to the positions, they can send the internal mails or contact the students through colleges after landing the management page. It is illustrated in Figure 3.

3.4.3 Design of management module
The main users of management module refer to the staffs in employment department and the administrators. The role of the staffs in employment department refers to publishing the latest

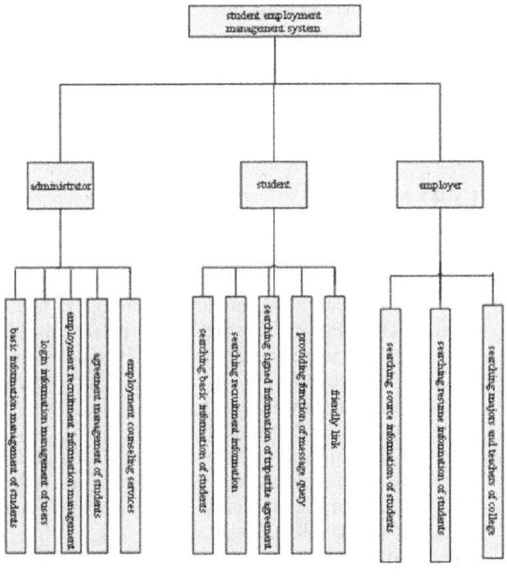

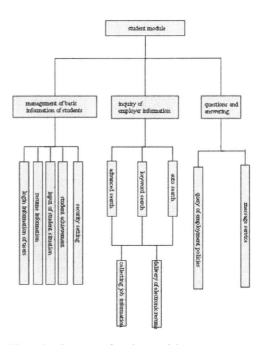

Figure 1. The whole structure of the student employment information management system.

Figure 2. Structure of student module.

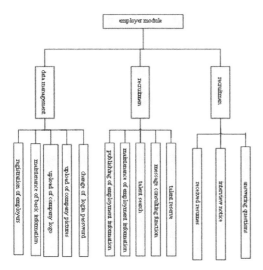

Figure 3.   Structure of employer module.

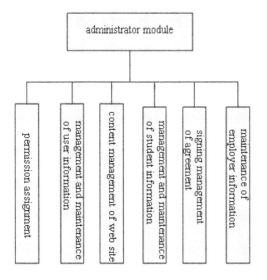

Figure 4.   Structure of administrator module.

employment information, the employment news, the graduate information, the job information and the employer information. At the same time, the online consulting, the query statistics and the employment management can also be realized. The administrators have the permissions to add, delete and modify the relevant information as well as the permission of data statistics and analysis operation. This function is of great importance to the student employment information management. Meanwhile, the administrators should pay more attention to the maintenance of system in the process of management so as to ensure the health and the stability of system. They should timely respond to the questions of users and delete the outdated information[5]. For the administrators, the main work is to set the basic information of system and implement the data backup, etc. It is shown in Figure 4.

4   CONCLUSION

In this paper, the employment information system has been discussed through analyzing the system demands, the system functions, the whole structure of system and the hierarchical design of system. In fact, the design of the employment management system must be based on the database, the web browser, the server and the network applications. The application program of server can be written on the basis of designing and creating database so as to gradually achieve the design function. The employment management system should be constantly improved through a variety of tests after being established. Afterwards, the scientific, efficient, strict and practical student employment management system can be built so as to achieve the fluent communication and meet the interaction requirements among employers, colleges and students. Then, the efficiency of college employment management work will be greatly improved and the standardization, the informatization and the systematization of employment management can be realized.

ABOUT THE AUTHOR

Yangyang Ge (1980), female, lecturer, the Graduate School, Harbin University of Science and Technology.

REFERENCES

[1] http://baike.so.com/doc/5381240.html [DB/OL] 360 data query.
[2] Yu Yanjie, Li Chengzhi. Common Technology of Comparative Data Paging Asp.net[J]. Journal of Harbin University of Science and Technology, 2010, (6)6–8.
[3] Li Jing. Implementation of Student Employment Management System Based on ASP[J]. Information and Computer, 2012, (9)84–86.
[4] Qiu Yue. Design and Implementation of Graduate Information Management System Based on ASP. NET[D]. University of Electronic Science and Technology of China, 2013: 20–30.
[5] Chang Jianyu. Design and Implementation of Employment Management System Based on web in Mandarin College[D]. University of Electronic Science and Technology of China, 2012: 24–26.

Sports Engineering and Computer Science – Luo (Ed.)
© 2015 Taylor & Francis Group, London, ISBN 978-1-138-02650-6

The study of DNS server configuration in heterogeneous environments

Xiaguang You
Electronic Information Department, Zhengzhou Electric Power College, Zhengzhou, China

Bingwen You
Customer Service Center, State Grid Zhengzhou Power Supply Corporation, Zhengzhou, China

ABSTRACT: The paper first introduces the principle of DNS; DNS is a hierarchical structure; there are two methods, Iterative Query and Recursive Query, to search DNS domain name database. It discusses the methods of configuring DNS Sever in Linux and Windows in heterogeneous environments. There are four important DNS configuration files in Linux environment: global configuration file; main configuration file; zone file and reverse resolution zone file. According to guidance of configuration in Windows environment, there are mainly forward lookup zone and reverse lookup zone. After completing configuration, Linux uses nslookup and dig to test and Windows uses nslookup to test.

Keywords: DNS; Linux; Windows; nslookup; dig

1   THE PRINCIPLE OF DNS

Computer technology has been developing fast since the first computer in the world ENIAC came into being in University of Pennsylvania in 1946. Computer network came into being because of the development of computer and communication. Computer network is divided into four stages [1]. Computer network is divided into WAN (Wide Area Network); MAN (Metropolitan Area Network) and LAN (Local Area Network) according to coverage area [2]. And Internet is the largest WAN. IP address is difficult to remember; people use easily-remembered domain names to surf the Internet, from the earliest English domain names to later Chinese names. But the computer itself can't distinguish domain names so that we need to resolve domain names into IP addresses which is completed by DNS (Domain Name System).

DNS is a hierarchical structure. There are two methods Iterative Query and Recursive Query to search DNS domain name database [3]. Iterative Query is suitable for Internet DNS and Recursive Query is suitable for local DNS.

There are kinds of computer systems on the Internet such as Linux and Windows. This paper discusses configuring DNS server in Linux and Windows in heterogeneous system. And we use VMware virtual machine to complete server configuring and testing.

2   DNS SERVER CONFIGURATION

2.1   *Linux red hat enterprise*

2.1.1   *Configuration*

First use rpm—qa|grep bind and rpm—qa|grep catching-nameserver to check whether bind and catching-nameserver packages are installed. If not installed, first install the two packages especially catching-nameserver package. There are four most important DNS configuration files in Linux environment: global configuration file; main configuration file; zone file and reverse resolution zone file. The four files copy and change names in the corresponding directory template. But we need parameters—a or—p when copying cp. Global configuration file and main configuration file are in directory/var/named/chroot/etc/ and zone file and reverse lookup zone file are in directory/var/named/chroot/var/named/.

a. Global configuration file named.conf
options {
    listen-on port 53 { any;};
    listen-on-v6 port 53 {::1;};
    directory "/var/named";
    dump-file "/var/named/data/cache_dump.db";
    statistics-file "/var/named/data/named_stat.txt";
    memstatistics-file   "/var/named/data/named_mem/stats.txt";
  query-source port 53;
  query-source-v6 port 53;

```
allow-query { any;};
};
logging {
    channel default-dubug {
    file "data/anmed.run";
    severity dynamic;
};
}
view a.net {
    match-clients { any;};
    match-destinations { any;};
    recursion yes;
    include "/etc/a.net";
};
```

b. *Main configuration file*
```
zone "." IN {
    type hint;
    file "named.ca";
};
zone "a.net" IN {
    type master;
    file "a.net.zone";
    allow-update { none;};
};
zone "0.168.192.in.addr.arpa" IN {
    type master;
    file "0.168.192.zone";
    allow-update {none;};
};
```

c. *Zone file a.net.zone[4]*

$TTL 86400

@	IN SOA	linux2.a.net.	root.a.net. (
42	; serial (d.adams)		
3H	; refresh		
15M	; retry		
1W	; expiry		
1D)	; minimum		
@	IN NS	linux2.a.net.	
@	IN MX 10	mail.a.net.	
Linux2.a.net.	IN A	192.168.0.132	
www.a.net.	IN A	192.168.0.142	
ftp.a.net.	IN A	192.168.0.152	
mail.a.net	IN A	192.168.0.162	
dns	IN CNAME	linux2	
bbs	IN CNAME	www	

d. *Reverse lookup zone file 0.168.192.zone*

$TTL 86400

@	IN SOA	linux2.a.net.	root.a.net. (
		42	; Serial
		28800	; Refresh
		14400	; Retry
		3600000	; Expiry
		86400)	; Minimum
@	IN NS	linux2.a.net.	
132	IN PTR	linux2.a.net.	
142	IN PTR	www	

152	IN PTR	ftp
162	IN PTR	mail

After server configuration, server has to be restarted. The restart command is service named restart.

i. *Testing [5]*

a. Use nslookup to do interactive testing

Testing host computer address A source record is shown in Figure 1. Testing reverse lookup pointer PTR source record is shown in Figure 2. Testing alias CNAME source record is shown in Figure 3. Testing mail

```
[root@linux2 ~]# nslookup
> www.a.net
Server:          192.168.0.132
Address:         192.168.0.132#53

Name:  www.a.net
Address: 192.168.0.142
> set type=a
> ftp.a.net
Server:          192.168.0.132
Address:         192.168.0.132#53

Name:  ftp.a.net
Address: 192.168.0.152
```

Figure 1.   Testing host computer address A source record.

```
> 192.168.0.142
Server:         192.168.0.132
Address:        192.168.0.132#53

142.0.168.192.in-addr.arpa     name = www.0.168.192.in-addr.arpa.
> set type=ptr
> 192.168.0.152
Server:         192.168.0.132
Address:        192.168.0.132#53

152.0.168.192.in-addr.arpa     name = ftp.0.168.192.in-addr.arpa.
```

Figure 2.   Testing reverse lookup pointer PTR source record.

```
> set type=cname
> dns.a.net
Server:          192.168.0.132
Address:         192.168.0.132#53

dns.a.net        canonical name = linux2.a.net.
> bbs.a.net
Server:          192.168.0.132
Address:         192.168.0.132#53

bbs.a.net        canonical name = www.a.net.
```

Figure 3.   Testing alias CNAME source record.

```
> set type=mx
> a.net
Server:          192.168.0.132
Address:         192.168.0.132#53

a.net    mail exchanger = 10 mail.a.net.
```

Figure 4.   Testing mail exchange machine MX source record.

```
[root@linux2 ~]# dig mail.a.net

; <<>> DiG 9.3.3rc2 <<>> mail.a.net
;; global options: printcmd
;; Got answer:
;; ->>HEADER<<- opcode: QUERY, status: NOERROR, id: 40701
;; flags: qr aa rd ra; QUERY: 1, ANSWER: 1, AUTHORITY: 1, ADDITIONAL: 1

;; QUESTION SECTION:
;mail.a.net.                  IN      A

;; ANSWER SECTION:
mail.a.net.           86400  IN      A       192.168.0.162

;; AUTHORITY SECTION:
a.net.                86400  IN      NS      linux2.a.net.

;; ADDITIONAL SECTION:
linux2.a.net.         86400  IN      A       192.168.0.132

;; Query time: 5 msec
;; SERVER: 192.168.0.132#53(192.168.0.132)
;; WHEN: Sun Jun 22 13:44:15 2014
;; MSG SIZE  rcvd: 81
```

Figure 5.   Testing forward lookup host address A source record.

```
[root@linux2 ~]# dig -x 192.168.0.162 @linux2

; <<>> DiG 9.3.3rc2 <<>> -x 192.168.0.162 @linux2
; (1 server found)
;; global options: printcmd
;; Got answer:
;; ->>HEADER<<- opcode: QUERY, status: NOERROR, id: 24381
;; flags: qr aa rd ra; QUERY: 1, ANSWER: 1, AUTHORITY: 1, ADDITIONAL: 1

;; QUESTION SECTION:
;162.0.168.192.in-addr.arpa.  IN      PTR

;; ANSWER SECTION:
162.0.168.192.in-addr.arpa. 86400 IN  PTR     mail.0.168.192.in-addr.arpa.

;; AUTHORITY SECTION:
0.168.192.in-addr.arpa. 86400  IN      NS      linux2.a.net.

;; ADDITIONAL SECTION:
linux2.a.net.         86400  IN      A       192.168.0.132

;; Query time: 6 msec
;; SERVER: 192.168.0.132#53(192.168.0.132)
;; WHEN: Sun Jun 22 13:45:20 2014
;; MSG SIZE  rcvd: 105
```

Figure 6.   Testing reverse lookup pointer PTR source record.

exchange machine MX source record is shown in Figure 4.

b. Use dig (Domain Information Groper) to do testing

Testing forward lookup host address A source record is shown in Figure 5. Testing reverse lookup pointer PTR source record is shown in Figure 6.

2.2   Windows Server 2008

2.2.1   Configuration

In default condition there is no DNS Server in Windows Server 2008. Open "Server Manager" → "Role" → "Add role" → "Next" → "DNS Server" → "Next" → "Next" → "Install" → "Exit" [6]. Open "start menu" → "tool" → "DNS" and we open DNS Manager. DNS Manager mainly has forwarded lookup zone and reverse lookup zone. Configure them respectively. Forward lookup zone is shown in Figure 7, and reverse lookup zone is shown in Figure 8.

2.2.2   Testing

Use nslookup to do testing. Nslookup is divided into Non-interactive and interactive, first use Non-interactive nslookup to do testing.
C:\Users\Administrator>nslookup www.a.net
Server: you10.a.net
Address: 192.168.0.132
Name: www.a.net
Address: 192.168.0.142
C:\Users\Administrator>nslookup ftp.a.net
Server: you10.a.net
Address: 192.168.0.132
Name: ftp.a.net
Address: 192.168.0.152
C:\Users\Administrator>nslookup mail.a.net
Server: you10.a.net
Address: 192.168.0.132
Name: mail.a.net
Address: 192.168.0.162
C:\Users\Administrator>nslookup 192.168.0.132
Server: you10.a.net

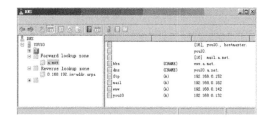

Figure 7.   Forward lookup zone.

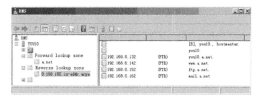

Figure 8.   Reverse lookup zone.

Address: 192.168.0.132
Name: you10.a.net
Address: 192.168.0.132
C:\Users\Administrator>nslookup 192.168.0.142
Server: you10.a.net
Address: 192.168.0.132
Name: www.a.net
Address: 192.168.0.142
C:\Users\Administrator>nslookup 192.168.0.152
Server: you10.a.net
Address: 192.168.0.132
Name: ftp.a.net
Address: 192.168.0.152
C:\Users\Administrator>nslookup 192.168.0.162
Server: you10.a.net
Address: 192.168.0.132
Name: mail.a.net
Address: 192.168.0.162
Then use interactive nslookup to do testing.
C:\Users\Administrator>nslookup
Default server: you10.a.net
Address: 192.168.0.132
> set type = a
> www.a.net
Server: you10.a.net
Address: 192.168.0.132
Name: www.a.net
Address: 192.168.0.142
> set type = ptr
> 192.168.0.142
Server: you10.a.net
Address: 192.168.0.132
142.0.168.192.in-addr.arpa name = www.a.net
> set type = cname
> bbs.a.net
Server: you10.a.net
Address: 192.168.0.132
bbs.a.net canonical name = www.a.net
www.a.net internet address = 192.168.0.142
> set type = mx
> a.net

Server: you10.a.net
Address: 192.168.0.132
a.net MX preference = 10, mail exchanger = mail.a.net
mail.a.net internet address = 192.168.0.162

3 CONCLUSIONS

DNS server is very important in the Internet. If there are problems in DNS, we can't surf the Internet. It's more complex in Linux to configuring DNS than in Windows. There are still many problems about DNS configuration such as DNS safety [7], accessorial DNS configuration, DNS clusters and dig usage and so on to be continually researched.

REFERENCES

[1] Wu Gongyi and Wu ying, "Computer Network Tutorial," 4th ed., Beijing: Publishing House of Electronics Industry, January 2007 pp. 1–2.
[2] Wu Gongyi and Wu ying, "Computer Network Tutorial," 4th ed., Beijing: Publishing House of Electronics Industry, January 2007 p. 12.
[3] Zhang Wen, Xing Shuqin and Yang Yanchang Transl. "Introduction to TCP/IP," 7th ed., Beijing: Publishing House of Electronics Industry, September 2003 p. 285 [Kenneth D. Reed, USA, WB47.0].
[4] Xie Shuxin, "Linux Network Server Configuration and Management Program Tutorial," Beijing: Science Press, 2011, pp. 177–178.
[5] Hu Lanlan, "A Study the DNS Server's Detection Based on BIND," Bengbu: Journal of Anhui Vocational College of Electronics & Information Technology, vol. 11, No. 62, October 2012.
[6] Zhang Wurong, Zhu Shengqiang and Tao An, "Windows Server 2008 Network Operating System," Beijing: Tsinghua University Press, 2011 p. 248.
[7] Wang Lixia, "Security Analysis of DNS Situation," Beijing: Computer Security, August 2011, pp. 65–68.

Sports Engineering and Computer Science – Luo (Ed.)
© 2015 Taylor & Francis Group, London, ISBN 978-1-138-02650-6

The design of Beijing Geographic Information System on Android

WenTing Guan & Jin Yu

Automation School, Beijing University of Posts and Telecommunications, Beijing, China

ABSTRACT: Nowadays the Geographic Information System (GIS) is widely used in various fields, and one of those is mobile devices. This project researches Beijing GIS based on Android platform. The user can search the index through menu, and then the application will automatically draw table or bar graph with data in SQLite databases. This application is going to offer scientists a convenient gateway to inquire information.

1 INTRODUCTION

With the popularity of mobile devices nowadays, Android, IOS, Windows phone operating systems occupy most of the smart phone market. The poor performance of traditional PDA processor based on WCE system leads to limited ability of the CPU to deal with threads and inability to do wireless communication. Therefore, we can only download data through USB. But in contrast, Android has also integrated various functions including GPS, orientation sensor and access to network via 3G. Android has become the leading platform in smart phone market.[1]

On Android devices, the data is stored in SQLite databases which applications access by constructing and executing queries, either directly or via Android content provider API calls. This paper talks about an application based on GIS which gives full expression to characteristic of GIS, the geographic information stored and worked have been coded into geographic mode and lots of geographic information which is provided to users is stored in database.

2 CONFIGURATION ENVIRONMENT OF ANDROID AND SQLITE

Android has two forms of development, SDK and NDK. SDK is a traditional development package which develops in Java and provides the tools and APIs necessary to begin developing applications. On the other hand, NDK is released by Google which is aimed to help develop coding in C++. However, Android doesn't support applications coded completely in C++ and the functions and bases offered by NDK are limited. At the same time, Java contains various classes and functions which can solve complex problems like multithreaded

programming and network programming easily, and the built-in Java garbage collection mechanism can free the memory used by objects which could avoid memory leak to some degree.[2] In summary, we choose Java in this project.

SQLite is an ACID-compliant embedded Relational Database Management contained in a small C programming library. SQLite implements most of the SQL standard, using a dynamically and weakly typed SQL syntax that does not guarantee the domain integrity.[3] In contrast to other database management systems, SQLite is not a separate process that is accessed from the client application, but an integral part of it. SQLite supports major operating systems like Windows, Linux and Unix, so it is widely used in embedded devices.

3 THE FLOW DESIGN

3.1 *The interface introduction*

The menu contains all the table names we need and the content is initialized by creatspinner() which is a new custom method. But two kinds of table are incompatible in the menu, one is android metadata included in SQLite database, the other is a table whose content is explanation of some Chinese in the database. If the two tables are accessed, the procedure will throw exception, so we read the unqualified table names and write them to array which can be tested and removed.

There are two other menus on the interface, table and bar. The menus content the indexes corresponding to the table name chosen. We need two associative containers to translate key into string and value into integer, also the containers store the specific attributes of table menu and bar menu. In this implementation, hashmap is chosen as associative containers because it supports fast access to key-value without sorted key.

Next, I use class created Button() to initialize the query button whose aim is to pass the indexes user chosen to other activity. An intent is an abstract description of an operation to be performed. It can be used with start an activity to launch an activity[4]. An intent contains full descriptions of an activity, including how to create an action and receive an action and some information attached on the intent. Here, I need the intent to transfer indexes user chosen. Once the query button is pressed, the corresponding table show or bar show will be awakened. If no indexes chosen, the program will prompt users to choose metrics first.

Figure 1 is the flow chart of the system.

3.2 The tables

There are two steps to draw a table.

1. I define the basic structure of the table in xml (Extensible Markup Language) document. A LinearLayout contains all interface elements of the form, next a HorizontalScrollView contains the final TableLayout, by which layout the user can horizontal slip the table in case too many columns or each column content too much.
2. Then I define a class named TableAdapter extended class BaseAdapter within Android,

which contains three important inner classes. The static inner class TableCell builds the width, height and content of every table cell, the static inner class TableRow is used to structure every line of the table. At last, the inner class TableRowView extended LinearLayout realizes the style of every line.

The column count is the number of indexes transmitted by intent. To fill the table row, we cycle to pass different length, width and contents into every cell. Every row stores in an ArrayList and every table is an ArrayList too. Take the filled row into the table, and the table is already finished.

3.3 The bars

The bar's layout is relatively simple consisting of one LinearLayout and a method drawChart(). In this method, first renew a custom database object extended from DBmanger, then transfer getResults() to pass indexes needed. To avoid needless duplication, do the repeated circulation to get the final indexes that will be abscissa of the bars. It needs four steps to draw a bar.

1. Set a SeriesRenderer to place the bar, and set the type face and font size of the axis, the background of the bar, division value and whether support viewport in SeriesRenderer.
2. Renew an object of bar and set its color.
3. Add the bar object to SeriesRenderer that is set already.
4. Start invoke method getBarChartView() from class ChartFactory, in which method draw a bar with the string array of abscissa and the number array of ordinate.

4 DATABASE MANAGEMENT

Android offers class SQLiteDatabase to do the operations, which offers some simple classes to manage the database. The method openDB() can easily open a database, conversely the method closed() is used to close the database.[5] And inquireDB() operates the database by some given SQLstatement. A new class DBmanager extended SQLiteDatabase is defined to manage associated operations with database.

There are four methods in class DBmanager:

1. getFields(), get original field names from a given table
2. getTypes(), get field types from a given table
3. getTitles(), translate original field names if they are in the form of code
4. getResults(), do the query that all the records of fields chosen from a given table

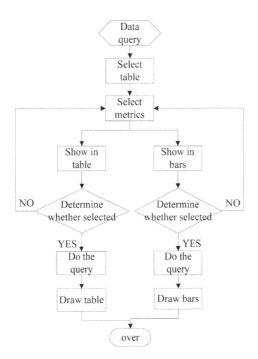

Figure 1. The flow chart of the application.

It is important to note that there are no functions defined to get indexes such as table names, field names and field types. But there exists a built-in table auto generated named sqlite-master in SQLite, and this table records original SQL sentence so that I can get the data through querying this table.

5 CONCLUSION

This program suits the MVC (Model, View, Controller) framework. The class DBmanager which we consider as M is used to encapsulate most of the database operations, it provides interface to process data. The V is interface layout that we implement it through XML and dynamic creation in Java. The control flow stands for C, which connects the upper interface and the underlying database. This project is object-oriented and releases high cohesion and low coupling. However, this study still leaves some problems for future discussion:

1. The interface design is not beautiful enough which needs professional designer.
2. The table is inflexible so the cell size couldn't adapt to the cell content.
3. The horizontal ordinate of the bar is kind of crowded.

In the process of working on designing and programming this thesis, I have gained much help from my tutor who has given me very important suggestions and guidance in critical time. In addition, this design has got a lot help and support from my friends. Thanks to their participation and accompany. I will work harder to explore new knowledge.

REFERENCES

[1] Wu Shyi-Shiou & Wu Hsin-Yi. 2011. The Design of an Intelligent Pedometer using Android. *2011 Second International Conference on Innovations in Bio-inspired Computing and Applications.*
[2] Cay, S Horstmann & Gary Cornell. 2011. *Cora Java.* Electronic Industry Press: 53–58.
[3] Karthick, S & Velmurugan. A. 2012. Android Su-burban Railway Ticketing with GPS as Ticket Checker. *2012 IEEE International Conference on Advanced Communication Control and Computing Technologies.*
[4] http://developer.android.com/reference/android/content/Intent.html.
[5] Grant Allen & Mike Owens. 2012. *SQL Definitive Guide.* Electronic Industry Press: 25.

Sports Engineering and Computer Science – Luo (Ed.)
© 2015 Taylor & Francis Group, London, ISBN 978-1-138-02650-6

Simulation design based on target detection and tracking control

HongCheng Zhou
College of Automation Engineer, Nanjing University of Aeronautics and Astronautics, Nanjing, China
Institute of Information, Jinling Institute of Technology, Nanjing, China

ZhiPeng Jiang
Institute of Information, Jinling Institute of Technology, Nanjing, China

ABSTRACT: Target detection and tracking simulation design of two-dimensional translational and three axis turntable are used to target the screen drag and three-dimensional angle attitude simulation of the control problem in essence belongs to the electric servo control. In terms of the design of the servo control, motion control platform including hardware design, the control law design of the control system and software design. Detailed describes the motion platform of mathematical modeling, control circuit and the control law design, through simulation and experiment verify the validity of the motion platform control system.

1 INTRODUCTION

Nature of moving target detection and tracking simulation platform control system for high precision servo system, is the control circuit, in which the shaft is in three-axis motion simulation turntable axis (pitch axis) platform as an example to describe the design of control system. The principle of high precision requirement is shown in Figure 1, as a current loop feedback (torque), speed loop and position loop[1].

According to the change of the current stable torque, the function of current loop is to adjust currency, to maintain a constant torque output. Speed loop can improve the stiffness of system, through the increasing of the damping coefficient of the system to improve the dynamic tracing ability. The main purpose of the position loop is to guarantee the precision of the controlling system[2]. The mathematical model of the dc torque motor is

$$\begin{cases} C_m = \dfrac{T_m}{I_a} = \dfrac{13}{7.2} = 1.81\,\text{N} \cdot \text{m/A} \\ C_e = \dfrac{U_e}{\omega} = \dfrac{U_e}{2\pi n_0/60} = \dfrac{24 \times 60}{2 \times 3.14 \times 120} = 1.91\,\text{V} \cdot \text{s/rad} \end{cases} \tag{1}$$

2 SPEED CONTROL LAW DESIGN

In engineering practice, the input signal is usually a low frequency signal, and the noisy signal of high frequency signals. If you want to meet the input signal, the output signal can be a very good reiteration and hopeful system has a larger bandwidth. Therefore, in order to make the turntable has a better performance, it is particularly important that controller is designed to make the turntable bandwidth in a reasonable range. Controlling system design usually requires speed ring band which is higher than the position of the loop band. In general, the definition of the closed-loop bandwidth is shown below:

$$\begin{cases} \Delta \left| \Phi(j\omega)_{0 \sim j\omega_s} \right| \leq 10\% \\ \Delta \phi(j\omega)_{0 \sim j\omega_s} \leq \pm 10° \end{cases} \tag{2}$$

$\Phi(j\omega)$ and $\phi(j\omega)$ respectively are amplitude-frequency characteristics and phase frequency characteristics of the closed-loop system. ω_s is the bandwidth of the closed-loop frequency response. Namely, $0 \sim \omega_s$ amplitude change in the range of $\pm 10\%$, the phase changes range $\pm 10°$.

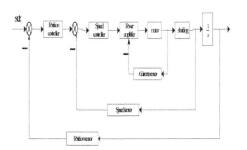

Figure 1. Three closed-loop control principle diagram.

The closed-loop system transfer function is[4]

$$\Phi_0(s) = \frac{176}{0.013s^2 + 8.336s + 177} \quad (3)$$

The basic structure of speed loop simulation platform servo controlling system block diagram is shown in Figure 2.

In the process of the simulation, PWM power amplifier $K_{pwm} = 10$, the sensor and press unit feedback processing can get open-loop transfer function which is[5]:

$$G(s) = \frac{18.1}{0.052s^2 + 34.321s + 4.1171}$$
$$= \frac{4.4}{0.013s^2 + 8.336s + 1} \quad (4)$$

Closed-loop transfer function is:

$$\Phi_0(s) = \frac{176}{0.013s^2 + 8.336s + 177} \quad (5)$$

According to the open loop transfer function and the closed loop transfer function, the bode plots of the open loop and closed loop frequency response diagrams are shown in Figure 3.

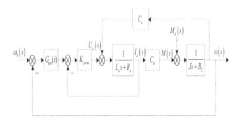

Figure 2. The structure diagram of speed loop servo system.

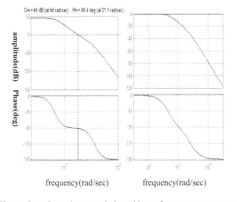

Figure 3. Open loop and closed loop frequency response Bode diagram.

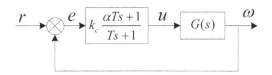

Figure 4. Advanced correction control chart.

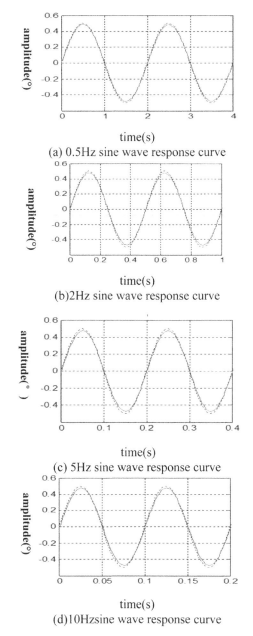

time(s)
(a) 0.5Hz sine wave response curve

time(s)
(b)2Hz sine wave response curve

time(s)
(c) 5Hz sine wave response curve

time(s)
(d)10Hzsine wave response curve

Figure 5. Correction sine wave response curve (input signal in the dotted line, output signal in the solid line).

The typical advanced correction controlling structure is shown in Figure 4[6,7].

When the input signal 0.5° sine wave of 0.5 Hz, 2 Hz, 5 Hz and 10 Hz frequency, the responses of the speed loop correction are shown in Figure 5 (a), (b), (c) and (d).

The formula can get the correction in advance of the transfer function[8]:

$$G_c(s) = \frac{1+0.42s}{1+0.014s} \tag{6}$$

After correction of the system open-loop transfer function

$$\begin{aligned} G_c'(s) &= G_c(s) \cdot G_0(s) \\ &= \frac{176(1+0.42s)}{(0.013s^2 + 8.336s + 1)(1+0.014s)} \end{aligned} \tag{7}$$

AS output signal amplitude and phase difference increase gradually, and the absolute value of the amplitude error is less than 10%, the absolute value of phase difference is less than 10°. The speed closed-loop bandwidth can meet the requirement of the dynamic performances.

3 THE POSITION CONTROL LAW DESIGN

The position control law can improve the dynamic response of the position loop. To complete the corrective controlling speed controlling circuit, basic structure diagram is shown in Figure 6[9].

Because of $G(s)$ the expression is complex, the physical implementation of full compensation conditions is very difficult to meet with the requirement of tracking precision part compensation mode. The closed loop transfer function is[10]:

$$\Phi(s) = \frac{\theta(s)}{\theta_0(s)} = \frac{G_{pos}(s)G(s)}{1+G_{pos}(s)G(s)} \tag{8}$$

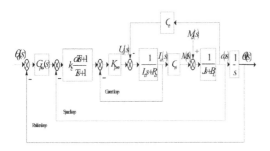

Figure 6. Position loop structure diagram.

According to the feedback compensation of input device, $G_f(s)$ closed-loop transfer function is

$$\Phi_f(s) = \frac{\theta(s)}{\theta_0(s)} = \frac{(G_{pos}(s)+G_f(s))G(s)}{1+G_{pos}(s)G(s)} \tag{9}$$

According to $E(s) = \theta_0(s) - \theta(s)$, system error transfer function is

$$\Phi_e(s) = \frac{E(s)}{\theta_0(s)} = \frac{1-G_f(s)G(s)}{1+G_{pos}(s)G(s)} \tag{10}$$

The current feedback compensation is

$$G_f(s) = \frac{1}{G(s)} \tag{11}$$

Comparison adding feedback compensation system and closed-loop system transfer function $\Phi(s)$ and $\Phi_f(s)$, the characteristic equation $1+G_{pos}(s)G(s) = 0$ of feedback compensation does not affect the whole system stability.

According to the speed loop control diagram, ignoring the non-dominant dipole and zero pole, it can get speed closed-loop transfer function[11]:

$$\Phi_V(s) = \frac{9.01(s+2.381)}{(s+2.159)} \tag{12}$$

Namely

$$G(s) = \frac{9.01(s+2.381)}{s(s+2.159)} \tag{13}$$

According to the formula (13), it can get the transfer function of feedback compensation

$$G_f'(s) = \frac{s(s+2.159)}{9.01(s+2.381)} = \frac{s(0.46s+1)}{9.94(0.42s+1)} \tag{14}$$

In order to suppress the high frequency noise, at the same time, the order of compensation link time fine-tuning, the feedback compensation in front of the transfer function can be increasingly inertia link, designed to increase the time constant of the inertial link is 0.001. So the resulting feedback compensation function can be expressed

$$G_f(s) = \frac{s(0.46s+1)}{9.94(0.42s+1)(0.001s+1)} \tag{15}$$

The frequency of 0.1 Hz, 0.5 Hz, 1 Hz and 2 Hz sine signal, the output signal are shown in Figure 7 (a), (b), (c) and (d).

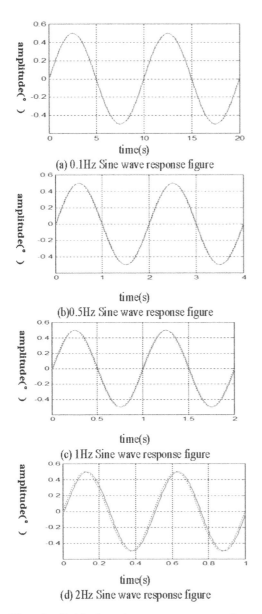

(a) 0.1Hz Sine wave response figure

(b) 0.5Hz Sine wave response figure

(c) 1Hz Sine wave response figure

(d) 2Hz Sine wave response figure

Figure 7.   Position loop sine wave response curve (input signal in the dotted line, output signal in the solid line).

The absolute value of amplitude error is less than 10%, the absolute value of phase difference within the scope of the 10°, can satisfy the requirement of the turntable position loop frequency band.

4   CONCLUSIONS

This paper introduces the moving target detection and tracking simulation platform of controlling system. Mathematical model of motion simulation turntable puts forward the control law of each control loop. The feedback compensation is designed on this basis. The digital simulation and table tests verify the correctness and effectiveness of the control strategy.

REFERENCES

[1] Alexandre S. Bazanella. Iterative minimizaation of H2 control performance criteria [J]. Automatica, 2008, 44(3):2549–2559.
[2] Cristian R. Rojas, Märta Barenthin, James S. Welshb and Håkan Hjalmarsson. The cost of complexity in system identification: The Output Error case [J]. Automatica, 2011, 47(9):1938–1948.
[3] G. Yin, Yu Sun and Le Yi Wang. Asymptotic properties of consensus-type algorithms for networked systems with regime-switching topologies [J]. Automatica, 2011, 47(7):1366–1378.
[4] G. Yin, Le Yi Wang and Shaobai Kan. Tracking and identification of regime-switching systems using binary sensors [J]. Automatica, 2009, 45(4):944–955.
[5] Henrik Ohlsson, Lennart Ljung. Identification of switched linear regression models using sum-of-norms regularization [J]. Automatica, 2013, 49(4):1045–1050.
[6] LennartLjung, Adrian Wills. Issues in sampling and estimating continuous-time models with stochastic disturbances [J]. Automatica, 2010, 46(5):925–931.
[7] Henrik Ohlsson, Fredrik Gustafsson and Lennart Ljung. Smoothed state estimates under abrupt changes using sum-of-norms regularization [J]. Automatica, 2012, 48(4):595–605.
[8] A.S. Bazanella, X. Bombois and M. Gevers. Necessary and sufficient conditions for uniqueness of the minimum in Prediction Error Identification [J]. Automatica, 2012, 48(8):1621–1630.
[9] L. Campestrini, D. Eckhard and M. Gevers. Virtual Reference Feedback Tuning for non-minimum phase plants [J]. Automatica, 2011, 47(8):1778–1784.
[10] M.C. Campi, M. Vidyasagar. Learning With Prior Information [J]. IEEE Transanctions on automatic control, 2011, 46(11):1682–1695.
[11] Maciej Niedźwiecki, Marcin Ciołek. Elimination of Impulsive Disturbances From Archive Audio Signals Using Bidirectional Processing [J]. IEEE Transanctions on Audio, Speech, and Language Processing, 2013, 21(5):1046–1059.

Sports Engineering and Computer Science – Luo (Ed.)
© 2015 Taylor & Francis Group, London, ISBN 978-1-138-02650-6

Speed control law design of moving target detection

HongCheng Zhou
College of Automation Engineer, Nanjing University of Aeronautics and Astronautics, Nanjing, China
Institute of Information, Jinling Institute of Technology, Nanjing, China

CunBao Chen
Institute of Information, Jinling Institute of Technology, Nanjing, China

ABSTRACT: In a servo system, the function of current loop is to adjust current according to the change of the current stable torque, in order to maintain a constant torque output. After the position controller output as input of speed loop, the amount of feedback is the value of velocity components after signal conditioning gets and speed signal common constitute a closed-loop speed control. Speed loop can improve the stiffness of system by the damping coefficient of the system to improve the dynamic tracing ability. The main purpose of the speed loop is to guarantee the precision of the control system.

1 INTRODUCTION

In the turntable control system where the encoder as the sensor is used, the measured pulse signal after conditioning gets feedback position, so as to realize a closed-loop position.

It will get the rate of position signal through differential signals and realizes the speed closed-loop control. Position controller in the system and by the control computer through the software implementation of speed controller, control law is introduced in detail below.

Section in according to the principle of three closed loop control and servo control system of the modeling of each component can get the system mathematical model. Among them, $G_{pos}(s)$ and $G_{spd}(s)$ respectively and the mathematical model of position controller and speed controller and current loop control model are set in the power amplifier[1,2]. Three-axis motion simulation turntable shaft of the motor selection which gets 160LYX45 dc torque motor specific parameters are as follows: the peak locked-rotor torque $T_m = 13\text{N} \cdot \text{m}$, stalling a peak current $I_a = 7.2\text{A}$, the armature resistance $R_a = 3.2\Omega$, the armature inductance $L_a = 20\text{mH}$, maximum no-load speed $n_0 = 120\text{r/min}$, rated voltage $U_e = 24\text{V}$, the rotational inertia $J = 2.6\text{kg/m}^2$ of the motor shaft, the damping moment coefficient $B_L = 0.05\text{N} \cdot \text{m} \cdot \text{s/rad}$. The mathematical model of the dc torque motor can be[3]

$$\begin{cases} C_m = \dfrac{T_m}{I_a} = \dfrac{13}{7.2} = 1.81\text{N} \cdot \text{m/A} \\ C_e = \dfrac{U_e}{\omega} = \dfrac{U_e}{2\pi n_0/60} = \dfrac{24 \times 60}{2 \times 3.14 \times 120} = 1.91\text{V} \cdot \text{s/rad} \end{cases}$$

(1)

2 SPEED CONTROL LAW DESIGN

In engineering practice, the input of a given signal is usually a low frequency signal, and the noise signal of high frequency signals which are to reduce the interference of the noise of the system, cannot be too big hope system bandwidth. But if you want to meet the input signal, the output signal can be a very good reiteration and the hope system has a larger bandwidth. So, in order to make the turntable have better performance, controller is designed to make the turntable bandwidth in a reasonable range which is particularly important. In control system design, it usually requires more speed ring band than high frequency band, the position loop in the project using the closed-loop transfer function characteristic of the system to define the bandwidth to represent the system. In general, the definition of the closed loop bandwidth is shown below[4]:

$$\begin{cases} \Delta \left| \Phi(j\omega)_{0 \sim j\omega_s} \right| \leq 10\% \\ \Delta \phi(j\omega)_{0 \sim j\omega_s} \leq \pm 10° \end{cases}$$

(2)

Φ $(j\omega)$ and ϕ $(j\omega)$ are respectively amplitude-frequency characteristic and phase frequency characteristics of the closed-loop system. ω_s is the closed loop bandwidth of frequency response, namely in $0\sim\omega_s$. System amplitude changes in the range of $\pm10\%$ and the phase changes range of $\pm10°$.

Research object moving target detection and tracking the basic structure of speed loop simulation platform servo control system block diagram are shown in Figure 1.

In the process of simulation, PWM power amplifier $K_{pwm} = 10$, is used as the sensor. Pressing unit feedback processing can get open-loop transfer function[5,6]:

$$G(s) = \frac{18.1}{0.052s^2 + 34.321s + 4.1171}$$
$$= \frac{4.4}{0.013s^2 + 8.336s + 1} \quad (3)$$

Namely

$$G_0(s) = k_c \cdot G(s) = \frac{176}{0.013s^2 + 8.336s + 1} \quad (4)$$

The system closed-loop transfer function is

$$\Phi_0(s) = \frac{176}{0.013s^2 + 8.336s + 177} \quad (5)$$

According to the calculation of the open loop transfer function, the closed loop transfer function can be system bode plots of the open loop and closed loop frequency response diagram, as shown in Figure 2.

The above system of open loop cut-off frequency is $\omega_c = 21.1\text{rad}/s$ and phase margin $\gamma = 88.4°$. Before correction closed loop frequency response, it is converted to frequency response $f_s = 0.59\text{Hz}$. Without calibration system narrow frequency band, it will not meet the requirements of dynamic. Open loop phase Angle at the same

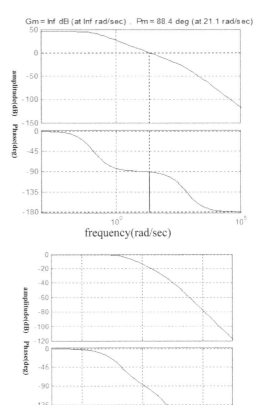

Figure 2. Open and closed loop frequency response diagram.

time spend big margin. From control theory point of view, it can improve the open loop gain phase margin to reduce the open-loop frequency band, but in the actual system, it is too large to open loop gain, which will be slightly high frequency noise amplification that leads to shafting vibration. So, it needs to use advanced correction to improve the frequency band of the system, reduce the phase margin and at the same time not to increase the open-loop gain of high frequency. Lead correction link transfer function can be expressed[7,8]

$$G_c(s) = \frac{\alpha Ts + 1}{Ts + 1} \quad (6)$$

When $\alpha > 1$, advanced network handover frequency $1/T$ and $1/\alpha T$. The typical advanced correction control structure is shown in Figure 3.

Adopting the advanced correction link of speed loop controller design, the formula can get

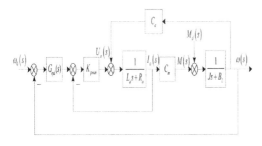

Figure 1. The structure diagram of speed loop servo system.

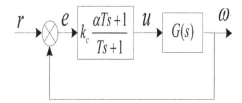

Figure 3. Advanced correction chart.

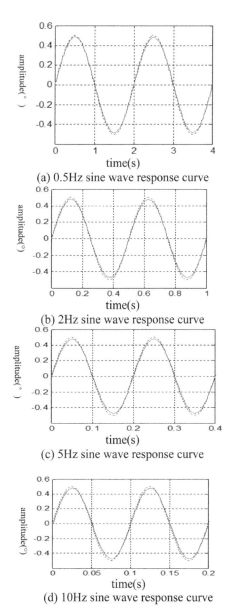

(a) 0.5Hz sine wave response curve

(b) 2Hz sine wave response curve

(c) 5Hz sine wave response curve

(d) 10Hz sine wave response curve

Figure 4. Correction sine wave response curve (input signal in the dotted line, output signal in the solid line).

correction link in advance of the transfer function for[9,10]:

$$G_c(s) = \frac{1+0.42s}{1+0.014s} \qquad (7)$$

After correction, the system open-loop transfer function is

$$\begin{aligned} G_c'(s) &= G_c(s) \cdot G_0(s) \\ &= \frac{176(1+0.42s)}{(0.013s^2 + 8.336s + 1)(1+0.014s)} \end{aligned} \qquad (8)$$

3 SIMULATION AND TEST

According to bode chart, closed-loop correction can get the speed loop bandwidth $\omega_s = 107 rad/s$, the corresponding frequency $f_s = \omega_s/2\pi \approx 17.04$Hz.

When the input 0.5° sine wave signal frequency is 0.5Hz, 2Hz, 5Hz and 10Hz, the response of the speed loop correction is shown in Figure 4(a), (b), (c) and (d).

By the picture above, after adding lead correction link, in the 0.5Hz ~10Hz range with the increase of frequency, output signal amplitude and phase difference also increase gradually. The absolute value of amplitude error is less than 10%. The absolute value of phase difference is less than 10°. The speed loop closed loop bandwidth can meet the requirements of dynamic performance of the turntable speed loop.

4 CONCLUSIONS

Firstly, this paper introduces the moving target detection and tracking simulation platform of control system. According to mathematical model of motion simulation turntable, it puts forward the control law of each control loop. On this base, the feedback compensation is designed. Digital simulation and table tests verify the correctness and effectiveness of the control strategy.

It firstly introduces the moving target detection and tracking simulation platform which controls system composition and the key components selection, and chooses the device established the mathematical model of the parts in the center axis of the three-axis motion simulation turntable. For example, "three closed loop control circuit is designed and puts forward the control law of each control loop and speeds loop adopting proportion and corrective control in advance. The position loop adopts integral separation and the proportional and integral control of saturated limiter.

On the basis of the feedback compensation, it is designed. The control strategy is verified by digital simulation and the turntable tests the correctness and effectiveness.

REFERENCES

[1] Shen, Q.K., Jiang, B., Cocquempot, V. Fuzzy logic system based adaptive fault-tolerant control for near-space vehicle attitude dynamics with actuator faults [J]. IEEE Trans. Fuzzy Syst. 2013, 21(2), 289–300.

[2] Cristian R. Rojas, Märta Barenthin, James S. Welshb and Håkan Hjalmarsson. The cost of complexity in system identification: The Output Error case [J]. Automatica, 2011, 47(9):1938–1948.

[3] Wang, R.R., Wang, J.M. Passive actuator fault-tolerant control for a class of overactuated nonlinear systems and applications to electric vehicles [J]. IEEE Trans. Veh. Technol. 2013, 62(3):972–985.

[4] G. Yin, Le Yi Wang and Shaobai Kan. Tracking and identification of regime-switching systems using binary sensors [J]. Automatica, 2009, 45(4):944–955.

[5] Henrik Ohlsson, Lennart Ljung. Identification of switched linear regression models using sum-of-norms regularization [J]. Automatica, 2013, 49(4):1045–1050.

[6] Lennart Ljung, Adrian Wills. Issues in sampling and estimating continuous-time models with stochastic disturbances [J]. Automatica, 2010, 46(5):925–931.

[7] Henrik Ohlsson, Fredrik Gustafsson and Lennart Ljung. Smoothed state estimates under abrupt changes using sum-of-norms regularization [J]. Automatica, 2012, 48(4):595–605.

[8] A.S. Bazanella, X. Bombois and M. Gevers. Necessary and sufficient conditions for uniqueness of the minimum in Prediction Error Identification [J]. Automatica, 2012, 48(8):1621–1630.

[9] L. Campestrini, D. Eckhard and M. Gevers. Virtual Reference Feedback Tuning for non-minimum phase plants [J]. Automatica, 2011, 47(8):1778–1784.

[10] Maciej Niedźwiecki, Marcin Ciołek. Elimination of Impulsive Disturbances From Archive Audio Signals Using Bidirectional Processing [J]. IEEE Transactions on Audio, Speech, and Language Processing, 2013, 21(5):1046–1059.

Sports Engineering and Computer Science – Luo (Ed.)
© *2015 Taylor & Francis Group, London, ISBN 978-1-138-02650-6*

The design and implementation of electricity charging system based on B/S

Zhen Zhao

Tianjin Bohai Vocational Technical College, Tianjin, China

ABSTRACT: At present, some problems exist in the power system charges, such as distributed network, charging is not convenient, and not conducive to statistics. In order to speed up construction of information in power industry needs to develop a system. Through this system, users can pay the electricity bill, current situation of the development of this system should be combined with the power industry. The system in addition to the basic function of the charges, a system administrator, can realize the management notice, new users, check the electricity and other functions. The system is developed with Java language application, based on the analysis of the demand from the system overview, system design, several aspects to elaborate.

Keywords: electricity; automation; charge; internet; Java

1 AN OVERVIEW OF THE PROJECT BACKGROUND AND SYSTEM

The electric power network covers from the city to the countryside. Power network coverage area is large, so the electric charge is inconvenient, the power net charges are used to the District, county as a unit, the tube member to the distributed network charges. The charging model making company statistics on all points of the state charges caused great inconvenience, at the same time, the safety of toll data have been charging transparency project is unable to be guaranteed. The user can only be a few days each month fixed fee, and not conducive to enterprise management and statistics. The power company for the electricity charge, automated demand payment has been imminent.

For each power company, charge is an important work. The growth of population, the increasing of users, has had great impact to the power company. For the power company, the management of large amounts of data is very important. Previously, people use the traditional artificial way to management document, which has many shortcomings, such as low efficiency, poor security. With the increasing number of users, the workload will increase greatly, which increases the workload and the labor intensity of workers. At present the company in China, there is still a considerable part of the datafile management still stays on the basis of paper, which waste a lot of manpower and material resources. With the development of science and technology, the computer has entered the human society in various fields of life and plays a more and more important role. The traditional manual management mode must take the computer as the foundation information management approach.

With the development of Internet, the application of B/S structure has become the mainstream of software architecture, and the development of B/S system is good at the Java language, the Java language development, the use of the system, on the front page of JSP and development, thanks to the cross platform Java, features safety, with PHP and ASP.NET do not have the advantages. At present, many industries such as telecommunications, finance, Service industry use the Java+JSP development. So this system is developed with JSP pages, the system also uses Spring, Ibatis, Javascript technology, these are the Java actual development of common framework and language, before I have the use of these technologies developed some practical system. There is no problem of this system.

Because of the cross platform of Java language, the system can run in Windows, Unix, Linux operating system, and this system interface is simple, intuitive, user and operator and the use of the system can get started quickly.

2 DEMAND ANALYSIS

The system function, through the investigation and analysis, this system mainly realize the following functions: administrator of new operators, management company announcement, the new users; the operator user payment; payment of a user, check your payment and other functions.

For system development, because the Windows operating system has the characteristics of easy use, high efficiency in development, this system was developed in Windows system. This system uses Spring MVC, Spring DAO module, and Spring AOP module directly in the aspect oriented programming function integrated into the Spring framework. You can easily make any Spring object management framework supporting AOP. Spring AOP module provides transaction management service application based on Spring. By using the Spring AOP, not dependent on EJB components, can let declarative transaction management integrated into the application.

The database uses the MySQL database, MySQL is a small relational database management system, which has the advantages of small volume, high speed, low total cost of ownership, especially the characteristics of open source, many small and medium sized websites in order to reduce total cost of ownership web site chose MySQL as the database, the system adopts MySQL database as the background database.

3 THE OVERALL DESIGN OF THE SYSTEM

The design goal of this system is to design a reasonable software structure, the user easy to use power charging system. System administrator management notice, new users, new operators, users pay the electricity bills and view the electricity function operator. Through this system, administrator can replace the manual payment by way of the current residents. Page intuitive, simple, not deliberately pursue page effects.

The system consists of the administrator module, the operator module, user module. The administrator module includes a notification management, create new users, create new operator, the operator module includes operator user monthly payment information; the user module includes user to pay for electricity and the user to view the tariff information.

According to the division of point of view, the system is divided into presentation layer, business logic layer, data access layer. The lower layer services, the lower is not dependent on the upper layer, invoked the provided interfaces.

The overall design of the page, as a typical application of Internet of the system, consider the user's experience, usually of Internet experience, many of the sites of different Webpage layout, style is consistent. This will give the user a more profound impression, is conducive to the site management and maintenance. Html frameset just to meet this need, all pages of this system is finally through main.jsp to display to the user.

System layout is divided into four parts: the top to display the name of the company, the announcement, the current date and the day's weather forecast; the bottom shows the company; when the user or operator is not logged in, the middle left display the login page, otherwise, display the corresponding menu options (there will be a demonstration); when a user selection option, right in the middle will display the corresponding content.

4 THE SYSTEM DESIGN IN DETAIL

4.1 *The database design*

The system has 5 entities, which are operator, user, notice, the payment record, address. The relationship between the operator and the announcement is a one to many relationship, an operator can issue a notice, a notice can only belong to one operator. The operator has the job number, name, password, is an administrator four attributes. Notice of a ID, title, content, release time, the operator of five attributes.

The relationship between the user and the place is a one to many relationship, a user can have multiple sites, a site can only belong to one user, the relationship between locations and payment records is a one to many relationship, a site can have multiple payment records, a record of payment can only belong to one. As shown in Figure.

4.2 *Logic design*

The system consists of 5 data tables.

Table is for recording system operator. The operator needs to login, so needs to have the job number and password, also needs to have a name. The operator is divided into administrator and operator, all have a used for identification of the column.

The announcement is for the recording system. Every advertisement must have a primary key of no practical significance, also will need the title, the specific content, release time, the publishing of one's work to ensure data consistency and uniqueness.

User table, used to record the users in the system. User must record the user's basic information, such as name, password, telephone, gender, birth date, ID number, the company. User also has a no actual meaning of the primary key to ensure data consistency and uniqueness.

Address table, which must have a primary key of no practical significance, ensure data consistency and uniqueness, but also a city, area, street, the user properties, in order to ensure the uniqueness of the street address, attribute cannot be repeated.

The payment record table, used to have a no practical significance of the primary key, and payment of year, month, day, address attribute, but also a sign to distinguish whether have already paid.

4.3 *The function module design*

The administrator module, the system administrator is the most basic role. A starting point is the system administrator module, the module design is critical to the accuracy of the data integration system, the administrator is the operation on the basis of data add, delete, change, check. The administrator can add, modify, query, delete announcement, when the user enters the system, will see consists of all notices of the scroll bar. Administrators can also add and delete users and operators.

In the operator module, the operator in the system of the duty payment situation views the user monthly. This function uses the Spring+Ibatis to be achieved. The seeAllCharge and seeAllChargeCount methods in the IchargeDao interface are to realize the query.

The user module, the user has 2 operation in the system, are to pay for electricity and the electricity. This function is realized by Spring+Ibatis. The seeCharge and selectChargeByUserCount methods in the IchargeDao interface to realize the query. Users pay for electricity, because between the user and the real estate is a one to many relationship, a user can have multiple properties, combined with the actual situation, the user as a house payment, a one-time property here must be put up to now all the unpaid bills paid, so the user clicks to pay for electricity, will see the corresponding to all the property and their unpaid electricity, a house can pay the electricity, also can pay all of the property of electricity.

5 THE OPERATION OF THE SYSTEM

5.1 *Installation position*

Ensure the server configuration to configuration requirements, and install the following software: JDK, MySQL database, Tomcat. After installation, environment variables, I JDK the installation position is: C:\Program Files\Java\jdk1.5.0_04, and then click my computer - > properties - > Advanced - > environment variables, add in the

system variables in the Path: C:\Program Files\Java\jdk1.5.0_04\bin; then the user variables in a new JAVA_HOME variable, value is: C:\Program Files\Java\jdk1.5.0_04.

5.2 *The system test*

The system is deployed to the server, open main.do in the browser to access the home page, the administrator login system, click "add new user" link to new users, normal and expected results view is consistent, click "add new operator operator" link, the normal view and expected results are consistent, click "add notification" and "view all notices" link, see the relevant operation and the expected results are consistent with the operator, log off the system, login system. Click the "view user fee" links, check the contents page display and background data in the database is consistent, log off the system, the user login system, click "electricity" links, check the electricity information and background data in the database is consistent, click on the "pay the electricity" link, follow the steps to pay for electricity, whether pay for electricity to view database data update has already been.

At present, all the modules and functions of this system have been tested many times, the system to normal operation.

6 CONCLUSION

This system can play an important role in the electric power enterprise's dailywork. This system can improve working efficiency, make the payment of electricity more convenient. However, this system also has some deficiencies in the design and implementation, in the latter part of the maintenance phase of the system will be perfect, meet the needs of enterprise development.

REFERENCES

[1] Sun Weiqin. Proficient in hibernate. Beijing: Publishing House of electronics industry, 2005.
[2] Liao Xuefeng Spring 2 core technologies and best practices. Beijing: Publishing House of electronics industry, 2007.
[3] Cai Yingli. Development trend of power automation management systems. Private science and technology, 2010, 24 (2): 1~8.

Sports Engineering and Computer Science – Luo (Ed.)
© 2015 Taylor & Francis Group, London, ISBN 978-1-138-02650-6

The design of high-power microwave power

Gang Zheng, Jian-Hua Wang & Dao-Dong Qin
Wuhan Institute of Technology, Hubei, China

ABSTRACT: With the increasing in wide application of microwave technology, microwave power supply attracts more and more attention; high output power, good stability and easy handling, etc. have got the excellent characteristics which a high-power microwave power supply must possess. With the improvement of people's technological requirements of downstream products, high-power microwave power supply emerged. The microwave power supply designed in this paper is controlled by ATmega16 L SCM, with power setting, feedback regulation, real-time power display, overcurrent protection and other functions.

1 INTRODUCTION

In order to achieve stable and efficient output power of microwave power supply, the author adopts ATmega16 L SCM as the control device of the whole electrical power unit; the SCM has high cost performance, 8-bit AVR microprocessor with high performance and low power and advanced RISC structure; its kernel has rich instruction set with 32 general working registers. All registers are directly connected to the Arithmetic Logic Unit (ALU), so that an instruction can access to two separate registers in one clock cycle. This structure greatly improves code efficiency and its data throughput is up to 10 times faster than ordinary CISC microcontroller. It is these characteristics that lay foundation for the perfect hardware design in the future.

The output power of microwave power supply $P = aUI$, where a is the conversion rate of the output power of magnetron, which relates to the module of magnetron and its own process performance; U is the negative high voltage applied to the cathode of magnetron; I is the anode current of magnetron[1]. In addition, the factors which affect the output power of microwave power supply include: magnetic field current I_m controlling the magnetic field strength, and filament current I_f. Experiment shows that the output power P of microwave power supply mainly depends on the cathode negative high voltage U and the field current I_m of the magnetron, while the filament current I_f has less impact on the output power, and the cathode negative high voltage U of the magnetron is mainly generated by the volatility of the magnetron cathode negative high voltage U primarily by the fluctuation of peripheral power grid. The design carries out purification for peripheral power grid to keep it stable, and which can stabilize the voltage; therefore, the output power of the microwave power supply can be stabilized as long as carries out associated control for the field current I_m and filament current I_f simultaneously.

2 CIRCUIT DESIGN

2.1 Design of overall system module

The microwave power supply adopts phase-control three-phase AC voltage regulator technology to control the magnetron and then controls the output power; the output power of the magnetron is set through the user carrying out operation for SCM as needed. The workflow of the system is as shown in Figure 1; 380V industrial electricity is sent to the step-up transformer primary after voltage regulator treatment, and it will be output via secondary after conversion to high voltage alternating current; and then carry out rectification, filtration and voltage stabilization treatment through rectifier circuit, filter circuit and regulator circuit respectively, and then apply the stable and adjustable DC in thousands of volts through treatment to the cathode of magnetron. Throughout the design, the role of A/D converter is that display the module display

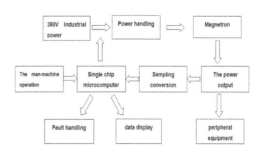

Figure 1.

through SCM after obtaining the data processing collected by the actual output power through the sampling circuit; meanwhile, contract the data with the feedback of output power set by us; control and regulate the cathode negative high voltage and anode current of the magnetron by SCM, in order to control the output power. The main functions of SCM during this process include: real-time adjustment for output power, disposal of equipment failures and display of preset power, output power and running time.

2.2 Start-up circuit

As shown in Figure 2, because the current generated in the case of instant close of the switch is too large and it is easy to cause greater impact on the circuit device, the start-up circuit of the high-power microwave power supply adopts soft-start mode with resistive subdivision. In the early startup of the device, make the relay JC2-1: 3 pull-in and JC3-1:3 open through pressing the key switch; each branch adopts resistance with 100 W100Ω to carry out voltage subdivision protection for the startup circuit; at the same time, ATmega16 L SCM enters into the initialization phase; after the device runs smoothly for certain period, disconnect JC2-1: 3, and pull JC3-1: 3, so that the protective resistance and the main circuit are detached, to avoid loss of resistance power, and enter the rectification, regulation and filtration circuit. At this moment, the device enters into the normal operation phase; the SCM is in place, and soft start-up of the switch circuit is complete. The soft start-up mode used in the design allows the device to run smoothly and stably, which can effectively limit the instant surge current produced during startup process; avoid unnecessary current impact on equipment and distribution network and can prolong the life of related equipment to a certain extent; what's more, the design itself has simple structure and is easy to operate[2–3].

2.3 Driving circuit and voltage division circuit

It is different from several previous microwave power supplies with small output power[4], the model of magnetron used in this experiment is CK-619; because the internal structure of the magnetron is different for different models, appropriate improvements and adjustments should be made for corresponding driving circuits and protection circuits.

Because the magnetic cores in magnetron used in previous microwave power supplies belongs to permanent magnetic cores, additional circuits are not needed to control it; while magnetic field of CK-619 magnetron is provided by external electromagnet, the driving circuit shown in Figure 3 is designed, in order to control the magnetic field intensity, and then achieve the purpose of anode current control. After the initial voltage in the circuit is through the step-up transformer, the voltage value is large, which can cause certain loss to the devices of magnetron and driving circuit or even decrease their life; therefore, distribute uniform voltage drop to each voltage subdivision

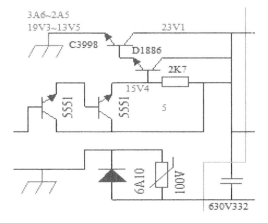

Figure 3. Driving circuit.

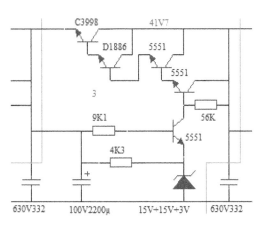

Figure 4. Voltage division circuit.

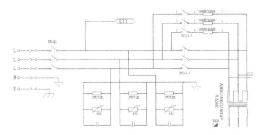

Figure 2. Start-up circuit.

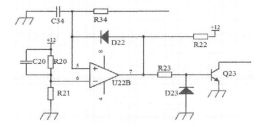

Figure 5. Over current protection circuit.

circuit by concatenating a plurality of voltage division circuits shown in Figure 4, which plays protective role.

2.4 Over-current protection circuit

In order to avoid short circuit of main circuit of microwave power supply or prevent excessive current of main circuit caused by some special factors damaging the device, especially damaging the magnetron which is the core device, overcurrent protection circuit is applied in the main circuit of the equipment, in order to prevent accidents. As shown in Figure 5, in the sampling circuit, the equipment can obtain the value of anode current through sampling of resistor R58; while in overcurrent protection circuit, however, the value of anode current is obtained through resistor R34. Meanwhile, C34 can carry out simple filtration treatment for the anode current. U22B is a voltage comparator, and in the design, we have adopted two resistors in series between HA17393, R20 and R21, which mainly provide reference voltage for the voltage comparator, and its value is about 2.8V, and the anode current converted is about 2A. During operation of the device, when the anode current is normal, i.e., less than the predetermined value of overcurrent, the output of voltage comparator is low level, and Q23 is turned off; in case of overcurrent of the device, the output of voltage comparator is high level, and Q23 is turned on; the SCM will receive overcurrent signal, and then reacts accordingly to cut off power supply of the device and stop running.

3 THE SOFTWARE PROGRAMMING

In the preparation of the program, taking into account the language of the application environment, the author uses C language, because the characteristics of the C language such as: easy to write, strong portability, high execution efficiency. And it also has the ability of strong data processing, so that the microcontroller can handle the sampling circuit

to get the data in time, and to make corresponding adjustments microwave power input. Finally we can obtain a stable output power.

Program used the hierarchical structure and modular development approach as a whole. The function of each process of single-chip microcomputer program is written individually separate into different modules, mutual interference between modules, which give priority to program calls. Mainly includes the function modules are: the main program module, initialization module, external processing module, equipment operation fault processing module, the watchdog module transform, sampling, data processing module and display module.

Void main(void) {}; Microcontroller running the main program, which responsible for circular calls to each subroutine modules.

Void init(void) {}; MCU initialization procedure, whose responsible is making sure internal registers are cleared initialization and reducing interference for the microcontroller input and output ports.

Void key(void) {}; External action handler can detect external button operation, response timely, which also includes button shake avoid misuse.

Void setjmp(void) {}; Equipment failure processing program, testing equipment operating conditions, making a adjust quickly or interruption when an exception occurred on the part of the operator.

Void watdog(void) {}; Watchdog program, ensure that the main function and interrupt function interlock, to prevent the system program run fly crashing.

Void A/D(void) {}; Sampling conversion program, access to the data sampling circuit is D/A conversion.

Void display(main) {}; Abnormal display program, including display, setting the power display, output power, gas and liquid flow display, operation time, the current display. The current display is divided into a filament current, magnetic field current and anode current.

4 CONCLUSION

Microwave power supply system is a complex system under extreme conditions, which determines its efficient controlling system is stable. With the improvement of technical requirements for products, the output power of microwave power and stability meet higher request. This paper mainly introduces several core parts of the microwave power supply system, only under the single chip microcomputer more efficient control, adjustment, feedback with more high-quality magnetron and safety protection circuit to form a perfect high power microwave power supply network.

REFERENCES

[1] Xiang dong, Wang Jianhua, Qin daodong, and other High-power microwave stable power supply design [J] Electrical Engineering Technology, 2008 (1): 26–27.

[2] Zhang shiquan Principles and Application of soft-start [J]. Mechanical and Electrical Technology, 2004 (2):11–15.

[3] Xu zhiwang Zheng liang Qin huibin New intelligent, high-current circuit breaker [J]. Mechanical and Electrical Engineering, 2012 (01):87–89.

[4] Zhang hailiang; Chen guoding; Xia deyin IGBT over-current protection circuit design electrical and mechanical engineering, 2012–08–20.

Sports Engineering and Computer Science – Luo (Ed.)
© 2015 Taylor & Francis Group, London, ISBN 978-1-138-02650-6

A highly efficient LLR-based algorithm for general spatially modulated multi-antenna wireless communication systems

Gang Li

School of Software and Communications Engineering, Jiangxi University of Finance and Economics, Nanchang, China

ABSTRACT: Multi-Input-Multi-Output (MIMO) systems are candidates for current and future wireless communication standards for their inherent high multiplex gain and diversity gain. They use multiple transmit antennas to send multiple data streams at the same time, and multiple receive antennas to demultiplex and demodulate these streams simultaneously. Spatially Modulated (SM) MIMO systems are modifications for traditional MIMOs, where only part of the transmit antennas is active for data transmission for each channel use, with the combination of transmit indices carrying information bits in the same time. The goal is to achieve higher energy efficiency versus spectral efficiency than in traditional MIMO systems. In this paper, a Log-Likelihood-Ratio (LLR) based detection algorithm is proposed to overcome the stringent computation requirement in optimal SM-MIMO detection algorithms such as the well-known Max-Likelihood (ML) algorithm. Analysis and simulation show that this algorithm is more computationally efficient than the optimal ML algorithm, yet at the cost of about 2-dBs in Signal-to-Noise power Ratio (SNR).

Keywords: Multi-Input-Multi-Output (MIMO); spatial modulation; Log-Likelihood Ratio (LLR); energy efficiency; computational efficiency

1 INTRODUCTION

1.1 *Background*

The MIMO techniques are very promising for the design of future wireless communication systems. They are capable of providing high data rates without increasing the spectrum utilization and the transmit power. Diversity gain can also be gained in MIMO, which lay very important foundations for physical-layer standards like WiMAX (Worldwide Interoperability for Microwave Access) (Andrews et al., 2007) and LTE-A (Long Term Evolution—Advanced) (Ghosh et al., 2010). In MIMO systems, all TAs are active at any time instance, resulting in SE optimization. Yet, this does not naturally lead to Energy Efficiency (EE) optimization (Hellings and Utschick, 2013), which is key to future communication systems requiring low energy-consumption techniques.

In (Mesleh et al., 2006), a novel MIMO scheme named SM MIMO was first proposed by Mesleh et al., in which only one antenna out of N_T transmit antennas is active every transmission, and the index of the active antenna conveys $\log_2(N_T)$ information bits alongside the $\log_2(M)$ bits (M is the size of signal constellation, or modulation order) carried on the activated transmit antenna for each channel use. This scheme has high energy efficiency, requiring only one RF link. The receiving algorithm can also be simplified greatly. In (Jeganathan et al., 2008b), Jeganathan et al. proposed an optimal detector for the SM. A Sphere Decoder (SD) was used for SM in (Younis et al., 2010a) which reduces the complexity of the SM optimal detector, yet can also achieve a bit error rate close to the optimal SM.

One disadvantage of the original SM MIMO is that it requires N_T (the number of TAs—transmit antennas) to be 2^p, where p is a positive integer. The other disadvantage is that the number of bits per channel use (bpcu) is in proportion to the base-2 logarithm of N_T, in contrast to $N_T \log M$ of the V-BLAST scheme. Thus SM MIMO requires much more TAs for the same number of bits per transmission as in V-BLAST (Bohnke et al., 2003). Meanwhile, SM can achieve only 1 degree of transmit diversity.

Many variants to SM MIMO have been proposed since (Mesleh et al., 2006), such as Generalized Space Shift Keying (GSSK) (Jeganathan et al., 2008a) and Space Shift Keying (SSK) (Jeganathan et al., 2009) proposed by Jeganathan et al., in which combinations of active antennas or the active antenna index are the only way to convey information bits.

In (Jinlin et al., 2010) and (Younis et al., 2010b), Generalized SM (GSM), which extended the SM idea to allow combinations of multiple active antennas to transmit the same signal symbols, were proposed. A comprehensive introduction of SM-MIMO was given in (Di Renzo et al., 2014).

1.2 *Paper organization*

In this paper, the author proposed a new MIMO transmission scheme to extend the GSM MIMO concept in (Jinlin et al., 2010) and (Younis et al., 2010b), and accordingly a computationally efficient algorithm based on LLR calculation to decide which antenna(s) are used for each transmission. The proposed scheme uses K active TAs out of the N_T (($N_T \le N_R$), with N_T, N_R being the number of transmit and receive antennas—RAs) available TAs to transmit different Signal Symbols simultaneously. As a result, the Spectral Efficiency (SE) can be much higher than SM MIMO at the cost of increased active RF links, thus reducing the number of TAs needed. The LLR-based detecting algorithm has a linear complexity with regard to the signal modulation order M, as compared to the exponential complexity in optimal detecting algorithms like ML-detection, which is in proportion to $N_A M^K$, where N_A is the number of applicable TA combinations.

The rest of the paper is organized as follows: The mapper and system model of the new GSM MIMO transmission scheme are presented in section 2. Section 3 presents the proposed detecting algorithm. Computation complexity analysis, simulation results and conclusion are provided in section 4, 5 and 6, respectively.

1.3 *Notations*

Throughout the paper, the following notations are used. Bold lowercase and bold uppercase letters denote vectors and matrices respectively. $[\cdot]^T$, Tr $[\cdot]$, $[\cdot]^*$, $[\cdot]^H$, and $[\cdot]^+$ are used to denote transpose, trace, conjugate, Hermitian and pseudo-inverse of a matrix or a vector, respectively. Furthermore, $\|\cdot\|^F$ are used to denote Frobenius norm of a matrix or a vector, and E $[\cdot]$ to denote the statistical expectation. We use n!, C (n, k) to denote factorial, binomial coefficient, respectively. Other notations will be explained in its first use.

2 SYSTEM MODEL

The system block diagram is given in Figure 1, where 2 out of 8 TAs are active (right column) to transfer 6 bits (2 8PSK symbols) on each active TA at one time.

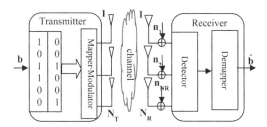

Figure 1.   Block diagram of the proposed GSM MIMO system.

In the proposed scheme, the random transmitted information bits are grouped into equal-length bit blocks. Each block contains p + K $\log_2$(M), where the combination of the K active TAs conveys the first p bits, M is the modulation order of signal symbols. In order to relax the requirement of linearity of amplifier, and to improve EE, MPSK are preferred. For example, consider a system with $N_T = 10$, K = 5, M = 2(BPSK), there are C(10,5) = 250 combinations available, in which we choose $2^7 = 128$ combinations to convey p = 7 bits using 5 active TAs. The other K$\log_2$(M) = 5 bits are transmitted on the 5 active TAs simultaneously using BPSK symbols. Here total 12 bits pcu can be got, as compared to 4 bits pcu for the original SM with 8 antennas, and 10 bits pcu for V-BLAST with the same configuration. With the $N_T = 20$, K = 10, M = 2, the proposed scheme offers 27 bits pcu, while V-BLAST can provide only 20 bits pcu. Thus the proposed scheme are more energy efficient as well as spectral efficient than the V-BLAST. Also to notice, when K = 1, it is the original SM. When K = N_T, it is a pure Spatial Multiplexing (SMX) scheme.

2.1 *Antenna mapping*

Different from the original SM scheme, where one of the 2^p TAs is used to send a modulated symbol, the proposed GSM scheme uses multiple active TAs to transmit different symbols simultaneously. The first problem for GSM is to choose a way to map numeric numbers ranging from 0 to N = 2^p-1 to distinct antenna combinations. Fortunately, combinatorial mathematics provides a one-to-one mapping between natural numbers and k-combinations, for all n and k (McCaffrey, 2004, Basar et al., 2013). For each n and k, C(n,k) can be presented by a sequence J of length k, which takes elements from the set {0,1,...,n-1}, according to the following equation:

$$Z = C(n_k, k) + \cdots + C(n_2, 2) + C(n_1, 1) \qquad (1)$$

where $n_1,...,n_k \in \{0,1,...,n-1\}$, and $n_i > n_{i-1}$ for all i ≥ 2 and i ≤ k and J = $\{n_k,...,n_2,n_1\}$. For example, when n = 8, k = 4, C(8,4) = 70. The following J sequences can be calculated:

$$69 = C(7,4) + C(6,3) + C(5,2) + C(4,1) \rightarrow J = \{7,6,5,4\}$$

$$68 = C(7,4) + C(6,3) + C(5,2) + C(3,1) \rightarrow J = \{7,6,5,3\}$$

$$\cdots$$

$$32 = C(6,4) + C(5,3) + C(4,2) + C(1,1) \rightarrow J = \{6,5,4,1\}$$

$$31 = C(6,4) + C(5,3) + C(4,2) + C(0,1) \rightarrow J = \{6,5,4,0\}$$

$$\cdots$$

$$1 = C(4,4) + C(2,3) + C(1,2) + C(0,1) \rightarrow J = \{4,2,1,0\}$$

$$0 = C(3,4) + C(2,3) + C(1,2) + C(0,1) \rightarrow J = \{3,2,1,0\}$$

In this way, a one-to-one correspondence of natural numbers to the k antenna indexes are established. Yet in this proposed scheme, only 2^p out of all $C(N_T, K)$ active TA combinations are used. For the example above, $p = 6$, $N = 2^p = 64$ combinations are used. That is,

$$N = \lfloor C(N_T, K) \rfloor_{2^p} \tag{2}$$

where p is the largest integer that satisfies

$$2^p \leq C(N_T, K) < 2^{p+1} \tag{3}$$

And $N = 2^p$.

In the proposed scheme, the number of bits per symbol transmitted by using the SM and the new scheme are calculated, respectively, as

$$\eta_{SM} = \log_2(N_T) + \log_2(M) \quad \text{bpcu} \tag{4}$$

$$\eta_{GSM} = \log_2(N) + K\log_2(M) \quad \text{bpcu}, \tag{5}$$

where bpcu means bit per channel use.

2.2 System model

The transmitter groups the incoming bits, **b**, into blocks of $\log_2(NM^K)$ bits, the first $p(=\log_2(N))$ bits are used to select the combination pattern of active TAs based on the combinatory method described previously, and the remaining $K\log_2(M)$ are mapped into a complex signal constellation vector $x = [x_1, x_2, \ldots, x_K]^T$ to be transmitted on the selected K active TAs, where $x_k (1 \leq k \leq K)$ is selected uniformly from an M-ary complex constellation such as MPSK etc., and is assumed to be independently identically distributed (i.i.d.) complex variables with covariance matrix $R_{xx} = E[x^H x]$. The average energy per transmission is limited as $E_x = \text{Tr}[R_{xx}]$, which is the power constraint at the transmitter

side. And the average symbol energy per timeslot is $\sigma_s^2 = Es/K$.

As with the traditional MIMO, the system model can be written as

$$y = Hs + v \tag{6}$$

where $y = [y_1, y_2, \ldots, y_{N_R}]^T$ is $N_R \times 1$ received sample vector, and **s** is an $N_T \times 1$ vector whose non-zero elements represent symbols transmitted on the antenna according to the index of the elements

$$s = [0, x_1, \ldots, 0, \ldots x_K, \ldots]^T$$

H is an $N_R \times N_T$ channel matrix between the transmit antennas and the receive antennas

$$H = \begin{bmatrix} h_{1,1} & h_{1,2} & \cdots & h_{1,N_T} \\ h_{2,1} & h_{2,2} & \cdots & h_{2,N_T} \\ \vdots & \vdots & \ddots & \vdots \\ h_{N_R,1} & h_{N_R,2} & \cdots & h_{N_R,N_T} \end{bmatrix} \tag{7}$$

where each item of H $h_{i,j}$ is a complex fading coefficient between the ith transmit TA and the jth RA and is modeled as an i.i.d. Zero Mean Complex Gaussian (ZMCG) variable with unit variance. We assume the channel is flat faded. **v** is the additive noise vector of size $N_R \times 1$, with i.i.d. elements each of which is also ZMCG with variance $N_0 = \sigma_n^2$.

3 LLR-BASED DETECTION ALGORITHM

3.1 The optimal ML algorithm

The optimal detector is the Maximum Likelihood (ML) detector as in (Jeganathan et al., 2008b), which estimate the combination index $\hat{m}$ and the transmitted symbol vector x jointly using the ML criterion

$$[\hat{m}, \hat{x}] = \underbrace{\arg\max}_{m,x} \{p(y|G_m, x)\}$$

$$= \underbrace{\arg\max}_{m,x} \{\|y - G_m x\|^2\} \tag{8}$$

where G_m is the channel matrix of the active antennas of the combination numbered m formed from K columns of **H**, and $p(y|G_m, x)$ is the likelihood function of **y** given G_m, x

$$p(y|G_m, x) = \left(\frac{1}{\pi\sigma_n^2}\right)^{N_R} \exp\left(\frac{\|y - G_m x\|^2}{\sigma_n^2}\right) \tag{9}$$

By iterating every m = 1,2,...,N, and every possible symbol vector x (totally M^K possibilities), the largest pair corresponding to the likelihood value is jointly detected as $\hat{m}$ and $\hat{x}$. However, this detector requires computations of exponential complexity in proportion to NM^K.

3.2 *Proposed LLR-based algorithm*

By applying Zero-Forcing to (6), we get

$$\mathbf{r} = \mathbf{x} + \mathbf{n} \qquad (10)$$

where $\mathbf{n} = \mathbf{Hv}$ is the new noise vector of length N_T, with zero mean and auto-covariance matrix as

$$
\begin{aligned}
\mathbf{R_n} &= \mathbf{E[nn^H]} \\
&= \mathbf{E[(H^H H)^{-1} H^H vv^H H(H^H H)^{-1}]} \\
&= \mathbf{(H^H H)^{-1} H^H [vv^H] H(H^H H)^{-1}} \\
&= \mathbf{(H^H H)^{-1} H^H [N_0 I] H(H^H H)^{-1}} \\
&= \mathbf{(H^H H)^{-1} N_0}
\end{aligned}
\qquad (11)
$$

And the new noise vector $\mathbf{n}$ is also ZMCG, with zero mean and variance as

$$\sigma^2 = N_0 \text{diag}\left[(H^H H)\text{-}1 \right] \qquad (12)$$

where diag($\mathbf{A}$) means the column vector formed with the diagonal elements of matrix A.

In our proposed GSM-MIMO system, as compared to the ML detector which has exponential computational complexity, which antennas are actually on for transmission can be estimated by computing their Log-Likelihood-Ratio (LLR) values after ZF operation on the received signal, using (10) and (12).

The LLR value for the α-th ($\alpha = 1,2,\ldots,N_T$) TA can be evaluated as

$$
\begin{aligned}
\lambda(\alpha) &= LLR \\
&= \ln \left[\frac{\sum_{i=1}^{M} P(x(\alpha) = s_i | r(\alpha))}{P(x(\alpha) = 0 | r(\alpha))} \right]
\end{aligned}
\qquad (13)
$$

where $s_i \in S$, S is the constellation set for the specified modulation scheme applied.

By applying Bayes formula, we get

$$
\begin{aligned}
\lambda(\alpha) &= \ln \left[\frac{\sum_{i=1}^{M} P(x(\alpha) = s_i | r(\alpha))}{P(x(\alpha) = 0 | r(\alpha))} \right] \\
&= \ln \left[\frac{\sum_{i=1}^{M} P(x(\alpha) = s_i)p(r(\alpha)|x(\alpha)) = s_i}{P(x(\alpha) = 0)p(r(\alpha)|x(\alpha) = 0} \right]
\end{aligned}
\qquad (14)
$$

Assume the K active TAs are taken uniformly, and the symbols transmitted on the antennas are uniformly selected from the constellation set S, we get

$$
\begin{aligned}
\lambda(\alpha) &= \ln \left[\frac{\sum_{i=1}^{M} P(x(\alpha) = s_i)p(r(\alpha)|s_i)}{P(x(\alpha) = 0 | r(\alpha))} \right] \\
&= \ln \left(\frac{K}{M(N_t - K)} \right) + \ln \left[\frac{\sum_{i=1}^{M} p(r(\alpha)|s_i)}{p(r(\alpha)|0)} \right] \\
&= \ln \left(\frac{K}{M(N_t - K)} \right) + \frac{|r(\alpha)|^2}{\sigma^2(\alpha)} \\
&\quad + \ln \left[\sum_{i=1}^{M} \exp \left(-\frac{|r(\alpha) - s_i|^2}{\sigma^2(\alpha)} \right) \right]
\end{aligned}
\qquad (15)
$$

By eliminating the common items in the LLR calculations by (15), the LLR can be modified as

$$\wedge(\alpha) = \frac{|r(\alpha)|^2}{\sigma^2(\alpha)} + + \ln \left[\sum_{i=1}^{M} \exp \left(-\frac{|r(\alpha) - s_i|^2}{\sigma^2(\alpha)} \right) \right] \qquad (16)$$

After calculating all N_T LLRs for TAs, the largest K values are picked out, and the corresponding indices are decided as the active TAs. By choosing the according items in (10), symbols on these TAs can be demodulated easily. Information carried by active antenna combinations can be detected simply by looking up table from section 2.1.

4 RECEIVER COMPLEXITY

This algorithm is highly efficient. First, The main computations needed of the proposed algorithm lies with the ZF operation, where the pseu-inversion needs $4 N_t^2 N_r - N_t(N_r+N_t) + O(N_t^3)$ flops (one flops means one complex multiplication or addition, real operation doesn't count in the analysis), and the Zero-Forcing needs $2 N_t N_r - N_t$ flops.

Second, the core of the algorithm, LLR calculation, needs $N_t(2M+1)$ flops, where the exp($\cdot$) and ln($\cdot$) can be simplified by table-lookup, or using Jacobian method, or even using MAX-LOG approximation.

Third, selecting the largest K LLRs requires sorting operation, which has order of $O(N_t \log_2(N_t))$ with $K < N_t$. The demodulation can be directly accomplished by hard detection.

In all, the proposed LLR-based algorithm has total computational complexity of

$$
\begin{aligned}
&N_t(N_r-1) + N_t^2(4 N_r-1) + N_t(2M+1) \\
&+ O(N_t^3) + O(N_t \log_2(N_t)) \sim O(NrN_t^3)
\end{aligned}
$$

In (Jeganathan et al., 2008b), computational complexity of ML algorithm is given as

$$\approx NM^K (N_r(2K-1)+Nr+1) \sim O(N_rKNM^K)$$

where N is the number of usable TA combinations.

By comparison, one can see that in high-rate transmission applications, where M and K are large, the proposed algorithm is far more efficient than ML algorithm. For example, when $N_R = N_T = 8$, $M = 16$, $K = 4$, then $N = 64$, so the part in the big-O of the proposed algorithm is $2^{12} = 4096$, while in ML, it is a humongous $2^{27} \approx 1342 \times 10^5$! Although this computation is rough (due to the neglecting of many items and constants) indeed, at least it shows the potential speed for the proposed scheme.

5 SIMULATION RESULTS

Two configurations of different parameters are simulated in the simulation. First, $N_T = 5$, $N_R = 8$, $K = 3$, $M = 4$ for QPSK modulation, so there are 9 bits per transmission. The result is shown in Figure 2. The second configuration is: $N_T = 8$, $N_R = 8$, $K = 4$, $M = 16$ for 16PSK modulation. So there are 22 bits per channel use. The result is shown in Figure. 3. For comparison, two sub-optimal algorithms based on ZF and MMSE are also simulated meanwhile.

As can be seen from both figures, with low SNR, the proposed algorithm is 4-dB worse than the ML detector. With SNR gradually increases, the gap is also narrowing down, yet it keeps around 2 dBs worse than the optimal ML detector. This is a price paid for the high computational efficiency of LLR-based algorithm, for the ML detector requires over 10 times more simulation time than the proposed LLR algorithm on the author's laptop.

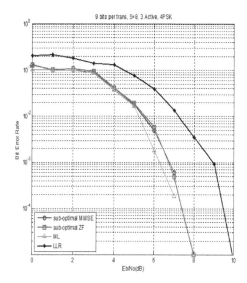

Figure 3. BER performance for 22 bpcu GSM MIMO.

6 CONCLUSION

A highly computationally efficient detecting algorithm based on LLR—calculation for generalized spatial modulation MIMO systems are proposed in this paper. With no extra spectral requirement, this algorithm requires far less computation time than the optimal ML algorithm, especially with large SE requirement where higher-order modulation and more combination numbers can be used to transmit large amount of bits at one time. This computational superiority comes at the cost of about 2-dBs degradation in BER performance.

Future work will focus on better improvement of BER performance to narrow the gap between the LLR-based algorithm and the ML detector.

REFERENCES

Andrews, J.G., Ghosh, A. & Muhamed, R. 2007. *Fundamentals of WiMAX: Understanding Broadband Wireless Networking,* NJ, USA, Englewood Cliffs.

Basar, E., Aygolu, U., Panayirci, E. & Poor, H.V. 2013. Orthogonal Frequency Division Multiplexing With Index Modulation. *IEEE Transactions on Signal Processing,* 61, 5536–5549.

Bohnke, R., Wubben, D., Ku, X., Hn, V. & Kammeyer, K.D. Reduced complexity MMSE detection for BLAST architectures. Global Telecommunications Conference, 2003. GLOBECOM '03. IEEE, 1–5 Dec. 2003. 2258–2262 vol.4.

Di Renzo, M., Haas, H., Ghrayeb, A. & Sugiura, S. 2014. Spatial Modulation for Generalized MIMO: Challenges, Opportunities, and Implementation. *Proceedings of the IEEE,* 102, 56–103.

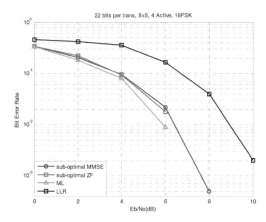

Figure 2. BER performance for 9 bpcu GSM MIMO.

Ghosh, A., Zhang, J., Andrews, J.G. & Muhamed, R. 2010. *Fundamentals of LTE*, Prentice Hall Press.

Hellings, C. & Utschick, W. Energy efficiency optimization in MIMO broadcast channels with fairness constraints. Signal Processing Advances in Wireless Communications (SPAWC), 2013 IEEE 14th Workshop on, 16–19 June 2013. 599–603.

Jeganathan, J., Ghrayeb, A. & Szczecinski, L. Generalized space shift keying modulation for MIMO channels. Personal, Indoor and Mobile Radio Communications, 2008. PIMRC 2008. IEEE 19th International Symposium on, 15–18 Sept. 2008 2008a. 1–5.

Jeganathan, J., Ghrayeb, A. & Szczecinski, L. 2008b. Spatial modulation: optimal detection and performance analysis. *IEEE Communications Letters*, 12, 545–547.

Jeganathan, J., Ghrayeb, A., Szczecinski, L. & Ceron, A. 2009. Space shift keying modulation for MIMO channels. *IEEE Transactions on Wireless Communications*, 8, 3692–3703.

Jinlin, F., Chunping, H., Wei, X., Lei, Y. & Yonghong, H. Generalised spatial modulation with multiple active transmit antennas. GLOBECOM Workshops (GC Wkshps), 2010 IEEE, 6–10 Dec. 2010 2010. 839–844.

Mccaffrey, J. 2004. Generating the mth lexicographical element of a mathematical combination. *MSDN Library online*.

Mesleh, R., Haas, H., Chang Wook, A. & Sangboh, Y. Spatial Modulation—A New Low Complexity Spectral Efficiency Enhancing Technique. Communications and Networking in China, 2006. ChinaCom '06. First International Conference on, 25–27 Oct. 2006 2006. 1–5.

Younis, A., Mesleh, R., Haas, H. & Grant, P.M. Reduced Complexity Sphere Decoder for Spatial Modulation Detection Receivers. Global Telecommunications Conference (GLOBECOM 2010), 2010 IEEE, 6–10 Dec. 2010 2010a. 1–5.

Younis, A., Serafimovski, N., Mesleh, R. & Haas, H. Generalised spatial modulation. Signals, Systems and Computers (ASILOMAR), 2010 Conference Record of the Forty Fourth Asilomar Conference on, 7–10 Nov. 2010 2010b. 1498–1502.

Sports Engineering and Computer Science – Luo (Ed.)
© 2015 Taylor & Francis Group, London, ISBN 978-1-138-02650-6

Physically disabled visual position tracking using skew lines model with head motion compensation

ZhengYe Guo & WenQiang Hu

School of Software and Engineering, Huazhong University of Science and Technology, Wuhan, China

ABSTRACT: In this paper, a new algorithm of physically disabled visual tracking is introduced. Some disabled individuals with moving difficulties can be solved using this method. In this algorithm, any head motion is allowed which is the most characteristic. The sight point estimation depends on midpoint of mid perpendicular for the two skew lines. In order to ensure the accuracy, the neutral network calibration and head motion compensation have been introduced. It shows a high accuracy and flexible method for physically disabled patients to fix the coordinate of target object.

1 INTRODUCTION

With the development of technology, improving the physically disabled life is urgent both in theory and practice. As we know tracking and estimating target objects has been an active research topic for many years. Computer vision tracking gives a new solution for this issue. Comparing to the visual tracking system, traditional methods use assisted mechanical to help disabled getting their target. However, it requires body movements of the disabled to handle the machine. A skew lines model solves this problem by tracking disabled persons' eyes. Using the three-dimensional coordinates of the target, they can use manipulator to get anything without move.

The skew lines model is a three-dimensional visual tracking method. Comparing to the two-dimensional tracking method, we take head motion compensation into consideration. It ensures the tracking accuracy with any head motion and it's more flexible for disabled to search their target. In order to cover the shortage of three-dimensional camera calibration, we use new ways with feedback regulation of neural network. It reduces calibration time and gives a high-accuracy data processing result. In general, this new method makes the tracking system simple, reliable and inexpensive for future engineering application.

2 VISUAL POSITION TRACKING CLASSIFICATIONS

Classify by the tracking methods, visual position tracking can be divided into two categories. One is two-dimensional visual tracking, another is three-dimensional visual tracking. As we know, two-dimensional mapping model apply to single camera visual tracking system, while three-dimensional visual estimate mainly apply to dual camera system.

Firstly we come to two-dimensional visual tracking. This tracking method uses a mapping function which has been calibrated to estimate the gazing direction. The import of the mapping function comes from a series of eye-movement characteristics for each eyes. And the export of the mapping function is the gazing direction of the sight point. The eye-movement characteristics changes with eye-sight direction. In order to get the mapping function, we need to calibrate each user on line. But two-dimensional visual tracking has two disadvantages. One is that the precision of the calibration decrease with the user's head motion, so we should stabilize the user's head. Another is that the calibration should test for many times.

Secondly, we come to three-dimensional visual tracking. Through space geometric model, this tracking method detects the 3d view parameters, and converted it into eye sight direction. From intersections of gazing direction and target plane, we can estimate the sight point. This method uses stereo vision, so it allows free head motion with errors. What's more, this method does not rely on eyes characteristic parameters and mapping function.

Therefore, comparing to the two-dimensional visual tracking, three-dimension tracking is more practical for physically disabled life.

3 THREE-DIMENSION VISION CAMERA CALIBRATION USING FEEDBACK REGULATION OF NEURAL NETWORK

3.1 *Neural network model*

In traditional ways, three-dimension vision calibration needs at least two cameras, but in this paper, it will show you a new way for calibration. Using feedback regulation of neural network, it makes calibration simple with simple equipments. With its outstanding performance for nonlinear approximation, feedback regulation of neural network can greatly reduce the error.

Using the result of preliminary image processing, we get the coordinate of angular points. First we send the coordinate to neural network and generalized equation. The expectations and the actual value come out. Secondly, we compare these two coordinate value to see whether it conform the expectation value. If it meets the demand, we will determine the distortion factor of nonlinear function g. If it does not meet the demand, we will use feedback regulation. At last, when the distortion factor become zero the calibration finished. See the structure chart of neural network shown in Figure 1.

3.2 The selection of parameter and function

First, it shows that three layer construction neural network will better approach the nonlinear function g. The buried layer function of the neural network is: $f_1(x) = (1/1 + e^{-x})$, Output layer function is: $f_2(x) = x$. Meanwhile, in order to improve the training speed, local minimum solution should be avoided (See the iterative algorithm in this paper):

$$w_{ji}(t+1) = w_{ji}(t) + \eta \sum_{p=1}^{2} \delta_{jp} x_{ip} + \alpha \Delta w_{ji}(t-1) \quad (1)$$

$$w_{kj}(t+1) = w_{kj}(t) + \eta \sum_{p=1}^{2} \delta_{kp} x_{jp} + \alpha \Delta w_{kj}(t-1) \quad (2)$$

w_{ji} is weight from input layer to buried layer and W_{kj} is weight from buried layer to output layer. δ_{jp}, δ_{kp} are error for buried layer node j and output layer node k when the input sample is p. See the calculation below:

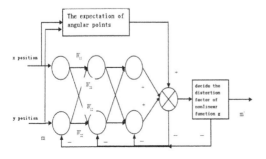

Figure 1. The structure chart of neural network.

$$\delta_{jp} = O_{jp} \times (1 - O_{jp}) \times \sum_{k=1}^{2} \delta_{kp} w_{kj}$$

$$\delta_{kp} = y_{kp} \times (1 - y_{kp}) \times (\hat{y}_{kp} - y_{kp}) \quad (3)$$

Secondly, from the mathematical modeling, we know that radial distortion can be described by few front parts of Taylor series. That is:

$$x_c = x(1 + k_1 r^2 + k_2 r^4 + k_3 r^6) \quad (4)$$

$$y_c = y(1 + k_1 r^2 + k_2 r^4 + k_3 r^6) \quad (5)$$

And tangential distortion can be described by other two parameters p_1, p_2. That is:

$$x_c = x + \left[2p_1 y + p_2 (r^2 + 2x^2) \right] \quad (6)$$

$$y_c = y + \left[p_1 (r^2 + 2y^2) + 2p_2 x \right] \quad (7)$$

So we can get the normalized image coordinate $p = (x, y)^T$ and the actual normalized image coordinate $p_d = (x_d, y_d)^T$, their relationship is as below:

$$p_d = \begin{bmatrix} x_d \\ y_d \end{bmatrix} = (1 + k_1 r^2 + k_2 r^4 + k_3 r^6) \begin{bmatrix} x \\ y \end{bmatrix}$$
$$+ \begin{bmatrix} 2p_1 xy + p_2 (r^2 + 2x^2) \\ p_1 (r^2 + 2y^2) + 2p_2 xy \end{bmatrix} \quad (8)$$

$r^2 = x^2 + y^2$ and k_1, k_2, k_3 are the radial lens distortion coefficient; p_1, p_2 are tangential distortion coefficient. We make $p_d = g(p)$, where g is the distortion transformation function.

Finally, due to the initial value of the weight decide the convergence speed of the program, we enactment right weight by calculation. As we know, $m' = g(k_1, k_2, k_3, p_1, p_2)$ is the function of $(k_1, k_2, k_3, p_1, p_2)$, its Jacobi matrix is: $J = [(\partial m'/\partial k_1) \ (\partial m'/\partial k_2) \ (\partial m'/\partial k_3) \ (\partial m'/\partial p_1) \ (\partial m'/\partial p_2)]$. With the least square method, we can get the initial value of the weight: $w_{ji}(0) = (J^T J)^{-1} J^{T(28)}$.

4 FEATURE EXTRACTIONS AND THREE-DIMENSIONAL RECONSTRUCTION

Recent studies show that Harris algorithm is a good method of corner detection. Because it's more suitable for matrix operation, high accuracy positioning and immune to noises. These make it the best way for feature extraction and get ready for target recognition, target tracking and three-dimensional reconstruction.

First step, we preprocess the two images we take from each camera. Then we use Canny operator to get the edge of the image.

Second step, it comes to the Harris algorithm for feature extraction. This method calculates the eigen value of squared-gradient for each pixel. And use the result to judge the feature points. See the mathematical expression below:

$$M = \begin{bmatrix} \left\langle \left(\dfrac{\partial I}{\partial x}\right)^2, W \right\rangle & \left\langle \left(\dfrac{\partial I}{\partial x}\dfrac{\partial I}{\partial y}\right), W \right\rangle \\ \left\langle \left(\dfrac{\partial I}{\partial x}\dfrac{\partial I}{\partial y}\right), W \right\rangle & \left\langle \left(\dfrac{\partial I}{\partial y}\right)^2, W \right\rangle \end{bmatrix} \qquad (9)$$

$$R = \det(M) - k * tr^2(M), (k = 0.04) \qquad (10)$$

Setting appropriate threshold value, compare R with the threshold value. If R is greater than the threshold value, it's the feature point.

Last step is three-dimensional reconstruction of the feature point. Along the polar direction, we find correspondence match points and calculate its parallax. See the parallax image shown in Figure 2.

Using the parallax, we can figure out depth information of three-dimension coordinate. See

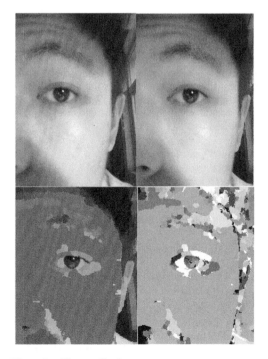

Figure 2. The parallax image.

the relationship between parallax and depth information below:

$$d = \frac{2ve\tan\left(\frac{\theta}{2}\right)z_0 z}{Ap_x(z - z_0)(v - l) + vez} \qquad (11)$$

p_x is the parallax, d is the depth information, v, e, θ, l, z, z_0, A can get from the calibration and actual measurement data. In conclusion, from the parallax we get the three-dimension coordinate of the feature point.

5 SKEW LINES MODEL WITH HEAD MOTION COMPENSATION

5.1 Eyeball optical axis calculate

In order to get the two skew sight lines we should know the coordinate of spherical center of the cornea O. So let's introduce the eyeball model first. In this model, we have known five points' coordinate. They are C_1, C_2, P_1, P_2, P_0 just as the Figure 3.

The three-dimension coordinate of C_1, C_2 come from the calibration. And the three-dimension coordinate of P_1, P_2, P_0 come from the feature extraction and three-dimensional reconstruction. So we get two equations of lines. If they are coplanar lines, intersection point will be the spherical center of the cornea. If they are skew lines, the mid-point of common normal will be the spherical center of the cornea. See the equation of the skew lines C_1P_1, C_2P_2 below:

$$\begin{cases} \dfrac{x - c_1 \cdot x}{c_1 \cdot x - p_1 \cdot x} = \dfrac{y - c_1 \cdot y}{c_1 \cdot y - p_1 \cdot y} = \dfrac{z - c_1 \cdot z}{c_1 \cdot z - p_1 \cdot z} = t_1 \\ \dfrac{x - c_2 \cdot x}{c_2 \cdot x - p_2 \cdot x} = \dfrac{y - c_2 \cdot y}{c_2 \cdot y - p_2 \cdot y} = \dfrac{z - c_2 \cdot z}{c_2 \cdot x - p_2 \cdot z} = t_2 \end{cases}$$

$$\qquad (12)$$

$$c_1 \cdot x - p_1 \cdot x = A_1, c_1 \cdot y - p_1 \cdot y = B_1, c_1 \cdot z - p_1 \cdot z = C_1$$
$$c_2 \cdot x - p_2 \cdot x = A_2, c_2 \cdot y - p_2 \cdot y = B_2, c_2 \cdot z - p_2 \cdot z = C_2$$

$$\qquad (13)$$

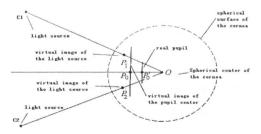

Figure 3. Eyeball model.

The coordinate of intersection point M_1 M_2 for the skew lines and their common normal are as below:

$$M_1 : \begin{cases} x_1 = c_1.x + A_1 t_1 \\ y_1 = c_1.y + B_1 t_1 \\ z_1 = c_1.z + C_1 t_1 \end{cases} \quad (14)$$

$$M_2 : \begin{cases} x_2 = c_2.x + A_2 t_2 \\ y_2 = c_2.y + B_2 t_2 \\ z_2 = c_2.z + C_2 t_2 \end{cases} \quad (15)$$

The equations of the common normal $M_1 M_2$ are as below:

$$\begin{cases} x = c_2 \cdot x - c_1 \cdot x + A_2 t_2 - A_1 t_1 \\ y = c_2 \cdot y - c_1 \cdot y + B_2 t_2 - B_1 t_1 \\ z = c_2 \cdot z - c_1 \cdot z + C_2 t_2 - C_1 t_1 \end{cases} \quad (16)$$

Because the common normal is perpendicular to each skew line, so:

$$(x\ y\ z)*(A_1\ B_1\ C_1) = 0, (x\ y\ z)*(A_2\ B_2\ C_2) = 0 \quad (17)$$

So we get the coordinate of the center of the cornea O:

$$\begin{cases} x_{\text{mid}} = (x_1 + x_2)/2 \\ y_{\text{mid}} = (y_1 + y_2)/2 \\ z_{\text{mid}} = (z_1 + z_2)/2 \end{cases} \quad (18)$$

Connect the spherical center of the cornea O and virtual image of the pupil center P_0, we get the direction of eyeball optical axis. In order to get direction of eye-sight, we need to know the compensation angle of the optical axis. That's the connection between eye-sight direction and eyeball optical axis direction.

5.2 The angle compensation and eye-sight direction calculate

The center of retina is not on the eyeball optical axis line. So the eye-sight direction has a certain angle with the eyeball optical axis. See the angle compensation model shown in Figure 4.

The X axis, Y axis and Z axis are the same as the world coordinate system. The angle between the eyeball optical axis and the X axis is θ. The angle between the eyeball optical axis and the Y axis is φ. So in the world coordinate system the eyeball optical axis' unit vector can be expressed by:

$$g = \begin{bmatrix} \sin\theta\cos\varphi \\ \sin\theta \\ \cos\theta\cos\varphi \end{bmatrix} \quad (19)$$

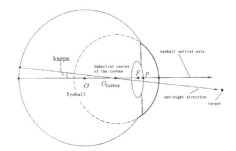

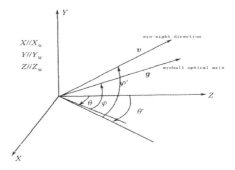

Figure 4.   Angle compensation model.

$$\begin{cases} \theta = \arctan(g_x/g_z) \\ \varphi = \arcsin(g_y) \end{cases} \quad (20)$$

Then we define α_{eye} the angle between the eye-sight direction and eyeball optical axis in horizon direction; define β_{eye} the angle between the eye-sight direction and eyeball optical axis in vertical direction. So the angle between eye-sight direction and the X axis is $\theta + \alpha_{eye}$, the angle between eye-sight direction and the Y axis is $\varphi + \beta_{eye}$. The vector v of the eye-sight direction can be expressed by:

$$v = \begin{bmatrix} \sin(\theta + \alpha_{eye})\cos(\varphi + \beta_{eye}) \\ \sin(\theta + \alpha_{eye}) \\ \cos(\theta + \alpha_{eye})\cos(\varphi + \beta_{eye}) \end{bmatrix} \quad (21)$$

$$\begin{cases} \alpha_{eye} = \arctan(v_x/v_z) - \arctan(g_x/g_z) \\ \beta_{eye} = \arcsin(v_y) - \arcsin(g_y) \end{cases} \quad (22)$$

α_{eye}, β_{eye} are the compensation angle.

5.3 The skew lines model and sight point estimation

The skew lines mean two straight lines in different planes. These two lines in our model are the two eye-sight direction lines. We define the two skew lines OP and $O'P'$, $M_1'M_2'$ is the common normal of the skew lines, M is the mid-point of the common

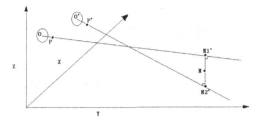

Figure 5. The skew lines model.

normal. In traditional ways, the sight point estimate relies on the intersection point between a straight line and a plane. But in this paper, we use the mid-point of the common normal of the skew lines to estimate the sight point. See the skew lines model shown in Figure 5.

From the results we calculate above. We can get the vector of the eye-sight direction for each eye, because two lines go through the spherical center of the cornea. So we can get the equation of the skew lines:

$$M_1'M_2': \begin{cases} \dfrac{x-O_x}{v_x} = \dfrac{y-O_y}{v_y} = \dfrac{z-O_z}{v_z} = t \\ \dfrac{x-O_x'}{v_x'} = \dfrac{y-O_y'}{v_y'} = \dfrac{z-O_z'}{v_z'} = t' \end{cases} \quad (23)$$

Use the method we have mention in the Eyeball optical axis calculate part. The coordinate of the mid-point M for the common normal $M_1'M_2'$ is the estimation of the target sight point.

5.4 Head motion compensation

The most characteristic of this method is allowing any head motion. So we should take head motion error into consideration. In this paper, we use support vector regression method to compensate error. Regression target is training the sample to get the function. This function makes the error lees than the assigned error. This system is multi input and single ended output. The input vector is: $g = [x\ y\ z\ r\ \theta\ P_x\ P_y]^T$, where $(x\ y\ z)$ are the coordinate of the sight point; r is the rate of the major and the minor axis of ellipse of the pupil; θ rotational angle of the ellipse of the pupil; $(P_x\ P_y)$ the coordinate of the center of the pupil in the image. The output of the system is the error $[d_x, d_y, d_z]$ between actual sight point and the ideal sight point.

The support vector regression includes kennel function selection, marginal coefficient selection and the loss function selection. In this paper, we use cross validation to choose optimal parameter

of the kennel function and marginal coefficient. First, we select the kennel function:

$$k(x_i, x_j) = \exp\left(-\frac{(x_i - x_j)^2}{2\delta^2}\right), \delta = 1, C = 200 \quad (24)$$

Through the experiment, we find the best termination precision is $\varepsilon = 0.01$, then we construct the optimization equation:

$$\min\left\{\frac{1}{2}\sum_{i=1}^{l} \begin{matrix} (\alpha_i - \alpha_i^*)(\alpha_j - \alpha_j^*)k(x_i, x_j) + \\ \varepsilon\sum_{i=1}^{l}(\alpha_i + \alpha_i^*) - \sum_{i=1}^{l} y_i(\alpha_i - \alpha_i^*) \end{matrix}\right\} \quad (25)$$

The constraint condition:

$$\sum_{i=1}^{l}(\alpha_i - \alpha_i^*) = 0, \alpha_i, \alpha_i^* \in [0, C] \quad (26)$$

Solving the equation, we get Lagrangian α_i, α_i^*, then we construct the regression function:

$$f(x) = \sum_{i=1}^{l}(\alpha_i - \alpha_i^*)k(x_i, x_j) + b \quad (27)$$

$$b = \frac{1}{l}\left\{\begin{matrix} \sum_{0<\alpha_i<C}\left[y_i - \sum(\alpha_j - \alpha_j^*)k(x_j, x_i) - \varepsilon\right] + \\ \sum_{0<\alpha_i^*<C}\left[y_i - \sum(\alpha_i - \alpha_i^*)k(x_j, x_i) + \varepsilon\right] \end{matrix}\right\} \quad (28)$$

From the regression function, we can figure out the error:

$$d_x = f_1(g), d_y = f_2(g), d_z = f_3(g) \quad (29)$$

We only train once when we build the system. In future, we don't need repeat training. Just use $f(g)$ to compensation the sight point estimation.

6 EXPERIMENTAL RESULTS AND ANALYSIS

Through the three step of the experiment, we let the user to watch the target point on the calibration plate. Calculate the sight point using skew lines model with head motion compensation. Compare the result with the theoretical value. See the data list in Table 1.

From the analysis of these data above, we find that the head motion compensation improve the accuracy for 10 mm–20 mm. It makes a great progress in sight point estimation. In order to

Table 1. Comparing the compensation result with the theoretical value.

Actual X (mm)	Actual Y (mm)	Actual Z (mm)	Calculated X (mm)	Calculated Y (mm)	Calculated Z (mm)	Compensated X (mm)	Compensated Y (mm)	Compensated Z (mm)
200	200	500	185.91	197.26	485.26	195.52	202.31	489.77
200	700	500	195.53	693.89	494.81	202.63	695.19	502.36
200	1200	500	203.47	1192.51	513.27	198.75	1192.96	506.64
700	200	500	699.56	201.38	486.55	697.81	189.66	487.21
700	700	500	697.23	689.82	507.40	703.18	702.59	500.09
700	1200	500	680.64	1195.91	477.47	688.77	1211.20	495.68
1200	200	500	1211.24	195.19	459.82	1192.41	198.13	488.57
1200	700	500	1208.56	715.26	475.95	1215.08	696.54	480.95
1200	1200	500	1220.17	1175.73	489.28	1209.19	1176.87	492.61

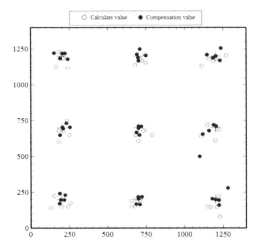

Figure 6. The comparison with head motion compensation.

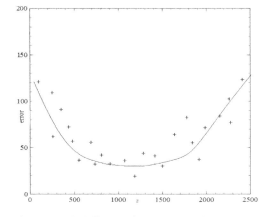

Figure 7. The influence of depth for Z axis.

So 500 mm–1500 mm is the best distance to use our system.

enable more intuitive data analysis, we project three-dimension coordinate to two-dimension coordinate. There are 9 calibrate points on the plate, we do the experiment 10 times, half of them is the sight estimation without head motion compensation, another half take the head motion into consideration. See Figure 6.

From Figure 6, we find head motion improve the accuracy of the sight point estimation. But we find another problem, the coordinate is also related to the Z axis. So we do experiment to find the connection between Z axis with the accuracy.

In this experiment we choose 25 Z value to test the relationship between coordinate accuracy and Z axis. In order to indicate the point distribution, we use log function, exponential function and quadratic function to create fit curve. The result shows that when Z value is within the limit of 500 mm to 1500 mm the accuracy is the best.

7 CONCLUSIONS

In order to improve the physically disabled life, we create this skew lines model with head motion compensation. This model use binocular visual position tracking to find any target in sight. With the help of skew lines model, any head motion is allowed. It becomes more flexible and comfortable for physically disabled to use. Also, in this method, we take high accuracy into consideration. The head motion compensation makes it more accurate than traditional visual tracking system. What's more, in this tracking system, we use a new calibration method. It reduces the demand for high precision instruments and makes it economical for popularization. I think this skew lines model with head motion compensation will change physically disabled life a lot.

ACKNOWLEDGMENTS

This work was financially supported by natural science fund of China granted No.51205145, National Basic Research Program of China (973 Program) granted No.2013CB035805.

REFERENCES

Baek, Y.M., S. Tanaka, K. Harada, N. Sugita, A. Morita, S. Sora, R. Mochizuki, and M. Mitsuishi. 2012. "Full state visual forceps tracking under a microscope using projective contour models," in Proc. IEEE Int. Conf. Robot. Autom., pp. 2919–2925.

Behrooz K.P. 1987. Minimization of the quantization error in camera calibration, proc. of image under Understand work shop, pp. 671–680.

Bleyer M., Gelautz M. 2007. Graph-cut-based stereo matching using image segmentation with symmetrical treatment of occlusions [J]. Signal Processing Image Communication, 22(2):127–143.

Chen H., Ye D., Che R.S.H., et al. 2006. Stereo camera calibration based on accurate control points [J]. Opt. Precision Eng., 14(5):903–909.

Chiang, Alpha C. 1984. Fundamental Methods of Mathematical Economics (3rd ed.), McGraw-Hill.

Devyver, M., A. Tsukada, and T. Kanade. 2011. "A wearable device for first person vision," in Proc. 3rd Int. Symp. Qual. Life Technol., pp. 1–6.

Gold, S., A. Rangarajan, C. Lu, S. Pappu and E. Mjolsness. 1998. "New Algorithms for 2D and 3D Point Matching", Pattern Recognition, (31):8,1019–1031.

Hongliang R., and P. Kazanzides. 2012. "Investigation of attitude tracking using an integrated inertial and magnetic navigation system for hand-held surgical instruments," IEEE/ASME Trans. Mechatronics, vol. 17, no. 2, pp. 210–217.

Janabi-Sharifi, F., Lingfeng Deng, and W.J. Wilson. 2011."Comparison of basic visual serving methods," Mechatronics, IEEE/ASME Transactions on, vol. 16, no. 5, pp. 967–983.

Kanade T., and M. Hebert. 2012."First-person vision," Proc. IEEE, vol. 100, no. 8, pp. 2442–2453.

Park, D.-H., J.-H. Kwon, and I.-J. Ha. 2011. "A novel position-based visual serving approach to robust global stability under field-of-view constraint," Industrial Electronics, IEEE Transactions on, vol. PP, no. 99, pp.

Sung-do C., and L. Soo-Young. 2012. "3D stroke reconstruction and cursive script recognition with magnetometer-aided inertial measurement unit," IEEE Trans. Consumer Electron., vol. 58, no. 2, pp. 661–669.

Tick, D.A., C. Satici, J. Shen, and N. Gans. 2013. "Tracking control of mobile robots localized via chained fusion of discrete and continuous epipolar geometry, IMU and odometry," IEEE Trans. Cybern, vol. 43, no. 4, pp. 1237–1250.

Wang, K., Y.H. Liu, and L. Li. 2013. "Vision-based tracking control of nonholonomic mobile robots without position measurement," in Proc. IEEE Int. Conf. Robot. Autom, pp. 5245–5250.

Wang Q.H., Tao Y.H., Li D.H., et al. 2007. 3D auto stereoscopic liquid crystal display based on lenticular lens [C]. Shanghai: Proc. of Asia Display, pp. 453–455.

Sports Engineering and Computer Science – Luo (Ed.)
© 2015 Taylor & Francis Group, London, ISBN 978-1-138-02650-6

Identification of comparative sentences in TCSL

LinNan Bai

Beijing Normal University, Haidian District, Beijing, P.R. China

ABSTRACT: Comparative sentence is one of difficulties in TCSL (Teaching Chinese as a Second Language) and the distribution of comparative sentences is various. In the textbooks of Teaching Chinese as a Second Language, there are about 10% comparative sentences. So the research on identification of comparative sentences is very significant. Previous studies were basically based on statistical methods, rare researches were done using the rule-based method. In this paper, a novel approach based on HNC (Hierarchical Network of Concepts) theory is proposed. We apply semantic analysis to write identification rules, and abundant linguistic features of comparative sentences are explored. Experiment on every category of comparative sentences verifies good performance of our method, and the parsing result of comparative sentence is a great help to Teaching Chinese as a Second Language.

1 INTRODUCTION

The comparative grammar item is a teaching focus in TCSL (Teaching Chinese as a Second Language), so how to teach Chinese comparative sentence is one difficult task for TCSL teachers. With the development of Natural Language Processing (NLP), sentence parsing is increasingly helpful in TCSL. Chinese comparative sentences have too many syntactic forms. Except for the most representative "BI sentence", there are many different syntactic forms expressing comparative meanings. For example, "YOU" could also be used like "BI" in this sentence "I'm not good at swimming, I don't swim as good as you". Moreover, "YOU" is often used after an adjective word to express comparative relation, for instance, "Among basic necessities of life, the food is more important than others."

As shown in Ding (2009), comparative sentences "contain comparative item, and express comparative meanings". In our research, we define comparative sentences as sentences which contain or imply four elements: comparative items, comparative mark, comparative aspect and comparative result. These four elements are essential to extract comparative relation. Liu (2004) studied the categories of comparative domain, not all sentences which contain comparative meanings belong to comparative category. We should exclude categorical judgment and election. In our work, we choose six categories of comparative sentences. They are: "BI" sentence (including the negative construction), You/ MeiYou/YouMeiYou+......+A, Zui/ Hai/Geng+A,A+YU,Xiang/He/Gen/Tong+......+ YiYang/ChaBuDuo(+A) and the "BuRu" type.

Comparative sentences are widespread in Chinese texts. In this paper, we research two corpuses, one is the textbooks of Teaching Chinese as a Second Language, and the other is the Chinese Penn Tree Bank. In the textbooks of Teaching Chinese as a Second Language, we choose 14 books and all HSK test texts, this corpus has 68946 sentences, and the Chinese Penn Tree Bank has 71433 sentences. About 6% sentences are comparative sentences in the textbooks of Teaching Chinese as a Second Language, and about 8% sentences are comparative sentences in Chinese Penn Tree Bank. But the distribution of comparative sentences is different.

The teaching texts for foreign students are progressively increasing day by day, TCSL also needs some advanced technology to identify linguistic phenomenon. In some present parser system, such as Stanford parser and LTP-Cloud, the identification and analysis of comparative sentence is not as good as other sentence patterns. So to automatically identify and analyze comparative sentences is our primary task.

In this paper, we proposed a novel approach based on semantic rules to identify the comparative sentences. Firstly, all comparative sentences are collected from the above two corpuses. Secondly, we divide comparative sentences into six different categories based on their syntactic features. At the same time, comparative words library is established. Then we set up the identification rules of Chinese comparative sentences. Next, we designed Sentence Semantic Analysis Engine to analyze and identify sentences. Lastly, we test the experimental results and evaluate them through precision and recall ration.

This paper focuses on automatic comparative sentences identification. The distribution of comparative sentences in the textbooks of Teaching Chinese as a Second Language is helpful to teaching Chinese. Our sentences parser is also helpful to teach and learn Chinese comparative sentences in TCSL.

2 RELATED WORK

Related work to this paper comes from both area of linguistics and natural language processing.

In the field of linguistics, researchers put their focus on the definition of comparative sentences and the interpretation of their syntax and semantic. Shang (2006) has systematically summarized researches about comparative sentences of some famous linguists, such as Ma Jianzhong, Li Jinxi, Gao Mingkai, Lv Shuxiang, Ding Shengshu. They only considered comparative meanings and expanded the edge of comparative sentences one by one. For example, Lv Shuxiang divided comparative sentences into 10 types by comparative degree.

As one type of special sentence pattern, comparative sentence is an important teaching point in the field of TCSL. Liu (2001) didn't define comparative sentence, but he divided comparing way into two types. The first one is "compare the similarity and difference of different things", it includes structures "A Gen B YiYang" and "A You B NaMe/ZheMe…"; while the second one is "compare the quality or degree", it includes "BI" sentence, "BUBI" sentence, "MeiYou" sentence and "BuRu" sentence. Ding (2009) gives the definition of comparative sentence, "comparative sentences are sentences which contain comparative items and express comparative meanings". He studies "less comparative sentence" and "equal comparative sentence". Except for patterns that Liu studied above, he also studies "Yu" sentence and "comparative sentence without marks". His classification is detailed and is helpful to teaching practice. Liu (2001) refers to different grammar teaching programmers and investigates frequency utilization of Chinese comparative sentence by Chinese and abroad students in China. Finally, they list 20 patterns about comparative sentence.

These researches are only from the perspective of linguistics, which are still far from the automatic identification. Despite their shortcomings, these researches can be useful to our work, such as tagging Chinese comparative sentences, building comparative words library and establishing identification rules.

It has also attracted a significant amount of researchers that studied the problem from the viewpoint of language information processing.

We have found that there are some researches on identifying and mining comparative sentences both abroad and at home. Nitin (2006a) research how to identify comparative sentences in English texts, they use algorithm of SVM and CSR to attain 84% precision and 83% recall. Nitin (2006b) propels Huang (2008) using LSR to extract comparative elements. In China, Huang et al. (2008) firstly paid attention to automatic identification of comparative sentences, they prove that SVM classifier based on sequential patterns can significantly outperform the traditional term-based classifier. Huang (2010) also took SVM as classifier but they then used CRF algorithm to identify entity and took the entity information as one feature.

Although some work has done to process comparative sentence, no research focuses on teaching Chinese comparative sentences as second language. The previous research basically adopted news corpus, they didn't consider the difference of different kinds of comparative sentences in the textbooks of Teaching Chinese as a Second Language. And the present parser about comparative sentences processing isn't applied to teaching practice. Different from the research above, our work focuses on the textbooks of Teaching Chinese as a Second Language, we will use HNC theory to study the identification and analysis about comparative sentences.

3 FEATURES

3.1 *Comparative mark (M)*

Comparative mark is the most important component of a comparative sentence. From our investigation, only three comparative sentences are no-mark sentences. In this paper, we didn't research no-mark comparative sentences. We summarize five types of comparative mark based on HNC theory, which is the core of comparative sentence study.

3.2 *Comparative result words (R)*

Comparative result is also essential to comparative sentence, and comparative result has abundant

Table 1. Comparative marks.

HNC symbol	The marks
L0	BI, BUBI, YOU, MEIYOU, YOUMEIYOU
EU	GENG, HAI, ZUI
HV	YU, GUO (过)
L1	XIANG, HE, GEN, TONG
EG	BURU

syntactic structure. There are 6 types of comparative results: A; A + complement; VP; (VP+) V+DE+NP+A; Verbs meaning increase or decrease; XIAN/HOU/ZAO/WAN...... +V.

3.3 Classes of comparative sentences

Through comprehensive consideration, we principally classified comparative sentences into 6 types by comparative marks. Meanwhile, comparative result was also taken into consideration. Because of particularity of negative comparative sentences, the comparative result types of "BUBI" sentence and "MEIYOU" sentence are different from "BI" sentence and "YOU" sentence, but the strategy of them is the same. As a result, our comparative sentences system contains 6 categories and 33 subcategories.

4 METHOD

Our approach is a combination of corpus for comparative sentences, comparative words library, and identification rules. We use Keyword-driven strategy and semantic rules to identify comparative sentences and eliminate non-comparative sentences.

4.1 The procedure

The procedure of comparative sentences identification is shown in the following figure. We firstly do some preprocessing to the corpus, including word segment, part-of-speech tagging, words library building. Then, we set up several models for different comparative sentences with different features. Next, semantic rules base is built. If a sentence could match any one of those rules, this sentence is a comparative sentence, otherwise, it's a non-comparative sentence. Meanwhile, the sentence parsing and syntactic analysis is done in this procedure.

4.2 Preprocessing

4.2.1 Comparative words library
We build comparative words libraries, in which we choose different categories of words, including adjectives, verbs, adverbs, prepositions. What's different is that our symbols stem from HNC theory, so do our semantic rules. Apart from these two libraries to identify comparative sentences, there is one HNC words library, in which we select words to analyze every sentence, so this library includes all kinds of part-of –speech of words. Let us see how the words are labeled in the comparative words library. Table 2 shows some examples of comparative mark words and result words.

Table 2. Comparative words library.

Chinese words	Attributes
YOU	\$ CC[v] GCC[V] LV[V] GBK_NUM[3] GXGY[GY] EPER[Y] \$ CC[hv] GCC[L] LV[HV] \$ \$ CC[l02] GCC[L] LV[L0] \$
XIANGTONG	\$ CC[u] CC[ug] GCC[U] LV[U] SC[jD000] FRAME_VALUE[0] \$
GAO	\$ CC[u] GCC[U] LV[U] \$

4.2.2 Disambiguation strategy
Many comparative words have more than one conceptual category, but in one specific sentence they only have one unique meaning. For instance, the word "BI" could be a verb in some sentences, it also is a preposition in other sentences. So we set up many exclusion rules to select the right attribute of key words.

4.3 Model

We build 5 models to identify different types of comparative sentences. These models distribute on the whole phases of our semantic parser.

4.3.1 L0 model
This is the most typical model of comparative sentences identification, and this model mainly process two types of comparative sentences, that are "BI" sentence and "You/ MeiYou/ YouMeiYou +......+A" structure. We used logical conception L0 to label "BI/BUBI/YOU/MEIYOU/YOUMEIYOU", which means these conception could introduce main semantic chunk.

Now we will give an example to show our processing procedure.

Chinese sentence: Is Mid Autumn Festival as lively as Spring Festival?

Phase 1: exclusionary rules
When "YOU" exists in comparative sentence, it is a L0 conception, as a result, we must eliminate its' other attributes. Here is the exclusionary rule of this sentence:

(0){CHN[YOU,MEIYOU,YOUMEIYOU]& LC_CC[v]}+(f){(1)CHN[ZHEME,NEME]+(2) LC_CC[u]} => !LC_SELECT(0,LC_CC,v)&!LC_SELECT(0,LC_CC,jlu)&!LC_SELECT(0,LC_CC,hv)\$

In this sentence, comparative mark is the activating point, which is set as node 0 in our system. CHN means Chinese string. If there is an adjective at the location closest to node 0 when searching forward, and a Chinese string "ZHEME" or "NAME" adjoining to this adjective before. attributes except L0 of "YOU" will be selected.

Phase 2: generating the core predicate
There is only one core predicate in a simple sentence from HNC theory. So to search the core predicate is the central task in our system. In this sentence, the predicate is an adjective.

(b){(-2)CHN[YOU,MEIYOU,YOUMEIYOU]}+
(-1)CHN[ZHEME,NAME]+(0){LC_
CC[u]&END%}=>LC_TREE(E,0,0)&PUT(fp,
LC_E_SCORE,E_U)$

An adjective is at the end of a sentence, we could search Chinese string "ZHEME" or "NAME" next to it ahead, and there is "YOU/MEIYOU/YOU-MEIYOU" before them. Then the adjective will be generated into an E tree, and the E was given the highest weight E_U.

Phase 3: generating L0
In this type of comparative sentence, the typical L0 is significant to analyze semantic trees.

(0){CHN[YOU,MEIYOU,YOUMEIYOU]&LC_
CC[102,10]}+(f){(1)END%&LC_CHK[E]&
LC_E_SCORE[E_U]}=>LC_TREE(L0,0,0)&
PUT(fp,LEVEL,1)$

In this sentence, "YOU" will be generated into a L0 tree whose level is 1, which means that it is the L0 of a sentence.

Pahse 4: matching L0 and core predicate
After generating L0 and E, there is one essential step to finally identify the comparative. That is weighted adjective predicative as E_FORMAT by L0. Only in this way, this adjective could truly be core predicate.

(b){(-1)CHN[YOU,MEIYOU,YOUMEIYOU]&
LC_CHK[L0]&!LEVEL[2]}+(0){LC_CHK
[E]&LC_E_SCORE[E_U]&END%}=>PUT
(-1,LEVEL,1)+PUT(0,LC_E_SCORE,E_FOR-
MAT)$

Ultimately, through those four steps above, our system could identify comparative sentences with L0 mark.

4.3.2 *ABK model*
Comparative sentence with "Xiang/He/Gen/Tong+......+YiYang/ChaBuDuo (+A)" structure is a little complicate. When a sentence has an adjective at the end, the adjective is always the predicate. If a sentence is without an adjective at the end, the comparative result words are the predicate. The second situation, we will use the L0 model. As to the first one, we will use other model-ABK model to process.
This model is similar to L0 model, the difference is that in ABK model the logical concept l1 is not introducing the main semantic chunk but the sun semantic chunk. It's easy to identify this kind of comparative sentence, because there is no need to match L1 and E. The task is just to identify EG and ABK chunk.
There are also some steps to identify the whole sentence. Many of these phases are same to L0 model, such as the exclusion rules and generating the core predicate. The key phase of this model is generating ABK. The following rule is an example to generate ABK.

(b){(1)CHN[YU,HE,TONG,GEN]}+(0)LC_CH
K[L1H]&CHN[XIANGBI,XIANGBIJIAO,YIY
ANG]=>LC_TREE(L1,-1,-1)+LC_TREE(ABK,
-1,0)$

4.3.3 *EG model*
L0 model and ABK model mainly process comparative sentence with preposition comparative mark. There are some sentences using verb to express comparative relation, such as "BURU/BUJI" and so on. This model is easier than the two models above. Firstly, these words have only one conception category. Secondly, the EG phase is the key step to identify comparative sentence, which is also the necessary step for every sentence.

4.3.4 *HV and EU model*
The structure "A+YU/GUO" and "HAI/GENG/ZUI+A" are special types of comparative sentence. Because there is no need to set special phases to process these kind of comparative sentences, the procedure is part of EG phase. "GUO/YU" and "HAI/GENG/ZUI" could be seen as the affix of the front adjective. "GUO/YU" is HV of the core predicate, which stands attribute component after verbs, while "HAI/GENG/ZUI" are adverbs before verbs, and we use EU to label them.

5 EXPERIMENTS

5.1 *Data sets*
Our test data come from the textbooks of Teaching Chinese as a Second Language. The data sets were all manually labeled. We have already defined the labeling specification of comparative sentences.
We have 12403 entries in our words library. The keywords are from the sentence we labeled, and they are also compiled into words library based on our labeling specification as shown before.

5.2 *Experiment and result*
We now give the precision and recall of identification of every type of comparative sentences.
From this table, we can see that the precisions for all the types of comparative sentences are more than 90%, and the recall ratios are more than

Table 3. The result of identification.

	Our system	
	Precision (%)	Recall (%)
BI/BU BI+NP+A	93.33%	84.48%
YOU/MEIYOU/YOU MEIYOU+......+A	92.98%	88.33%
ZUI/HAI/GENG+A	97.63%	86.31%
A+YU	94.38%	89.62%
Xiang/He/Gen/Tong+ ......+YiYang/ ChaBuDuo(+A)	98.11%	88.14%
Bu Ru type	96.72%	80.82%

80%. We analyzed the reasons: some comparative chunks are complex than comparative sentences; there are various types of complement in simple comparative sentences, which still need to be complemented; some labeling mistakes and segmentation errors are inevitable.

6 CONCLUSION AND FUTURE WORK

This paper proposed strategy based on HNC rules of identifying comparative sentences in the field of TCSL. We give a clear categories based on syntactic feature and the different models to identify every type of them. Our semantic system has high precision and better stability. Our system is not only helpful to TCSL, but also available to future research about comparative sentences. In our future work, we will expand our data sets and improve the precision of our system. On this basis, we will do comparative relations extraction.

ACKNOWLEDGMENT

This paper is supported by the National High Technology Research and Development Program of China (No. 2012 AA011104).

REFERENCES

Chen, J. & Zhou, X. 2005. The Ranking and Selection of the comparative syntactic item. Language Teaching and Reseach, 02, 22–33.
Ding, C. 2009. A Course for Mandarin Chinese Grammar. Peking University Press, pp. 81–95.
Huang, X. & Wan, X. & Yang J. & Xiao, J. 2008. Learning to Identify Chinese Comparative Sentences. Journal of Chinese Information Processing, (05).
Huang G. & Yao T. & Liu Q. 2010. Mining Chinese Comparative Sentences and Relations Based on CRF Algorithm. Applications Research of Computers, (06).
Liu, Y. 2004. The Difference of Comparative Category and Adjacent Categories. Journal of Xuzhou Normal University: Philosophy and Social Sciences Edition, 29(4), 57–60.
Liu, Y. & Pan, W. & Gu, W. 2001. Practical Modern Chinese Grammar. The Commercial Press, pp. 833–851.
Nitin, J. & Bing, L. 2006. Identifying comparative sentences in text documents. Proceedings of SIGIR, 244–251.
Nitin, J. & Bing, L. 2006. Mining Comparative Sentences and Relations. Proceeding of AAAI.
Shang, P. 2006. The Summarize of comparative sentences research. Journal of Applied Linguistics, 02, 58–6.

Construction and application of enterprise private cloud

Xia Yang & Kai Wang
Tianjin Bohai Vocational Technology College, Tianjin, China

ABSTRACT: For enterprises, strong ability of data service is essential. In pursuit of green and efficient today, cloud computing has become the inevitable choice for enterprise. I hope this can help companies choose to build private cloud solutions to reduce the difficulty of the enterprise users deploy cloud computing, so as to promote cloud computing into the enterprise.

1 THE CONCEPT OF CLOUD COMPUTING

Cloud computing is a scalable platform that can be deployed and configured dynamically as needed. This platform can contain huge computing resources and network storage capacity is determined by the physical presence of the massive server hardware, storage, network, network facilities and safety equipment consisting of cloud computing platforms.

Development of cloud computing is built on a virtualized resource management model, in support of virtualization technology, the huge hardware resources are integrated into a huge resource pool. Users can easily and quickly get computing and storage resources they need from the resource pool.

2 CLOUD COMPUTING IS AN INEVITABLE CHOICE FOR THE COMPANY'S FUTURE DEVELOPMENT

Enterprise managers have long recognized the importance of corporate computer networks, a high degree of information for enterprises, each of the daily activities of enterprises rely on network systems to complete. Therefore, throughout the enterprise network system has become an important system of enterprises.

However, with the traditional model of network development to the present, the bottleneck on its architecture has been revealed. Rely on expanding the scale does not bring efficiency, network resources cannot be effectively allocated, large data center is only a very low actual performance, ninety percent of the server can only play about 10% of the work efficiency. Server itself and the power consumption of air conditioning systems tend to become huge companies have to bear a huge burden. Nevertheless, many enterprises in order to meet the ever-increasing number of new systems and getting huge database, every year the need to constantly invest a lot of budget funds used to purchase more servers, storage devices and related systems. Brings higher efficiency and lower energy consumption after increased scale.

Diversity of business development, and changes in corporate management structure, and further exacerbated the contradictions of traditional network architecture and is now fast and flexible allocation of resources needed. Enterprise restructuring in accordance with market changes, and often have limited traditional network structure, the complexity of the configuration process resources and slow, has been unable to meet the modern enterprise in a rapidly changing market environments.

Cloud computing is an innovative IT use model, which provides businesses with a new possibility. Through enterprise cloud computing platform can be a lot of server hardware abstraction become a huge resource pool, and then based on different business systems, different business units for rapid and effective distribution. Those are usually the waste of resources that can be fully utilized; computing and storage resources can be allocated based on the actual needs of the mission. Data center cloud computing architecture of its work efficiency can reach more than 85%, which compared with the traditional architecture of the data center is a huge increase in efficiency. The era of cloud computing companies can lower cost, access to far more than the previous computing resources and storage capacity, and its flexible resource allocation, business migration and integration capabilities of modern enterprise desired.

Companies to enter the cloud computing era of the two selections are public and private cloud. Public cloud is undoubtedly the most ideal choice for business. In the public cloud, enterprises can completely abandon the huge complex infrastructure and associated upgrading and management of maintenance work. Instead it is to buy and get the appropriate computing resources and application services according to their needs from the public cloud. Currently the major cloud service providers have Amazon, Google, Windows, IBM, etc., they can already provide affordable cloud computing services for users.

However, along with the rapid development of cloud computing platform there are some safety problems, especially in recent years have occurred since the accident on a public cloud platform, but also caused a lot of loss of individual users and business users' important data. Most business managers are still willing to retain their core business in their own internal private network systems, taking into account the factors of reliability, safety, controllability on corporate use of public cloud or a great concern.

Taking into account the safety and security of the network infrastructure upgrade business systems, many companies will locate their target in a more realistic option, construct their own private cloud.

Unlike public clouds, private clouds is based on the company's own facilities, and is constructed of a single enterprise customers use, so it can provide the most effective control of the data, security and quality of service. Enterprises have the infrastructure, and you can control the deployment of applications on this infrastructure. More importantly, many companies have established relatively complete hardware facilities, as long as the necessary upgrading and transformation, these hardware resources can be fully utilized in the construction of a private cloud. Building enterprise private cloud is to maximize retain their investment in IT systems. In addition, under the cloud computing environment to improve the utilization of the servers will greatly improve the work efficiency of the data center, more flexible application deployment also brings improve management effectiveness.

3 CLOUD COMPUTING INFRASTRUCTURE

Enterprise private cloud data centers to facilitate the enterprise internal control and coordination, to avoid privacy issues and public cloud data center availability issues exist, but there are many other problems. Enterprises should be the security and stability of the data center, flexibility, scalability, comprehensive consideration, in order to ensure safe and stable data storage. Internal cloud computing center will be in the near future an important part of the enterprise IT infrastructure, before there are some basic questions need to be addressed, including security issues, standards and cost three aspects.

3.1 Security issues

Cloud Computing Center primary issue is self-managed enterprise security and reliability requirements, environmental management, safety equipment, energy supply, disaster management and other data center to meet the basic needs of the future long-run, especially in its core data services after the completion of the data migration, security and stability are the most fundamental requirements of business-to-storage devices for enterprise data loss is unacceptable. In the case of mixed old and new systems will become more complex, IT departments need to introduce professionals to the corporate data center monitoring and management, in particular, to establish a strict management procedures, as well as respond to respond to complex emergencies when the treatment plan, to avoid human error operation caused greater losses.

3.2 Construction standards

Determine the overall standard is the basis for the integration and use of information resources in the enterprise cloud computing center construction process is even more complicated. Enterprise system to their own original full investigation, including future data system clear and specific data standards, building standards clearly network infrastructure, data center equipment configuration standards, data center reliability and security standards, business system resource allocation criteria, and a series of related to the old and new systems integration, specific standard line, operational management issues on the new system.

3.3 Costs

The small and medium-sized enterprise because of cost issues more choice public cloud services. Public cloud services to facilitate the daily management of data resources for the small and medium-sized enterprise, reducing the energy consumption of its equipment expenditures, but also generate a data storage security, cost of user privacy, safety supervision and other security issues, as well as storage and migration. For large enterprises, will all operations are transferred to the cloud, you may pay a higher cost. Establishment costs within the enterprise private cloud data centers, but also worthy of attention and consideration. According to Microsoft's White Paper "cloud computing economics" findings, fewer than 100 servers for small and medium companies use far more than the cost of a private cloud public cloud, and for server installations reached about 1,000 units of large enterprises, the private cloud is feasible. James Staten, principal analyst with market research firm

Forrester Research, said that only 5% of the world's large enterprises to run an internal cloud computing, in which only half of the companies may actually have an internal cloud computing. Seen from the above data, the establishment of the enterprise data center is costly and conditions are harsh.

4 VIRTUALIZATION CONSOLIDATION OF DATA SERVICES

In the enterprise cloud computing construction process, companies need to combine their characteristics more scientific long-term planning. Pursue short of cloud computing companies can not solve the problem for a long time will have to face, enterprises in upgrading their IT systems in the process, totally have more choices to meet the needs of enterprises.

Private clouds high construction and maintenance costs will make many enterprises cannot afford, therefore choose some low-cost virtualization solution also can effectively improve the work efficiency of the company's existing network infrastructure. Physical server virtualization, storage virtualization coupled to meet the actual needs of many companies, and at a lower cost, and will run on business systems on separate physical server migration to a virtual machine, you can greatly strengthen the efficiency of a single server. In future, the conditions will ripe, virtualization future business systems can more easily migrate to a public or private cloud computing platform.

5 SUMMARY

IT companies to abandon traditional development mode, select efficient intensive cloud computing platform is the inevitable technological progress. Stable and progressive development model is accepted by most businesses. In cloud computing integration solutions, companies can easily build their own cloud infrastructure. Private cloud will become the most powerful assistant, enterprises a more flexible and convenient means of resource allocation, greatly expanding the business capabilities of enterprises.

REFERENCES

[1] Deng H., Qiu Q., Wang T., et al. Design and implementation of service bus prototype for RFID Logistics-Customs Clearance Service Platform [J]. 2nd International Conference on Information Science and Engineering (ICISE2010), 2010: 1971–1975.
[2] Armbrust M., Fox A., Griffith R., et al. Above the clouds: A Berkeley view of cloud computing [R]. U.C. Berkeley: EECS Department, Feb 2009.
[3] Wang L., Laszewski G.V. Cloud computing: a perspective study [J]. Journal of New Generation Computing, 2010, 28 (2): 137–146.
[4] Mell P., Grance T. The NIST Definition of Cloud Computing [R]. National Institute of Standards and Technology, Information Technology Laboratory, 2009.

Sports Engineering and Computer Science – Luo (Ed.)
© *2015 Taylor & Francis Group, London, ISBN 978-1-138-02650-6*

The analysis to guide and encourage college graduates to the grass roots-oriented employment

MeiYang Li

Xi'an University of Architecture and Technology, Shaanxi, China

ABSTRACT: The popularization of high education makes college graduates become ordinary workers. The grass-roots level is the main channel to recruit the graduates and is an effective way to ease the employment pressure. For the concepts, policies, systems or some other reasons, the grass-roots level employment boom has not been formed. This paper begins with the importance to guide the graduates to accept the grass-roots level employment, and analyses the existence of the bottlenecks of grass-roots level employment, and then proposes the establishment of the long-term systems to guide the graduates to employ in the grass-roots level.

Keywords: college graduates; grass-root employment; long-term system

1 INTRODUCTION

The report of the 17th National Congress of the Communist Party of China clearly points out: "Accelerating social development with the focus on improving people's livelihood give priority to education and turn China into a country rich in human resources." Thus, employment is not only vital to people's livelihood, but also the foundation of the country' peace and the basis of the country's progress. In the severe employment situation, employing in the grass-roots level becomes an effective way to ease the employment pressure. Only when the society, government, schools, and graduates cooperatively change our concepts, and implement the policies, guiding the graduates to contribute to the grass-roots level may it be achieved.

2 TO GUIDE AND ENCOURAGE COLLEGE GRADUATES TO THE GRASS-ROOTS-ORIENTED EMPLOYMENT IS AN INEVITABLE TREND OF DEVELOPMENT OF THE TIMES

With the popularization of high education, college graduates has increasingly become common laborers and become an important part of new labor force, the grass-roots level must become the main channel to absorb the employment of graduates. General Office of the CPC Central Committee and State Council on "To guide and encourage the employment of college graduates to the grass-roots level" views that: Actively guide and encourage college graduates to the grass-roots-oriented employment is conducive to the healthy growth of young talents and to improve the structure of the grass-roots talent is conducive to promote the balanced development of urban and rural economic, as well as conducive to build a socialist harmonious society and consolidate the party's ruling status. Therefore, to guide and encourage college graduates to the grass-roots-oriented employment is a great project to benefit the country.

To guide and encourage college graduates to the grass roots-oriented employment is directly related to the major problems of the new historical conditions on "Cultivating what kind of people, How to cultivate people". Cultivating what kind of people is the most important issue in the educational field. The party's educational policy clearly states that education is to commit the socialist modernization and develop the socialist builders and successors. Former President Hu Jintao once pointed out: Masses of young people should consciously link their personal destiny with the destiny of the motherland and the nation,should closely link to their personal ideals with the attaining the goal of building a moderately prosperous society in all respects, should consciously serve the motherland and the nation, should selflessly contribute to society and work hard, and also should continuously make progress, in the positive social practice to create the youth with eternity and without regrets.

The grass-roots level is indeed the way for the development of the youth, a good place to understand the nation's conditions, sharpen youths' will and character, enhance the feelings with the working people, proves a good stage to get through the toughs, devote intelligent, and achieve successful career. Therefore, guiding the graduates to face the grass-roots employment connects with the fundamental issues on whether the builders and successors are qualified.

To guide and encourage college graduates to the grass roots-oriented employment is a fundamental way to solve structural contradictions in the employment of graduates. Talent resource is an important national resource. In the need of the socialist modernization, the number of the college students is not big; they are still in great need. As the employment of college graduates over-centralized and the southeast coast and economically developed areas, many graduates face dual pressures of employment. So the structural contradictions are very prominent. In order to solve this contradiction, we need to bring forth new ideas, change concepts, and actively guide the graduates to start their own business, and do knowledge-based entrepreneurship to the grass-roots level. The grass-roots, extremely rich in content, is an extensive concepts, including the vast rural areas, and the city's communities. It covers not only the county-level party government organs, enterprises, institutions and social groups, organizations, but also a variety of economic organizations, even includes non-public organizations, small and medium enterprises, contains both the self-owned businesses, self-employment, and a strenuous industry and a hard job. Thus, the grass-roots provide broader space and many chances for progress. It has a long-term strategic significance to alleviate the current and future periods of stress for employment.

To guide and encourage college graduates to the grass roots-oriented employment is the objective need to build a new socialist countryside and achieve the balance in China's economic and social development. At present, we are in the new period to further implement the scientific concept of development, promote technology and educational strategy and speed up for building a moderately prosperous society. The college graduates are valuable national talent resource, and their employment is a major issue involved in the overall society, not only related to the vital interests of the masses, but also directly affects the coordinated development of economy and society. The construction of new socialist countryside not only needs the leaders of management with the good master of science and technology, but also bears the task to provide sustainable development of the reserving able teachers for building new socialist countryside. Guiding and encouraging college graduates to the grass roots-oriented employment provides talent reserve and intellectual support to the new socialist countryside build. It has important practical significance for the coordinated development of urban and rural, and the socialist harmonious society.

3 THE DIFFICULTIES THAT GRADUATES FACE TO THE GRASS-ROOTS EMPLOYMENT

Along with university's unceasing increased enrollment, population of the graduates grows largely. The graduates' employment has become the focal point that the society pays attention to. To encourage and guide university graduates to face the grass-roots employment, and to establish the persistent effect mechanism, we should adapt with the socialist market economic system for the graduates to face the grass-roots employment, which is the need to speed up the new socialism countryside reconstruction, and thoroughly apply the scientific outlook on development as well as the harmonious socialism society construction. All levels of the government and the universities have made many efforts and attempts. I believed that there are three big bottlenecks that affect the university graduates to grass-roots employment.

The graduates do not understand why they go to the grass-roots. As a result of the traditional high education's influence, going to college is considered the way to change their fates. So the only goal of going to college is becoming the "city slicker" With the university enrollment population unceasingly increased, graduating population increases rapidly, the employment pressure increases, the university graduates fall from the outstanding ones to the common labor, at this time college educational popularity has the serious contradiction with the college students' outstanding person's concept. According to the investigation, the graduates in choice of the job, above 80% graduates choose the National work Unit, state-owned Large and middle scale Enterprises, foreign enterprises or high new enterprise. In the unit region's choice, some 85% graduates choose the southeast coast, the economical developed area, the provincial capital city and so on. Thus as it can be seen, the majority of graduates tend to think highly of high stability, high income and good environment. They turn a blind eye to the demand of the grass-roots. Facing the huge employment pressure, many graduates would rather wait for jobs, than accept the grass-roots employment, they do not realize: It is very important to adapt social need, the choice of enterprise development targets, takes the correct growth path. At present, the social most urgent need is the basic unit talented person. Take Shanxi as an example, in 2007 our province only in countryside compulsory education stage, teacher' post achieves 1964, and will assign 600–800 people every year to the villages and towns. The graduates' enthusiasm is not high.

The graduates do not know what kind of graduates that in the grass-roots are most needed. The market regulation is an efficient path to achieve the talent resources' reasonable disposition. Along with the socialist market economic system's further deepening, this is the trend for more graduates to choose to face the grass-roots employment, but the reform of university's educational system is relatively slow. The school curriculum takes the discipline as a center, lacks of the improvement of students' adaptiveness and the application ability. The curriculum comes apart from actual need, theory content lag behind realistic situation. Moreover, in their improvement process, they attach more importance to the knowledge and theory than the practice and skill, in addition, parts of the graduates can't bear hardships and stands hard work, which causes the graduates to adapt slowly in grass-roots work, the seat of honor was slower, thus affects the grass-roots' enthusiasm to receive graduates.

The grass-roots level doesn't know how to keep graduates. To guide and encourage the graduate to the grass-roots employment, we need the policy and the laws and regulations' support and it is a systematic project. The Party and the government takes this work seriously, but because of the restrictive provincial economy development, many less-developed areas need large quantities of outstanding talents, on the other hand, they actually have no establishment and the fund to attract talented person; And also we have many examples such as the social distribution systems, the social security systems, the lateral economic ties, and the local financial circumstance, the talented person who develops mechanisms, system problems and relative backwardness in environment.

REFERENCES

[1] Hu Jintao, Hold High the Great Banner of Socialism Chinese Characteristics And Strive to Complete the Building of A Moderately Prosperous Society in All Respects—Report to the Seventeenth National Congress of the Communist Party of China.
[2] Suggestion on Guiding And Encouraging the Employment of College Graduates towards Basic Level—Proclaimed by General Office of the CPC Central Committee.
[3] Zhou Ji: Implement the Central Spirit And Complete A Solid Employment Work—Speech on Work Employment Conference of of College Graduates from All Country in 2006.

Sports Engineering and Computer Science – Luo (Ed.)
© *2015 Taylor & Francis Group, London, ISBN 978-1-138-02650-6*

The analysis of real estate industry credit system and evaluation system

MeiYang Li

Xi'an University of Architecture and Technology, Shaanxi, China

ABSTRACT: In the process of building a market economy, credit system plays a vital role, and it is fundamental to a market economy. For enterprises, the analytical evaluation of credit has taken an important part of building a credit society. As for real estate industry—one of the "special forces" in our socialist market economy, a credit system is indispensable to its sustainable development. However, building a real estate market credit system has lagged far behind. Therefore, it is necessary to establish a comparatively perfect real estate market credit system, in order to promote the healthy development of the industry. Studying and establishing a feasible real estate enterprise's credit system and evaluation method is not only a basic way to promote healthy sustainable development of the industry but also an important part in building credit system in the whole real estate industry. In this paper, the status of the real estate industry development, real estate industry credit system and evaluation system are analyzed and discussed.

Keywords: real estate enterprise; credit; evaluation system; existing problem; implication; preliminary conception

1 THE CONNOTATION AND DENOTATION OF CREDIT AND CREDIT SYSTEM IN REAL ESTATE ENTERPRISE

Management definition of credit is: credit is a kind of strategic assets. Partners bound with trust can improve the odds of successful cooperation. In fact, someone believes that if both partners can trust each other, the time needed for successful cooperation is the shortest. In enterprise strategic view, credit is often described as: valuable, rare, difficult to imitate, and cannot be replaced. Therefore, an enterprise to be considered trustworthy will have a competitive advantage when forming and adopting cooperation with others.

Credit is a sense of responsibility to keep promises, it is also the moral behavior that a person who is responsible for the consequences of their own actions. In the narrow sense, credit refers to the credit behavior generated in bank lending behavior, later it's extended to every social domain. The nature of credit is unilateral movement of value based on the premise of repay; and the essence of the credit system is "commitment". According to famous economists Jinglian Wu, credit is a kind of ability that fiduciary does not need to pay cash to obtain goods, services or money based on credit grantor's trust.

An impeccable industrial credit system mainly includes three components: the national credit management system, enterprise credit risk management and prevention system and credit risk reveal and evaluation system. From current point of view, the industrial credit system which can meet the demand of the market economy has not yet been established in our country. To establish the real estate industrial credit system is a root-curing measure to improve and standardize the real estate market, and it is also an important aspect to reform and develop the real estate industry.

Real estate industry credit mainly involves two aspects. On one hand, it is the consideration of risks that real estate enterprises as borrowers in the credit market, their ability and willingness to repay the debt and interest. To investigate credit in this particular area, we should focus on enterprises' future profitability and cash flow, and analyze deeply to enterprises' solvency and insolvency. Here, it is worth noting that, first of all, the real estate enterprises has its own features compared with other industrial and commercial enterprises, real estate industry is vulnerable to government regulations and policy control, such as city planning, land sales, financial investment, and tax policy which can directly impact the real estate industry. Secondly, the real estate industry also has a significant relationship with regional economic development and economic cycle. Thus it shows large fluctuation in their asset quality, financial status and cash flow. Chinese real estate enterprises (especially residential development enterprises) are more willing to focus on new-property sales, while old—properties rarely enter the market. It is also very rare that real estate companies do rental business, because their income mainly depends on the sale. This mode of business causes the unstable financial status of the real estate business, especially in cash flow aspect. This part of credit often in corporate with finance loans, in this paper we call it "capital credit".

On the other hand, when we talk about the credit problem in the real estate industry, we actually refer to how the real estate building enterprises and related companies fulfill the agreement on their products (such as commercial housing sales), that can also be seen as whether or not contracting parties can honestly fulfil their obligations. This part of credit problem mainly focuses on the seller's side, including the authenticity of promises in the advertising, the price (price increases), shrinking area, changing of design and functionality, building qualification, property certificates, the quality of the house, the quality and cost of property management, and many other aspects. In this paper, we call this "product credit".

Since the real estate enterprise credit system has just started, it's immaturity and imperfection leaves much room to be improved and strengthened. And currently its popularization is very narrow; many provinces are still in the bud. So, the Ministry of construction has issued several documents on establishing the real estate credit system, strengthening the government policy support, and providing guidance to real estate enterprises, so that the real estate market can be standardized and orderly. The basic characteristics of the real estate enterprise credit can be summarized as follows:

Integration: Real estate enterprise credit involves financial accounting, the market economy, construction engineering, marketing management, and other various factors, all of which need to be combined to carry out a comprehensive credit evaluation.

Evolution: With the development of the real estate industry and construction industry, real estate enterprise credit evaluation system cannot always be fixed. It must change along with social and natural environment, human needs and social economic status, in order to adapt to the continuous development of the real estate market.

Quantifiable: credit should be quantified as credit rating, credit rating marks the integrity status of credit stakeholders. Creditable and discreditable are opposite sides of credit rating. Credit evaluation must be quantified, only in this way we can provide reasonable rating and sorting to enterprise's credit status.

2 THE MAIN PROBLEMS EXISTING IN THE CREDIT EVALUATION IN OUR COUNTRY

Our country's credit rating was developed after the 1980s, many of the credit rating systems just serve for financial enterprises, and other enterprises' credit evaluation systems are still in a relatively blank stage. It is just the beginning to use it as a means of market regulation, therefore, there are inevitably many problems, among which the most outstanding problems are:

2.1 Lack of credit rating awareness

In current social environment, the value of social credibility has not yet been fully revealed, as a result, some of the organizations, enterprises, and individuals who need to conduct credit rating are still lack of awareness, lack of attention, or simply do not want to participate in the credit rating. Some of the organization, enterprises, and individuals, who need to know the credit rating of their counterparts, do not know how to search or use the credit rating information. Some of them who know their counterparts' credit rating don't even believe or have never take it as the main decision making factor. This situation accounts for the demand of the credit rating market.

2.2 The credit rating agency system is not perfect

Right now, there are several agencies doing credit rating, we have institutions mainly focus on credit rating, and credit rating department in banking system, but the work in each of those agency did not widely, deeply and systematically carried out, instead they all doing their own job, lack of contact and communication. Missing authoritative rating companies like Moody's and Standard & Poor's, the credit rating industry in China is not driven by the market, because the size of the market is too small, so the government plays an important role in pushing forward of developing credit rating. In that case, the rating has some political color, so the credibility of the credit rating has not been widely accepted.

2.3 Lack of scientific rating index and methods

At present there is no unified service standard and rating system in China. The rating standards are also not unified or comprehensive, and there is no scientific rating method, so that the rating results are lack of objectivity, impartiality and accuracy. Even more, in some circumstances, the rating results are not assessed based on the actual situation, but on how much money has been paid.

Publishing false information in listing companies are collective illegal activities, which involve the issuer, accounting and auditing organizations. Thus it is impossible to have a good credit rating system if there are no standardized credit rating agencies.

3 THE PRELIMINARY CONCEPTION OF ESTABLISHING A REAL ESTATE CREDIT SYSTEM

Credit is a kind of ability that fiduciary does not need to pay cash to obtain goods, services or money based on credit grantor's trust.

According to enterprise strategic management theory, the external environment has great effect on the growth and profitability of the enterprise, the enterprise external environment can be divided into three main layers: the overall environment, industry environment and competitive environment. The overall environment includes all factors in the broader social environment which can affect one industry or enterprises in such industry. As of real estate enterprises, credibility has permeated the whole external environment. It is very important and cannot be ignored. An impeccable industrial credit system mainly includes three components: the national credit management system, enterprise credit risk management and prevention system, and credit risk reveal and evaluation system. By looking at the situation in our country, there is no credit system that can truly meet the requirements of the market economy. To establish the real estate industrial credit system is a root-curing measure to improve and standardize the real estate market, also it is an important aspect to reform and develop the real estate industry. There are several aspects to be considered when establishing the real estate industry credit rating system:

3.1 Create credit record and issue credit status

According to the book—Management information systems for the information age: "Today we have entered an era of information, an era that knowledge and information becomes productivity." There is no doubt that information has become more and more important. Information has turned into a product. We can determine people's needs of information from the three dimensions of information: time, space and form. Also according to the strategic management theory, due to the asymmetry of information, it has caused serious losses to consumers. Especially for the booming real estate industry, information asymmetry phenomenon is more serious, information exchange, transfer, expression can cause various problems, and it has

brought a lot of problems to the healthy development of our province and even to the whole real estate industry. So it is necessary to establish a complete credit evaluation system to release and communicate information. In this way, consumers, enterprises, and government can know the industry and enterprises in this industry.

A complete credit record should include all kind of credit information related to the referring subject in a certain period of time, such as: the execution of justice status, administrative control status, status of on-going contracts, fair competition status, intellectual property rights, business reputation, whether or not the referring subject is people-oriented, and respect social ethics and many other aspects. The publicity of credit rating can be divided into: outstanding credit rating, high credit rating, good credit rating, fair credit rating and general credit rating. To a certain extent, this can reflect the credibility of market behavior, the default risk degree, market operation activates affected by adverse economic environment, and on-going business's risk level, etc.

3.2 Establish credit system of rewards and punishments

According to the theory of management, enterprises aiming at long-term and balanced development must have an perfect incentive mechanism. The system of rewards and punishment is one of the most effective. Especially in current situation that real estate market is lack of credibility, it is necessary to adopt reasonable incentive mechanism of rewards and punishment.

As of bad influence, produce and sale counterfeit goods, key market, key area and key cases, fully mobilize the enthusiasm of law enforcement, judicature, news media and other aspects of the strength, increase punishment, and focus on the difficulties to breakthrough. Then gradually establish a self-disciplined, public supervised, widely participated credit system, which should not only have the feasibility and operability, but also has good social effect.

3.3 The establishment of credit evaluation system

Management theory said: A perfect management mechanism, not only has the incentive mechanism, but also should have a proper evaluation system in order to provide objective results, promote the enthusiasm of employees, so as to promote the sustainable and healthy development of the enterprise. Speaking of real estate credit, it is more important to establish a proper evaluation system.

The main body of evaluation should involve three parties together: government, sales representatives and consumer representatives. Set up a special committee to regularly assess the integrity of enterprises, publish the assess result to the public, deal with those cases, and finally remove the dirt from the market.

REFERENCES

[1] Zhang Li, Wang Wei: The Study on the Target System of Credit Rating in real estate development enterprise, [J]Chinese Real Estate 2013(8).

[2] Sheng Juan: The Study on Chinese Real Estate's Credit Loss And the Construction of Credit System, Nanjing Jilin University.

[3] Jia Renpu: The Construction And Realization of Credit Rating System of Real Estate, Journal of Yangzhou University 2009(2).

Sports Engineering and Computer Science – Luo (Ed.)
© 2015 Taylor & Francis Group, London, ISBN 978-1-138-02650-6

A brief analysis of the problem of undergraduate entrepreneurship education and its countermeasures

MeiYang Li

Xi'an University of Architecture and Technology, Shaanxi, China

ABSTRACT: With the expansion of college enrollment, the number of graduates has increased year by year, the employment of graduates has gain more attentions of the whole society. To strengthen the education of college students' entrepreneurship and promote employment has become the basic way to alleviate employment pressure. Under the current system of higher education, there still exist problems like lack of thinking and understanding, incomplete course system, incomplete construction of teaching and academic staff, and weak cultural atmosphere. In order to promote in-depth development of college students' entrepreneurship education, this paper brought up two innovative aspects in college students' entrepreneurship education and realization, so as to cultivate college students' entrepreneurial spirit and entrepreneurial ability, and change passive employment to active employment.

Keywords: undergraduates; entrepreneurship education

1 INTRODUCTION

Being an organic part of quality education, college students' entrepreneurship education centering on developing their entrepreneurial comprehensive quality, exploiting their potential and molding talents with great potential, competitiveness, and good social adaptability, which is also aiming at creating more employment opportunities via encouraging entrepreneurship. The new education mode promoting the building of an innovation-oriented country, and realizing high degree of unification of individual and social values. [1] With the rapid development of higher education, and the continuous deepening of reform of college students' employment system, the difficulties of employment in college graduates has been the main factor influencing the stability of the Chinese society, and hindering the economic development. The 17th Nation Congress of the Communist Part of China (CPC) noted, "Employment is the fundamental to the people's livelihood. And we must insist on implementation of proactive employment policies, strength the guidance of the government, perfect the market employment mechanism, enlarge the employment scale, and improve the employment agencies. We also have to perfect the policies to encourage entrepreneurship and self-employment, and strengthen the education of employment ideas to help more workers become entrepreneurs". College students are carrying the historic responsibilities of building China into a powerful socialist country, while strengthening the entrepreneurship education of college students, and promoting employment via encouraging entrepreneurship are not only the main way to alleviate the difficulties of employment in college students, but also the way to practice scientific concept of development.

2 THE CURRENT SITUATION OF ENTREPRENEURSHIP EDUCATION OF COLLEGE STUDENTS IN CHINA

The time for developing entrepreneurship education in college students is short in China. By carrying out the activities such as college graduates' venture contests, venture forums, and so on, and in conjunction with media's attention and propaganda to the entrepreneurship of college graduates, especially publishing reports of cases with college undergraduates' entrepreneurial successes, have well aroused college students' entrepreneurial enthusiasm. To promote the rapid development of entrepreneurship education of college students in China, the Ministry of Education has taken a series of measures in 2002, the Ministry of Education determined 9 universities as the pilot ones for entrepreneurship education, such as Renmin University of China, Qinghua University, Beijing University of Aeronautics and Astronautics, Heilongjiang University, etc., which marks the beginning of the study and practice of college entrepreneurship education in China. From then on, these colleges, began holding various forms of activities, and made beneficial attempts in entrepreneurship education. Amongst those activities, the most influential ones are: Qinghua University half-price rented its office buildings to the graduates to found their companies; cooperating with Zhangjiang High-Tech Park in Pudong, Fudan University implemented the action plan on students' scientific innovation, and set up dedicate venture fund; Wuhan University carried out the "San Chuang" education, integrating innovation, creation and entrepreneurship. Meanwhile, the colleges have made relevant attempts in the establishing of organizations for entrepreneurship education. For example, Beijing University of Aeronautics and Astronautics set up a college for the training of entrepreneurship and management for non-degree education purpose, and Heilongjiang University launched a college for entrepreneurship education, and set up a learning system in relation with entrepreneurship, including elective courses, subsidiary majors, etc., and compiled relevant teaching materials, unfolding the entrepreneurship education in Chinese universities.

"Driven" by the government and "motivated" by the market, a great number of colleges quickly promoted the further development of the work. Henan University of Science and Technology established Chinese entrepreneurship research institute, started four selective courses including entrepreneurial management, etc., and set up an incubation fund for ventures to provide field and technological guidance for students' ventures. Zhejiang Sci-Tech University started intensive classes on entrepreneurship education. Anshan University of Science and Technology founded an association to promote the employment and entrepreneurship of college graduates. East China Normal University and Donghua University set up selective courses such as "entrepreneurship education course", "entrepreneurship and risky investment", etc. Although many colleges didn't conduct entrepreneurship education, they are gradually accepting the concept of entrepreneurship education in educational teaching practices, and these attempts lay a good foundation for college students' entrepreneurship education and promote the education to be conducted over a wider range.

3 BARRIERS RESTRICTING THE DEVELOPMENT OF COLLEGE STUDENTS' ENTREPRENEURSHIP EDUCATION IN CHINA

3.1 Lack of thinking and understanding

As one of the most major entrepreneurship education places, colleges should positively guide and motivate college students to develop good entrepreneurship consciousness, and create undertaking practice opportunities for college students. At present, colleges, college students, and the Chinese society are lacking of thinking and understanding of entrepreneurship education in different degrees. First, colleges are lacking of understanding of entrepreneurship education. Many colleges have a low importance impression towards entrepreneurship education. They haven't formed a systematic course system yet, and they haven't taken entrepreneurship education as a part of employment guidance, focus on guidance of entrepreneurship tactics to their students, and naturally ignored the development of the graduates' entrepreneurial consciousness, spirits, and abilities. Second, college students have biased understanding of the entrepreneurship education. In the process of accepting entrepreneurship education, college students normally not use their subjective initiative, and having weak active consciousness for entrepreneurship, thus making the entrepreneurship education have little value. On one hand, they just take entrepreneurship education as a second choice, and only when they are unable to find a decent, stable, and high-income job, they took entrepreneurship into their consideration in order to solve their food and clothing problems. On the other hand, many students believe they are lacking of entrepreneurship qualities and conditions, and they think entrepreneurship is just the matter of the few people, so they have little enthusiasm towards entrepreneurship. Third, the Chinese society is lacking of understanding to the entrepreneurship education. Most parents have a contra-attitude towards graduates' entrepreneurship, and psychologically reject the funds and risks that are bound with entrepreneurship, and regard entrepreneurship as a way of avoiding duties and running no gainful business.

3.2 Incomplete course systems

Education serves the social and economic development. With the coming of intellectual economy, to better meet the needs of the society, entrepreneurship education also must keep up with the times, and the building of the course system must be strengthened. Currently, one major problem existed in college students' entrepreneurship education in China is that it is separated from discipline and specialty education, and a series of courses focusing on entrepreneurship education are missing. Because there is no major entrepreneurship education in the specialty catalog issued by China's Ministry of Education, there are only small number of colleges are running courses in entrepreneurship education, and among which only several colleges taking entrepreneurship education as a part of employment guidance to conduct the trial. Entrepreneurship education has not integrated in colleges' teaching systems, nor associated organically with discipline and specialty education. The result of these are entrepreneurship education is separated from disciplines and specialties, and entrepreneurial practices are lacking of support from specialty knowledge, which in return directly resulted that the entrepreneurship education only has a name, without substantial content inside, the enterprising qualities of college students are hard to enhance, and even more, it's unable to form rich campus atmosphere of innovation and entrepreneurship in this way.

3.3 Incomplete construction of teaching and academic staff

The premise and basis of implementation of entrepreneurship education is that the educators must have consciousness and thinking of innovation and entrepreneurship, while the entrepreneurship education in China is still in its starting stage, which determines the teaching staff is in shortage. When carrying out entrepreneurship education activities, colleges mainly rely on courses or speeches run by management specialty teachers, and extracurricular practice activities mainly rely on entrepreneurial motivation, speeches, and entrepreneur contests, etc., organized by student affair divisions, youth league committees, and counselors. Entrepreneurship instructors in colleges not only lack entrepreneurial consciousness, spirits, knowledge, and abilities, but also lack entrepreneurial experiences and experience. Therefore, the enlightenment of entrepreneurship education in most colleges is done by integrating entrepreneur resources in society, and inviting them to come and make speeches, which are far from satisfying the needs of students for entrepreneurship education.

3.4 Cultural atmosphere is weak

Cultural atmosphere has a cultivation effect upon the formation of students' entrepreneurial concepts. Currently, the problem of weak cultural atmosphere commonly exists in colleges' entrepreneurship education in China. First, college students' entrepreneurship education lack of good campus entrepreneurial culture. Under the influence of Chinese traditional education system, the campus culture lack of entrepreneurial atmosphere of brave exploration, courageously marching forward, daring to innovate, and daring to outperform others. As a result, the entrepreneurial enthusiasm and fighting will of college students can't be activated effectively. Second, college students' entrepreneurial education lack good social

entrepreneurial environment. Entrepreneurship education is rooted in campus culture, but is not a school behavior solely, the implementation of which is even a social program. Due to the flourishing of middle course in traditional Chinese culture, and the incomplete policy system, the social resources for the implementation of entrepreneurship education is pretty lacked, and the difficulties in raising venture funds, plus not enough support from entrepreneurship policies, as well as lack of good entrepreneurial environment, etc., have greatly influenced the initial and positivity of college students' entrepreneurship.

4 THE FULFILLING APPROACHES OF COLLEGE STUDENTS' ENTREPRENEURSHIP EDUCATION IN CHINA

College students' entrepreneurship education is a complicated systematic engineering, which needs the common efforts of society, schools, students, etc., to work together. We should change the single mode of entrepreneurial education in the past that only through class teaching, and we should enhance college students' entrepreneurial abilities, by arousing their entrepreneurial enthusiasm and make entrepreneurship education penetrate through their whole studying process.

4.1 Guide positively, and identify future's development direction

Because of diverse psychological characteristics of individual student, their entrepreneurial potential and possibilities vary. Therefore, before implementing entrepreneurial education, we need to assess and guide our training objects to reinforce training effects. Concretely speaking, based on educational concepts of teaching students in accordance with their aptitudes, and personality development, starting from growth rules of talents, conducting common "psychological assessment on career tendency", and setting up courses on career planning, holding dedicate speeches on "specialty development prospects", etc., we can make students gradually identify their individual development directions, and lay foundation for the implementation of entrepreneurship education.

4.2 Teach the students in accordance with their aptitudes, and teach specialty knowledge

The development of students' entrepreneurial consciousness, spirits, abilities, must be based on the basic theory knowledge and under the influence of rich study atmosphere. Therefore, in entrepreneurship education, class teaching plays a leading role. We should carefully select fruitful experts, scholars, enterprise management personnel, and successful entrepreneurial personnel in entrepreneurship education area as teachers in entrepreneurship education class, set up common knowledge course for students, teach basic knowledge for students' entrepreneurship, cultivate their entrepreneurial consciousness, arouse their entrepreneurial interests, and motivation, in order to make them learn and master the basic rules in entrepreneurial activities, and the most fundamental approaches to conduct entrepreneurial practices. On that basis, relying on "collegiate system", we should open dedicated classes, to conduct deep-level entrepreneurship education for the students we selected who has great interests and potential in entrepreneurship education, and gather them together for certain aspects of entrepreneurship, and help them accumulate knowledge needed in entrepreneurship. Because these students are more enthusiasm in practicing and much higher possibility of future entrepreneurship than ordinary students, this kind of entrepreneurship education in form of dedicated classes can provide them with deep-level entrepreneurial knowledge and richer atmosphere.

4.3 Relying on specialties, and reinforce specialty education background

At the same time of opening selective courses to students, and holding dedicated training classes, the specialty teaching should also take entrepreneurship education into their teaching contents in coordination with their own practical situations, for example, run entrepreneurial courses combining their own related specialties, organize the students to carry out entrepreneurial practice activities related to their specialties, establish cooperation relationship with enterprises in related specialty areas to provide practice opportunities for students, etc. On that basis, we should positively explore and implement new approaches combining the specialties, engage experienced social personages, encourage brave reform and innovation of the teachers in teaching content, methods, and assessment form, establish "demonstrating sites in entrepreneurial education", further explore to the combination of theories and practices of entrepreneurship education, gradually form course platforms constituted by entrepreneurial courses and dedicated speeches in specialty areas, combining practical practices in the specialty areas, inspire the students to integrate organically the entrepreneurship with the specialties learnt, and at the same time learning the specialty knowledge, and add related entrepreneurship education concepts in, to lay a foundation for entrepreneurship in the specialty area.

4.4 Practice bravely, and cultivate entrepreneurial abilities all round

Specialty education advocates the combination of theories and practices. For the entrepreneurship education, this principle has more meaning, because the development of entrepreneurial consciousness, spirits, and abilities can't be separated from the accumulation of necessary knowledge, and the practical application and practices can't be ignored either, and it's in the practices that the study abilities of students can be examined and activated, to prompt they turn their entrepreneurial knowledge into entrepreneurial abilities, and finally become successful entrepreneurs. Therefore, we should constantly encourage the students take part in the second-class activities in and out of campuses creatively, and they can improve their willpowers, judgments, and organization and coordination abilities. We should hold "venture contests" to encourage the students to carry out entrepreneurship in simulation, and make the students deepen their learning of foundation of enterprises in the process under the guidance of experts from enterprises, risky investment circle, and improve their teamwork cooperative abilities. For the relatively armature teams in ventures, contact social institutions to help students realize their ventures, and provide entrepreneurial help for them, and help them walk on the way of entrepreneurship indeed.

4.5 Strengthen the guidance, and perfect the tutoring system for entrepreneurship

In the concrete entrepreneurial practices, especially at the stages of entrepreneurship simulation and incubation, engaging experienced personnel in entrepreneurship as the tutors is quite necessary, that is to say furnish "entrepreneurship tutors". On one hand, we should encourage schools' teachers to practice in enterprises, and communities, and after gaining practical experience, they can be the tutors of students' entrepreneurial teams. On the other hand, we should fully consider the desire of social personages "focusing on the youth, sharing experience", and engage successful entrepreneurial persons, enterprise senior management staff members, and venture capitalists as entrepreneurial tutors, to give effective guidance to students in particular entrepreneurship programs. Through these, we could not only let the students draw more successful experience and modes from entrepreneurial practices, and learn entrepreneurial knowledge and tactics more closely related to entrepreneurship, but also during the contact with entrepreneurship tutors, we can perceive their rare and valuable entrepreneurial qualities, set up entrepreneurial ideals under the direction of examples, and refine our own entrepreneurial qualities.

REFERENCES

[1] Luo Meiping: The Current Situation and Counter-measures of College students' in China. Journal of Zhejiang University of Industry and Commerce, 2006 (2).

[2] Cai Dan, Li Junhong, Wang Meiyan: The Study And Exploration of entrepreneurship education in College and Universities, Ecomist.

[3] Wu Jianqiang: Entrepreneurial Education Model and Practical Exploration, Journal of Jiujiang Technical Institute, 2006 (1).

[4] Yue Fengli: The Exploration, Scientific And Techonollogical Pross, And Countermeasures of the Construction Mode and realization approach of College students' Entrepreneurial Education, 2005, (7).

Sports Engineering and Computer Science – Luo (Ed.)
© 2015 Taylor & Francis Group, London, ISBN 978-1-138-02650-6

Running test system based on STM32

RongChun Zou, Lei Zhou, XiaoJu Liu & SuoDong Sun
Institute for Electronic Information, Hangzhou Dianzi University, Zhejiang, China

ABSTRACT: This paper presents a wireless data transceiver system with STM32F103 and STM32F107 of ST Company as the company's main chips. It controls the NXP Company's highly integrated card reader chip MFRC522 and is combined with high-power wireless module UTC4432. The system functions of record, summary and backup are realized through the background logical processing combined with the socket of java to get data from the fixed network port initiatively.

Keywords: wireless data transceiver; STM32; running test system; java

1 INTRODUCTION

With the science processing, nowadays lot of devices of the Internet of things tend to be more intelligent. This paper presents a new design system of sports test. It includes a reader terminal part, data base station receiving section and a web server part. The web server part has the characteristics of long transmission distance, low packet loss rate, stability, automatic statistical score and online inquiry and so on.

2 THE REALIZATION OF HARDWARE

2.1 *The principle of the overall design*

There are the requirements that the students in Zhejiang Province must complete a certain number of long-distance running events in each semester. According to it, an idea to design a system of intelligent long-distance running test has been generated. The ideas of the overall design are: When testing, the students go to the card terminal and brush with the campus card. Then after the data base station receives the wireless signal which is launched by the reader terminal, the signal is transmitted to the web server for processing. The system can record the long-distance running distance, frequency and time automatically and accurately. As well as, it can be inquired via the network. The students' running space cannot be limited to the stadium, and it also can conduct on the campus roads. Simultaneously, it greatly reduces the PE teachers' work intensity and possesses high practicability. At last, it is easy to grasp the students' long-distance running completion. The system diagram is shown in Figure 1.

2.2 *The design of the card reader terminal*

The main functions of reader terminals are reading the student number of the student card (Mifare1 card). It shows the student number in the liquid crystal with the buzzer sounding and transmitting the data to the data base station.

In the card reader terminal, the master chip adopts the Italian semiconductor company's embedded processor STM32F103VET6 based on the Cortex-M3 kernel. The processor supports a variety of communication bus, including the two I²C interfaces, three SPI interfaces, 5 USART serial interfaces and other interfaces, whose maximum operating frequency can attain 72 MHz. Meanwhile, the processor also has 80 general I/O, 16-bit timer, A/D converter and real-time clock. The control chip is used to read the data on the student card. It is MFRC522, which is the highly integrated contactless read-write card chip.

The transmission module utilizes the principle of modulation and demodulation and integrates them into all kinds of contactless communication methods and protocols. MFRC522 can connect MCU through the three following interfaces: SPI

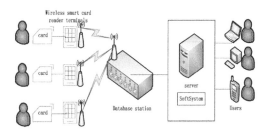

Figure 1. The system diagram.

interface, serial UART interface and I²C interface. The design uses a four-wire spi and its communication clock signal is generated by MCU. Thus, MFRC522 sets to the slave mode for receiving data from MCU[1]. The diagram of reader terminal is shown in Figure 2 and the interfaces of MCU and MFRC522 are shown in Figure 3.

After reading the data, the data is displayed on the LCD, and is sent to the data base station via a wireless module. Here the used transmission module is the highlight and innovation of this design.

This design uses a wireless module that adopts the control chip Si4432. It is introduced by the Silicon Labs Company and is a wireless transceiver chip of EZRadioPRO series with highly integration, low-power and multi-band. Si4432 has always been the "the king through the wall" in the world. The transmission distance is up to 2000 m in an ideal environment. Besides, Si4432 possesses lower utility costs and can compose a transceiver system with high-reliability and high stability only requiring a crystal and small amounts of components[4]. The system designs a wireless transmission module of medium-distance transmission using high-frequency amplification in the RF front-end with high sensitivity, and long-distance transmission capacity. And the wireless module supports serial port configuration[5].

Throughout the MCU online configuration, the transmitted power of wireless modules, air transmission rate and baud rate can be configured. Finally, MCU and the interface of this wireless module are presented in Figure 4.

Thus, the design of the card reader terminal is completed.

2.3 The design of the database station

The database effect is: The data which are sent by the data card reader terminal is received by the wireless module of the database. In this condition, the data is sent to the web server through the network and allow the web server for further processing.

The master chip which is adopted by the database is still TM32F107 produced by ST Company. Unlike STM32F103, processors of STM32F107 series has self-bring MAC of IEEE802.3. which only needs to increase the PHY physical layer chip to connect to TCP/IP networks. The processor supports two standard interfaces to connect to the physical layer module: Media-Independent Interface (MII) defined by the IEEE802.3 protocol and simplified media-independent interface (RMII). This design uses RMII interface mode, because a RMII port only needs seven data lines, and MII requires 14; RMII data transmission requires only two lines, MII takes four.

In the case to meet this design, the RMII mode is used. This design uses a network chip

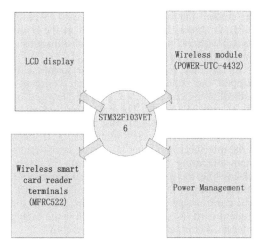

Figure 2.   Reader terminal diagram.

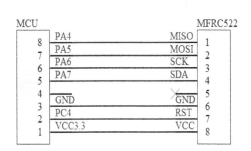

Figure 3.   The interfaces of MCU and MFRC522.

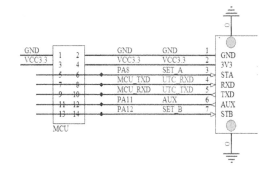

Figure 4.   MCU and the interface of this wireless module.

424

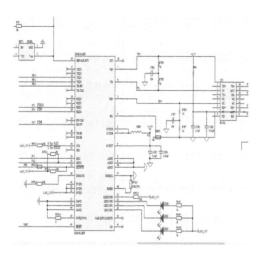

Figure 5.   Network interface module.

DM9161 AEP, which is a fully integrated and cost-effective single-chip Fast Ethernet PHY and is the 0.25 um 10/100M adaptive Ethernet of transceiver adopting the small craft. DM9161 AEP connects to the MAC layer via variable voltage MII or RMII standard digital interface to support HP Auto-MDIX †. The network interface module diagram is shown in Figure 5.

COL pin (36 pins) is a high level, indicating that the interface mode is RMII. In the condition of RMII interface, the TXD0, TXD1, RXD0 and RXD1 are only utilized. 26–29 feet and 35 feet CRS/PHYAD4 are used to determine the physical address DMA9161 AEP. By the LED blinking, the network can determine no problems.

LED blinks once every 500 ms after the reset by the power-on, or by writing to the PHY register. Three LED pins are bidirectional pins which can be configured as high effective or low effective ones. The pin is pulled to the high, and then after reset it expresses low effective. The pin is pulled to the low, and then after reset it expresses high effective.

Similarly, the wireless module of the data station and wireless module in the reader terminal are the same types, which can communicate with each other through configuration parameters.

3   REALIZATION OF SOFTWARE

3.1   *The realization of card reader by software*

The type of the reader module identification cards is Mifare1 card (referred to as M1 card), M1 card is divided into 16 sectors, each sector is composed with four blocks (block 0, block 1, block 2, block 3). Here these 64 blocks of 16 sections are numbered

from 0 to 63 according to absolute addresses[2][3]. Reader's Pseudo-code programs are as follows:

```
int flag = 0;
unsigned char i;
unsigned char j = 0;
  char status;
    status=PcdAuthState(PICC_AUTHENT1A,
BlockAddr, Password, pSelSnr);
  if (readstatus ! = OK) {
    return ERROR;
    }
  status = read_card_block(BlockAddr, pbuffer);
  if (readstatus ! = OK) {
    return ERROR;

    }
Read MI card;
return OK;
```

3.2   *The transplantation of Lwip protocol stack in data base station*

The main work of the network is to achieve the TCP/IP protocol stack based on the hardware platform. But the TCP/IP protocol is so complicated that many features are useless in this system. Therefore, the system selects the mature TCP/IP protocol stack for transplantation and LwIP is such a mature TCP/IP protocol stack.

LwIP is the TCP/IP protocol stack developed by the Swedish Institute of Computer Science for embedded systems. It is a Light Weight IP protocol and can work with or without operating system. LwIP has the main functions of TCP/IP protocol stack, and on this basis reduces the occupancy of RAM, which can work with just a dozen KB of RAM and about 40 KB of ROM[6]. That is why LwIP is suitable for low-end embedded systems. The initialization procedure of LwIP is as follows:

```
char macaddress[6];
mem_init();
memp_init();
IP4_ADDR(&ipaddr);
IP4_ADDR(&netmask);
IP4_ADDR(&gw);
Set_MAC_Address(macaddress);
netif_add(&netif, &ipaddr, &netmask, &gw,
NULL, &ethernet_init, &ethernet_input);
netif_set_default(&netif);
```

After the transplantation of Lwip system, the data base station could communicate with the web server through a simple TCP/IP protocol to ensure the stability of data transmission[7][8].

3.3   *The execution process of software system*

The enabling of the serial ports and timer interrupt are opened after completing initialization. The

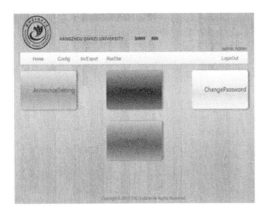

Figure 6.　The effect drawing of the network.

interrupt priority of timer is higher than serial ports. The timer will be in a break state every 3 seconds and collect the data sent by the wireless card reader terminal. The timer will also deliver it to the next polling signal (the next interrupt number). After that, the data is sent to the data processing terminal through a network interface. It counts on the next intelligent statistics and displays the real-time dealed information immediately on the website.

3.4　Software realization of web server

In the aspect of network connection, the java's matured socket technology is used and the data of fixed port under the fixed IP address is fixed by the large join of the server. The information can manage to be transformed in a certain form. After ensuring no packet loss with checksum verification, the packet is decoded to obtain the original data. It is converted into useable data and backup data to prevent data loss. The database makes use of the commit to prevent database crashing in the recording process. The effect drawing of the network is shown in Figure 6.

4　SYSTEM TESTING

4.1　Packet loss testing

In order to test the packet loss rate of the system, we take the following tests: The wireless date base station sends a polling signal to 9 wireless data card terminals. They are placed around the base and the polling signals that the wireless data card terminals matched are different. For each polling signal, the wireless data card terminals will send the data to the data base station. Then the date base station will send the data to the network interface after the data is received. The operating frequency of

wireless modules for the test is 468 MHz, the air transmission rate is 9600 bps and the serial port baud rate is 115200 bps. In order that all modules can communicate with each other, local host and destination address are set up as 0×3 A and 0×3 A respectively. After the test, the statistic is shown in Figure 7. The data packet loss rate is 0.5% or less, in condition that high accuracy is not required, such a deviation is acceptable.

4.2　Overall system test

To test the feasibility of the whole system, we can take the following tests: 3 card terminals put around the date base. After the system is turned on, the testers will take the long-distance running test in accordance with the installing of the website. Then they can find the information about long-distance running. Test results are shown in Figure 8.

As it can be seen from the results, according to the way of swing card, we can see the distance that the person runs and the running speed. Thus we can calculate that the long-distance running is valid or not.

	Received	Lost
A	994	0
B	888	4
C	1010	0
D	843	4
E	952	4
F	844	0
G	1885	0
H	1901	8
I	1898	4

Figure 7.　The packet loss test result.

Run Info				
Num	Date	Miles	Ave-Speed	Y/N
1	2014-03-10 AM	1600	1.41	Y
2	2014-03-10 PM	1600	1.18	Y
3	2014-03-13 PM	1600	1.30	Y
4	2014-03-17 AM	1600	1.32	Y
5	2014-03-17 PM	1600	1.09	Y
6	2014-03-18 AM	1600	1.30	Y

Figure 8.　The test results.

5 CONCLUSION

The system combines a variety of Internet of Things technologies and Internet technology and designs this long-distance test system, which is practical and convenient. Slightly modified, the system can be applied to more occasions which require wireless date transmission. According to the result of the test, the system is stable and intelligent. For teachers and students, it provides a more convenient way of management which can save a lot of manpower and resources.

ACKNOWLEDGMENT

In this paper, the research was sponsored by Key scientific and technological innovation team project of Zhejiang Province "The key technology of intelligent identification and application" (Project No. 2010R50010).

REFERENCES

[1] Philips. MFRC522 contactless read IC product datasheet (Rev3.6) [EB/OL]. [2012–08–27]. http://www..nxp.com/documents/datasheet/MFRC522.pdf.

[2] Zhu Yi, Wang Gang, Wang Hong-Jin. Design of RFID Reader Antenna at 13.56 MHz Frequency [J]. Journal of Microwaves, 2008(10): 22–26.

[3] NXP Inc. Mifare ISO/IEC 14443 PICC Selection (Rev1.0) [EB/OL]. 2006. http://www.nxp.com.

[4] Silicon Laboratories. Si4432 ISM transceiver datasheet [EB/OL]. [2009–02–20]. http://www.silabs.com/Support%20Documents/TechnicalDocs/Si443X_Short.pdf.

[5] Zhang ling. Design of high-performance wireless transceiver application platform based on SI4432 [J]. Computer Technology and its applications, 2010, 36(12):124–127.

[6] Adam Dukels. Design and Implementation of the LwIP TCP/IP Stack [J]. Swedish Institute of Computer Science. 2001(2).

[7] Kong Dong, Zheng Jian-hong. Transplant and Application of LWIP in ARM Platform [J]. Communications Technology, 2008, 41(6):38–40.

[8] Feng Shuang, Jiang Nian-ping. Wireless Data Transmission Applications Based on STM32 [J]. 2012, 21(9):228–231.

Sports Engineering and Computer Science – Luo (Ed.)
© 2015 Taylor & Francis Group, London, ISBN 978-1-138-02650-6

Students involved in e-sports discussion on factors affecting bodily functions

Xue-Min Han
School of Applied Sciences, Hainan University, Hainan, China

Meng Qian
Hainan University, Hainan, China

ABSTRACT: As an emerging competitive sport, the e-sports have developed into a sports video game with a modern competitive sports spirit between people. However, the existence of people gaming with the lack of awareness and mainstream media coverage will be playing electronic games with the ordinary confusion. In this paper, there may be a study about that time factor of the e-sports affects physical and bodily functions, and the conclusions and recommendations are designed to provide a reference for the development of China's electronic sports.

Speaking of e-sports, especially when it comes to domestic area where exists anti-gaming phenomenon, many teachers, parents, and students regard the games as the scourge both physically and mentally, with strongly opposing to the next generation of contacting with the games. As a sports and entertainment means, electronic game is a double edged sword. Its impact on people's physical function is very strong. It can become a healthy sport leading fashion, but the key is to properly guide and standardize the rational e-sports market, develop fashion philosophy of happy sports, electronic sports movement and cultural spirit, truly make e-sports become a healthy, active, and progressive sports project, being consistent with China's national condition. This paper analyzes the characteristics of Hainan factors, bodily functions of middle school students who involved in e-sports. And it is designed to provide a theoretical basis for the sound development of China's electronic games.

Keywords: middle school students; electronic games; physical skills

1   SUBJECTS AND METHODS

1.1   *Study factors of electronic games affecting middle school students' bodily functions*

1.2   *Research object select 15 high schools' students in Hainan for the survey (Table 1)*

Table 1.   Survey secondary type list.

Type of school	School name	Quantity
City focus on school	Hainan High School Middle School Teachers Sea Guoxing High School Jiaji Middle School	4

(Continued)

Table 1.   (*Continued*)

Type of school	School name	Quantity
General secondary school	Double Island Academy Haikou Haikou City The Eighth Middle School High School, Haikou China Haikou Gold Plate Experimental School	5
Total		15

1.3   *Research methods*

1.3.1   *Literature*
Use Hainan University College Library to view the condition concerning e-sports carried out through the library network retrieval centers, and this paper is closely related to the theory of six.

1.3.2 Interview

Hainan SoZ interviews e-sports Club CEO Mr. Fang Tianhua and gaming players to gain first-hand e-sports reference of youth involvement.

1.3.3 Questionnaire

According to the purpose and content of this thesis, the students designed a questionnaire. Hainan high school students were surveyed.

Questionnaire's validity and reliability test 1) validity: expert evaluation method. Invited nine experts of Hainan University (of which three professors, and six associate professors). The validity of the questionnaire can be tested, and questionnaires are valid (Table 2).

Reliability test: the use of "re-measurement method" test of its credibility. Selected Hainan University undergraduates and conducted every 16 people to have a try, and the result of the correlation coefficient is 0.901, twice before, indicating good reliability survey.

Recovery of questionnaires distributed deliver the questionnaires personally to the heads of relevant sports department and school-based physical education teachers and let them to distribute. 2750 copies of questionnaires, delete invalid ones, the valid ones was 2383, and questionnaire effective rate was 86.7% (Table 3).

1.3.4 Mathematical Statistics Act

The main use of SPSS11.5 statistical software is the use of interactive chi-square test, and frequency analysis on survey data and related data for statistical processing.

Table 2. Validity of survey results.

Index	Very effective %	Effective %	The evaluation results
Student			
Questionnaire content validity	22.2	77.8	Effective
Student			
Questionnaire content validity	11.1	88.9	Effective

Table 3. Questionnaires issued the recall.

Grant of questionnaires	2750
Recover of questionnaires	2565
Recovery %	93.3
The number of valid questionnaires	2383
The effective rate	86.7

1.3.5 Comprehensive study method

In the overall understanding of the study, based on the research carried out systematically summed up the problem analysis, the impact factors of physical function corresponding countermeasures theory and high school students to participate in e-sports.

2 RESULTS AND ANALYSIS

2.1 Time factors affecting bodily functions

Table 4 shows that: Students participating in the e-sports life of Hainan to less than 1 year are mostly concentrated in the period of holidays or weekends, the time involved in every 1 to 2 hours based. Lives of men, women and grades the level of features and differences in the timing of each participation are highly significant.

Students involved in the case in Table 4 e-sports time Kainan.

Miss He Huixian said: different from traditional arcade games and online games, e-sports is a confrontational game between people.[1] E-sports and traditional sports items as needed according to the physical condition of each exercise reasonable control over the size of the load, carried out under scientific training time control systems in order to achieve the effect of physical and mental development. These results indicate: E-sports is too concentrated on one time period (such as weekends or holidays) of high-intensity combat electronic games, and it is clearly contrary to the underlying principles of sports training (training systems, suitable load, etc.). It will not be conducive to the development of participants' bodily functions. The interviewee of e-sports players said: since e-sports has become China's official sports, school sports should be the creation of electronic curriculum to popularize. E-sports expert Mr. Qin Gan believes that e-sports campus is imperative that we organize e-sports tournament in Hainan University, and the purpose is to make schools become electronic sports talent cultivation base, and promote the development of e-sports conscience cycle. Hubei Provincial Sports Bureau Shaw Aishan believes that the introduction of e-sports has offered professional e-sports and met the needs of the times. People playing electronic games comply with nature. As long as with the correct guidance and management, it will not only not affect the students' academic, but also promote the overall development of students.[2] Ling Gang (2005) pointed out that e-sports has been made to the campus of Nanjing Zhongshan College, and opened China's first professional electronic sports. Professional training through the school system in favor of e-sports enthusiasts proper guidance e-sports as a powerful tool to adapt to the modern information society,

Table 4.

		Sex/%		Grade/%		X2	
		Male	Female	Junior high school	Senior middle school	Sex	Year
Life	Within 1 year	26.3	46.2	44.9	27.4		62.508**
Feature	1~2a	24.0	28.9	25.2	25.7		
	2~3a	22.2	13.8	13.6	22.1		
	3a Above	27.5	11.0	16.3	24.8		
Time	Weekend	43.6	49.1	46.5	44.8		4.991
Feature	Usually at night	5.8	7.2	7.3	5.8		
	Day after school	4.3	2.8	3.0	4.2		
	Holidays	46.4	40.9	43.2	45.3		
Time	Within 1h	18.1	35.4	29.9	21.0		380517**
Feature	1~2h	40.7	34.2	40.3	37.8		
	2~3h	24.8	15.5	16.8	24.0		
	3h Above	16.4	14.9	13.1	17.2		

Note: ** indicates highly significant difference ($P < 0.01$), the following tables.

so students get a better physical and mental development.[3] Combining research results at home and abroad, it is recommended every time junior high school students and women to participate in e-sports control within 45 min, high school male control in less than one hour is better. In addition, if high school students combine the electronic sports and outdoor sports participation it would be a good guide strategy.

2.2 Effect of bodily functions that may exist

Table 5 shows that: the differences between men and women born in the flexibility of brain activity, skeletal deformities significant aspects, and the lack of sleep leads to decreased immunity differences in terms of highly significant; low grades in hand-brain coordination ability, skeletal deformities areas the difference was significant, differences in the rapid response capability, and loss of appetite regarded highly significant.

The differences of grade level are mainly caused by different physiological functions. Scientific research shows that: high school students (16–20 years old) in human physiology and development of the second peak of the students at this stage of bone hardness enhancement, increased inorganic constituents, endurance levels were significantly increased, and rapid response capabilities and coordination capabilities are reach adult levels. Rapid response capabilities and the ability to hand—brain coordination with youth participation in e-sports is mainly reflected in the sensitivity of the mouse and keyboard control, which is what we usually call "hand to force and mind to bear." Tian Lei (2000) states: "Healthy children involving in guiding appropriate electronic games may be beneficial, and it can improve the coordination of hands, eyes, ears and brain." Zonghao (2005) considers electronic sports can exercise and improve participants' thinking skills, response capacity and coordination ability, and promote all-round development of young people.[5] It is clear that high school students in this capacity is much better than junior high school students. And no matter in body immunity or logical thinking ability, high school students are better than junior high ones. And the differences of the brain in hand with coordination and rapid response capability between high school and junior high school students lie in the loss of appetite.

These differences between boys and girls mainly due to physiological function and psychological motivations between male and female students. Usually at the same age, male body immunity is significantly stronger than female, which determines the highly significant differences of the boys and girls in terms of decreased immunity. In addition, the smaller a women's bone density is, easier the bone deforms; currently e-sports project requires a high degree of coordination battle class-based for hand the brain; psychological and behavioral characteristics in men are more lively and active than women, and these factors have contributed to male and female students in the brain activities flexibility, skeletal deformities, lack of sleep leading to decreased immunity. Of course, any movement should be a good grasp of the issues, if excessive participation in e-sports, not only does not increase the flexibility of the system of the brain, it will lead to brain damage. Japan's leading expert Senzhao' latest male brain science research, said: being addicted to video games can lead to brain function decline.[6] "The British Medical News" (February 2, 2002) reported a 15- year-old boy,

Table 5. Factors case of bodily functions.

Sex/%		Year/%		X2	
Male	Female	Junior high school	Senior middle school	Sex	Year
Positive impact					
Nerve flexibility	55.5	51.1	51.1	54.1	4.455*
Hand-brain coordination	70.4	68.8	66.0	71.2	6.425*
Rapid response capability	72.8	71.1	66.2	74.6	17.898*
Negative impact					
Decreased immunity	68.1	83.8	74.3	76.8	79.385**
Loss of appetite	32.5	31.8	38.4	29.3	19.399**
Skeletal deformities	28.0	31.8	33.0	28.5	4.097*
					4.837*

Note: * indicates significant difference ($P < 0.05$).

playing the game due to time up to 7 hours a day and put on a "brain—Arm System shaking syndrome" that Raynaud's mechanical syndrome. According to the temperature change, except the pain, there is a color change.[7]

3 CONCLUSIONS AND RECOMMENDATIONS

3.1 *Conclusion*

1. The time high school students participating in e-sports sessions were concentrated on weekends and holidays, and it did not promote the healthy development of the student body.
2. E-sports factors for students bodily functions depends primarily on the amount of time involved in e-sports and personal fitness.
3. Caused by different physiological functions involved in e-sports men and women born in decreased immunity aspects of the difference was highly significant; Grade level differences in rapid response capability, and loss of appetite regard highly significant.

3.2 *Recommendation*

1. Strictly controlling of time to participate in e-sports is an important way to keep health and development of middle school students. Under the proposal serious homework teen premise, weekly participation in e-sports 3 to 4 times, each time to accompany children at home, time playing electronic games strictly controlled within 45 min, high school male control within one hour.
2. Establish electronic games are "sports" consciousness, rationally guide young people to participate in e-sports. Parents and children participate in e-sports together, and not only the children can be reasonable and effective using

education management, and more importantly it enhance the feelings between child through this e-sports entertainment tool, and promote the healthy development of the child's body and mind. To the electronic sports expert, it is also a professional university medical physics and engineering doctoral jian ying YOU 2008 won the "International Women's Tournament of Warcraft" runner-up to our profound revelation, his father You Changyu has no objection to her daughter's playing online games. He felt his daughter becomes lively, cheerful, learned to express her, and how to deal with success and failure by playing the game. Dad You Changyu was profound about the experience, "the game is not terrible, how the parents handle it would decide the game become a devil or an angel."[8]

REFERENCES

[1] He Huixian Let mathematical interpretation of the sports Magnifique [J]. Sports Culture Guide, 2004, 8: 3–7.
[2] Zhou Jian electronic games does not mean online games—love of Hubei Sports Bureau Shaw Hill Interview [N] Wuhan Sports Report 2004-05-05-14.
[3] Ling Gang. Wang Fengxian research campus of the development of e-sports [J]. Wuhan Institute of Physical Education, 2005, 39 (80):110–112.
[4] Tian Lei, computer games and intellectual trajectory [M] Changsha: Hunan Science and Technology Press, 2000:1–2.
[5] Zonghao, Li Bo, Wang E-sports Introduction [M] Beijing: People's Sports Publishing House, 2005:5.
[6] Zhang date Hui, Zhou Qingping, Han Jin away from the "video game brain" [M] Shenyang: rolls Publishing Company, 2005.
[7] [Law] Jacques. Hainaut, translation of video games [M] Chengdu: Sichuan Art Publishing House, 2004:75.
[8] "play" into the North doctoral students "Warcraft" Global Runner [EO/OL]. Http://news.xinhuanet.com/viedo/2009–04/content-11197881.htm.

Sports Engineering and Computer Science – Luo (Ed.)
© 2015 Taylor & Francis Group, London, ISBN 978-1-138-02650-6

Based on the measurement of sports building stadiums and drawing course website construction

LanJun Ma
Department of P.E., Xi'an University of Architecture and Technology, Xi'an, China

ABSTRACT: In this paper, we used of the methods about the literature, system analysis and computer of website construction and programming. From the course of website running environment, it used the function design and the structure of framework, multimedia courseware manufacture technology, dynamic interaction technology, Open Database Connection (ODBC) and BBS and timely discussion system. Meanwhile, we set up the course of introduction, electronic lesson plans, teaching outline, online testing and online FAQ, architecture gallery, gallery surveying, analysis of different sports venues gallery, online lab homepage module, each layer building in the development of the course website.

Keywords: sports building stadiums; measuring and drawing; the course of website; building

1 INTRODUCTION

At the beginning of the 21st century, Xian university of Architecture & Technology, set up the new subject, the first sports architecture management to fill in the gap in the field of higher education in China. We know that sports architecture management as a new crossing subject, it combines sports science, architecture, management disciplines as a whole, After ten years of hard exploration, the system has been constructed and will be widely recognized by the industry. This is reason about Chinese social progress and economic growth, the development of undertakings of physical culture and sports. The sports organizations run requires sports architecture, all need efficient sports architecture management, more in need of a large number of high-level sports architecture management personnel. Therefore, we should take care and support for sports architecture and management. It is one of today's sports higher education justice not honor responsibility. The measurement of the sports building venues as well as the illustration is sports architecture management professional, excellent courses is an introduction to students' learning sports architecture, building environment and equipment, construction planning and architecture management, sports venue construction craft, construction and operation management, sports information and foundation of computer applications, and many other course. In view of the sports building venues survey and drawing "course particularity, need to apply the computer technology and network technology to the teaching content such as images, text, sound, form input computer, combines text, graphics, images, animation, sound, applying Authorware, PowerPoint, FroutPage2000, Autocad, Dreamweaver, Flash and other software to make vivid network teaching software. Build the sports building venues survey and drawing course website, can widen the teaching space, so as to improve the teaching quality, and can fully exert the high quality teaching resources, from this angle has important practical significance.

This article used the literature method, system analysis method, computer methods of website construction and the programming, construction management in our school sports professional students in grade 2008–2013 as the research object, discussed about the sports building venues survey and drawing of this course web site building, the site in the trial run in our campus.

2 "SPORTS BUILDING STADIUMS MEASURING AND DRAWING" COURSE WEBSITE DESIGN OF THE BASIC IDEAS AND OPERATION ENVIRONMENT

2.1 *The sports building venues survey and drawing of the basic ideas of course website design*

The course of "Sports building stadiums measuring and drawing" used mostly according to the traditional teaching methods and blackboard writing and slide projector, color chart, building models and other auxiliary teaching means. Even though the application of teaching media increased the intuitiveness of classroom teaching and vitality, the frequent switching between a variety of teaching media is more time-consuming and laborious, which certainly will affect the effective teaching time to be compressed, the classroom teaching of information to be limited. Teachers in the classroom teaching are unable to fully discuss and communication, it is difficult to carry out student-centered teaching activities. Design aiming at the status quo, application of multimedia network technology to build the sports building venues survey and drawing "course website, must reflect openness, flexibility, high efficiency and interactivity, autonomy and participatory principles. The course website needs to cover are: text, image teaching, online reading, online practice and test and network remote teaching and real-time content such as video on demand and way.

According to the basic idea of the course website design, web site management module is designed, it is mainly for teachers, students and visitors to the different registration and landing; Course network courseware function module for teachers, students and visitors to the different people to download lesson plans, teaching outline and the correlative chapters content; Online laboratory function module, facilitate the experimenter choose experimental section, browse the experiment purpose, content, time requirements and submit the online experiment report; Network question bank and the online test function module, facilitate students and related visitors choose the test question types, difficulty coefficient of each chapter, answer the questions on time, and scoring, flexible organization students online testing, timely submit answers; Student management function module, easy to verify the validity of the student's name, password, and students' course selection, learning, discussion, information query, modify and submit assignments and homework and personal information; BBS and chat function module, facilitate visitors use of bulletin boards and chat rooms talk to each other, published their own problems and view, teachers can also be posted on the bulletin board, facilitate the timely communication and discussion between teachers and students; Better educated and practical function module, convenient for visitors to browse the basic knowledge of sports building stadiums system, architecture gallery, gallery surveying, analysis of the various sports venues gallery, website of reading books and laboratory online content, develop the students' ability of independent observation and analysis. The measurement of the sports venues as well as the illustration of the three dimensional animation design and streaming media playback function modules: use the three dimensional animation design and streaming media broadcast function module can play sports action stone fragments lens and the slow motion,

display at the 3D animation of the venue and the analysis of different sports stadiums.

2.2 The sports building venues survey and drawing course website running environment

With Windows 7 as the platform, image acquisition card, CD burners, scanner, digital camera as the main hardware configuration, with Internet Explorer browser combining multimedia technology to build a completely open manageable teaching website system. The course website using the client or network multimedia classroom/web server mode, can be realized in network coverage area of image, voice, text and other multimedia information sharing, storage, management and search. The sports building venues survey and drawing of the development of the course website mainly adopts the multimedia courseware manufacture technology, dynamic interaction technology, open database connection (ODBC) and BBS and timely discussion system.

3 THE SPORTS BUILDING VENUES SURVEY AND DRAWING OF THE CONSTRUCTION OF THE MAIN INTERFACE OF COURSE WEBSITE

Course website frame structure of the main interface is according to the website function module of the empty structure combined with the theme of node space structure design, in a second main interface site navigation framework with introduction, electronic teaching plan, teaching syllabus, online testing, on-line question-answering, architecture gallery, surveying gallery, online galleries, sports venues analysis lab homepage module in each layer of hot spots, such as the mouse click to enter the corresponding lower home page. In the framework of communication with the address, zip code, domain name, E-mail address and digital device and related content.

4 "SPORTS BUILDING VENUES SURVEY AND DRAWING" COURSE WEBSITE CONSTRUCTION OF THE TEACHING PLAN AS A WHOLE ARRANGEMENT OF INTERFACE

"Sports building stadiums measuring and drawing" course website is from the teaching content, management content, machine learning and interaction between teachers and students, teaching content and related resources website operation of the software and hardware support environment. The teaching content of the course website design includes chapters, divided into questions after class, stage testing, and based on the arrangement of the different levels of learning content. Because of the courseware, informative, courseware management system should have a friendly user interface and information retrieval browsing system, can meet the students in the greatest degree of browsing, retrieval, access and download. Measurement of different sports venues and the emphasis and difficulty in drawing, students are difficult to set up the correct space structure in mind. We can use hypertext, hyperlinks, 3D animation, virtual simulation and streaming video on demand, etc way, some abstract plane structure image solid ground, facilitate student learning, observation and control.

5 "SPORTS BUILDING STADIUMS MEASURING AND DRAWING" COURSE WEBSITE ONLINE READING ROOM OF THE BUILDING

"Online reading room" can help students to the teaching content of strengthening memory; Equipped with electronic books related to architecture, surveying the url hotspots, online sports teaching courseware, electronic construction survey map urls hotspots, facilitate students convenient link to browse.

6 "SPORTS BUILDING VENUES SURVEY AND DRAWING" COURSE WEBSITE OF ONLINE LAB BUILDING

"Sports building stadiums measuring and drawing" course website should not only have a friendly man-machine interface, it is more important between teachers and students can be realized through the network of teaching and learning interaction. The course system and teaching content convenient teacher for adjustment and replacement. Site dynamic learning resources can be introduced through hyperlinks, and other ways, and make the teachers and students through discussion and cooperation in the form of a learning task. Teachers can through the web for students learning tasks and problems, teachers can through the network to monitor student learning space, to guide and help students at any time. In this kind of teaching model, students can also interact between sports venues such as the different design style of understanding and mastering, issued a different point of view of the individual, agreed by mutual discussion. Teachers are no longer the mentor but active participants, so that the students' subjective initiative raises the student to find the problem, the ability to analyze and solve problems. Online

testing interface node structure design based on network chapter and through ASP (Active Server Pages) and question bank practice to implement the dynamic interaction of websites online test. Mainly reflected on the home page design practical, dynamic interaction, quickly obtaining, self test and evaluation, etc. After the completion of the course website to integrate high quality teaching resources, to overcome the lack of teaching instruments, expand the students' learning space, improve the overall quality of students majoring in sports architecture management and learning, management ability is of great significance.

7 CONCLUSIONS

We have built the course of network about "sports stadium construction measurement and drawing". It is not only provides a new learning platform for subject teachers and students, but also to reform the traditional teaching model to achieve integration and information technology disciplines, The important way to gradually establish an information literacy to students. Meanwhile, the site for the establishment of teaching has become an important part of the campus network, and its completion will largely improve the utilization of the campus network.

REFERENCES

[1] Wang Dewei. Sports management: principles and methods—college sports architecture management professional compile for universal use materials [M]. Beijing: People's sport publishing house, 2009: 334–353.
[2] Yang Shao.xian. Media teaching software design and development, institutions of higher learning education technology series materials [M]. Beijing: higher education press, 2008:278–296.
[3] Zhao Huang, Li Fujun. Sports venues automation management-college sports architecture management professional compile for universal use materials [M]. Beijing: people's sport publishing house, 2009: 187–195.
[4] MinRui. The development of the network multimedia CAI courseware and application [J]. Journal of PLA university of science and technology (natural science edition), 2009, 3(5):32–35.
[5] Zhang Xiaoyan. Network multimedia technology [M]. Xi'an: xi'an university of electronic science and technology press, 2009:163–177.
[6] Tang sixin. Based on Web design and creation of Web standards—China's institutions of higher learning basic computer education curriculum system for teaching [M]. Beijing: tsinghua university press, 2009:213–237.
[7] Liu Yuanhang Liu Wenkai. Principle and build modern remote education system [M]. Beijing people's posts and telecommunications press, 2008:241–263.7.

Sports Engineering and Computer Science – Luo (Ed.)
© 2015 Taylor & Francis Group, London, ISBN 978-1-138-02650-6

The research and application of Computer Adaptive Testing based on logistic model

Qingchao Jiang & Bin Zhao
Hebei Software Institute, Baoding, Hebei, China

ABSTACT: Based on the Item Response Theory (IRT), Computer Adaptive Testing technology (CAT) is now the main direction of the computer test technology. Among all the IRT models, the logistic model is the most widely used one. The existing CAT systems are mostly developed from this model. In this paper, a test is implemented between computer science and non-computer science students. Logistic model and the classic CTT model are all implemented in the test. Though comparing the data, we can see that the advantage of logistic model is more clarified and has stronger ability in distinguishing similar data.

1 IRT AND LOGISTIC MODEL

As the development of information technology, CAT (Computerized Adaptive Testing) has become the main developing direction of educational measurement which is based on the Item Response Theory (IRT). It is an integrated product of the modern educational and psychological measurement theory and computer technology. The core idea is "person apt Surveying." That is to say, people with different levels of ability can accept a group of subjects with their own characteristics suited to the level of questions.

The basic idea of Item Response Theory (IRT) is that there is a relationship between some latent traits when tested with their response (correct answer probability) for the project, and this relationship can be represented by a mathematical model. By mathematical models, item response theory established the relationship between the ability who was tested and project parameters as well as the correct answer probability. It can not only shorten the test time effectively, but also estimate the ability who was tested with the least subjects.

Logistic model (Logistic Model) proposed by Birnbaum is the most widely used IRT model. According to the parameter numbers of different characteristic function, the model can be divided into one-Parameter Logistic Model (1PLM), two-Parameter Logistic Model (2PLM) and three-Parameter Logistic Model (3PLM). Three Parameters Logistic Model (3PLM) is showed as follows:

$$P_i(\theta) = c_i + \frac{1 - c_i}{1 + e^{-Da_i(\theta - b_i)}}$$

Among them, $P_i(\theta)$ represents the probability which can answer questions correctly of number i subject with the ability level of θ. θ represents the subject's level of ability. b_i represents the difficulty of the questions i, e denotes the natural logarithms $e = 2.71828$, D represents Scale factor $D = 1.7$, a_i represents the discrimination of questions, c_i represents the guess parameters of questions i.

The characteristic chart of three-parameter logistic model is s-shaped curve. It describes the relationship between different levels of tested and the probability of a correct response of the questions. The chart is showed in Figure 1.

2 LOGISTIC EXPERIMENT

The purpose of this experiment was to study the student's changes in terms of logical thinking who study in vocational colleges. In the experiment,

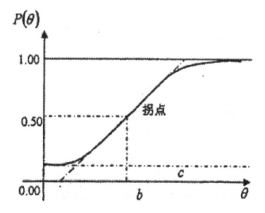

Figure 1. Character curve of logistic model (3PLM).

they all have finished professional learning after three years of studying. For comparison, it selected software programming majors and non-computer-related major students in this test.

2.1 Tested

The tested was made by a total of fifty people in this experiment. They were all juniors of vocational colleges, with twenty-five majoring in software programming professionals and twenty-five majoring in non-computer-related professional. The tested people were selected randomly, rather than according to their score. Software programming professional tested completed nearly two years of programming learning. Most non-computer-related professional tested did not accept the programming aspects of professional learning.

2.2 Experimental design

A total of 30 items in this test are tested from the aspects of "divergent thinking, logical reasoning and creative design" respectively, studying three main areas of logical thinking ability. Each project is assigned the score of 0 to 3, respectively; that 0 point and 1 point are poor, and the ability of 2 points and 3 are preferable.

2.3 One-dimensional detection

The method uses principal component factor analysis SPSS11.0 software analysis results data, the results show that the KMO test of sphericity set questions you can use principal component analysis ($\chi^2 = 2769.53$, p <.000). The set of questions are constituted by the four dimensions, the first load (44.2%) common factor greater than 20% and more than a second amount of load factors common factor (4.2%), the slope Figure the first inflection point was a factor in the location, thus can be considered to meet this test unidimensional IRT model assumptions.

2.4 Based on two levels of IRT item parameter estimation and project analysis model

Based on the characteristics of the subject itself, we will score of 0 points and 1 point of the project a unified recorded as 0 points, score 2 points and 3 points of the project referred to as a sub-unified. This part of the research program uses BILOG-MG3.0 maximum likelihood method to estimate the parameters of the data, and based on a, b, c three kinds of parameters on the questionnaire items were tested and evaluated.

2.5 Subjects score analysis

The purpose of this test is to test the impact of the questionnaire three years of vocational education for students thinking ability to provide students with a reference for the future in terms of whether the computer's choice of profession. CTT theoretical basis, purpose can be achieved by different groups were measured by the contrast between scoring and sharing. After the IRT model a, b, c parameters are also able to measure the ability of those affected (in this case, the behavior is consistent with the degree of subjects with autism symptoms) were estimated to be discriminating on the project under the two models the ability to compare.

3 TEST DATA ANALYSIS

3.1 The estimation and analysis of two parameters IRT model

From Table 1, we can see that in addition to a parameter discernment 27th extraneous items are high, can effectively identify high and low scorers. Through the projects that fit test, 27th questions with a total score of little scoring column related to 0.047, lower than the standard higher than 0.3 is generally required in the process of selecting the questions, while the likelihood ratio chi-square for the project the extent of the fit test showed that the model is only 24.5%, while the degree of model fit other projects in between 47.2% and 98.7%. The first 27th questions is an evaluation of the ability of divergent thinking, the view that through long-term oriented education, developmental divergence ability students are subject to a certain amount of influence, the results of the test project parameters supported this view.

Difficulty value of all the items is in the vicinity of zero b value, and is slightly larger than 0.

Table 1. Two IRT item parameter estimation model.

No.	a	b	No.	a	b	No.	a	b
1	1.127	0.228	11	1.169	0.282	21	1.38	0.242
2	1.007	0.207	12	1.672	0.431	22	1.836	0.413
3	0.746	0.179	13	1.703	0.42	23	1.67	0.324
4	1.197	0.248	14	0.736	0.155	24	1.61	0.376
5	1.306	0.283	15	0.877	0.163	25	1.134	0.283
6	1.317	0.264	16	1.028	0.221	26	1.603	0.315
7	1.978	0.485	17	1.206	0.232	27	0.315	0.356
8	1.612	0.387	18	1.997	0.519	28	1.576	0.322
9	1.77	0.396	19	1.475	0.305	29	1.51	0.333
10	1.24	0.24	20	2.146	0.561	30	1.234	0.279

The follow shows the definition of information functions.

$$I_i(\theta) = \frac{a_i^2 (1 - c_i)}{\left[c_i + e^{1.7a_i(\theta - b_i)} \right]\left[1 + e^{1.7a_i(\theta - b_i)} \right]}$$

This survey shows that the amount of information reaches maximum when the Testeds' ability value near the middle range. Questionnaire reaches the best accuracy when measuring the behavior of those living in the middle or ambiguous identification of subjects. Figure 2 shows that the capacity value about zero, which the maximum amount of information, standard deviation is minimized.

3.2 *Part Tested score analysis*

Comparing the scores of the tested in the CTT and IRT models, we can see that the relevant of the two

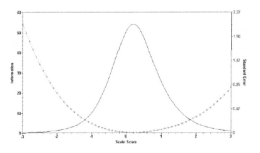

Figure 2. Amount of information curve.

Table 2. CTT score and IRT score comparison table.

Tested	CTT score	Stander error	IRT score	Stander error
1	57	1.34	0.445	0.008
2	63	1.34	−1.596	0.452
3	54	1.34	−0.475	0.161
4	63	1.34	0.445	0.002
5	107	1.34	2.146	0.419
6	24	1.34	−1.15	0.344
7	78	1.34	−1.398	0.296
8	76	1.34	−1.597	0.452
9	67	1.34	−0.518	0.245
10	92	1.34	1.333	0.048
11	81	1.34	1.18	0.336
12	77	1.34	−0.807	0.437
13	88	1.34	−1.077	0.411
14	56	1.34	−0.447	0.006
15	62	1.34	−1.346	0.27
16	57	1.34	0.448	0.01

scoring methods reached .970 (p <.000). It means that the two models in identifying logical thinking ability and decision-making criteria are the same. But the standard error of ability value which IRT model estimates is smaller, it is more precise in identification of tested. Especially when the ability of tested in the middle of the value, IRT information function reaches its maximum, with such tested identification standard error is very small. This once again shows the IRT model in the diagnosis of borderline subjects plays a tremendous role. On the other hand, we can also see that the parameter of IRT model on discrimination issues provides more useful information. The CTT model due to the small sample size, the distinction of the many topics is the same, and is out of dated, it does not meet the actual situation. It is difficult to promote from the sample overall, and can only provide a qualitative judgment basis.

4 CONCLUSION

Currently, computer adaptive test applications based on logistic model are very extensive. However, there are some challenges and difficulties in the practical application of computer-based adaptive test based on item response theory: the selection of item response theory model parameter; parameter estimation algorithm; the decision of development strategy. To resolve these difficulties, it often combines with the specific circumstances of the practical application. The design and implementation of choice must fit the project itself.

REFERENCES

Birnbaum, A. (1968). Some latent trait models and their use in inferring an examinee's ability. In F.M. -Lord & M.R. Novick (Eds.), Statistical theories of mental test scores (pp. 395–479). Reading, MA:Addison-Wesley.

Bock, R.D., & Aitken, M. (1981). Marginal maximum likelihood estimation of item parameters: Application of an EM algorithm. Psychometrika, 46, 443–459.

Chang, H., & Ying, Z. (1996). A global information approach to computerized adaptive testing. Applied Psychological Measurement, 20, 213–229.

Chang, H., & Ying, Z. (1999). A-stratified multistage computerized adaptivetesting. Applied Psychological Measurement, 23, 211–222.

Sports Engineering and Computer Science – Luo (Ed.)
© *2015 Taylor & Francis Group, London, ISBN 978-1-138-02650-6*

University education management system safe system research based on B/S–C/S model

Yue Sun

Dean's Office, Northeast Dianli University, Jilin, China

ABSTRACT: In this paper, based on the advantages and disadvantages of B/S (Browser/Server) model and C/S (Client/Server) model, a variety of security issues and the causes of the present educational administration system are analyzed and a university education management system model based on B/S–C/S model is proposed, which can offset defects when they are used alone. In the meantime, to avoid risks that arise from server configuration is undeserved and defects in designer code, then role authorization management mechanism is applied to the pattern, achieving the correspondence of system permissions and roles. Finally, a university education management system safe system model based on the model of combining B/S with C/S is designed, so that to ensure the safety of educational management system's message, also assure the safe and stability operation of the system.

Keywords: education management system; safe system; B/S model; C/S model; role authorization management

1 INTRODUCTION

Education management system can meet all kinds of needs from colleges and universities in current and future to information resource collection, storage, processing, organization, management and use. Realize the high integration, sharing, unify management and scheduling of information resources. It proposes the accurate, timely information for various management, and it is on the analysis of these information and fast processing; it also provides a fast and efficient electronic platform for the exchange of information, educational management; eventually to improve the management level and work efficiency, and reduce the burden of work. However, with the expansion of network applications, something negative also gradually revealed, such as computer virus, Trojan horse, hacker attacks and so on. They cause the system paralysis, data corruption, the site being attacked, home page had been tampered. Because of the particularity of the educational management system, there are a lot of important and confidential data, once destroyed, the consequence will be unimaginable, so it is particularly important to research and establish education management system safe system research [1].

2 THE MODE ANALYSIS OF THE SYSTEM PLATFORM

University educational administration management system is a large complex computer network information system, it's usually to construct university network educational administration management system by B/S or C/S application architecture [2], it makes the university educational administration management realize remote office, break the space constraints. Specific data flow is shown in Figure 1.

2.1 B/S model

System based on B/S model is a three-layer distributed application system by dividing the server into two parts: data server and the WEB server (see Fig. 2).

This model greatly simplifies the client's design, they can log in the server as long as able to browse webpage, which greatly convenient for teachers, students and other users. For system expansionary, as long as this model can extend application of data server, and the remote user need to start the WEB server application. But for a large amount of data interaction system or higher requirements for the speed of the system, the model is not very suitable, such as curriculum arrangement system, performance management system. At the same time as a result of the limitation of the browser, it is difficult to develop a highly complex and powerful subsystem; In addition, because the model is based on a highly open network environment, the security is not high, vulnerable to malicious attacks. Therefore, for high security requirements of educational administration system, it must establish a complete security system as a guarantee.

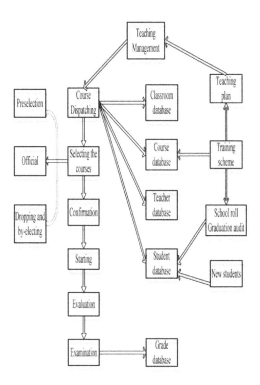

Figure 1. The data flow diagram of educational administration management system.

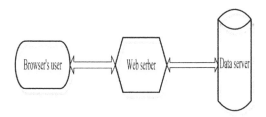

Figure 2. B/S model.

2.2 C/S model

System based on C/S model is a two-layer distributed application system by connecting the client and data server directly (see Fig. 3).

The advantage of this model lies in: (1) it can be very convenient to access and processing a variety of data through the network; (2) users perform client complete program, so strong interactivity, obvious advantages in dealing with a large amount of data; (3) occupying less network resources, transmission speed; (4) the model is based on LAN, can set the appropriate permissions for different users, so security is higher, and high security is needed in the educational administration system. But with

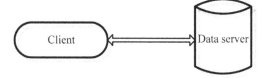

Figure 3. C/S model.

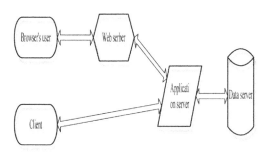

Figure 4. The educational administration management system model based on B/S and C/S model combination.

the rising number of the client, its defect is obvious: First, the client application must be installed and set up the corresponding parameter to access the server, so the demand of the user's operation level is higher, and the subsequent maintenance workload is bigger; Second, if the client number is more, easy to appear the phenomenon of the heavy load on the server, and causing paralysis of the server; Third, unable to realize information public release of student achievement, teacher's schedule.

2.3 Education management system model based on B/S–C/S model

B/S model interface is simple, easy to operate, and has strong expansibility and high sharing, so for the network rapidly developing today, B/S model will bound to get better development and application, however, for the current technology and the characteristics of the educational administration system, simple using B/S model is not necessarily the best choice. Strong interactivity and high security are the main characteristics of C/S model, for a large amount of data and high security requirement educational administration management system, C/S model must be the best choice, but for the release of the student's grade, teacher's schedule query and so on, B/S model is bound to more convenient than C/S model. So according to the characteristics of the educational administration system and the advantages and disadvantages of two models, we eventually adopt B/S combined with C/S model to design the educational management system model, shown in Figure 4.

3 EDUCATION MANAGEMENT SYSTEM SAFE SYSTEM

3 layer or multilayer distributed application system model largely avoids the disadvantage of B/S and C/S mode, take full use of the advantages by combining these two models and greatly improve the efficiency of the system, however, it is not difficult for us to see. if the server is configured incorrectly, or web design code defects, so the whole system still exists a lot of risks in the operation process, the security of the system is still fragile. For this reason, we should do permissions allocation and management to all user access to the system, do some encryption measures on sensitive data and information to ensure only authorized users can access.

3.1 RBAC model

At present the most popular is based on RBAC (role based access control) model [3–4], RBAC technology was originally proposed in 1992 by NIST. The technology main research content is to divide users into the different roles of corresponding to its organizational structure system, thus reducing the complexity of authorization management. RBAC's core idea is the mapping relationship between the authority and role, in the university educational administration management system according to the needs of different working to create the corresponding different role, at the same time the user assigned to the corresponding role, so that we can use the roles to associate users and permissions, unlike the traditional access control technology to access directly allocated to the users. Thus the access authorization is divided into 2 stages: the first stage is to role authorization; the second stage is to make the role assigned to the corresponding user. RBAC includes users, roles and permissions three entities, as shown in Figure 5.

The user can represent human, can also be a machine, it is also the main of perform operations on data in the system and access to other resources. In university educational administration management system users probably can be divided into: students, teachers, teaching management and system administrator 4 parts. Role refers to the post or specific work in the educational management system. It includes students, dean of teaching, teaching secretary, teaching affairs officer and provost, and so on.

Authority refers to all kinds of data in educational administration system, information access

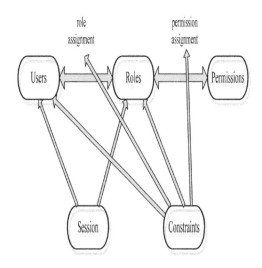

Figure 5. RBAC model.

permission and operational rights. For student users it includes accessing announcement information, coursing selection, checking schedule and results, doing an objective evaluation for teachers; for teacher users it includes querying schedule and coursing selection list, inputting student achievement, carrying on the examination paper analysis, accessing to notice, inputting and submitting the teaching calendar; system administrators have permissions to assign permissions, monitor other users. Session refers to allocate corresponding role according to user's department and the division of work. For example, a user belongs to the college teaching secretary, so his corresponding role teaching secretary, his authority is the teaching secretary's authority, thus avoiding doing repeat distribution to permissions of the same category user. Constraint is a very important concept in the RBAC model. For example: inputting, submitting and reviewing student performance can't be the same user, in the process of role assignment once grant the two roles to one user, it will out of the oversight. Therefore, through the effective set of constraints to limit the distribution of role permission.

3.2 Education management system safe system model based on RBAC mode

As mentioned above, the role can easily to map the structure of university education management to the educational administration system. Due to the small probability of changing work and working content, we make the authority and role corresponds directly, rather than the user itself. So once permissions initialized distribution, we don't need

443

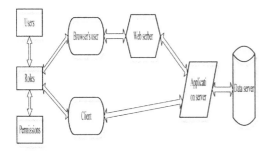

Figure 6. The educational administration management security architecture model based on RBAC.

to do any larger adjustment, the mobility of staff wouldn't largely affect the rights management mechanism. RBAC access entrust mechanism supporting itself, this makes it more suitable for the application of university educational administration system. The above mechanism, we design the following system model (see Fig. 6).

4 CONCLUSION

With the development of information technology and network technology, automatic course scheduling, online course selection, student management and teaching management, all of these need educational management system to complete. However, how to guarantee the safety of information in the transmission process has become key issues in the educational administration management system. Therefore the university educational administration system is facing great challenges, more improvements are needed in security, efficiency and stability, and so on. This paper analyzes the current problems in educational management system and the advantages and disadvantages of B/S and C/S model, designs the combination model of educational administration system, inserts role authorization management mechanism into the security system, so that the efficiency of the educational administration management system has been greatly improved, in terms of the security of the system has higher security.

REFERENCES

[1] Niu Haoyu, 2013.03. Design and implementation of the educational administration management system for security design and realization of enhanced. University of Electronic Science and Technology of China.
[2] Zhao Hong, 2013.05. Design and implementation of the university course scheduling system based on the B/S structure. University of Electronic Science and Technology of China.
[3] Lei Hao et al. 2004,15(11):1680–1688, Threshold access control scheme based on quantified permissions [J]. Journal of Software.
[4] Shan Zhiyong & Sun Yufang, 2004,41(2):287–298, A RBAC model applied to the operating system and its implementation [J]. Journal of Computer Research and Development.

Sports Engineering and Computer Science – Luo (Ed.)
© *2015 Taylor & Francis Group, London, ISBN 978-1-138-02650-6*

E-learning design and steering of real didactic stands

D. Kowalik
University of Technology and Humanities of Radom, Poland

M. Siczek & J. Wojutyński
Institute for Sustainable Technologies—National Research Institute, Radom, Poland

ABSTRACT: The paper presents real didactic stands for e-learning courses in virtual didactic laboratory for the "Mechatronics Technician" profession. The e-learning programmes were designed based on the European and the National Qualification Frameworks, educational requirements and labour market. The didactic stands are for learning of PLC (Programmable Logic Controller) and HMI (Human Machine Interface) panel. The learner can freely steer a real object, programme research tasks and verify the correctness of the tasks undertaken. The stands are independent to each other, autonomous and equipped in security controller, what does not allow for to make the stand destroyed. Each didactic stand is equipped with an industrial computer, which is a kind of a measurement server. The main task of the server is to maintain the communication between the PLC controller and the user with the help of tooling programmes and compression and the transfer of the video transmission. The authors present didactic tasks undertaken on one of the stands, which is related to the design and the steering of the technological transport line. The outcome of the learning (in the virtual laboratory already designed) is to acquire the skills of undertaking vocational tasks in real work environment.

1 INTRODUCTION

The vocational education model includes tasks related to the acquirement of qualifications in real and laboratory conditions. e-learning is an alternative way of education in order to be "on time" regarding the development of new technologies in industry and the lack of the equipment (innovative didactic resources) in schools and laboratories of universities. The use of IT in education influences the decrease of costs (organisational, investment, modernisation) of the university and the student on the didactic materials [1]. Vocational and continuing e-learning education should be treated as priority as it reduces the unemployment and the poverty [2]. The research on the use of interactive media shows great profits for students of vocational technical and engineering schools through independency in solving the own technical ideas [3]. The role of the use of virtual didactic laboratories is bigger in vocational e-learning. The student in a real time undertakes experiments and laboratory tasks on distance with the use of the real research materials [4,5]. Special software and cameras control the correctness of the tasks undertaken [6]. The course for vocational and continuing education must include the requirements of the labour market on the actual needs of the vocational qualifications and educational standards

and the standards of vocational competencies [7]. The authors designed the real didactic stands and elaborated the course based on the qualifications searched on the labour market.

2 ASSUMPTIONS OF THE DESIGN OF A VIRTUAL LABORATORY

2.1 *Educational and vocational requirements*

In order to design the virtual laboratory with e-learning real didactic stands for vocational and continuing education the European and the National Qualification Frameworks [8,9], vocational education standards [10,11], standards of the equipment of ateliers and laboratories in schools [12,13], e-learning educational system, the software for the educational process and the selection of educational staff were into consideration. The authors assumed that the didactic stands for the "Mechatronic Technician" profession direct to the achievement of the K3 qualification (see Table 1): The design and the programming of mechatronic devices and systems. The outcome of the education is to acquire the following skills:

- to design the technical documentation for mechatronic devices and systems;
- to design mechatronic devices and systems;

Table 1. Qualification for the mechatronic technician profession [10].

| Profession | Vocational qualification | |
	No	Name
Mechatronic technician	K1	Assembly of mechatronic devices and systems
	K2	The use of mechatronic devices and systems
	K3	The design and programming of mechatronic devices and systems

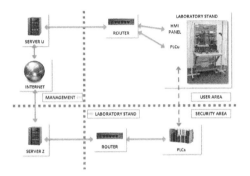

Figure 1. The conception of the stand for e-learning of PLC programming.

• to programme mechatronic devices and systems.

The Table 1 presents the qualifications separated in vocational standards for the "Mechatronic Technician" profession.

2.2 The concept of the measurements and steering didactic stand

The conception of a didactic stand for e-learning of PLCs programming, consists of the user controller (PLCu) and paralleled protection controller (PLCsafety). The conception was illustrated in Figure 1 [3]. The controller, CCD camera and optional HMI panel interface are equipped with RJ45 connectors (Internet) and plugged into the router to the server. The user of the stand enters a program to the PLC and by using a CCD camera observes the way the program runs. Specialised LCMS software placed on the user server manages an access to both the stand and learning process.

The protection controller PLCs controls the interoperability of the user program in the user controller PLCu and detects dangerous for the stand conditions, e.g. activation of limit switches, exceeding the temperature limit values, emergencies, and so on. The occurrence of an unsafe condition is signalled with a red light and the possibility to heat the heater is blocked, e.g. when the temperature is excessed. Therefore, the protection controller detects critical conditions in controller and blocks the user controller activity. The protection controller PLCs impacts are invisible to the user during the program execution when unsafe conditions were not registered. In case of emergency, the user can, for example, retract the pneumatic actuator from his program or monitor program software provided with the stand.

Figure 1 also shows a hardware and computer structure of the stand for e-learning in the virtual laboratory.

2.3 Technical requirements for didactic stands

Problems with stands for the virtual laboratory are associated with different access levels to the stand, i.e. the stand is seen differently by an educational service provider, the recipient, and the manufacturer.

From the point of view of educational service providers, the stand for e-learning should reach didactic aims determined for established teaching standards and learning levels (packages of exercises for specific occupations and competence levels). The LCMS software should provide appropriate teaching aids (exercise instructions, control's documentation, runtime software, and so on), register introduced to the PLCu controller software, store monitoring recordings from the CCD camera, and assist the evaluation of exercises.

The design of the stand should ensure safety when an incorrect program is introduced by the user. This is the main role of the protection controller that detects critical conditions in the controller and blocks its operation. Figure 2 shows the stand from the point of view of the user.

The hardware and computer organisation of the stand is not important for the user who sees only

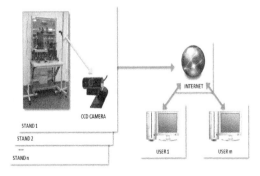

Figure 2. The stand for e-learning of PLC programming (the user view).

the stand. The user should have access to the following elements:

- electrical design of the stand without any visible circuits of the protection controller;
- the structure of the controller addresses that have been used;
- documentation of all sensors and actuators that have been used;
- documentation of the PLC and operator panel HMI, and;
- sample programs;
- educational packages including didactic material, exercises and projects [15].

3 CONSTRUCTION OF THE E-LEARNING DIDACTIC STAND ON THE EXEMPLARY TECHNOLOGICAL TRANSPORT LINE

3.1 *A model of technological transport line*

In order to perform a didactic task, within K3 qualification, a model of technological transport line was selected. A model of technological transport line with two conveyor belts set in the 2U layout was selected for implementation after in-depth analyses. The conveyor belt shown in Figure 1. Sorting elements 1 and 2 as rotating modules with stepper motors and encoders have been replaced with sliding compartments controlled with mini pneumatic actuators that move sorted items between the lines. Sorting elements 3 and 4 have been replaced by a permanent redirecting element. Moreover, developed actuators protect against jamming of sorted items.

Original line drives with synchronous motors M1 and M2 were replaced by DC motors 24VDC with PWM speed controllers—this enables one to easily control the speed of the line. Sorted items in a cylinder shape are placed in standard pallets.

Items are arranged according to the following criteria: colour, height, and type of material.

3.2 *Performance of tasks on the didactic stand*

The user (student) after logging in the e-learning management system can see: exemplary programmes, scheme of a stand, documentation of the PLC and HMI panel and the instruction for doing exercises.

Figure 4 shows the user view of the e-learning stand for PLCs programming of technological transport.

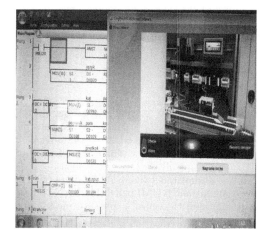

Figure 4. User interface—e-learning of PLC programming of technological transport (view monitor user).

Figure 5. e-learning stand for PLC programming of technological transport.

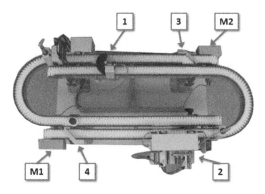

Figure 3. Model of the technological line of the 2U type.

The user can see a laboratory stand with the PLC and HMI operator panel on the screen; access programming software for the PLC and HMI that can be installed on a user's computer or shared by the management system.

The stands are actually surrounded with a much greater amount of hardware and software. On Figure 5 a protection controller PLCS, invisible for the user, detects critical conditions in the user controller PLCU and blocks the user controller.

The stands for e-learning of PLC programming of technological transport enable remote:

- observation of the experiment (audio-video);
- visualisation of the experiment in real time;
- control of the experiment course—remote control of the experiment, and;
- access to apparatus located in another place (e.g. institution).

4 CONCLUSIONS

The main objective of the virtual laboratory designed is to enable e-learning in Mechatronic Technician profession and e-learning improvement of qualifications, which are searched on the labour market in the qualification course systems.

It was assumed that stands for learning of PLC programming and a HMI panel (Human-Machine Interface) have to control a physical model of a device/process. A special attention was paid to the development of physical models that represent elements found in industrial practice with the use of various types of sensors—inductive, laser, optical, vibration, ultrasonic and actuators. Such a stand enables to learn how to use PLC as the measurement and steering system.

The presented real didactic stand ensures a student to get familiar and to understand the techniques of programming of the PLC and the steering elements applied in practice.

The stands can be applied by PLC producers in order to organise trainings for service staff during the introduction of new products on the market.

Scientific work executed within the Strategic Programme "Innovative Systems of Technical Support for Sustainable Development of Economy" within Innovative Economy Operational Programme.

REFERENCES

[1] Kraemer, R. 2014. Advancing without new resources, Educause Review online 49(4).

[2] Bappa-Aliyu, M. 2012. Integrating e-Learning in Technical and Vocational Education: A Technical Review. International Journal of Academic Research in Business and Social Sciences 2(5).

[3] Karahocaa, D., Duldaa, I., Karahocaa, A., Yücela, A., Gulluoglua, B., Arifoglua, E. 2010. Interactive e-content development for vocational and technical education. Procedia—Social and Behavioral Sciences 2(2): 5842–5849.

[4] Khan, B.H. 2001. Veb-Based Training. Englewood Cliffs, New Jersey: Educational Technology Publications.

[5] Foss, B., Oftedal, B.F., Løkken, A. 2013. Rich Media e-Compendiums: A New Tool for Enhanced Learning in Higher Education, European Journal of Open, Distance & E-Learning 16(1): 102–114.

[6] Horton, W. 2000. Designing Web-Based Training: How to Teach Anyone Anything Anywhere Anytime, New York: Wiley.

[7] Kowalik, D. 2013. Polish vocational competence standards for the needs of adult continuing education and the European labour market. In F. Zheng (ed.), 2013 International Conference on Advanced Information Engineering and Education Science: 95–98. Paris-Amsterdam-Beijing: Atlantis Press.

[8] Declaration of the European Ministers of Vocational Education and Training, and the European Commission. 2002. Copenhagen: Convened in Copenhagen.

[9] Development of proposals for substantive and institutional implementation of the National Qualifications Framework and the National Qualification Register for learning throughout life. Project. 2010–2014. Warsaw: Educational Research Institute. (in Polish).

[10] A basic curriculum for professions. 2012. Warsaw: Ministry of National of Education. (in Polish).

[11] Recommended equipment of ateliers and schools workshops. Electric and electronic area. 2013. Warsaw: National Centre for Supporting of Vocational and Continuing Education. (in Polish). http://new.koweziu.edu.pl/wyposazenie-pracowni [cit. 2014-06-30].

[12] Recommended equipment of ateliers and school workshops for mechatronic technician profession. 2013. Warsaw: National Centre for Supporting of Vocational and Continuing Education. (in Polish). http://wyposazenie.koweziu.edu.pl/technik_mechatronik.docx [cit. 2014-06-30].

[13] The Polish Classification of Occupation in Vocational Education. 2010. Warsaw: Ministry of National of Education. (in Poland).

[14] Wojutyński J., Dobrodziej J., Siczek M., Kaczyński J. 2011. The development of a model stand for programming of measurement and steering systems in the convention of the virtual laboratory. Research task. Radom: Institute for Sustainable Technologies—National Research Institute, Radom. (in Polish).

[15] Kowalik D, Kowalik M. 2009. The design of multimedia educational packages for e-learning, In: Jastriebow A. (ed.), Information of technology in the era of the XXI century: 338–343. Radom: Technical University of Radom. (in Polish).

Section 4: Materials science

Sports Engineering and Computer Science – Luo (Ed.)
© 2015 Taylor & Francis Group, London, ISBN 978-1-138-02650-6

Study on properties and composition of garment material on pressure generated by sportswear

Hongxia Chen, Huawei Yang & Xiaohong Li
Xi'an University of Architecture and Technology, Xi'an, China

ABSTRACT: To enhance athletes' performance and recovery speed, Sports Compression Garments (SCG) have been used by them. Two different knitted fabrics with different physical properties and elastic performance were chosen. The pressure generated by sleeves was measured using pressure-measuring device. The results showed that different material composition of fabric assemblies influenced the pressure delivery of garment differently. However, no clear relationship between the fabric percentage in assembly composition and the generated pressure was established.

1 INTRODUCTION

Sports Compression Garments (SCG) have been used by athletes to enhance their performance and speed of recovery[1–3]. Many commercial branded sportswear are claimed to provide the wearers with enhancing blood flow, better muscle oxygenation, reduced fatigue, faster recovery and reduced muscle oscillation, etc[1]. Therefore, SCGs are becoming increasingly popular in a multitude of sporting activities.

The degree of pressure produced by a compression garment is determined by the following principle factors: the construction and fit of the garment, structure and physical properties of its materials, the size and shape of the part of the body to which it is applied and the nature of the sporting activity undertaken[1]. The function of SCG is claimed to be improved through various garment design and engineering factors. In addition the compression has the aerodynamic advantage of reducing the frontal area and thus the total drags[5]. Often these suits are composed of different high-modulus elastic fabrics of different physical performance attributes. However there is a lack of systematic research capable of linking, the positive attributes of the SCGs to the material properties and choose design patterns in the inhomogeneous cases.

In this paper, some things were studied, which were the experimental issues related to measuring the distribution of pressure exerted by the fabric, the influence of tensile properties of experimental knitted fabrics and their material assemblies with their different orientations to stretch upon the distribution and magnitude of the pressure generated by them on a cylindrical body.

2 EXPERIMENTAL

2.1 Materials

Two different knitted fabrics with different physical properties and elastic attributes were chosen. The fabrics were joined by three thread cover stitch. Fabric sleeves were made of different combination of fabrics A and B (Table 2). Fabric sleeves were of the dimensions providing 25%, 50% and 75% strain around the circumference of the cylinders of 90, 130 and 160 mm in diameter, respectively, when positioned over cylinders. Fabric sleeves were 500 mm in length and of the width required for each strain, plus 20 mm for seam allowance where they were sewn along their length.

2.2 Characterization

For tensile test, 30 cm × 5 cm samples with different composition of fabrics A and B in course direction were prepared in strip formation (Table 1). All fabrics and sleeves were conditioned and tested in standard temperature atmosphere of 20 ±2°C and 65 ±2% relative humidity. Fabric sleeves were made of different combination of fabrics A and B (Table 2). Samples composed of fabrics A and B

Table 1. Samples' composition in course direction.

Composition	Fabric A (%)	Fabric B (%)
C0	0	100
C1	20	80
C2	50	50
C3	80	20
C4	100	0

Table 2. Sleeves composition.

Sleeves	Fabric A (%)	Fabric B (%)
S0	0	100
S1	20	80
S2	50	50
S3	80	20
S4	100	0

Table 3. Fabrics properties.

	Fabric A	Fabric B
Fiber composition (%)	Nylon 70 Elastane 30	Nylon 70 Elastane 30
Mass (g/m²)	225	190
Courses/cm	22	24
Wales/cm	26	26
Thickness (mm)	69	66

in strip formation were tested for determination of strain at specified force in course directions. The tensile test was conducted according to ASTM D3107-2007, using Instron Tensile Tester. The pressure generated by sleeves was measured using Salzmann pressure-measuring device. The measurement was conducted on the cylinders with diameter comparable to the size of the wearer's leg.

The measurement on the cylinder was conducted as follows: the cylinder was positioned vertically on its holder so that the cylinder did not move. The sensor used was the short sensor of 330 mm with 4 measuring points. The sleeve was positioned on the cylinder over the Salzmann sensor. The sleeve fabric was spread evenly so there was no fold, kink and air bag between the cylinder, sensor and the fabric. The result could be read on the MST and on the software. The measurement of pressure induced by the sleeve on the cylinder was repeated. The result was analyzed using descriptive statistical methods.

3 RESULTS AND DISCUSSION

The test results of physical properties of fabric used in this study are given in Table 3 and the elongation of fabrics at 25 N forces was given in Table 4. The result of MST Salzmann calibration test result was given in Table 5.

Figure 1 shows the generated interfacial pressure by sleeves of different material composition in three different cylinder diameters. It could be seen that every composition of sleeves generated different pressure delivery. The result of analysis of variance test could be viewed at Table 6. Almost all one way analysis of variance test resulted in the decision to reject H0 (p Value < 0.05) which means that the sleeves with different composition generated different pressure delivery.

From the tensile test result, it could be seen that the strip with different composition has different elongation. The statistical results showed that the average of C0-C4 was different. So that when these different composition fabrics formed into sleeves and positioned over a cylinder, they were expected to generate different pressure on

Table 4. Stress and strain of samples in course direction.

Composition	Stress (N)	Strain (mm)
C0	26	153.22
C1	26	146.67
C2	26	146.89
C3	26	140.21
C4	26	147.20

Table 5. MST Salzmann calibration result.

Point of measurement	Pressure on 25 g (mmhg)	Pressure on 45 g (mmhg)
b	4.8	7.6
b1	5.0	7.8
c	4.7	7.9
d	4.6	7.9
f	4.5	7.4
g	4.0	6.0
Average	4.6	7.43
St dev	0.340	0.728

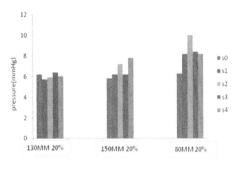

Figure 1. Pressure delivery of different composition of sleeves on different cylinder diameters.

the underlying cylinder. From the pressure measurement, it is evident that the sleeves of different compositions resulted in different generated pressure. This was confirmed with result of one way analysis of variance. However, as it can be seen from Figure 1, there was no clear relationship between the percentage of sleeve composition and the generated pressure, as the pressure result did not exhibit a similar trend. The generated pressure of sleeves on 90 mm cylinder had the similar trend with the generated pressure of sleeves on 160 mm cylinder however, it had different trend with the pressure delivery of sleeves in 130 mm cylinder.

4 CONCLUSIONS

Different material composition of comprising fabric influenced the generated interfacial pressure by fabric sleeves. The pressure generated by the sleeve formed from two fabrics with different attributes was different from the pressure generated by sleeves of the single constituent material. There is no clear relationship between percentages of material composition with the pressure delivery generated, probably because the elastic attributes of the experimental fabrics at the specified force were not considerably different, and the equipment used to measure the generated pressure in this study was not more suitable for the measurement of the distribution of interfacial pressure.

REFERENCES

[1] Troynikov, O & Ashayeri, E. 2010. Factors influencing the effectiveness of compression garments used in sports. 8th Conference of the International Sports Engineering Association (ISEA), Procedia Engineering 2; pp. 2823–2829.
[2] Nusser, M & Senner, V. 2010. High-Tech-Textiles in Competition Sports. 8th Conference of the International Sports Engineering Association (ISEA), Procedia Engineering 2; pp. 2845–2850.
[3] Dascombe, B & Osbourne, M. 2010. The physiological and performance effects of lower-body compression garments in high-performance cyclists. 8th Conference of the International Sports Engineering Association (ISEA), Procedia Engineering 2; pp. 2856–2860.
[4] Liu, R & Little, T. 2009. The 5Ps Model to Optimize Compression Athletic Wear Comfort in Sports. Journal of Fiber Bioengineering and Informatics; vol. 2, no. 1, pp. 41–52.
[5] Oggiano, L. & L. Sætran. 2012. Experimental analysis on parameters affecting drag force on speed skaters. Sports Technology; 3(4): pp. 223–234.

Sports Engineering and Computer Science – Luo (Ed.)
© *2015 Taylor & Francis Group, London, ISBN 978-1-138-02650-6*

Tool wear of aluminum/chromium/tungsten-based-coated cemented carbide in cutting hardened steel

Tadahiro Wada
Nara National College of Technology, Nara, Japan

Hiroyuki Hanyu
OSG Corporation, Aichi, Japan

ABSTRACT: An aluminum/chromium based coating film, called (Al,Cr)N coating film, has been developed. This coating film has a slightly more inferior critical scratch load and micro-hardness. Therefore, to improve both the scratch strength and micro-hardness of the (Al,Cr)N coating film, the cathode material of an aluminum/chromium/tungsten target was used. It added the tungsten (W) to the cathode material of the aluminum/chromium target. To clarify the effectiveness of the aluminum/chromium/tungsten-based coating film, we measured the thickness, micro-hardness and critical scratch strength of aluminum/chromium/tungsten-based coating film. This film was formed on the surface of a substrate of cemented carbide ISO K10 formed by the arc ion plating process. The hardened steel ASTM D2 was turned with the (Al,Cr,W)N, (Al,Cr,W)(C,N), (Al,Cr)N and the (Ti,Al)N coated cemented carbide tools. The tool wear of the coated cemented carbide tools was experimentally investigated. The following results were obtained: (1) the micro-hardness of the (Al,Cr,W)N or (Al,Cr,W)(C,N), (Al,Cr)N coating film was 3110 HV0.25 N or 3080 HV0.25 N, respectively. (2) the critical scratch load of the (Al,Cr,W)(C,N) coating film was 123 N, which was much higher than that of the (Al,Cr)N or (Ti,Al)N coating film. (3) in cutting the hardened steel using (Al,Cr,W)(C,N) and (Ti,Al)N coated carbide tools, the wear progress of the (Al,Cr,W)(C,N) coated carbide tool was almost equivalent to that of the (Ti,Al)N coated carbide tool. The above results clarify that the aluminum/chromium/tungsten-based coating film, which is a new type of coating film, has both high hardness and good adhesive strength. And it can be used as a coating film of WC-Co cemented carbide cutting tools.

1 INTRODUCTION

Hardened steels used for dies or molds are widely cut as a substitution for grinding. Polycrystalline cubic boron nitride compact (cBN) tools are used for cutting hardened steels, due to their higher hardness and higher thermal conductivity. However, in higher feed rate turning or discontinuous cutting, e.g. milling, drilling and tapping, main tool failure of the cBN occurs easily by fracture. It is because the cBN has poor fracture toughness. In these cutting hardened steels, coated cemented carbide tools, which have good fracture toughness and wear resistance, are effective tool materials. The Physical Vapor Deposition (PVD) method, which is a coating technology, is widely applied to cutting tools. The reason is that the PVD method can be coated at a lower treatment temperature and a higher adhesion of the deposition to the substrate. In this case, titanium based films (e.g. TiN, (Ti,Al)N) are generally used as the coating film [e.g. 1, 2].

An aluminum/chromium-based coating film, namely (Al,Cr)N coating film, has recently been developed. An aluminum/chromium-based coated tool was evaluated through the machining of sintered steel, and showed greatly improved performance [3]. Furthermore, it was clarified that the (Al,Cr)N coated cemented carbide is an effective tool material in cutting hardened sintered steel [4]. However, according to our study, the critical scratch load, which is the measured value by scratch test, of the (Al,Cr)N coating film is 77 N. And the micro-hardness is 2760 HV0.25 N. Therefore, in order to improve both the scratch strength and the micro-hardness of the (Al,Cr)N coating film, cathode material of an aluminum/chromium/tungsten target was used in adding tungsten (W) to the cathode material of the aluminum/chromium target.

In this study, to clarify the effectiveness of aluminum/chromium/tungsten coating film for cutting hardened steel, tool wear was experimentally investigated. The hardened steel was turned with the aluminum/chromium/tungsten-based coated tool according to the Physical Vapor Deposition (PVD) method. Moreover, the tool wear of the aluminum/chromium/tungsten-based coated tool was compared with that of the (Al,Cr)N and (Ti,Al)N coated tools.

2 EXPERIMENTAL PROCEDURE

Coating deposition was performed by an arc ion plating system (KOBE STEEL, LTD. AIP-S40). Various coating films were deposited on WC-Co cemented carbide ISO K10.

We measured the thickness, hardness and scratch strength (critical scratch load measured by scratch tester) of various coating films. These films were formed on the surface of a cemented carbide ISO K10 substrate formed by the arc ion plating process.

The work material used was hardened steel (ASTM D2, 60HRC). The chemical composition of the hardened steel is shown in Table 1. The tool material of the substrate was cemented carbide, and four types of PVD coated cemented carbide were used as shown in Table 2. Namely, the coating films used were (Al,Cr,W)N, (Al,Cr,W)(C,N), (Al,Cr)N and (Ti,Al)N coating film. The (Al,Cr,W)N or (Al,Cr,W)(C,N) is a new type of coating film whereas (Al,Cr)N or (Ti,Al)N is a conventional and commercial type. The configurations of the tool inserts were ISO TNGA160408. The insert was attached to a tool holder MTGNR2525M16. In this case, the tool geometry was (−6, −6, 6, 6, 30, 0, 0.8 mm). The turning tests were conducted on a precision lathe (Type ST5, SHOUN MACHINE TOOL Co., Ltd.) by adding a variable-speed drive. The driving power of this lathe is 7.5/11 kW and the maximum rotational speed is 2500 min^{-1}. Hardened steel was turned under the cutting conditions shown in Table 3. The tool wear was investigated.

Table 1. Chemical composition of the hardened steel. (AISI D2, 60HRC) [mass %].

C	Cr	Mo	Mn	Si	V
1.47	11.5	0.82	0.37	0.32	0.20

Table 2. Tool material in turning of AISI D2.

Tool type	Tool material
Coated tool	Substrate: Cemented carbide ISO K10 Coating layer: (Al,Cr,W)N, (Al,Cr,W)(C,N), (Al,Cr)N, (Ti,Al)N

Table 3. Cutting conditions.

Cutting speed	0.50 [m/s]
Feed speed	0.1 [mm/rev]
Depth of cut	0.1 [mm]
Cutting method	Dry

3 RESULTS AND DISCUSSION

To clarify the formation of (Al,Cr,W)N coating film on the substrate of cemented carbide, Scanning Electron Microscope (SEM) observation was conducted on the cross section of the insert. The result is shown in Figure 1. This insert has a (Al,Cr,W)N coating film 4.4 μm thick, and the thickness is almost constant. Moreover, there is flaking of the (Al,Cr,W)N coating film, and the (Al,Cr,W)N coating film and the ISO K10 cemented carbide substrate strongly adhere.

Figure 2 shows the SEM observation of the tool wear in turning hardened steel with the four types of coating film. This is at a cutting speed of 0.50 m/s, a feed rate of 0.1 mm/rev and a depth of cut of 0.1 mm. In this Figure, "L" is the cutting distance. In the case of the two types of aluminum/chromium/tungsten-based-coated cemented carbide tools, namely the (Al,Cr,W)N and the (Al,Cr,W)(C,N) coated tools shown in Figure 2(b) and Figure 2(c), respectively, there is a crater on the rake face. And there is no remarkable adhesion on either the rake face or flank. No remarkable flaking of the coating layer is found either. Furthermore, no remarkable flaking of the coating layer is found. In the case of the (Al,Cr)N and (Ti,Al)N coated tools shown in Figure 2(a) and Figure 2(d) respectively, the wear pattern of the two types of coated tool is the same as that of the two types of aluminum/chromium/tungsten-based-coated cemented carbide tool.

The above results indicate that the main tool failure of the four types of coated tools was the flank wear. It was within the maximum value of the flank wear width of 0.2 mm. Therefore, the maximum value of the flank wear width (VBmax) was measured with a microscope.

In cutting the hardened steel using the four types of coated tools, the wear progress was investigated. The wear progress is shown in Figure 3. Although the wear progress of the (Al,Cr,W)N coated tool is faster than that of commercial type (Ti,Al)N, the wear progress of the (Al,Cr,W)N coated tool is slower than that of commercial type (Al,Cr)N. In particular, there is little difference in wear progress between the (Al,Cr,W)(C,N) and (Ti,Al)N coated tools. This indicates that the (Al,Cr,W)(C,N) can be used for cutting the hardened steel. Thus, adding

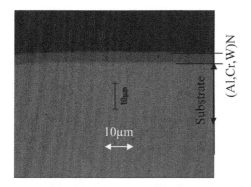

Figure 1.   Cross section of (Al,Cr,W)N coating film.

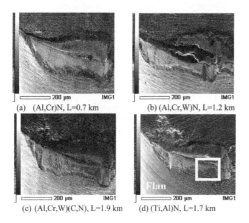

(a)   (Al,Cr)N, L=0.7 km

(b) (Al,Cr,W)N, L=1.2 km

(c) (Al,Cr,W)(C,N), L=1.9 km

(d) (Ti,Al)N, L=1.7 km

Figure 2.   Tool wear at cutting speed of 0.50 m/s, feed rate of 0.1 mm/rev, depth of cut of 0.1 mm and cutting method of dry cutting.

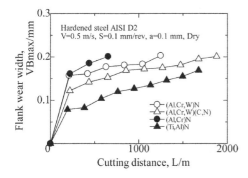

Figure 3.   Wear progress.

tungsten (W) to the aluminum/chromium-based-coating film is effective for improving the wear-resistance.

The above-mentioned results show that the (Al,Cr,W)(C,N) can be used as the coating film for

cutting hardened steel. Therefore, SEM observation and Energy Dispersive X-ray Spectrometer (ED) analysis were conducted on the worn surface.

Figure 4 shows the SEM observation on the worn surface of the four types of coated tools. This figure shows the details of "A" shown in Figure 2. In Figure 4, the adhesion is found on the abraded surface of the coating film for all coated tools. Moreover, in the region enclosed by the rectangle, many striae scratched by a hard material are found on the abraded surface of the coating film, too.

Figure 5 shows the EDS mapping analysis on the worn surface of the four types of coated tools shown in Figure 2. In Figure 5, the EDS analysis for the iron (Fe) and oxygen (O) mapping on the cutting part is shown. As compared with the iron element on the worn surface of the four types of coating film shown in Figure 5(a), there is little difference in the oxygen element among all types of coating films. Moreover, as compared with the oxygen element on the worn surface of the four types of coating films shown in Figure 5(b), there is little difference in the oxygen element among all types of coating films.

Therefore, the main wear mechanism of the four types of coating films is both abrasive wear and adhesion wear. For abrasive wear, the wear-resistance of the coating film often depends on the hardness of the coating film. That is, a coating film with higher hardness has good wear-resistance. For adhesion wear, the wear-resistance of the coating film often depends on the scratch load between the substrate and the coating film. That is, a coating film with higher scratch load has good wear-resistance. Therefore, the characteristics of the coating films were investigated.

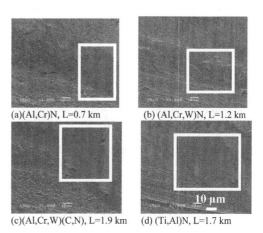

(a)(Al,Cr)N, L=0.7 km

(b) (Al,Cr,W)N, L=1.2 km

(c)(Al,Cr,W)(C,N), L=1.9 km

(d) (Ti,Al)N, L=1.7 km

Figure 4.   Details of "A" shown in Figure 2.

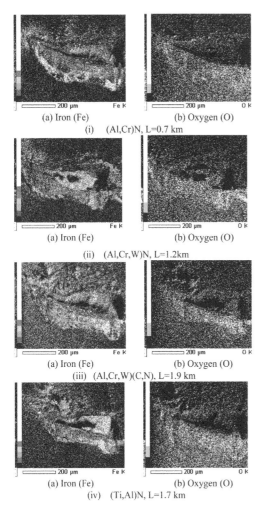

(a) Iron (Fe) (b) Oxygen (O)

(i) (Al,Cr)N, L=0.7 km

(a) Iron (Fe) (b) Oxygen (O)

(ii) (Al,Cr,W)N, L=1.2km

(a) Iron (Fe) (b) Oxygen (O)

(iii) (Al,Cr,W)(C,N), L=1.9 km

(a) Iron (Fe) (b) Oxygen (O)

(iv) (Ti,Al)N, L=1.7 km

Figure 5. EDS mapping analysis on the worn surface (i) (Al,Cr)N, (ii) (Al,Cr,W)N, (iii) (Al,Cr,W)(C,N) and (iv) (Ti,Al)N coated tool. EDS analysis of (a) iron and (b) oxygen is shown in Figure 2.

The film characteristics were then observed. Table 4 shows the characteristics of the coating films. The four types of coating films are compared. The 4.4 μm or 3.3 μm thickness of the two types of aluminum/chromium/tungsten-based coating film is thicker than that of (Al,Cr)N or (Ti,Al)N coating films' 3.0 μm. However, it is considered that an aluminum/chromium/tungsten-based film with sufficient thickness has been formed. The microhardness of the two types of aluminum/chromium/tungsten-based coating film about 3100 $HV_{0.25 N}$ is higher than that of the (Al,Cr)N coating film 2760 $HV_{0.25 N}$ or the (Ti,Al)N coating film 2710 $HV_{0.25 N}$.

In order to evaluate the adhesion between the substrate and the aluminum/chromium/tungsten

Table 4. Characteristics of the coating films.

Coating material	Thickness of film [μm]	Micro-hardness [$HV_{0.25N}$]	Critical scratch load* [N]
(Al,Cr,W)N	4.4	3110	81
(Al,Cr,W)(C,N)	3.3	3080	>130
(Al,Cr)N	3.0	2760	77
(Ti,Al)N	3.0	2710	73

*: Measured value by scratch test.

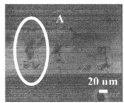

(a) (Al,Cr,W)(C,N) coating film (b) (Al,Cr)N coating film
(Load 130N) (Load 77N)

Figure 6. Microscopic photographs of the scratch track by a scratch tester on (a) (Al,Cr,W)(C,N) coated cemented carbide and (b) (Al,Cr)N coated cemented carbide.

coating film, a scratch test was conducted for the four types of coated cemented carbide tools. This test is typically used to evaluate the adhesion force of thin films. Figure 6 shows microscopic photographs of the wear track in the scratch test. Figures 6(a) and (b) show the (Al,Cr,W)(C,N) and (Al,Cr)N coating film, respectively. The pattern of progressive scratch load is one whereby cracks on the coating surface form and flaking of the coating layer occurs. First, the cracking on the surface of the coating film occurs with the increase of the load. Next, the catastrophic failure happens with the increase of the load at portion "A" shown in Figure 6(b). For the (Al,Cr)N coating film shown in Figure 6(b), the catastrophic failure occurs at the load 77 N. However, for the (Al,Cr,W)(C,N) coating film shown in Figure 6(a), the catastrophic failure does not occur until the load increases to 130 N.

In Table 4, the critical scratch load of the (Al,Cr,W)N coating film is slightly larger than that of the (Al,Cr)N and the (Ti,Al)N coating film. And, the critical scratch load of the (Al,Cr,W)(C,N) coating film is over 130 N. The critical scratch load of the (Al,Cr,W)N coating film was 81 N. It is higher than that of the (Al,Cr)N coating film at 77 N or the (Ti,Al)N coating film at 73 N. As compared with the critical scratch load of the two types of aluminum/chromium/tungsten-based coating film, the critical scratch load of the

(Al,Cr,W)(C,N) coating film over 130 N is higher than that of the (Al,Cr,W)N coating film 81 N. Therefore, by comparing the (Al,Cr,W)(C,N), the (Al,Cr,W)N and the (Al,Cr)N coated tools, the (Al,Cr,W)(C,N) coating film has both higher hardness and good adhesive strength. The wear progress of the (Al,Cr,W)(C,N) coated tool is slower than that of the (Al,Cr,W)N and (Al,Cr)N coated tools as shown in Figure 3. It is because the (Al,Cr,W)(C,N) coating film has good wear resistance.

As mentioned above, there is little difference in the wear progress of the (Al,Cr,W)(C,N) coating film and (Ti,Al)N coating film shown in Figure 3. Therefore, (Al,Cr,W)(C,N) coated cemented carbide is considered an effective tool material and hardened steel can be cut by this tool.

4 CONCLUSION

In this study, to clarify the effectiveness of aluminum/chromium/tungsten-based coating film, we measured the thickness, micro-hardness and critical scratch strength of aluminum/chromium/tungsten-based coating film. This coating film was formed on the surface of a cemented carbide ISO K10 a substrate formed by the arc ion plating process. The hardened steel ASTM D2 was turned with the (Al,Cr,W)N, (Al,Cr,W)(C,N), (Al,Cr)N and the (Ti,Al)N coated cemented carbide tools. The tool wear of coated cemented carbide tools was experimentally investigated.

The following results were obtained:

1. The micro-hardness of (Al,Cr,W)N or (Al,Cr,W)(C,N), (Al,Cr)N coating film was 3110 HV0.25 N or 3080 HV0.25 N, respectively. And the micro-hardness of two types of aluminum/chromium/tungsten-based coating film was higher than that of both the (Al,Cr)N coating film 2760 HV0.25 N and the (Ti,Al)N 2710 HV0.25 N.

2. The critical scratch load of (Al,Cr,W)(C,N) coating film 123 N was much higher than that of (Al,Cr)N coating film 77 N or (Ti,Al)N coating film 73 N.

3. In cutting the hardened steel using (Al,Cr,W)(C,N) and (Ti,Al)N coated carbide tools, the wear progress of the (Al,Cr,W)(C,N) coated carbide tool was almost equivalent to that of the (Ti,Al)N coated carbide tool.

The above results clarify that the aluminum/chromium/tungsten-based coating film, which is a new type of coating film, has both high hardness and good adhesive strength. And it can be used as a coating film of WC-Co cemented carbide cutting tools.

ACKNOWLEDGMENT

This work was supported by JSPS KAKENHI Grant Number 24560149 (Grant-in-Aid for Scientific Research (C)).

REFERENCES

[1] K. Sakagami, G. Yongming and T. Yamamoto: Pro. 4th Int. Conf. on Progress of Cutting and Grinding, (1998), p.38.
[2] H. Nakagawa, T. Hirogaki et al.: Pro. 6th Int. Conf. on Progress of Machining Technology, (2002), p.81.
[3] T. Wada, K. Iwamoto, H. Hanyu, and K. Kawase, "Tool wear of (Al,Cr)N coated cemented carbide in cutting sintered steel", Journal of the Japan Society of Powder and Powder Metallurgy, vol. 58, pp.459–462, 2011 (in Japanese).
[4] T. Wada, M. Ozaki, H. Hanyu, and K. Kawase," Tool Wear of Aluminum-Chromium Based Coated Cemented Carbide in Cutting Hardened Sintered Steel", International Journal of Engineering and Technology, Vol.6, No.3, pp.223–226, 2014.

Sports Engineering and Computer Science – Luo (Ed.)
© 2015 Taylor & Francis Group, London, ISBN 978-1-138-02650-6

Differences of bacterial contamination of the materials in hospital medical care

Y.F. Su
Infection Control Team, St. Joseph's Hospital, Taiwan

L.C. Lu
Information Management Department, National Yunlin University of Science and Technology, Taiwan

ABSTRACT: Quantities of bacterial contamination of the materials in hospital medical care are the critical reason for in-hospital infection. This study observed 53 different in-hospital medical care materials. Each material was observed before and after the health care worker finished a complete care procedure. The bacterial quantities were calculated from the main medical materials and workers' hands in 3 different areas. Regression analysis was used for estimating the relationships among variables. The result shows that bacterial quantities increased dramatically on ungloved hands, respiratory pipes, respiratory masks, and fluid tubes ($p < 0.01$). Hand washing and the proper use of surgery gloves should be the primary concern to reduce the bacterial infection in hospital.

1 INTRODUCTION

1.1 *Bacterial infections*

All hospitals have infection control procedures and policies. And all hospital workers should follow the SOP to avoid infections. However, the risk of infection can never be completely eliminated. And some materials, which are used in the hospital, may increase more bacteria than other material items and thus should be paid more concern to Figure 1, the example shows the contaminated surfaces increase cross-transmission on different materials (Kaiser 2007). Studies show that at least 30% of all nosocomial infections are preventable (Haley 1985). Important factors for nosocomial infections

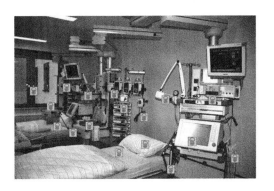

Figure 1. Contaminated surface increases cross transmission (Norbert 2007).

could be the contaminated surface of the materials, health care staff, and medical machines.

1.2 *Purpose of the study*

Bacterial contamination can be prevented by cleaning the surface of medical materials or adopting medical materials whose surface is not easy for bacteria to survive or grow. To know about the most effective material for controlling nosocomial infection, we observed and studied the quantities of bacterial contamination of different medical materials in daily hospital operations.

2 LITERATURE REVIEW

2.1 *Medical materials hygiene*

In a survey of physicians (Wester 1999), 45% poor hand washing practices are an important cause of bacterial contamination in hospitals. In various patient care scenarios, medical staffs have practiced appropriate medical material surface in only 25% or even lower opportunities (Vernon, 2000). Table 1 shows the hospital personnel self-reported and observed cleansing rates.

2.2 *Alcohol-based cleansing*

Alcohol-based cleansing can help to overcome the time problems of washing the material surface and actually improve surface condition

Table 1. Hospital personnel self-reported and observed cleansing rates.

Category	N	%
Self reported	123	85%
Right	63	51%
All other	48	28%

*Summary from CARP (Vernon et al, 2000).

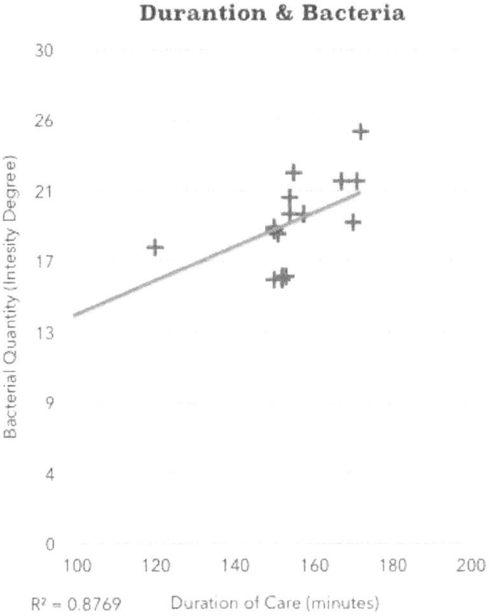

Figure 2. Relationship between duration of care and bacterial contamination quantity.

(Pittet et al. 1999; Boyce 2001). An alcohol-based cleansing can de-germ material surface in less than 30 seconds and enhance the killing of transient bacteria.

The use of alcohol for material surface could be costly. In a study comparing alcohol cleansing and water-based cleansing, Boyce et al. have commented that alcohol cleansing is accessible and convenient to use, and has a pleasant fragrance.

Alcohol-based cleansing may require ongoing educational reinforcement, compliance monitoring, and feedback to personnel. However, the huge resistance of personnel practice, the adoption of germ or bacteria resisted medical material may be a more preferable infection control strategy.

3 STUDY METHODS AND ANALYSIS

3.1 Medial materials observations

A total of 53 medical materials were conducted. Bacteria quantities were calculated before and after the medical treatment procedures, for a total of 420 minutes of observation. Types of medical materials consisted of ungloved hands in 8 cases (15%), respiratory pipes in 12 cases (23%), respiratory masks in 14 cases (26%), machine surface in 9 cases (17%), and flow tubes in 10 cases (19%).

3.2 Study site

An area teaching hospital participated in this study, and the main observations were performed in intensive care units, surgical and respiratory care wards. Types of patient care activities that were observed represented a wide variety of patient care. During each observation, hospital medical professionals executed different types of medical service.

3.3 Types of medical care

Duration and types of medical care were positively associated with the quantity of bacterial contamination in different types of medical material. It was shown in Figure 2. All care activities except drug service activities were positively associated with

bacterial quantity in any medical material. Subsequent analyses were therefore restricted to the observations of medical materials and the relation with bacteria contamination.

3.4 Types of medical material

All medical materials were positively associated with bacterial contamination. Association was significant with the material or contact with material. Average number (Fig. 3).

Increasing bacterial contamination over time was seen in every medical material. However, ungloved hands were the most obvious reasons.

3.5 Regression analysis

Regression analysis of ungloved hands, respiratory pipes, respiratory mask, machine surface, and flow tubes are the independent predictors of bacterial contamination. Five types of medical materials were independently associated with bacterial quantity intensity degree. Observations made in the medical procedure show that hands without any cleansing or protection have high association degree.

3.6 Analysis result

The analysis result identified several predictors of bacterial contamination of medical materials in the

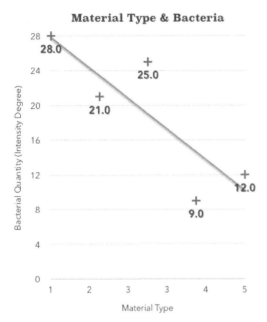

Material Type & Bacteria

Figure 3. Relationship between material type and bacterial contamination quantity intensity.

hospital. Contamination progressively increased during the time of patient care. Generally, medical givers who did not have any material or cleansing method to protect their hands had the highest intensity of germ. Patient care activities and medical materials associated with higher rates of contamination were unprotected hands, respiratory pipes and masks. These materials were associated with moisture contact with patients. A practical recommendation resulting from this observation is that hand antisepsis immediately after these activities should be mandatory. Observing data from this study support that the hand cleansing after dirty contacts or contacts with infected sites is required.

The bacterial contamination for different medical materials was highest on unprotected hands but lowest on machine surface. This study is the first to relate bacterial contamination and medical machine surface.

Our findings are consistent with other evidence. Consequently, alcohol treated hands may be less likely to transfer bacteria than washed hands. In a study reported (Ehrenkranz 1991), hand washing did not prevent the transfer of aerobic gram-negative bacilli by medical care workers' hands from heavily colonized patient groins to urinary catheters. Use of un-medicated soap may not prevent transferring gram-negative bacilli to medical machines or devices from a heavily contaminated

source. Our observations extend these findings by providing quantitative assessment of contamination of the medical materials.

4 CONCLUSIONS

The adoption of anti germ medical material requires aggressive implementation of several strategies (Weinstein & Kabins 1981), for example: ongoing surveillance of resistance and molecular typing of isolates. Monitoring adherence of healthcare workers to control measures and feedback anti germ mathematical modeling could be used to judge the value of medical materials of infection control. In these calculations and screening, the most effective control measures are hand cleansing and protection method.

Current infection control strategies are aimed at the hygiene and anti germ materials. We must harness technology to improve and direct adherence to these strategies. Future approaches may adopt these anti germ medical materials profoundly that underlie evolution of resistance.

REFERENCES

Boyce, J. 2001, Antiseptic technology: access, affordability, and acceptance, Emerg Infect Diseases, 7(2): 231–233.

Ehrenkranz N.J., & Alfonso, B.C. 1991, Failure of bland sop hand wash to prevent hand transfer of patient bacteria to urethral catheters, Infect Control Hosp Epidemiol, 12: 654–662.

Haley, R.W., Culver, D.H., White J.W., et al. 1985, The efficacy of infection surveillance and control programs in preventing nosocomial infections, Am J Epidemiol 121: 182–205.

Norbert Kaiser File:Intensivstation, 2007, Licensed under Creative Commons Attribution-Share Alike 3.0 via Wikimedia Commons.

Pittet, D., Mourouga, P., & Perneger, T.V. 1999, Compliance with handwashing in a teaching hospital, Ann Intern Med, 130: 126–130.

Vernon M.O., Trick W.B., Schwartz D., Welbel S.F., Wisniewski M., Fornek, M.L., et al. 2000, Marked variation in perceptions of antimicrobial resistance (AR) and infection control (IC) practices among healthcare workers (HCWs), Proceedings of APIC 2000.

Weinstein, R.A, Kabins, S.A. 1981, Strategies for prevention and control of multiple-drug resistant nosocomial infections, Am J Med, 70: 449–454.

Wester, C.W., Durairaj L, Schwartz, D., Husain, S., Martinez, E., & Evans, A.T. 1999, Antibiotic resistance: who cares? Physician perceptions of antibiotic resistance among inpatients: its magnitude, causes, and potential solutions, Proceedings of the 37th Annual Meeting of the Infectious Diseases Society of America.

Sports Engineering and Computer Science – Luo (Ed.)
© 2015 Taylor & Francis Group, London, ISBN 978-1-138-02650-6

A major perspective of the characterization and quantification of soft magnetic nanocrystalline ribbons

V. Tsepelev & V. Konashkov
Ural Federal University, Ekaterinburg, Russia

O. Zivotsky & A. Hendrych
VSB-Technical University of Ostrava, Ostrava, Czech Republic

Yu. Starodubtsev
Gammamet Research & Production Enterprise, Ekaterinburg, Russia

ABSTRACT: The concept of the quasi-chemical model of the liquid micro-non-uniform composition has been taken into account. Besides, a research has been made on the physical properties of the Fe-based melts being crystallized. So the unique technology of the melt time-temperature treatment has been developed. Amorphous ribbons produced using this technology require optimal annealing temperatures to be specifically selected. The results of studying nanocrystalline magnetic core' properties and their structure in the course of annealing at temperatures below and above the optimal ones are presented.

1 INTRODUCTION

The concept of the quasi-chemical model of the liquid micro-non-uniform composition is being developed under the supervision of B.A. Baum in our laboratory. According to it, the metal melt consists of space areas (groups, sibotaxes or clusters), within which the atom arrangement is characterized by certain ordering—short-range order. Due to moderately intense particles' heat movement, the clusters have no clear-cut boundaries. The atoms' ordered arrangement is being continuously replaced by another one, moving away from the cluster's core. For the same reason, the time of the given cluster's existence is limited and depends upon the chemical bonds' energy. At the same temperature, it is possible for two or more cluster ordering types to co-exist.

With the aim of determining the interrelation and interaction of real systems' hard and liquid phase properties, the liquid metal state has been studied. Therefore, preferences have to be given to the models based on the quasi-crystal description concept as compared with the variety of liquid's models.

Quasi-crystallizing implies modeling the liquid structure on the basis of actual inter-particle inter-actions clearly manifests themselves in a crystal body (in the absence of heat motion). This approach is based on the unity of the condensed substance state, the unity of inter-particle attraction forces on either side of the melting interval. Taking into account the above concept and research made on

the physical properties of the metal—and cobalt-based melts being crystallized, the unique technology of the melt time-temperature treatment has been developed. Amorphous ribbons produced, using this technology, require optimal annealing temperatures to be specifically selected.

2 EXPERIMENTAL

The kinematic viscosity ν is one of the important characteristics of metal melts and the main values measured in case torsional vibrations are recorded. Determining these values by torsional vibrations are based on measuring the vibration factor δ. The number of models and programs is designed to calculate ν from the δ known.

The amorphous ribbons were obtained by melt quenching on a rotating disk (ribbons were 20–25 μm thick and 10 mm wide). Toroidal specimens of the 32 mm outer diameter and 20 mm inner diameter were wound out of these ribbons. The studies of Finemet type specimens were carried out.

Static hysteresis loops, initial magnetic permeability $\mu_{0.08}$ and specific magnetic dissipation $P_{0.2/20}$, as well as magnetic hysteresis squareness ratio K_S were measured at annealing temperatures for 1 hour of ageing $P_{0.2/20}$ at frequency of 20 kHz and induction 0.2 T. Losses were estimated by means of dynamic hysteresis loops.

Atomic Force Microscopy (AFM) is based on the formation of images using the probe that scans

465

the specimen. By approaching the probe to the sample surface and moving in a raster scan, line by line, the resulting image is constructed. The force interaction acting between the tip and the specimen is represented as a function of position. Magnetic Force Microscopy (MFM) is a technique for imaging the distribution of magnetic field emanating from the magnetic sample. We used the Nntegra Prima platform (NT-MDT) operated in semi-contact regime. The two pass methods have been utilized to visualize the topography contour of the specimen as well as the magnetic contrast. The tips involved in measurement were coated with Co-Cr magnetic layer. However, we presented only the topography of measured samples because no magnetic response was observed.

Magneto-Optical Kerr Effect (MOKE) methods for surface hysteresis loops and magnetic domains observations were used. Hysteresis loop arrangement uses the light from the red laser diode operated at 670 nm, that is linearly polarized (s- or p-polarization) and incidents the shiny surface of the sample. The specimen is located within the air coil generating well defined external magnetic field along the ribbon axis. After the reflection, the beam is elliptically polarized. It continues through the Wollaston prism that splits the light into two mutually orthogonal laser beams captured by two PIN photodiodes. The differential intensity of those signals is proportional to the magneto-optical angle of Kerr rotation θ_K. Hysteresis loops are obtained by measuring θ_K as a function of external magnetic field applied in the sample plane and plane of light incidence (longitudinal configuration). The light penetrates into the depth up to 20 nm. And the information is obtained from local area restricted by the laser spot of 300 μm in diameter.

The specially designed polarization Kerr microscope is utilized for magnetic domain observations. This technique is based on the magneto-optical Kerr microscopy and uses the method of almost crossed polarizers. The best magnetic contrast is obtained by subtraction of two surface images. The first one is stored in memory of the computer while the saturation magnetic field is applied. The second one is the image when the magnetic field is gradually decreased towards to zero. In the paper we present the magnetic domain patterns of ribbons in remnant state with detection of longitudinal magnetization component.

3 RESULT AND DISCUSSION

The kinematic viscosity (Fig. 1) of nanocrystalline magnetic alloy of Finemet type has been studied. The first regime of heating was up to the temperature below t_c, supplemented by quenching.

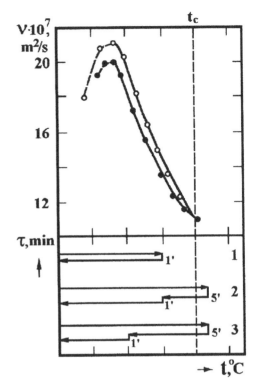

Figure 1. The polytherm of kinematic viscosity of the amorphous Finemet type alloy: • is heating; o is cooling; t_c is critical temperature. The numbers near the curves denote the thermal diagrams of the three regimes.

The second regime was heating up to the temperature above t_c supplemented by quenching as in the first regime. The third regime is heating up to the temperature above t_c followed by overcooling and further quenching.

The physical characteristics of amorphous ribbons were tabulated. The best service properties of ribbons were obtained in the third regime.

The physical properties versus the annealing temperature are shown in Table 1.

The optimal properties are obtained by annealing at temperatures ranging from 542 to 572°C. The maximum relative permeability and minimum coercive force are obtained during annealing at 542°C for the nanocrystalline material of a magnetic circuit.

Figure 2 represents the AFM observations of Finemet samples annealed at 300°C, 440°C, and 520°C. The topography presented for the specimens annealed at 300°C and 440°C is quite similar. Small clusters are dispersed almost randomly in amorphous matrix. The frequency of spherical formation is in strong correlation with the annealing temperature indicating the nanocrystallization process.

Table 1. Physical properties values, depending upon the annealing temperatures.

Annealing temperatures, T_a, °C	$\mu_{0.08}$	μ_{max}	K_R	H_c, A/m
482	7300	147000	0.69	2.15
502	14400	330000	0.77	1.25
522	25000	394000	0.65	0.90
532	35000	540000	0.63	0.66
542	52000	713000	0.63	0.41
552	91000	663000	0.59	0.45
562	98000	664000	0.63	0.46
572	120000	688000	0.61	0.51
582	105000	588000	0.58	0.56
592	69000	430000	0.56	0.56
602	61000	237000	0.59	1.58

it can be attributed to the FeSi nanocrystallites surrounded by the amorphous matrix.

Figures 3 and 4 represent the results of MOKE measurements. The sample annealed at 300°C exhibit still a lot of stresses coming from the preparation process. This is manifested by magnetic domain pattern consisting of two types of domains. Narrow curved domains indicate the regions influenced by the presence of tensile stress while the fine finger print domains refer to the planar compressive stress. Measured hysteresis loop shows low remanence-to-saturation ratio and high anisotropy field ($H_a \approx 60$ Oe). It is typical for a magnetostrictive alloy with random internal stresses introduced during the rapid solidification.

However, the shape of magnetic domains and hysteresis loops is changing due to the increase of

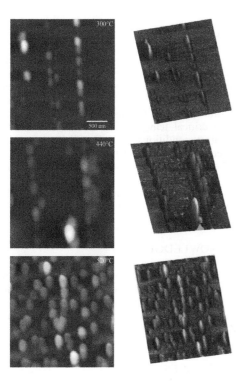

Figure 2. Atomic force microscopy image from the shiny side of ribbons annealed at 300°C (upper subplots), 440°C (middle subplots), and 520°C (lower subplots). In the left panels the topography of surface is shown, while the 3D visualization is represented in the right panels.

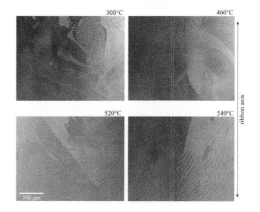

Figure 3. MOKE hysteresis loops observed on the shiny surface of ribbons annealed at 300°C, 520°C, 540°C, and 560°C.

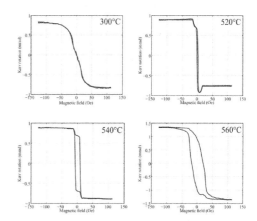

Figure 4. Magnetic domain patterns of Finemet samples measured using the MOKE microscopy annealed at different temperatures.

Contrary, the samples annealed at 520°C exhibit different surface morphology. The composition of particles appears to be homogeneous over the whole sample surface with the size up to 100 nm in average. In accordance with Ref. (Salinas et al. 2000), annealing temperature, when stress relaxation followed by the nanocrystal formation on the surface is expected. At annealed temperatures of 520°C and 540°C wide strip domains with fast reversal indicate enormous decrease of anisotropy field. The presence of finger print domains in specific regions refers to the fact that the stresses are not entirely relaxed. This is in good agreement with measured hysteresis loops showing the anisotropy field only about 15 Oe. Moreover, asymmetric loops were observed at annealing temperatures of 500°C (not shown) and 520°C. In accordance with Refs. (Zivotsky et al. 2012, Zivotsky et al. 2013) we expect that such asymmetries are the consequence of Quadratic MOKE (QMOKE). Surface magnetic properties are strongly inhomogeneous. Therefore some places exhibit the large QMOKE contributions while at the other places it vanishes. Such behaviour is explained by the random orientation of nanocrystals or amorphous clusters on the ribbon surface. The change of orientation and size of grains, manifested by increasing of the coercive field, is probably responsible for vanishing of QMOKE at temperature of 540°C. At symmetric loop we distinguish two magnetic phases, the first one with fast reversal (almost rectangular loop) and the second small contribution close to the saturation magnetic field. Their origin is connected with the formation of FeSi nanocrystals embedded in the amorphous matrix. Further increase of annealing temperature (560°C) leads to the enlargement of crystals size, to their gradual penetration into the ribbon bulk. And it marked deterioration of soft magnetic properties of the ribbons (increase of the anisotropy field and the coercive fields of both phases).

Figure 5 shows histograms of the grain sizes at different heating levels at t 522, 552 and 60 2°C.

The average grain size at these temperatures has not changed and was equal to 7, 9 and 8 nm, respectively. With the temperature increases, however, the distribution of grain size changes. Low temperature of 522°C is significant for the maximum grain size of 2 nm. These grains, with no clear boundaries, are, most likely, the nuclei (clusters) of the crystalline phase in the amorphous matrix.

At the temperatures of 552 and 602°C the grain size distribution is virtually identical. At a higher temperature compared with the temperature of 522°C, the proportion of grains is larger than 10 nm from 20.5% to 33.5%.

In the nanocrystalline magnetic alloys Finemet type, low initial permeability corresponds to the structure, the share of which in the residual non-crystallized amorphous matrix is significant. The coercive force of the alloy at the annealing temperature of 522°C is very low, i.e., in this area a low coercive force does not result in the high initial permeability. The highest initial permeability occurs when the structure is characterized by the largest volume of the ordered phase Fe_3Si. Increasing the coercive force and decreasing the initial permeability at high temperatures are attributed to the formation of the magnetically hard phase Fe_2B.

4 CONCLUSIONS

The scientific as well as applied approach to addressing the problem of the particle's liquid state structure should rely on experimental data related to the particular liquid; take into account the temperature interval of its existence.

The critical temperature values are set under laboratory conditions through studying the melt's properties temperature dependences.

The results of studying nanocrystalline magnetic core' properties and their structure in the course of annealing at temperatures below and above the optimal one are presented.

ACKNOWLEDGEMENT

This work was supported by IT4 Innovations Centre of Excellence project reg. no. CZ.1.05/1.1.00/02.0070 and Nanotechnology-basis for international cooperation project reg. no. CZ.1.07/2.3.00/20.0074.

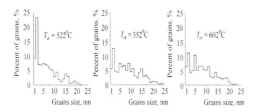

Figure 5. Frequency distribution of grain by size detected after heat treatment at 522°C, 552°C and 602°C. Investigated alloy is Finemet type.

REFERENCES

Salinas D.R., García S.G., Bessone J.B., Pierna A.R., Nanocrystallization process of the $Fe_{69.5}Cu_1$ $Nb_3B_9Si_{13.5}Cr_4$ FINEMET-type alloy: and AFM study, Surf. Interf. Anal. 30, 305 (2000).

Zivotsky O., Jiraskova Y., Hendrych A., Matejka V., L. Klimsa V., Bursik J., Influence of annealing temperature and atmosphere on surface microstructure and magnetism in FINEMET-type FeSiNbCuB ribbons, IEEE Trans. Magn. 48, 1367 (2012).

Zivotsky O., Klimsa L., Hendrych A., Jiraskova Y., Bursik J., A new phenomenon on the surface of FINEMET alloy, J. Supercond. Nov. Magn. 26, 1349 (2013).

Section 5: Biomedical

Sports Engineering and Computer Science – Luo (Ed.)
© 2015 Taylor & Francis Group, London, ISBN 978-1-138-02650-6

Experimental study of effects of different combinations of acupuncture points to the index of myocardium in exercise-induced fatigue rats

ChunMing Zhu

Department of Physical, Binzhou University, Binzhou, Shandong, China

ABSTRACT: In this research, we used experimental method to observe activity of SOD, GSH-Px and MDA content after swimming training. Besides, this method also observes implementation of different combinations of acupuncture points to SD rats in 5 groups. The results showed that acupuncture treatment made a greater impact in SOD, GSH-Px, MDA on myocardial rats. This indicated that acupuncture points had some effect on the elimination of exercise-induced fatigue in rats. By comparing the differences of data between groups, we could deduce the conclusion that acupuncture treatment (Dazhui plus Housanli) had the most obvious effect for eliminating athletic fatigue, following by acupuncture treatment (Sanyinjiao plus Housanli). However, acupuncture treatment of Shenyu plus Housanli is ineffective. The mechanism is not clear why reverse trends of certain cardiac index appeared, and this demands further studies.

Keywords: acupuncture treatment; combinations of acupuncture points; exercise-induced fatigue; myocardium; physiological and biochemical index

1 INTRODUCTION

As sport performance continues to be refreshed in international competition, the exercise training load is also increasing. This leads lack of timely cardiac recovery of accumulation fatigue to athletes during the training. Their hearts are constantly bearing too much burden to adapt and maintain the established exercise intensity. When it exceeds its tolerance, some ultra-structures of myocardial cells will irreversible change as a consequence. Numerous cases of sudden death are related to movement, which raise higher demand in the study of fatigue recovery. Abundant experiments show that the body is in a state of fatigue after a large load exercise. Some indicators of myocardial antioxidant system occur sensitive and corresponding changes: activity of SOD, GSH-Px decrease, MDA content increases and so on.

Therefore, these indicators after exercise and changes after acupuncture stimulation may reflect the impact of exercise on the body and the effect of acupuncture stimulation. We can use scientific, convenient, effective and non-toxic recovery means after exercise training. Then it will impact a significant and profound significance in improving the effect of exercise training. Acupuncture theory is based on the balance of Yin and Yang in theory. The theory is a unique Chinese medical treatment, and it possesses a favorable and two-way adjustment function. It would eliminate fatigue mechanisms through the effect of dredging the meridians, relaxing the muscles and stimulating the blood circulation, invigorating vital energy and replenishing blood, and balancing Yin and Yang. Previous studies have found that stimulate with a combination of different points will impact different effects on fatigue to body recovery. But which combination of acupuncture is better, and is more conducive to the elimination of accumulated fatigue, yet is to be verified.

This study is based on the theoretical foundation of the profound Chinese medicine. Actually, it is carried out to stimulate acupuncture points of Sanyinjiao, Dazhui, Shenyu, Zusanli of rats which are after heavy load exercise training. By it, we can observe changes in myocardial antioxidant capacity of each index. The experimental results gain effects of stimulating different combinations of acupuncture points to fatigue recovery, and reflect the degree of muscle fatigue elimination. It will help to screen out a combination of acupuncture points which is evident, scientific, rational and simple. At last, it can provide new ideas of eliminating exercise training fatigue.

2 SUBJECTS AND METHODS

2.1 *Experimental materials*

Experimental Subjects and Groups: Clean male healthy adult SD (Sprague-dawley) rats 60, weighing 250 ± 30 g. They are provided by the

Experimental Animal Center of Shanghai University of Traditional Chinese Medicine. With the national standard rodent diet, it also frees diet. The breeding environment temperature is $20 \pm 2°C$, with 12 hours light duration and the relative humidity of 40% to 55%.

As a result, 60 rats were randomly divided into five groups: sedentary control group, the training control group, the training acupuncture group 1, 2, 3 (Zusanli plus Sanyinjiao group, Dazhui plus Sanli group, Shenyu plus Zusanli group), n = 14.

2.1.1 *Experimental reagent*

Malondialdehyde (MDA) kits, superoxide dismutase (SOD) kit, glutathione peroxidase (GSH-Px) kit (provided by Nanjing Jiancheng Bioengineering Institute)

2.2 *Experimental controlling conditions*

Exercise environmental conditions:

The Pool is of iron sink, with smooth inner wall, and diameter 50 cm, and depth of about 60 cm which is more than 2 times the length of rat. The water temperature is controlled at $30 \pm 1°C$.

Exercise training program:

What the training program used is mainly adopted Zhao Liping [1] exercise fatigue model with minor changes. Sedentary control group fed routinely without any intervention, and did no exercise at ordinary times. Exercise control group underwent a 4 days adaptive swimming training, which is once a day with the training time of 30 min. Subsequently, the exercise fatigue model was established by swimming training for 10 days. The swimming training time was 90 min in the first 1–4 days, and the swimming training time was 120 min in the next 5–10 days. Actually, the steps are:

1. Pick it up promptly when the rat was found exhaustive during the training process;
2. Put it into the cage and dried its hair with a hair dryer;
3. Picked it up for a 5 min rest for the rats of short time exhaustion, and continued to swim training;
4. Use a stick to drive it in order to prevent floating and to maintain its motion state when it was found floated in the water.

Rat weight, athletic ability and activity state were observed and recorded in detail before and after swimming training. The training programs of sports acupuncture group and exercise control group were entirely consistent. At last, pick up the rat out of the water each time for 15 minutes rest to dry after training, and did its acupuncture on the treatment stage immediately.

Select and Collocation of Points: In the past experimental studies in rats with acupuncture, it has been widely confirmed that acupuncture Zusanli acupoints would alleviate exercise-induced fatigue. This experiment is based on the common points of attending functions, and adjunct acupuncture points with the principle of enhancing the efficacy of mutual collaboration. We used Zusanli (Housanli), coordinated with other synergistic points as Sanyinjiao, Dazhui, Shenyu to compare which combination would be more effective. We selected points according to Huaxing Bang [2] et al The development of the rat point map.

Acupuncture Methods: We used hand acupuncture to stimulate. Expect of not retaining the needle on points of Dazhui and Shenyu, the acupuncture time of other points is 1 min, with twisting frequency of 30 s a time. In order to allow all trained rats suffered consistent stimulus conditions, and thereby maximize the reliability of the experimental results. The exercise controls should also be fixed in the treatment stage, with the fixed time equal to time of acupuncture treatment in exercise acupuncture group.

2.3 *Sampling*

The rats were immediately sampled after the last training and acupuncture treatment. The rats are anaesthetized with intraperitoneal injection of 20% urethane (5 ml/kg). Directly cut the chest to pick out the heart first, and then rinse with ice saline. Place it in the –40°C ultra-low temperature freezer after dropping OTC embedding medium.

2.4 *Test indicators and methods*

Malondialdehyde (MDA), SOD, and GSH-Px, in strict accordance with the relevant provisions of the operation of each test kit.

2.5 *Statistics processing*

Experimental data of sample indicators are expressed in the way of Mean $\pm$ Standard deviation. We used SPSS13.0 statistical software package to do a single factor analysis of variance analysis. The significance level $\alpha = 0.05$.

3 RESULTS

3.1 *Comparison of myocardial SOD activity between groups*

Acupuncture treatment had an obvious impact on myocardial SOD activity of rats which had been trained with large loads exhaustive swimming.

a. The training control group was significantly lower than the sedentary control group (P < 0.05);
b. Training acupuncture group 1, 3 and the control group were higher compared to the training control group, in which training acupuncture group 3 was significantly increased (P < 0.01);
c. While training acupuncture group 1 compared with the exercise control group, it had tended to increase, but had no statistical significance (P> 0.05);
d. Training acupuncture group 2 compared with the training control group, tended to decrease (P> 0.05). (See Table 1).

3.2 Comparison between groups of myocardial GSH-Px activity

Acupuncture treatment had an obvious impact on myocardial GSH-Px activity of rats which had been trained with large loads exhaustive swimming.

a. The training control group was significantly lower than the sedentary control group (P < 0.05);
b. Training acupuncture group 1,3 and the control group were higher compared to the training control group, but there is no statistical significance (P > 0.05);
c. While training acupuncture group 2 compared with the training control group, it had tended to increase, but had no statistical significance (P> 0.05);
d. A comparison between the three training acupuncture group, the mean is 3 > 1 > 2, wherein there is a highly significant difference between the 2 and 3. (See Table 2).

Table 1. List of changes of myocardial SOD activity in each group ($\bar{x} \pm s$).

Group	N	SOD (U/mg prot)
Sedentary control group	10	31.30 ± 0.59
Training control group	11	28.89 ± 1.08*
Training acupuncture group 1	12	30.51 ± 0.50
Training acupuncture group 2	13	28.19 ± 0.62^Δ
Training acupuncture group 3	13	30.867 ± 0.55^{#ΔΔ}

Note: *represents:
a. There is significant difference comparing with sedentary control group, P < 0.05;
b. #i represents that there is significant difference comparing with training control group. (P < 0.05);
c. ^Δrepresents there is few significant difference (P < 0.05) comparing the latter training acupuncture group with the former training acupuncture group;
d. ^ΔΔrepresents there is highly significant difference (p < 0.01) comparing the latter training acupuncture group with the former training acupuncture group.

Table 2. List of changes of myocardial SOD activity in each group ($\bar{x} \pm s$).

Group	N	GSH-Px activity (U)
Sedentary control group	10	171.02 ± 13.44
Training control group	11	157.71 ± 9.12*
Training acupuncture group 1	12	164.30 ± 13.87
Training acupuncture group 2	13	149.18 ± 13.02^Δ
Training acupuncture group 3	13	169.53 ± 21.88^{ΔΔ}

Note: ditto.

Table 3. List of changes of myocardial MDA content in each group ($\bar{x} \pm s$).

Group	N	MDA content (nmol/mg prot)
Sedentary control group	10	1.25 ± 0.14
Training control group	11	1.46 ± 0.11*
Training acupuncture group 1	12	1.27 ± 0.09^{##}
Training acupuncture group 2	13	1.40 ± 0.17^Δ
Training acupuncture group 3	13	1.13 ± 0.09^{##ΔΔ▲}

a. Represents the mean was significantly different (P < 0.05), comparing with sedentary control group;
b. #represents the mean was significantly different (P < 0.05), comparing with training control group;
c. ##represents there was highly significant difference, comparing with training control group;
d. ^Δrepresents the mean was significantly different, comparing a group of the previous group are significantly different (P < 0.05) the number of the training acupuncture;
e. ^ΔΔrepresents there was highly significant difference (P < 0.01), comparing the latter training acupuncture group with the former training acupuncture group;
f. ▲represents there was highly significant difference (P < 0.01), comparing training acupuncture group 1 with training acupuncture group 2.

3.3 Comparison between groups of myocardial MDA content

Acupuncture treatment had an obvious impact on myocardial MDA content of rats which had been trained with large loads exhaustive swimming.

a. The significance was significantly lower in the control group than sedentary control group (P < 0.05);
b. Compare acupuncture training groups with control training group, they were all decreased, in which group 1, 3 were significantly decreased (P < 0.01). In addition, group 2 tended to decrease with no statistical significance (P > 0.05);
c. Compare each group between the three training acupuncture groups, the mean in each group is

3 < 1 < 2, wherein there was significant difference between 1 and 2 ($P < 0.05$). Besides, there were significant differences between groups 1,3 and groups 2,3. (See Table 3).

4 ANALYSIS AND DISCUSSION

4.1 *Effects of acupuncture to myocardial SOD activity of rats which are after big load exercise*

SOD is an important enzyme in the body's free radical scavenging systems, and distributes in the cytoplasm and mitochondria. Besides, it reflects the level of activity of the functional status of free radicals in the system at a certain extent. Most studies have shown that chronic aerobic exercise can vary degrees to improve SOD activity. This may be because that lipid peroxidation is relevant with exercise intensity and exercise duration. The exercise time of rats is short, and the body is far from exhaustive functional state. So there is no body anti-oxidative capacity shortage or relative lack of capacity shortage phenomenon after exhaustive exercise. On the contrary, SOD activity is activated due to the short-term effects of exercise on the mobilization of the tissues and organs of the body. Han Liming (1995) [3] reported that 70 min swimming significantly reduced myocardial MDA content in myocardium, liver and kidney, and increased cardiac SOD activity tendency.

The acute strenuous exercise and exhaustive exercise can cause decreased myocardial SOD activity. Zhang Jun found that MDA content in myocardial mitochondria was significantly increased in rats after exhaustive swimming. Moreover, SOD activity and GSH-Px activity was significantly reduced. Jenkins et al [4] reported that SOD activity was inhibited in the fatigued muscle cell. Wang wenxin et al [5] reported that SOD activity in myocardial mitochondria decreased after exhaustive exercise.

But there are also opposite results found. Zhang Yunhun (1995) [6] found that SOD levels fell in mouse heart, liver, and skeletal muscle tissue after exhaustive swimming, while SOD activity increased in the brain. And comparing with the control group, there was no significant difference found. And myocardial SOD activity was also found an increasing trend. The reason may be the effect of the delayed activation of antioxidant mechanisms. This phenomenon has also been similarly reported by Dai Yi (2001) in the study of the relationship between the central mechanisms and peripheral mechanisms of fatigue.

Acupuncture Zusanli could improve the overall health of the mice, improve the swimming endurance of the mice and improve SOD, GHS-Px

activity in blood plasma, liver, muscle, and decrease MDA content. It is shown in the study of acupuncturing Zusanli at swimming training mice. This indicates that acupuncture Zusanli can improve exercise capacity in mice. The mechanism is related to correcting imbalance of free radical metabolism in sports mice. Zhu Zhaohong et al have found that SOD in myocardial tissue and GSH-Px activity in liter tissue is all raised. MAD content in myocardial tissue are decreased in acupuncture mice group. This indicates that acupuncture could improve the metabolism of free radicals in multiple organs of mice.

The study found that myocardial SOD activity was significantly decreased ($P < 0.05$) in training control group comparing with the sedentary. In addition, this indicates that the ability of large load movement of generating free radicals in rats exceed the ability of scavenging SOD. SOD activity decreased on account of consumption of accumulated free radicals to SOD which led to body fatigue in rats. It may be because that exhausting exercise enables organizations to generate a lot of free radicals, which is in order to resist the attack of free radicals. And the myocardial tissue activated an antioxidant defense system. On one hand, SOD was consumed in scavenging free radical. On the other hand, intermediates of $H2O2$ produced from free radical disproportionation produced its own "killing" effect. It suppressed its own function, and therefore decreased enzyme activity.

There is an increasing trend but no statistically significance ($P > 0.05$), comparing the training acupuncture group 1 with the training control group. There is significant increase, comparing the training acupuncture group 3 with the training control group. As a result, it suggests that acupuncture treatment of (Sanyinjiao plus Housanli) and (Dazhui plus Housanli) increases SOD activity, and plays a certain effect to the recovery of fatigue. The mechanism may be because that acupuncture has a good regulation to mitochondria and cytoplasm of the body cells. And this may enable the depressed SOD resulted from the former state of fatigue to regress to normal levels. However, comparing with training control group, training acupuncture group 2 emerged to a reduced phenomenon which aggravate the body fatigue. That is acupuncture (Shenyu plus Housanli) may have the opposite effect, and the reason is not clear at present, it needs further study.

4.2 *Effects of acupuncture to myocardial GSH-Px activity of rats which are after big load exercise*

Glutathione peroxidase (GSH-Px) is widespread in the body and is an important antioxidant enzyme, which can catalyze reduction reactions of reduced glutathione (GSH). Thus protect integrity of the cell membrane structure and its function [8]. And its activity can be measured to reflect the degree of fatigue of the body.

FuYu Ping [9] reported that SOD and GSH-Px activity was significantly reduced (compared with the non-exercise group) in myocardial mitochondrial of rats which are after big load exercise. Chen Yajun [10] reported that GSH-Px activity had significantly decreased in two organizations of skeletal muscle and cardiac muscle of rats which are after acute exercise. Then it still could not fully recover in 48h. This could be related to increased free radical production after exercise, and too much consumption of GSH-Px. This indicates that acupuncture has some effect on improving SOD and GSH-Px activity in the body.

In addition, Xu Yuming et al [11] have found filiform needle and Magnetic pick-up needle has an obvious function in promoting the recovery of body fatigue. Acupuncture can stimulate the body's own potential to play. Zhai Daodang and Jin Wenquan et al [12] found that moxibustion could eliminate exercise fatigue by improving immunity and affecting metabolism.

This research found that GSH-Px activity was significantly decreased (P < 0.05) in sedentary control group compared to the training group. This indicates that free radicals of rats which are generated from large load exercise exceeded the GSH-Px scavenging capacity. Large accumulated and chemically active free radicals reacted with CHO, proteins, nucleic acids and lipids in the body. The things happened as follow steps:

1. caused damage and destruction of cell function and structure;
2. led to inactivation of GSH-Px;
3. blocked the catalytic reduction reaction of H2O2;
4. decreased myocardial contractility
5. finally, led to fatigue to rats.

This indicates that acupuncture treatment (Sanyinjiao plus Housanli), and (Dazhui plus Housanli) of these two points in combination can effectively stimulate the body's antioxidant enzyme system function, and improve cardiac GSH-PX activity. Then play a certain effect of eliminating sports fatigue. The mechanism may be that acupuncture has played a protective role in maintaining the integrity of cell membrane structure and function. Acupuncture can reduce the damaging effects of H2O2 and lipid peroxidation to the cell membrane. Consequently it enhances the ability of the body's antioxidant enzyme systems, eliminate sports fatigue and enhance the athletic ability. However, the training acupuncture group 2 comparing to the training control group has emerged to reduce the phenomenon, and enhanced body fatigue. That is, acupuncture (Shenyu plus Housanli) may have the opposite effect. The reason is not clear at present, it needs further study.

4.3 Changes of myocardial malondialdehyde (MDA) content of rats which are after acupuncture

When exercise-induced fatigue of the body occurs, the oxygen free radicals cause cell damage not only through peroxidation of polyunsaturated fatty acid in the biological membrane, but also through lipid hydroperoxide decomposition products (such as MDA, etc.)

Therefore, test MDA content in the myocardial antioxidant system could reflect the extent and degree of cell damage in vivo lipid peroxidation. In addition, the content can indirectly determine whether the body has a sport fatigue.

Fu yuping [13] found that MDA content increased significantly in myocardial Mitochondria after exhaustive exercise. Lipid peroxidation in post-exercise myocardial mitochondria was significantly elevated after exhaustive training which may be the main mechanism of myocardial injury. Guo yongli, Zhang jun et al [14] found that serum MAD content was significantly elevated when the rats swam to exhaustion. The research showed that after the rats experienced exhaustive exercise of moderate intensity, superoxide anion generated from skeletal muscle and MDA content were significantly higher than pre-exercise. Zhang jun [15] found that after the rats experienced exhaustive swimming. MDA content in myocardial mitochondria was significantly increased. The above reports indicate: MDA content in skeletal muscle and cardiac muscle was significantly higher in exhaustive rats than they were called sedentary rats before. Our experiment got the same results as the above reports.

Xu lanfeng et al [16] observed the role of health moxibustion in improving the body's SOD activity and decreasing the main degradation product malondialdehyde (MDA of lipid peroxidation). It indicates that acupuncture could reduce the effectiveness of MDA, and indirectly reflect that it could improve the body fatigue.

This study found (see Table 3) that MDA content in increased significantly (P < 0.05) in the training control group than in the sedentary control group. This indicates that myocardia have a good antioxidant capacity when it is in normal state. The myocardial antioxidant system is not sufficient to balance the oxygen free radicals after heavy load exercise. When the free radical scavenging system of the heart muscle can not balance the large number of free radicals, the damaging effects of free radicals begin to occur. The free radical scavenging systems becomes more and more difficult to balance the growing number of free radicals. The harmful effects would be more and more serious. Thereby it would affect the exercise capacity, and lead to exercise-induced fatigue.

a. Compared with training control group, the training acupuncture groups were significantly decreased;
b. Wherein training acupuncture group 1, 3 were significantly decreased ($P < 0.01$);
c. Training acupuncture group 2 tended to decrease, but without statistical significance ($P > 0.05$);
d. Compare between any two groups of the three, their means were 3 <1 <2, in which comparison between group 1 and group 2 was significantly different ($P < 0.05$);
e. Comparison between group 1 and group 3, and comparison between group 2 and group 3 were highly significant ($P < 0.01$).

This indicates that three kinds of combinations of acupuncture points have some effect on elimination exercise-induced fatigue, the average number is 3 < 1 < 2. This suggests that acupuncture treatment (Dazhui plus Housanli) have the best effect on elimination exercise-induced fatigue. Based on the theory of traditional Chinese medicine and acupuncture point function, the mechanism is probable that the rats are more sensitive to Dazhui acupuncture. Dazhui belongs to governor meridian points, and is the intersection of the hand channel, foot channel and Six Yang meridians which is commonly known as 'confluence of all yang-channels'. It is the main point of health care. Because acupuncture this point will have the effect of boosting yang, activating yang and relieving exterior, and expelling wind and cold pathogens, and Sedating fever. Match acupoint with Zusanli (Housanli), the effect of acupuncture treatment (Sanyinjiao plus Housanli) is followed, and acupuncture treatment (Shenyu plus Housanli) the most general.

5 CONCLUSION

Myocardial SOD rats acupuncture treatment could have an impact on SOD, GSH-Px, MDA in rats of each group. This indicates that acupuncture points could have some effect of the elimination of exercise-induced fatigue in rats. By comparing the data differences between groups, we found that acupuncture treatment (Dazhui plus Housanli) had the most obvious effect of eliminating the exercise-induced fatigue, acupuncture treatment (Sanyinjiao plus Housanli) followed. At last, the effect of acupuncture treatment (Shenyu plus Housanli) is inconspicuous.

The conditions why the trend of myocardial indicators appeared reverse changes, such as SOD reduced after acupuncture (Shenyu plus Housanli), is not clear. It needs further study.

In conclusion, acupuncture can be used as a safer and more effective means to restore the body's cardiac fatigue. It can reduce the extent of myocardial injury fatigue, and protect cardiomyocytes.

REFERENCES

[1] Watts R.L., Zimmerman J., Towards a Positive Theory of the Determination of Accounting Standards, The Accounting Review, pp. 112–134.
[2] Healy P.M, The Effect of Bonus Schemes on Accounting Decisions, Journal of Accounting and Economics, April, 85–107.
[3] Hopwood A.G., Towards an Organizational Perspective for the Study of Accounting and Information Systems, Accounting, Organizations and Society (No. 1, 1978) pp. 3–14.
[4] Collins, D.W., Kothari, S. P, An Analysis of Intertemporal and Cross-Sectional Determinants of Earnings Response Coefficients, journal of Accounting & Economics, pp. 143–181.
[5] Easton P.D, Zmijewski M.E, Cross-Sectional Variation in the Stock Market Response to Accounting Earnings Announcements, Journal of Accounting and Economics, 117–141.
[6] Beaver, W.H., 1968, The Information Content of Annual Earnings Announcements, Journal of Accounting Research, pp. 67–92.
[7] Holthausen R.W., Leftwich R.W., The Economic Consequences of Accounting Choice: Implications of Costly Contracting and Monitoring, journal of Accounting & Economics, August, pp77–117.
[8] Patell J.M, Corporate Forecasts of Earnings Per Share and Stock Price Behavior: Empirical Tests. Journal of Accounting Research, Autumn, 246–276.

Sports Engineering and Computer Science – Luo (Ed.)
© *2015 Taylor & Francis Group, London, ISBN 978-1-138-02650-6*

An experimental study of Planta Pedis' pressure distribution caused by different gaits of coednas

Jie Du, Baoying Yu & Suqing Yin
Hebei Institute of Physical Education, Shi Jiazhuang, Hebei Province, China
Project of Science and Technology Agency of Hebei Province, China

ABSTRACT: By means of the sports biomechanics, the thesis employs Belgium plat system of Footscan insole and tests the maximum pressure of the tenth subarea of coednas's Planta Pedis when they walk normally and jog in three different treads: normal walk, toe out and toe in. By employing the method of random sampling, the study takes 30 coednas as the research target and probes into the features of the deformed gaits. The aim is to make contributions for researchers to know about the shifting mechanism of human gait pattern.

Keywords: coednas; Planta Pedis; pressure

1 INTRODUCTION

It is not difficult to find that young students show up different degree of toe out, toe in, kicking the ground and tiptoe in their gaits. If young women develop the lower limb malformation, joining the army and work are to be affected, thus it has a bad effect on their lives. Some of them may be depressive, even worse, pessimistic and world-weary. It is known that plantar pressure is the pressure distribution between Planta Pedis and the bearing surface. It is the counterforce of the ground of the shoe sole directly received by human Planta Pedis in the motor process. Thus plantar pressure is the concurrent result of Planta Pedis and shear force. By employing the accurate regularities of distribution of Planta Pedis, the study analyzes the features of plantar pressure for coednas' different gaits, and probes into the motion features of two deformed gaits and normal gait. The aim of the study is to provide significant research basis for the assessment to motor function of human body and the design of orthopedic sneakers and artificial sport limbs.

2 RESEARCH OBJECT AND METHOD

2.1 *Research object*

Following the principle of simple random sampling, the study chooses 30 coednas as the research object. Their joint motion is normal and the average age is 22 years old (20–23). Their average weight is 54.1 kg (46–59 kg). The objects are without foot deformity, abnormal gait and foot trauma. See the general condition of subjects in Table 1.

2.2 *Research method*

The methods used in the thesis include Document literature, Measurement and Mathematical statistics.

2.2.1 *Consult document literature*

The study consults some important documents, theoretical works, and the research results of plantar pressure and gait feature, thus can make people fully understand some basic principles, methods and the status of the same research at home and abroad at present. It will avoid repetition of former research, find out the predecessor's deficiency and the study's innovation. At the same time, by selecting the method of gait analysis in advance and the index, it will provide powerful theoretical support for the success of the study.

Search strategy: Consult Wanfang database from the library of Hebei Institute of Physical Education and related thesis from December 1997 to June 2014 from CNKI. The index word of the thesis' title is gait and the language is limited to

Table 1. General condition of subjects.

Number	Age	Height (cm)	Weight (kg)	BMI (kg/m^2)
30	22.3 ± 1.6	162.6 ± 1.3	54.1 ± 6.4	20.5 ± 3.4

Chinese, getting 395 thesis, 39 of them are academic dissertation. The other index word of the thesis' title is plantar pressure, getting 48 thesis, 4 of them are academic dissertation. The researchers have skimmed and intensively read these data, and selected those data that meet the criteria and looked up the full article of them.

2.2.2 Measurement

Test instrument: Employ Belgium plat system of Footscan insole. Measurement frequency is 100 Hz, sensor thickness is 2.2 mm, sensor density is 4/cm², extraction thickness is 1.5 mm, pressure range is 1–60 N/cm², minimum resolution is 25 g, consistency is ±25 g.

2.2.3 Mathematical statistics

Data acquisition and analysis employs bundled software footscan software 7.0 with simple data statistics. The result is shown by $\bar{x} \pm S$, and the comparison among groups adopts one-way analysis of variance. The researchers divide planta pedis into 10 areas during analyzing: the first metatarsal bone, the second metatarsal bone, the third metatarsal bone, the fourth metatarsal bone, the fifth metatarsal bone, middle part of the foot or arch, heel medial, and heel lateral. The researchers calculate three times and measure average max force, contact area and impulse of each area. Besides, all the subjects are measured the related messages in height, weight, foot length and foot breadth etc. and examined to see if all of them have illness or not in their planta pedis before examination.

3 RESULTS AND ANALYSIS

3.1 Counterforce curve of the ground

The force received during sports is complex; it includes both internal force and external force. Inside the body, it has continuously changing myodynamia that gives rise to the counter-force of the ground, as well as the resistance caused by the tightening of myolemma, ligament, muscle tendon and the sarcous glutinousness. Outside the body, there is gravity, air friction, and the force acted on human body by the ground. The force acted on human body by the ground is the counterforce of the ground. It has three directions: vertical direction, front and back direction and left-right direction. As the measuring apparatus can test the counterforce in the vertical direction, the plantar pressure curve reflects the forces received by foot in the vertical direction. When the heel touches the ground, peak pressure appears in the heel and quickly drops. By the support of the middle part of the foot, the second peak appears in the front of the metatarsal bone of the front sole or toes. During sports, if the relatively higher plantar pressure appears, the tissue of planta pedis is injured. The maximum stress value had better appear in the forefoot before leaving the ground in order to enhance the stability and reduce the impact. Besides, the bigger difference between the numerical value of the maximum stress when the half sole leaving the ground and the heelpiece down to the ground, the more stable the control of the foot.

3.2 Centrode of plantar pressure

The arch of foot is composed of inside longitudinal arch, outside longitudinal arch and transverse arch. It is solid, light and stretchy. It is good for lasting standing and supporting bigger pressure. When walking, it can buffer the impact caused by the supporting counterforce. The inside longitudinal arch is high and has a big span. It is from calcaneus, passes through astragalus, navicular bone, cuneiform and is composed of the first metatarsal bone, the second metatarsal bone and the third metatarsal bone. The astragalus is the highest and at the top of inside longitudinal arch. Human centrode of plantar pressure exactly conforms to the inside longitudinal arch in stress diagram in dissection. It is the rule of normal human physiological movement. When people walk in the gait of toe out or toe in, centrode of plantar pressure

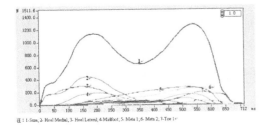

Figure 1. Counterforce curve of the ground.

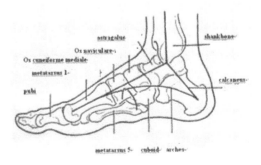

Figure 2. Structure of the arch of foot.

confines to the heel and the first metatarsal bone, the second metatarsal bone and the third metatarsal bone. It conforms to the rule of normal human physiological movement. If the pressure centrode is curved, it shows that human body's range of vacillation from left to right is big. If the shifting speed of pressure centre is too fast and asymmetrical, it shows that the control of the middle part of foot's touching the ground is bad. If the front sole is in the area of dense pressure center, it shows that the front sole's pedal and stretch is delayed, the touchdown is too long, and the pressure center is too dense. If the heavy load stimulates the second metatarsal bone and the third metatarsal bone for a long time, callosum is easily formed. The training of ligament in the planta pedis should be strengthened. Enhancing the control of the middle part of foot can make the pressure center move steadily and the speed more evenly. If the touchdown is moderate, the pressure in the front sole is decreased.

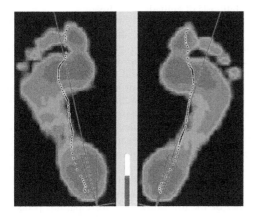

Figure 3.   Centrode of plantar pressure.

3.3   Pressure analysis of the tenth subarea in planta pedis

3.3.1   The distribution of planta pressure when walking in three-step gait

Seen from Table 2, when the subjects walk in normal gait, the maximum pressure appears in the second metatarsal bone of the front sole. The maximum pressure of each part of body in descending order is as follows: the first metatarsal bone, the third metatarsal bone, the fourth metatarsal bone, arch, the first phalanx, the second to the fifth phalanx, the fifth metatarsal bone, and the regularities of distribution of both feet is almost in consistency. When walking in the gait of toe out, the maximum pressure appears in the heel. The maximum pressure of each part of body in descending order is as follows: the second metatarsal bone, the first metatarsal bone, the third metatarsal bone, the first phalanx, the fourth phalanx, the second to the fifth phalanx, arch, the fifth metatarsal bone, and the regularities of distribution of both feet is almost in consistency. When walking in the gait of toe in, the maximum pressure appears in the heel. The maximum pressure of each part of body in descending order is as follows: the second metatarsal bone, the third metatarsal bone, arch, the fourth metatarsal bone, the first phalanx, the fifth metatarsal bone, the second to the fifth phalanx, and the regularities of distribution of both feet is almost in consistency. When the subjects walk in the gait of toe out and toe in, the heel peak is higher than that of normal gait, the discrepancy is of statistical significance ($P < 0.01$). As the pressure in the plantar pedis from the heel to and entire foot and front sole is not in good transition, the buffering effect is not well received, making the heelpiece get bigger pressure. When walking in gait of toe out and toe in, the pressure of the front sole is less than that

Table 2.   The maximum pressure of the tenth N subarea in planta pedis when walking in three-step gait.

Body parts	Plants pressure in the left			Plants pressure in the right		
	Normal gait	Toe out	Toe in	Normal gait	Toe out	Toe in
Toe 1	48.0±11.3	50.1±14.2	27.9±4.1*	51.9±4.2	81.4±13.7	12.9±4.1**
Toe 2–5	13.7±3.6	20±9.1	22.7±8.3	30.9±50.9	24.4±6.4	12±3.5
Meta 1	79.3±6.8	86.1±19.4*	27.9±7.2**	70.7±12.4	84.9±15.3	29.1±7.2*
Meta 2	127.3±31.2	99.9±21.6*	87.9±13.2*	131.4±26.8	96.9±18.6*	99.4±25.3*
Meta 3	6.4±9.4	83.6±15.3	102.9±24.1	102±29.5	45.0±6.3	69.9±17.4
Meta 4	20.1±4.8	30.4±7.3	35.1±8.4	37.7±8.3	13.7±4.6	26.1±3.6
Meta 5	8.6±1.2	1.2±3.5	2.1±3.8**	9±1.4	5.6±1.5	16.3±1.4
Midfoor	9.9±2.1	3.4±1.4	43.3±13.1**	14.1±2.7	16.3±3.9	33.4±4.8
Heel medial	115.4±21.3	142.3±32.4*	112.3±38.6	123.4±23.6	192.9±35.7**	135±26.9*
Heel lateral	122.6±26.1	150.3±29.7**	188.3±32.9*	87.4±28.6	140.6±30.4**	132.4±31.5**

Note: *Compared with normal gait, $P < 0.05$, **Compared with normal gait $P < 0.01$.

Table 3. The maximum pressure of the tenth N subarea in the planta pedis when jogging in three-step gait.

Body parts	Plants pressure in the left			Plants pressure in the right		
	Normal gait	Toe out	Toe in	Normal gait	Toe out	Toe in
Toe 1	46.3±8.1	75±8.7	48.4±7.5	104.1±9.7	93.4±7.3	80.1±7.4
Toe 2–5	20.1±3.5	8.1±1.3*	40.7±4.2	33.4±2.5	4.3±1.0	48.9±2.8
Meta 1	135±18.4	185.6±145**	124.7±11.4*	201.9±19.4	273.4±23.8**	153±12.4*
Meta 2	150±23.7	141±19.4	152.6±15.7	215.1±24.3	224.6±21.5	194.1±16.5*
Meta 3	114.4±18.4	97.7±8.3*	115.7±9.5	108.9±8.0	89.1±7.4	93.4±8.2
Meta 4	42±3.2	34.7±6.9	46.7±7.4	39.4±3.9	28.3±2.0	33.9±7.0
Meta 5	18.9±4.2	15.9±2.7	36.4±4.9	18.4±2.1	9.0±1.3	20.1±2.7
Midfoor	24.4±7.6	35.1±4.6	87±9.2	24.9±3.8	42.9±7.5	29.1±4.3
Heel medial	99±10.5	165±21.3**	213.9±249**	131.6±12.0	190.3±16.4**	133.3±14.2
Heel lateral	103.7±17.8	121.7±16.5*	189.9±19.6**	124.3±11.7	141.9±9.5**	163.4±110**

Note: *Compared with normal gait, $P < 0.05$, **Compared with normal gait $P < 0.01$.

of the heelpiece, it is not good for the stability of walking.

3.3.2 The distribution of planta pressure when jogging in three-step gait

Seen from Table 3, when the subjects jog in normal gait, the maximum pressure appears in the second metatarsal bone of the front sole. The maximum pressure of each part of body in descending order is as follows: the first metatarsal bone, the third metatarsal bone, heel, the first phalanx, the fourth metatarsal bone, the fifth metatarsal bone, arch, the second to the fifth phalanx, and the regularities of distribution of both feet is almost in consistency. When jogging in the gait of toe out, the maximum pressure appears in the first metatarsal bone. The maximum pressure of each part of body in descending order is as follows: the second metatarsal bone, heel, the third metatarsal bone, the first phalanx, arch, the fourth metatarsal bone, the second to the fifth phalanx, the fifth metatarsal bone, and the regularities of distribution of both feet is almost in consistency. When jogging in the gait of toe in, the maximum pressure appears in the second metatarsal bone. The maximum pressure of each part of body in descending order is as follows: the first metatarsal bone, heel, the third the first metatarsal bone, the first phalanx, the second to the fifth phalanx, arch, the fourth metatarsal bone, the fifth metatarsal bone, and the regularities of distribution of both feet is almost in consistency. When the subjects jog in the gait of toe out and toe in, the heel peak is higher than that of normal gait, the discrepancy is of statistical significance ($P < 0.01$). As the pressure in the plantar pedis from the heel to and entire foot and front sole is not in good transition, the buffering effect is not well received, making the heelpiece get bigger pressure. When jogging in the gait of toe out and toe in, the pressure in the first and second metatarsal bone is higher than that of normal gait, the discrepancy is of statistical significance.

Pressure peak is an important indicator in reflecting the pressure distribution of plantar pressure. The number of pressure peak and body parts has a significant influence on the accumulation of foot weariness and occurrence of damage. The study tests the distribution of plantar pressure when the subjects walk in toe out, toe in and normal gait, which is different from that of foreign races. It shows that different races have different plantar pressure. At the same time, people of different region and life style and mannerism are different in plantar pressure. Thus when employing it, researchers shouldn't copy or use foreign data and normal value. When normal person walks, the order of the touchdown of entire plantar pedis is as follows: heel, arch, metatarsal bone and phalanx etc. When different groups of people walk, features of gait are different, which shows the pressure interphase is different in the plantar pedis' touchdown.

4 CONCLUSION

When the subjects walk in the gaits of toe out and toe in, the peak heel is higher than that of normal gait, the discrepancy is of statistical significance ($P < 0.01$). As the pressure in the plantar pedis from the heel to and entire foot and front sole is not in good transition, the buffering effect is not well received, making the heelpiece receive very great pressure. When the subjects walk in the gaits of toe out and toe in, the pressure of front sole is less than that of back sole, it is not good for the stability of walking. When the subjects jog in toe out and toe in, the pressure in the first metatarsal bone and second metatarsal bone is bigger that of normal gait, the discrepancy is of statistical significance.

REFERENCES

Betts R.P, Frank C.I, Dnckworth T.J, et al. 1980, Static and Med dynamic foot pressure measurement in clinical orthopaedics, Biol Eng Comput. 18(9): 674–684.

Hongfeng Huo, Yanxia Wu, Liming Fu, Huanbin Zhao, 2009, The features of plantar pressure when young woman walking in abnormal gait. Chinese Journal of rehabilitation Medicine. 9(24): 841–843.

Hua Zhou, 2007, The study of the athlete's gait of middle distance race in intermediate run and the distribution of plantar pressure. Shan Dong Institute.

Lord M, Reynolds D.P, Hughes, J.R.1986. Foot pressure measurement: A review of clinical findings. Biomed Eng, 8(10): 283–294.

Xiangjiang Rong, Hongen Yao, Weiqiang Wang etc. 2004, A quantitative research of the posture and cycle of hemiplegic gait. Journal of TianJin Institute of Physical Education.19(2):56–58.

Xiaolan Zhu, Fang Zhao, Xinglong Zhou. 2006, An analysis of the features of aged person's gait and the initial establishing of its evaluation system. Journal of Beijing sport university. 29(2): 201–203.

Sports Engineering and Computer Science – Luo (Ed.)
© 2015 Taylor & Francis Group, London, ISBN 978-1-138-02650-6

Instrumented evaluation of upper extremity motor function for stroke rehabilitation applications

K. Daunoravičienė, J. Žižienė, J. Griškevičius & A. Linkel
Vilnius Gediminas Technical University, Vilnius, Lithuania

A. Juocevičius & I. Raudonytė
Vilnius University Hospital Santariskiu Klinikos, Vilnius, Lithuania

ABSTRACT: The present work introduces to another method of evaluation of stroke (ST) patients Upper Extremities (UE) movements. ST is a third leading cause of dead in the most countries of the world. More than half ST of the suffered patients become disabled for short time and 10% need care forever. An assessment of effectiveness of any proposed rehabilitation method requires evaluating objectively the recovered movement of upper extremity. This article explores the possibilities of quantitative evaluation of upper extremity function via instrumented analysis, thus proposing an additional tool that could facilitate the monitoring of rehabilitation process and with larger accuracy. Defined by the physician subjective values according to Wolf Motor Function Test (WMFT) scale corresponds to objective intervals of UE segments movement's amplitudes of joint angle and angular velocity.

1 INTRODUCTION

ST is a leading cause of death in a most countries of the world after cardiovascular and cancer diseases (Andrejevaite, 2009, Valaikiene, 2007). Lithuania suffers 2 times worst situation compare with the Europe ST position because 3000 ST patients of 20000 die every year (HPML 2014). In the course of time ST patients mortality decreases but number of residual effects of movements after ST tendentiously increases (Edwards, 2002). During first 30 days after ST about 17–34% patients die (Juocevicius, 2009a). More than 50% of the ST experienced patients become disabled temporarily or forever, 20% are able to come back to their previous work and 10% require care in the future (Juocevicius, 2009b). Statistically it is shown that 15% of the ST suffered patients die because of infarct (Janoniene, 2004). ST patient's expenditure of treatment in the Britain represents 5% of the National Health protection budget (Juocevicius, 2009a). In Sweden treatment of one ST patient per year costs 79000 US dollars (Juocevicius, 2010). In general ST is important economic and social problem of all countries in the world.

As early as possible (in first 90 days after stroke) and individualized rehabilitation after ST is important for successful recovery of biomechanical UE function (Dorothy, F. 2012). It is important to evaluate ST patient's affected movements of UE or rehabilitation effectiveness by one of the methods:

Action Research Arm (ARAT), Box and Blocks (BB), Jebsen-Taylor Hand Function, Nine-Hole Peg, Chedoke Arm and Hand Activity Inventory (CHANAI) Tests, WMFT and so on. (Catheryne, 2013). WMFT is one of the most popular and contemporary methods to evaluate UE movement quality in early stage after ST and during rehabilitation to observe the progress of movement recovery (Steven, 2001).

Nowadays rehabilitation physicians use combined methods for the assessment of patients' motor functions and use various scoring systems such as International Classification of Functioning, disability and health (ICF), Barthel index of activities and daily living, Wolf motor function test and so on. Despite the fact, that application of WMFT is plausible due to its' high reliability (David, 2001) and it is comparable with other methods similar to Cronbach alpha (Rinske, 2010), there is still a risk of subjectivity because WMFT evaluation scale ranges from 0 to 5 and different physicians can score the same movement with different scores. Therefore, the main purpose of our research is to explore the possibilities of quantitative evaluation of upper extremity function via instrumented analysis, thus proposing an additional tool that could facilitate the monitoring of rehabilitation process and with larger accuracy. The basic task of this article is to introduce the numerical method and kinematic testing to obtain objective and sensitive quantification of UE movement during combined WMFT and ICF exercises.

2 METHODS

2.1 Subjects

All measurements were carried out at the Vilnius University Hospital Santariškių Klinikos Center of Sports Medicine and Physical Rehabilitation. A total of fourteen ST patients (5 women and 9 men, mean age was 60.8 ± 12.5 (mean $\pm$ SD)) were recruited in the study. None of participants had any other injuries or disease affecting movement or coordination and all provided informed consent prior to participating in the study. Measurements were made for all participants before rehabilitation and some of them repeated provided tasks after rehabilitation.

2.2 Experimental setup, motor tasks and parameters

Four portative wireless inertial measurement units (Shimmer Research, Dublin, Ireland) each capable measuring linear acceleration, angular velocity and magnetic heading in 3D were used for measurements. Three sensors were fixed on the arm, forearm, and hand and last one on the chest by the straps. The data from the sensors was received via a Bluetooth connection at a sampling frequency of 51.2 Hz and stored on the computer and processed using LabVIEW, MATLAB and Excel software (see Fig. 1). Subjects were sitting on the chair near the table. Each ST subject was asked to perform 11 motor tasks specifically designed by the physiotherapist for the assessment of UE function according to WMFT and ICF guidelines. Maximal angular velocities (ω_{max}) of each arm's segment, movement duration (t) were measured and the range of motion of shoulder and elbow joints (A_{sh} and A_{el}) with maximal joint amplitudes (A_{max}) was calculated and chosen for further evaluation of motor performance.

Out of total 11 motor tasks only three tasks were executed by all participants. Other exercises were performed by stronger participants with less ST damage or after rehabilitation. During each task the physiotherapist rated each subject's performance according to WMFT scale (0 to 5). Each task was repeated 3 times. Successfully completed motor tasks by the subjects were chosen for further analysis. According to WMFT and ICF, completed tasks were as follows:

1. IIA: extend elbow task, requiring assessing functional ability of the elbow, when patient attempts to reach across the table by sliding hand palm-down placed on textile cloth forwards and second time backwards.
2. The second and third tasks were performed in the same sitting position, only the subject was able to push the arm side-on (elbow extension in horizontal plane) and after back towards himself (elbow flexion in horizontal plane).

Probably after rehabilitation these ST subjects could perform provided tasks in better quality with more evaluable rates, but at this stage only amplitudes of the UE motion and the task duration (t) could be assessed and compared with the rates set in WMFT score. MATLAB software was used to process raw measurements data and calculate the above-mentioned parameters.

3 RESULTS

According to WMFT 0 points score means that there is no movement during the task. 1 point means very small functional activity of the damaged UE and any movement needs a lot of efforts. Subjects evaluated 0 or 1 did not show any significant changes of amplitudes after the measurements. However, those subjects who were evaluated in 2 and more points showed meaningful enough results of the ST affected UE. The amplitudes of the shoulder and elbow were increasing and only the task was performed very slowly with all efforts. The Figures 2 and 3 presents A_{sh} and A_{el} (dotted lines), and the average amplitude (bold lines changes over the movement cycle of ST patients rated in 2 points and the same dependencies for patients rated by 5 point according to WMFT (see Fig. 3).

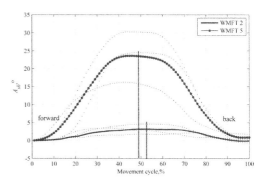

Figure 2. Shoulder joint amplitudes during first task of ST patients rated by 2 and 5 according WMFT.

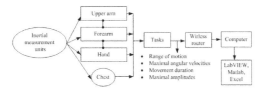

Figure 1. Measurements scheme.

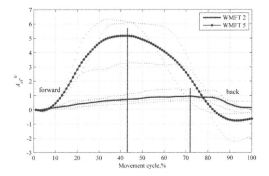

Figure 3. Elbow joint amplitudes during first task of ST patients rated by 2 and 5 according WMFT.

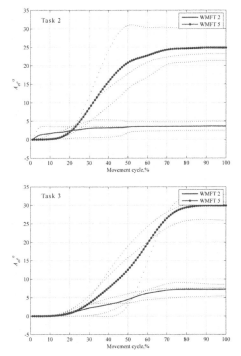

Figure 4. Elbow joint amplitudes during second and third tasks of ST patients rated by 2 and 5 according WMFT.

It was noticed from the Figure 4 that maximal A_{sh} exceeded in the end of first task stage (pushing the arm forward or at shoulder extension), approximately in the middle of the movement cycle. The amplitudes are very small in comparison with variations of arm segments amplitudes during cycle received form the ST patients rated by WMFT in 5 points.

This rate means that the movement is normal and the task was performed very well. It should

be noted that only ST patients after rehabilitation could perform such quality movements and the highest rate for ST patients before rehabilitation was 4 or less. It is observed from Figures 2 and 3 that A_{sh} and A_{el} maximal values are reached not at the same time; more synchrony ST damaged UE movement noted in results of ST patients with higher results by WMFT. This explained the visual observation as ST patients with lower WMFT score performed weaker, slower and not coordinated UE segments movement, when one UE segment compensate another's motion.

Levels of A_{sh} and A_{el} presented in the Figure 5 show the differences of values and relation with WMFT score. In all tasks in higher WMFT rate the growing values of amplitudes were found and the movement time tended to decrease. This proves the fact that more precise and qualitative movement is performed more quickly in comparison with weak and not coordinated UE motion.

It was noted, that all ST subjects doing exercises tried to move their damaged UE first by moving and swinging the body and after that the movement of the arm began from the shoulder joint. The role of the hand in the task was minimal or equal to zero. Intervals of A_{sh} and A_{el}, and movement time according WMFT are presented in Table 1.

In general, maximum angular velocities tend to increase approximately from minimum 2° to

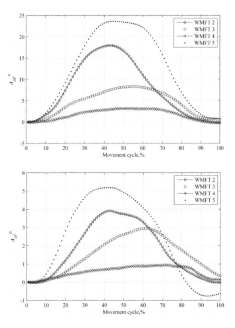

Figure 5. Shoulder and elbow joints amplitudes during first task of ST patients rated by 2 and 5 according WMFT.

Table 1. Parameters.

WMFT	2	3	4	5
Task 1				
$A_{sh,}$ °	2.11÷5.39	7.72÷10.20	11.79÷23.35	16.23÷30.25
$A_{el,}$ °	0.62÷1.45	1.60÷4.76	2.50÷5.72	3.29÷6.31
t, s	7.62÷10.03	5.62÷9.03	2.86÷5.38	2.51÷4.40
Task 2				
$A_{el,}$ °	2.47÷5.36	5.67÷7.82	6.86÷25.26	21.37÷31.04
t, s	5.04÷7.91	2.35÷8.40	1.84÷4.90	1.46÷3.34
Task 3				
$A_{el,}$ °	5.47÷11.24	11.36÷13.8	16.30÷21.51	26.15÷32.88
t, s	3.28÷10.70	3.10÷6.89	2.03÷3.63	1.9÷2.52

maximum 32° in the scale of WMFT from 2 to 5 points in all performed motor tasks. In every level of WMFT they became about 2 times bigger and movements were performed faster (*t* was increasing).

From all eleven motor tasks only three of them were fully performed by the patients and considered of good quality. Therefore, these motor tasks could be used for the evaluation of ST damaged UE function before rehabilitation and repeated post-rehabilitation. After analyzing results of each ST subject the relation between measured values and usual WMFT rates was clearly seen. It was found that every higher Wolf scores match higher amplitudes of the upper extremity segments and the task was made more actively, accurate and faster. Therefore it was decided to collect all measured parameters meaningful to the WMFT score, and to show that the movements of patients could be evaluated not only by visual sight but also by some parameters that could allow to select and to provide effective means of rehabilitation to individual ST person.

4 CONCLUSIONS

The close relation between kinematic parameters and WMFT rate was found and alternative evaluation of ST patients UE movements was proposed. This proposed kinematic testing could be used to obtain objective and sensitive quantification of UE movement during WMFT and ICF exercises. ST patients could be evaluated not from the visual sight by WMFT scale, but more detailed analysis could be made by assessing biomechanical parameters of the ST affected UE. Such methodology could be used for all the tasks provided in this article to ST patients and help to identify the target of rehabilitation. However, this presented research has to be improved by larger number of tested ST patients before and after rehabilitation to get reliable enough method for UE movement recovery evaluation.

This is important for our further work towards bio-mechanical model-based rehabilitation.

REFERENCES

Andrejevaite, V. et al. 2009. Seimos medicina. Vilnius: Nacionalinis medicinos mokymo centras.

Catheryne, E. 2013. Assessment of upper extremity impairment, function, and activity following stroke: Foundations for clinical decision making. Journal of hand Therapy 26(2): 104–115.

David M.M. 2001. The Reliability of the Wolf Motor Function Test for Assessing Upper Extremity Function After Stroke. Arch Phys Med Rehabil 82: 750–755.

Dorothy, F. et al. 2012. An Evaluation of the Wolf Motor Function Test in Motor Trials Early After Stroke. Arch Phys Med Rehabil. 93(4): 660–668.

Edwards, S. 2002. Neurological physiotherapy. A problem—solving approach. Churchill Livingstone 35(63): 120–50.

Health Protection ministry of Lithuania (HPML): Web page www.sam.lt attended 2014-09-06.

Janoniene, D. 2004. Ligoniu kompleksinės reabilitacijos efektyvumas po galvos smegenų kraujotakos sutrikimų. Sveikatos mokslai 1:36–40.

Juocevicius, A. et al. 2009a. Sergančių galvos smegenų insultu pacientų, reabilituotų trijose reabilitacijos paslaugas teikiančiose įstaigose, charakteristika. Gerontologija 10(4): 214–222.

Juocevicius, A. et al. 2009b. Sergančių galvos smegenų insultu pacientų funkcinės būklės pokyčiai ir jų prognozė antrame reabilitacijos periode. Gerontologija 10(4): 238–248.

Juocevicius, A. et al. 2010. The influence of feedback on balance in stroke patients. Gerontologija 11(4): 233–239.

Rinske, N. et al. 2010. A Comparison of Two Validated Tests For Upper Limb Function After Stroke: the WMFT ARAT. Rehabil Med 42: 694–696.

Steven, L.W. et al. 2001. Assessing Wolf Motor Function Test as Outcome Measure for Research in Patients After Stroke. Stroke 32: 1635–1639.

Valaikiene, J. 2007. Galvos smegenų insultas: etiopatogenezė, paplitimas, diagnostikos metodai ir jų vertė parenkant optimalią gydymo taktiką. Radiologija 13(3): 225–231.

Sports Engineering and Computer Science – Luo (Ed.)
© 2015 Taylor & Francis Group, London, ISBN 978-1-138-02650-6

Biomechanical markers for Parkinson's Disease and essential tremor discrimination

J. Griškevičius, J. Žižienė, K. Daunoravičienė & D. Lukšys
Vilnius Gediminas Technical University, Vilnius, Lithuania

V. Budrys, R. Kaladytė-Lokominienė & D. Jatužis
Clinics of Neurology and Neurosurgery, Faculty of Medicine, Vilnius University, Vilnius, Lithuania
Vilnius University Hospital Santariskiu Klinikos, Vilnius, Lithuania

P.M. Aubin
University of Washington, Seattle, USA

ABSTRACT: Analysis of Rapid Alternating Movements (RAM) can be used to detect movement disorders such as Parkinson's Disease (PD). The present work applies the principles of RAM analysis for extracting potential (or sensitive) biomechanical markers that would enable the discrimination of PD from Essential Tremor (ET). PD and ET have similar symptoms making differential diagnoses sometimes challenging. A total of 30 subjects were divided into three groups: 10 PD, 10 ET and 10 healthy control subjects. Wireless inertial sensors were used to measure angular velocity and acceleration during multi-joint arm motion performing rapid alternating movements of flexion and extension of an elbow joint in sagittal plane. Several parameters representing biomechanical performance of upper extremity were extracted from the RAMs. The results showed that amplitude of the first peak of shoulder and elbow angular velocities and time until the first peak are the potential metrics that can discriminate between PD and ET subjects.

1 INTRODUCTION

Parkinson Disease (PD) is a chronic, progressive neurodegenerative disorder characterized by motor symptoms of rest tremor, rigidity, bradykinesia and gait disturbance. Despite the fact, that PD is a common disorder, accurate diagnosis remains challenging especially in early stages of the disease. As stated in the study of Suchowersky et al. (2006), currently the best methods for diagnosis confirmation appear clinical examination with long term follow-up. Typically, PD is diagnosed clinically using Unified Parkinson's Disease Rating Scale (UPDRS). There are other rating scales, although they have not been fully evaluated for validity and reliability (Ramaker et al. 2002, Edersbach et al. 2006). The diagnosis can be difficult, especially when differentiating it from Essential Tremor (ET), a "benign" disorder characterized by action tremor and postural tremor without latency. The study of Jankovic et al. (2008) showed that during early onset the manifestation of PD symptoms may be similar to essential tremor making a clinical diagnosis difficult. Three cardinal clinical features of Parkinsonism are tremor at rest, rigidity, and

bradykinesia, in addition with postural instability and freezing in the late stages. Hughes et al. (1992) reported that 69% of the patients with PD have rest tremor at disease onset.

According to several studies for diagnosis of Parkinson's disease various markers are needed to differentiate neurodegenerative disorders and predict PD before the onset of symptoms (Robichaud et al. 2009, Shtilbans & Henchcliffe 2012, Haas et al. 2012). While many PD features are relying on motor PD markers, studies showed that, for example rapid eye movement behavior disorder can precede clinical motor symptoms associated with PD (Haas et al. 2012). Among these there are other clinical biomarkers such as bowel dysfunction, olfactory deficits and mood disorders.

There has been encouraging work that demonstrates PD subjects can be distinguished from healthy control subjects based on a patient's upper extremity kinematic (Chan et al. 2010) or electro-myographic (Robichaud et al. 2009) data. Different technologies have been used to record movements in 2D via digitizing pads (Fimbel et al. 2005), 3D tracking systems (Poizner et al. 1998), diado-chokinesimeter for rapid alternating movements of wrists (Deprés et al. 2005) and six degrees of freedom magnetic motion tracker and rotational sensors for simultaneous recording of wrist and forearm movements (Ghassemi et al. 2006). While analyzing rapid alternating movements, various metrics have been proposed to quantify recorded data. Some authors applied information theory methods to evaluate performance of periodic movements in a horizontal plane using mutual information metrics (Oliveira et al. 2011). Other authors use similarly called metrics such as irregularity of RAM cycle amplitude (Ghassemi et al. 2006) and index of irregularity (Fimbel et al. 2005) to assess motor performance and lack of smoothness of motion. Despite named similarly by using the term "irregularity", irregularity of RAM cycle amplitude evaluates standard deviation of linear envelope of absolute amplitude values, while index of irregularity correspond to the presence of jerk singularities or number of upwards zero-crossings in second derivative of limb's angular velocity.

The present work applies the principles of RAM analysis for extracting potential (or most sensitive) biomechanical markers that would enable discrimination between PD and essential tremor. Our long-term goal of this research is to discover biomechanical markers of PD and ET in order to facilitate clinical diagnosis and monitoring of the disease progression. In this study we collected upper extremity kinematic data from PD, ET and healthy control (CO) subjects and investigated possible RAM features and metrics that may aid in classification.

2 MATERIALS AND METHODS

2.1 Subjects

Research was carried out on volunteers who were divided into three groups—10 control (CO) subjects (5 men, 5 women, aged: 64.5 ± 8.0 (mean ± SD)), 10 Parkinsonian (PD) subjects (4 men, 6 women, aged: 64.5 ± 9.8 (mean ± SD)) and 10 Essential Tremor (ET) subjects (1 man, 9 women, aged: 68.1 ± 8.3, (mean ± SD)). None of the participants had any other injuries or disease affecting movement or coordination other than PD or ET and all provided informed consent prior to participating in the study.

2.2 Experiment setup

Three wireless inertial sensors (Shimmer Research, Dublin, Ireland), each able to measure linear acceleration, angular velocity and magnetic heading in three dimensions were attached to each patient's most affected hand, forearm and arm. The data from the sensors were received via a Bluetooth wireless connection at a sampling frequency of 51.2 Hz and stored on the computer.

Subjects were sitting on the chair with their affected arm next to a semitransparent white screen on which a multimedia projector displayed from behind the screen a stick figure's arm and target as a green circle. The stick figure arm was used as the reference starting position for the subjects arm and forearm. The screen was positioned parallel to the sagittal plane of the subject. A sound cue was used to inform subjects about the pre-start of the measurement and after one second delay the green colored frame appeared on the screen as a signal to start performing a motion until a visual stop signal was presented.

2.3 Motor tasks

Each subject performed eight sagittal plane upper extremity movement tasks in randomized order. Three motor tasks required increasing amounts of shoulder joint flexion were performed with and without a dual contralateral open-close hand task (Aubin et al., 2012). Two motor tasks were RAM of flexion and extension of elbow in sagittal plane at approximately 45° amplitude, one time without simultaneous contralateral open-closing hand motions, and a second time with simultaneous open-closing hand motions. The subjects were asked to perform the motion as many times and as fast as comfortable during the allowed 5 to 7 seconds. Each trial was performed 6–7 times and a random three trials were selected for analysis.

2.4 Metrics and statistical analysis

The following metrics for the analyses were selected: average frequency (f) of the motions, average time period (see Fig. 1) between maximum peaks of angular velocity (T_i), average angular velocity (ω_{avg}), time until first peak (t_{1stpk}), average amplitude of first peak (A_{1stp}), average time period between absolute angular velocity peaks

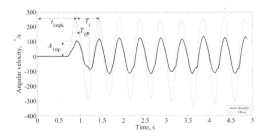

Figure 1.   Angular velocities of shoulder and elbow joints in sagittal plane recorded during the RAM measurement.

(T_{yp}), index of irregularity (upwards zero-crossings of the second derivative of the angular velocity, i.e. angular jerk), the rate of zero-crossings (R_{zc}) and approximate entropy of angular velocity and acceleration. Each metrics were calculated both for elbow and shoulder joints.

MATLAB software was used to process the raw measurement data and calculate the above-mentioned metrics. Statistical analysis of the metrics was performed using IBM's SPSS v22 software. A one-way ANOVA with a significance level of $\alpha = 0.05$ was used to test the null hypothesis that the means of RAM parameters are the same between the PD, ET and CO groups. A paired-samples t-test was used to compare RAM metrics from the motor task without contralateral open-closing hand motions to the motor tasks with contralateral open-closing hand motions.

3 RESULTS

The full set of metrics for the comparative analysis consisted of 23 parameters (due to limited space, Table 1 shows an abbreviated parameter set).

A one-way between subjects ANOVA was conducted to compare the RAM metrics for discrimination of CO, PD and ET groups. When analyzing motor task without contralateral hand opening and closing during the rapid alternating movements, we found that there was a statistically significant

difference in average amplitude of first peak of shoulder angular velocity [$F(2, 27) = 23.875$, $p = 0.0000$], time until first peak of angular velocity for shoulder [$F(2, 27) = 8.696, p = 0.0012$] and elbow [$F(2, 27) = 5.089, p = 0.0133$]. Post hoc comparisons using the Tukey's HSD test indicated that the mean score for PD group of average amplitude of first peak of shoulder angular velocity (mean $\pm$ SD = 188.46 $\pm$ 62.56) was significantly different from CO and ET groups (CO: mean $\pm$ SD = 64.07 $\pm$ 34.07, PD: mean $\pm$ SD = 71.90 $\pm$ 32.01). However, there was no significant difference between CO and ET groups. The mean score for ET group of time until first peak of shoulder angular velocity (mean $\pm$ SD = 0.9375 $\pm$ 0.1754) as well as the time until first peak of elbow angular velocity (mean $\pm$ SD = 0.8739 $\pm$ 0.1573) was significantly different from CO and PD groups (CO: mean $\pm$ SD = 0.7100 $\pm$ 0.1049, PD: mean $\pm$ SD = 0.7142 $\pm$ 0.1255).

When analyzing motor task with contralateral hand opening and closing during the rapid alternating movements, we found that there was a statistically significant difference in frequency of movements [$F(2, 27) = 3.626, p = 0.0403$], average amplitude of first peak of shoulder angular velocity [$F(2, 27) = 47.028, p = 0.0000$] and average amplitude of first peak of elbow angular velocity [$F(2, 27) = 3.485, p = 0.045$], average time period between angular velocity peaks: for shoulder [$F(2, 27) = 4.779, p = 0.017$], for elbow [$F(2, 27) = 4.799, p = 0.016$], time until first peak of angular velocity for shoulder [$F(2, 27) = 8.880, p = 0.001$] and elbow [$F(2, 27) = 4.073, p = 0.028$].

Post hoc comparisons using the Tukey's HSD test indicated that the mean score for ET group of frequency of movements (mean $\pm$ SD = 0.9606 $\pm$ 0.4849) was significantly different from CO and PD groups (CO: mean $\pm$ SD = 1.4658 $\pm$ 0.3894, PD: mean $\pm$ SD = 1.3127 $\pm$ 0.4104, see Fig. 2).

Table 1. RAM parameters.

Parameters	Groups	Mean	SD
Frequency of	CO	1.5038	0.2992
movements, Hz	PD	1.6308	0.6782
	ET	1.1849	0.5774
Average amplitude of	CO	64.07	34.07
first peak of shoulder	PD	188.46	62.56
angular velocity, °/s	ET	71.90	32.01
Average amplitude of	CO	194.45	78.77
first peak of elbow	PD	115.19	46.67
angular velocity, °/s	ET	129.17	94.10
Average time period	CO	0.6040	0.1404
of elbow movement	PD	0.6080	0.2327
cycle, s	ET	0.9900	0.5763
Average amplitude of	CO	169.35	70.63
shoulder absolute	PD	180.37	63.70
angular velocity, °/s	ET	151.14	95.45
Time until first peak	CO	0.7170	0.0837
of shoulder angular	PD	0.7312	0.1216
velocity, s	ET	0.9375	0.1754
Approximate entropy	CO	0.4870	0.0965
of shoulder angular	PD	0.4491	0.0613
velocity	ET	0.5043	0.1289

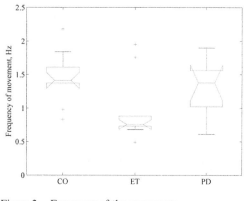

Figure 2. Frequency of the movements.

Average amplitude of first peak of the shoulder angular velocity (mean ± SD = 197.95 ± 40.68) for the PD group was significantly different from the CO and ET groups (CO: mean ± SD = 75.80 ± 26.34, ET: mean ± SD = 68.23 ± 32.14). The average amplitude of first peak of elbow angular velocity (mean ± SD = 175.87 ± 80.80) for CO group was significantly different from PD group (mean ± SD = 93.09 ± 54.19), but not from ET group (mean ± SD = 112.02 ± 76.70). Average time period between the peaks of shoulder angular velocity (mean ± SD = 0.994 ± 0.331) for ET group was significantly different from CO (mean ± SD = 0.635 ± 0.177) group. No significant difference found between PD (mean ± SD = 0.726 ± 0.279) and ET groups. Average time period between the peaks of elbow angular velocity (mean ± SD = 0.997 ± 0.348) for the ET group was significantly different from the CO (mean ± SD = 0.637 ± 0.0172) group. No significant difference found between PD (mean ± SD = 0.721 ± 0.267) and ET groups. The mean score for the ET group of time until first peak of shoulder angular velocity (mean ± SD = 1.072 ± 0.275) was significantly different from the CO and PD groups (CO: mean ± SD = 0.773 ± 0.1103, PD: mean ± SD = 0.7838 ± 0.098). The time until first peak of the elbow angular velocity (mean ± SD = 1.0116 ± 0.251) for the ET group was significantly different only from the PD group (mean ± SD = 0.7728 ± 0.118).

Comparing ANOVA results for both motor tasks, we can see that with the introduction of additional task for a contralateral hand the set of statistically significant biomechanical metrics also increases, i.e. from 3 to 7.

A paired-samples t-test was conducted to compare motor task without contralateral open-closing hand with motor task with open-closing hand. There was a significant difference in the frequency of movements in PD group ($t(9) = 2.574$, $p = 0,029973$) and in ET group ($t(9) = 2.838$, $p = 0,019457$). There was a significant difference in the average amplitude of absolute elbow angular velocity in ET group ($t(9) = 2.564$, $p = 0,0305$) and in PD group ($t(9) = 2.838$, $p = 0,0195$). No statistically significant differences found in CO group between two motor tasks.

In general, the maximum angular velocities of both the shoulder and elbow tend to decrease (for example, CO: from 81.17 °/s to 80.38 °/s, PD: from 86.47 °/s to 77.32 °/s and ET: from 75.02 °/s to 71.995 °/s for shoulder) with the introduction of opening-closing hand motion. This acts such as disturbance and for some of the subjects it was a challenge to maintain opening-closing hand motion without starting to move their whole contralateral arm in the same fashion as the arm being measured. The time until first peak of the shoulder and elbow angular velocities tends to increase in all three groups, i.e. subjects reacted slightly slower while performing the motor task with contralateral hand opening and closing. Based on this reaction time parameter all subjects started to move their forearm (flexion of elbow) a bit sooner than arm (flexion of shoulder). None of the metrics described and used successfully in the detection of movement disorders like PD by other author's (i.e. index of irregularity and irregularity of RAM cycle amplitude) were found significantly different.

4 CONCLUSIONS

The most sensitive metrics to distinguish between PD and ET subjects were the amplitude of the first peak of shoulder and elbow angular velocities just after the subject started to move and time until the first peak (i.e. reaction time). However, it also makes difficult to apply in clinical setting since this metric does not separate ET patients from healthy control subjects. Thus clinical discrimination between PD and ET might be easier than between ET and CO subjects. Collecting data from more patients and discovering additional discrimination metrics based on the upper extremity multi-joint motion tasks may improve the misclassification rate in the future.

REFERENCES

Aubin, P.M., Serackis, A., Griškevičius, J. 2012. Support vector machine classification of Parkinson's disease, essential tremor and healthy control subjects based on upper extremity motion. Biomedical Engineering and Biotechnology (iCBEB), IEEE Conference Publications: 900–904.

Chan, J., Leung, H., Poizner, H. 2010. Correlation among joint motions allows classification of Parkinsonian versus normal 3-D reaching. IEEE transactions on neural systems and rehabilitation engineering 18(2): 142–149.

Despres, C., Richer, F., Roberge, M-C., Lamoureux, D., Beuter, A. 2005. Standardization of quantitative tests for preclinical detection of neuromotor dysfunctions in pediatric neurotoxicology. NeuroToxicology 26: 385–395.

Ebersbach, G., Baas, H., Csoti, I., Müngersdorf, M., Deuschl, G. 2006. Scales in Parkinson's disease. Journal of neurology 253(4): iv32–35.

Fimbel, E.J., Dubary, A.S., Philibert, M., Beuter, A. 2003. Event identification in movement recordings by means of qualitative patterns. Neuroinformatics 1: 239–257.

Fimbel, E.J., Domingo, P.P., Lamoureux, D., Beuter, A. 2005. Automatic detection of movement disorders using recordings of rapid alternating movements. Journal of Neuroscience Methods 146: 183–190.

Ghassemi, M., Lemieux, S., Jog, M., Edwards, R., Duval, C. 2006. Bradykinesia in patients with Parkinson's disease having levodopa-induced dyskinesias. Brain Research Bulletin 69: 512–518.

Haas, R.B., Stewart, H.T., Zhang, J. 2012. Premotor biomarkers for Parkinson's disease—a promising direction of research. Translational neurodegeneration: 1–11.

Hughes, A.J., Daniel, S.E., Kilford, L., Lees, A.J. 1992. Accuracy of clinical diagnosis of idiopatic Parkinson's disease: a clinic-pathological study of 100 cases. Journal of neurology, neurosurgery & psychiatry 55: 181–184.

Jankovic, J. 2008. Parkinson's disease: clinical features and diagnosis. Journal of neurology, neurosurgery & psychiatry 79: 368–376.

Oliveira de Elias, M., Menegaldo, L.L., Lucarelli, P., Andrade, B.L.B., Buchler, P. 2011. On the use of information theory for detecting upper limb motor dysfunction: an application to Parkinson's disease. Physica A 390: 4451–4458.

Poizner, H., Fookson, O., Berkinblit, M.B., Hening, W., Fledman, G., Adamovich, S. 1998. Pointing to remembered targets in 3-D space in Parkinson's disease. Motor control 2: 251–277.

Ramaker, C., Marinus, J., Stiggelbout, A.M., Johannes van Hilten, B. 2002. Systematic evaluation of rating scales for impairment and disability in Parkinson's disease. Movement disorders 17: 867–876.

Robichaud, J.A., Pfann, D.K., Leurgans, S., Vaillancourt, D.E., Comella, L.C., Corcos, M.D. 2009. Variability of EMG patterns: a potential neurophysiological marker of Parkinson's disease? Clinical neurophysiology 120(2): 390–397.

Shtilbans, A., Henchcliffe, C. 2012. Biomarkers in Parkinson's disease: an update. Current opinion in neurology (25): 460–465.

Suchowersky, O., Reich, S., Perlmutter, J., Zesiewicz, T., Gronseth, G., Weiner, W.J. 2006. Practice parameter: diagnosis and prognosis of new onset Parkinson disease (an evidence-based review): report of the quality standards subcommittee of the American academy of neurology. Neurology 66: 968–975.

Sports Engineering and Computer Science – Luo (Ed.)
© 2015 Taylor & Francis Group, London, ISBN 978-1-138-02650-6

Potassium metabolism of steel workers exposed to high-temperatures

Zhenzhen Tian
School of Public Health, Hebei United University, Tangshan, Hebei, China

YongMei Tang
School of Public Health, Hebei United University, Tangshan, Hebei, China
Hebei Coal Mine Health and Safety Laboratory, Tangshan, Hebei, China

Yingxue Li
School of Public Health, Hebei United University, Tangshan, Hebei, China

ABSTRACT: Objective of the study is to understand the high temperature steel workers potassium intake, sweat consumption and levels in the body, to provide the basis for improving the health level of high temperature operation workers, reducing the harm caused by high temperature operation research. Methods: select a steel corp of 226 workers exposed to high temperature, non high temperature 39 workers, proceed environmental meteorological condition test, the general survey, dietary survey, sweat and sweat loss of potassium determination and in vivo detection. Results: high temperature workers intake of potassium is 2450.1 mg/d, medium and heavy physical labor intensity of workers intake under high temperature operation suggestions quantity; with in the high temperature physical labor intensity of workers, compared with the intake differences had no statistical significance ($p > 0.05$); sweat in the amount of the loss of potassium increased with temperature, and high temperature is greater than the amount of workers lost; serum potassium concentration of 103.2 mg/L, lower than the normal range. Conclusions: high temperature workers did not increase the intake of potassium due to high temperature labor, and did not reach the high temperature operation groups recommended intake so that high temperature workers should improve the intake of potassium, in order to improve the life quality and labor efficiency.

Keywords: high-temperature steel workers; potassium; metabolism

1 INTRODUCTION

Workers exposed to high temperatures were one of the special operations groups. Working under high temperature conditions, metabolism enhanced, sweat excretion increased, people will lose a lot of minerals with the increasing sweat evaporation. If these minerals were not supplement timely, it will affect the normal physiological functions of the body, thereby affecting the health of workers[1-7]. Potassium and heart activity were closely linked, potassium concentration decreasing prone to heart arrhythmia. In addition, potassium metabolism disorders may lead to disruption of water distribution and acid-base balance disorders. In this state, people prone to fatigue, dizziness, loss of appetite, decreased blood pressure and other symptoms[8-10]. Therefore, the investigate on the intake and consumption of sweat and lever of potassium of high-temperature steel workers, explore the special nutritional needs of workers exposed to high temperatures,

providing a scientific basis for the development of interventions.

2 SUBJECTS AND METHODS

2.1 Study subjects

We selected 226 high-temperature workers from steelmaking workshop and steel rolling workshop in a steel company, and selected 39 non high-temperature workers in the same workshop.

2.2 Research methods

By way of face to face inquiry, surveying general survey of workers and three days of dietary intake; collecting fasting blood of workers, the urine during the work, the sweat of forehead, chest and back of steel workers during work (the excretion of sweat during work was determined by weight difference method, sweat excretion = the weight before work and after urinating – the weight after work and after urinating + the weight before and after drinking + the weight before and after eating – the weight before and after defecation – the weight before and after urinating), to prepare for laboratory test[11–13].

2.3 *Environmental determination*

The temperature, humidity, heat index and heat radiation intensity were determined by heat index instrument, and then calculated the wet bulb globe temperature index (WBGT index). The work whose average WBGT ≥ 25°C called high-temperature operation[14].

2.4 *Statistical analysis*

Building a database in Excel 2003, doing statistical analysis in SPSS13.0, compare the mean of two sets with t test, and make multigroup comparisons with variance analysis and q test, count data with chi-square test.

3 RESULTS

3.1 *General conditions of steel workers exposed to high-temperatures*

The age range of steel workers exposed to high-temperature was 22 to 54 years in the survey, the mean age was 36.9 ± 5.9 years old; length of service ranges from 1 to 33 years, the average length of service was 13.6 ± 6.4 years; workers with moderate physical strength accounted for 39.7% of the total high-temperature workers, workers with severe physical strength accounted for 60.3% of the total number of high-temperature workers; junior middle school education, high school and technical secondary school education, college and above education workers accounted for 10.2%, 59.3%, 30.5% of the total high-temperature workers. Reference in Table 1.

3.2 *Potassium intake of steel workers exposed to high-temperatures*

Daily intake of potassium range was 1103.0~5494.0 mg/d among steel workers exposed to high-temperatures, the average intake was 2450.1 ± 633.6 mg/d. Refer to the recommended intake of hot work crowd, their potassium intake was lower than the recommended intake; under the medium physical labor level, daily intake of potassium was no difference between high-temperature workers and non-high-temperature workers (*p*>0.05); under the heavy physical labor level, daily intake of potassium was no difference among high-temperature

Table 1. General conditions of steel workers exposed to high-temperatures.

Item		n	Ratio (%)
Age (year old)	22_	47	20.8
	32_	135	59.7
	42_56	44	19.5
Length of service (year)	1_	56	24.8
	10_	124	54.9
	20_42	46	20.3
Physical activity level	Moderate	89	39.4
	Severe	137	60.6
Education	Junior middle school education	23	10.2
	high school and technical secondary school education	134	59.3
	College and above education	69	30.5

workers with different operation temperature (*p*>0.05). Reference in Table 2.

3.3 *The excretion of potassium of steel workers exposed to high-temperatures*

The excretion of potassium range was 349.3_991.2 mg among steel workers exposed to high-temperatures, the average excretion was 528.0 ± 146.8 mg; under the medium physical labor level, The excretion of potassium of high-temperature workers was higher than non-high-temperature workers (*p*<0.05); under the heavy physical labor level, the excretion of potassium was difference among high-temperature workers with different operation temperature (*p*<0.05 for all), the excretion of potassium increased with the rising of WBGT index, all had a statistical significance. Reference in Table 3.

The excretion of potassium was accounted for 90.0% of the total excretion in sweat and urine, sweating is the main way for potassium; under the medium physical labor level, the ratio of excretion potassium accounted the total discharge of high-temperature workers was higher than non-high-temperature workers (*p*<0.05); under the heavy physical labor level, the ratio of excretion potassium accounted the total discharge were not difference among high-temperature workers with different operation temperature (*p*>0.05). Reference in Table 4.

3.4 *Levels of potassium in the body of steel workers exposed to high-temperatures*

Serum potassium concentration range was 88.4_237.7 mg/L among steel workers exposed to

Table 2. Potassium intake of steel workers exposed to high-temperatures.

Item	WBGT (°C)	Labor intensity	n	K (mg/d)
High-temperature	30_	Medium	89	2505.6 ± 500.6
		Heavy	14	2417.1 ± 508.0
	35_	Heavy	85	2441.2 ± 455.8
	40_43	Heavy	38	2358.9 ± 523.2
Non-high-temperature	20_25	Medium	39	2471.3 ± 598.1

Table 3. The excretion of potassium of steel workers exposed to high-temperatures ($\bar{x} \pm s$).

Item	WBGT (°C)	Labor intensity	n	K (mg)
High-temperature	30_	Medium	89	438.7 ± 66.8[a**]
		Heavy	14	468.1 ± 42.9
	35_	Heavy	85	560.1 ± 101.9[b**]
	40_43	Heavy	38	687.2 ± 172.5[b**c**]
Non-high-temperature	20_25	Medium	39	128.5 ± 20.1

Where a represents the comparison with non-high-temperature workers, b represents the comparison with heavy manual workers in 30_34°C, c represents the comparison with heavy manual workers in 35_39°C; $^*p<0.05$, $^{**}p<0.01$.

Table 4. The excretion constitution of in sweat and urine of high-temperature workers.

Item	WBGT (°C)	Labor intensity	n	The output of potassium in sweat and urine (μg)	Potassium/ sweat (%)
High-temperature	30_	Medium	89	816.2 ± 110.9[a**]	85.7[a**]
		Heavy	14	869.6 ± 139.6	87.8
	35_	Heavy	85	1015.5 ± 109.2 [b**]	91.8
	40_43	Heavy	38	1207.5 ± 246.2[b**c**]	94.3
Non-high-temperature	20_25	Medium	39	260.3 ± 38.1	38.6

Where a represents the comparison with non-high-temperature workers, b represents the comparison with heavy manual workers in 30_34°C, c represents the comparison with heavy manual workers in 35_39°C; $^*p<0.05$, $^{**}p<0.01$.

high-temperatures, the average concentration was 137.2 ± 29.2 mg/L, which was lower than normal serum concentrations of adult; under the medium physical labor level, the average serum potassium of high-temperature workers was lower than non-high-temperature workers ($p<0.05$); under the heavy physical labor level, the average serum potassium was difference among high-temperature workers with different operation temperature ($p <0.05$), the average concentration of serum potassium of workers in 30_34°C and 35_39°C was greater than workers in 40_43°C, the difference all had statistical significance ($p<0.05$ for all). Compared with the normal adult range of serum potassium concentration, serum potassium deficiency of steel workers exposed to high-temperatures was 58.8%; under the medium physical labor level, serum potassium deficiency of high-temperature workers was higher than non-high-temperature workers ($p<0.05$). Reference in Tables 5 and 6.

4 DISCUSSION

High-temperature operation means the WBGT index of workplace is equal or greater than 25°C. For high-temperature workers, metabolism enhances and the consumption of water increase, people will lose a lot of minerals, if it cannot be timely supplement, result in reduced levels of the body's nutrition, then make the body more easily transition from physiologic changes to the subclinical state, so that contribute to pathological changes[1–7]. In addition, the environment of high-temperature and high intensity of labor load may also reduce the brain cortex function and abate to

Table 5. Serum potassium concentration of steel workers exposed to high-temperatures ($\bar{x} \pm s$).

Item	WBGT (°C)	Labor intensity	n	Serum potassium concentration (mg/L)
High-temperature	30_	Medium	89	$157.8 \pm 28.0^{a**}$
		Heavy	14	136.7 ± 16.6
	35_	Heavy	85	125.5 ± 11.0
	40_43	Heavy	38	$115.6 \pm 15.3^{b**c**}$
Non-high-temperature	20_25	Medium	39	185.2 ± 32.7

Where a represents the comparison with non-high-temperature workers, b represents the comparison with heavy manual workers in 30_34°C, c represents the comparison with heavy manual workers in 35_39°C; *$p<0.05$, **$p<0.01$.

Table 6. Lack of potassium levels in the body of steel workers exposed to high-temperatures.

Item		WBGT (°C)	Labor intensity	n	Lack n	Lack %
K	High-temperature	30_	Medium	89	23	25.8^{a**}
			Heavy	14	9	64.3
		35_	Heavy	85	63	74.1
		40_43	Heavy	38	34	89.5
	Non-high-temperature	20_25	Medium	39	4	10.3

Where a represents the comparison with non-high-temperature workers; **$p<0.01$.

adapt ability, resulting in a decline in sensitivity and coordination[15].

Steel workers as one of the main groups of high-temperature workers, in this study, the survey included potassium intake, the amount lost in sweat and levels in the body of high-temperature operation of the steel workers, understand the current situation of potassium loss of workers exposed to high temperatures, then by comparing the difference between the amount of discharge of high-temperature workers and non-high-temperature workers in the same intake level, explore the special needs of workers exposed to high temperatures, in order to provide the reference for the study of putting forward the corresponding prevention, control, and complementary measures.

Adult normal in vivo hold potassium 15640 mg, 80% distributes in intracellular fluid, accordingly potassium plays an important role in sustaining intracellular fluid and bulk. Simultaneously, potassium ion has a close connection with metastasizing Chlorine ion, potassium metabolism disorders may lead to disruption of water distribution and acid-base balance disorders. The concentration of sanguis-potassium has a close connection with cardiomotility, the heart facility generates premature systole and other arrhythmias when it decreases, when human body is short of potas-

sium, people prone to fatigue, dizziness, loss of appetite, decreased blood pressure and other symptoms. Under the high-temperature operation, owing to profuse sweating, except losing mass of sodium chloride with sweat, as well provoking the notable lose of potassium. Li Yujiu[16] found, high-temperature workers lose 392 mg potassium through sweat, 42 mg sodium through urine, sweat lost sodium is 20.5 times of urine. He Yingjiang[17] reported such as the high temperature environment, calculated at 4 L sweat, sweat in potassium loss amount is 1003~1780 mg. Ning Hongzhen[18] reported, the high temperature coal mine workers sweat potassium excretion was 1480 ± 410 mg, urinary potassium excretion was 340 ± 79 mg. At present, excessive potassium loss may be one of the causes of heat, so attention should be paid to the problem of high potassium operator. High-temperature operation workers not only because of sweat loses large amounts of potassium, but in glomerular cells in feelings of profuse sweating, reducing the blood sodium, blood volume decreased after the stimulation of renin secretion of aldosterone, the renin-angiotensin-aldosterone system increased secretion, excretion of urinary potassium increased significantly, up to 1798.7~2932.6 mg/d, so the sweat urinary excretion of potassium, the total will exceed intake can cause negative balance. High temperature physical activity, because the potassium excretion increased as well as the reference standard general dietary potassium intake is low, there will be a negative potassium balance, resulting in low serum potassium, higher body temperature after work. The potassium in high temperature environment of human endurance and plays an important role in preventing heatstroke, so the high temperature environment of workers compensation should be a variety of electrolyte containing potassium salts, minerals, rather than merely added sodium chloride.

The survey showed that, the iron and steel workers exposed to high temperature potassium intake in the corresponding recommendations below, can not meet the needs of labor under high temperature environment; intake and non high temperature exposed workers was not significantly different, and with the operation temperature has no obvious change. Increased loss of potassium in sweat with the increase of temperature, and more than non workers exposed to high temperature loss; compared with the WBGT index operation environment 35_40°C and 30_35°C, the average discharge workers sweat potassium increased. The investigation object of potassium in vivo levels lower than the general population reference range; concentration equal labor intensity of workers exposed to high temperature and serum potassium is lower than that of workers exposed to high temperature, concentration of the same physical labor intensity of different working temperature of workers exposed to high temperature plasma potassium increased with temperature are different; iron and steel workers exposed to high temperature in the average level of potassium low, lack of rate higher than that of non exposed to high temperatures.

In conclusion, working under high temperature conditions, through the sweat loss of potassium, loss increased with the increasing of temperature, and potassium in average level is low, so it should be according to the temperature changes of different labor intensity of workers, different temperature operating temperature of potassium supplement corresponding, in order to improve the health of workers exposed to high temperature, reduce heat exposure, while avoiding the adverse effects caused by excessive supplement.

REFERENCES

[1] Zhang Guoqiang, He Hanzhen, Zhang Wei. High temperature physical and health [M]. Shanghai: Shanghai science and Technology Press, 1989: 83–91.

[2] Pan Lijie, Yan Xige. Glass products industry high temperature effects on physiological functions of workers Chinese [J]. Journal of industrial medicine, 2007, 20 (5): 325–326.

[3] Sharman Rk et al. effect of high ambient temperation induced heat stress on the functions of small intestine and colon in albino rats [J]. Indian Res, 1982(75): 593–606.

[4] Sharman Rk et al. simultaneous measurement of gastric-small intestine motility absorption of proline in rats exposed to heat [J]. Indian JMed Res, 1982(81):330.

[5] A. Sengupta, R.K. Sharma et al. Acute heat stress in growing rats:effect on small Intestinal morphometry and in vivo absorption [J]. Journal of Thermal Biology, 1993(18)3:145–151.

[6] Lebedeva NV, Alimova ST, Efendiev FB. Study of mortality among workers exposed Heating microclimate (epidemiological study) [J]. Gig Trprof Zabol, 1991, (10):12–5 (In Russian).

[7] Moulin JJ, Wild P, Mantout B, Fournier Betz M, Mur JM, Smagghe G. Mortality form Lung cancer and cardiovascular dwaseases among stainlesssteel producing workers [J]. Cancer Causes Causes Control, 1993, 4(2):75–81.

[8] Wu Fuping, Li Songqiang. The health of human body and chemical elements [J]. Guangdong chemical industry, 2005, 4: 33–35.

[9] Li Shouchun. The biological role of inorganic elements in the body [J]. Journal of Nanchang Education College 1999, 3: 50–54.

[10] Gu Kun, Shi Zhenxun. The chemical elements and health [J].the Yunnan environmental science,. 1999, 18 (3): 51–55.

[11] Li Yujiu, Zu Guodong, Yang Jun, et al. Study on high temperature exposed workers in vitamin B1, B2, C and inorganic nutrition [J]. Chinese public health, 1998,14 (8): 467–470.

[12] Nassim Hamouti, Jian Del Coso, Juan F ortega, et al. Sweat sodium concentration during exercwase in the heat in aerobically trained and untrained humans [J]. Eur J Appl Physiol, 2011, 111: 2873–2881

[13] Stephen J, Genuwas Detlef Birkholz, Ilia Rodushkin, et al. Blood, Urine, and Sweat (BUS) Study: Monitoring and Elimination of Bioaccumulated Toxic Elements [J]. Arch Environ Contam Toxicol, 2011, 61:344–357.

[14] "meteorological conditions in the hot work environment determination method" GB/T 934–2008.

[15] Jin Tai Yi. Occupation occupation health and medical [M. 6 edition. Beijing: People's Medical Publishing House, 2007: 14.

[16] Li Yujiu, Zu Guodong, Yang Jun, Jia Lihong, et al. The workers exposed to high temperature of vitamin B1, B2, C and inorganic salt on the nutrition of [J]. China public health, 1998, 14 (8): 468–470.

[17] He Yingqiang, Chen Jidi, Yang Zeyi. Thermal environment research work and sweat losses of water and inorganic elements [J]. China sports Med, 1987, 6 (2): 70–74.

[18] Ning Hongzhen, Guan Weijun, Ren Lei, et al. Mine workers investigation on nutrition and water salt metabolism [J]. Chinese Journal of industrial medicine, 2002, 15 (5): 280–281.

Author index

Printed and bound by CPI Group (UK) Ltd, Croydon, CR0 4YY

18/10/2024

01776219-0011